USEFUL INTEGRATION FORMULAS

$$\int \sin^2 t \, dt = \frac{1}{2}t - \frac{1}{4}\sin 2t + C$$

$$\int \cos^2 t \, dt = \frac{1}{2}t + \frac{1}{4}\sin 2t + C$$

$$\int \frac{dt}{a^2 + t^2} = \frac{1}{a}\tan^{-1}\frac{t}{a} + C$$

$$\int te^{at} \, dt = (t - 1) \, e^t + C$$

$$\int e^{at} \sin bt \, dt = \frac{e^{at}}{a^2 + b^2}(a \sin bt - b \cos bt) + C$$

$$\int e^{at} \cos bt \, dt = \frac{e^{at}}{a^2 + b^2}(a \cos bt + b \sin bt) + C$$

$$\int t \sin bt \, dt = \frac{1}{b^2}\sin bt - \frac{t}{b}\cos bt + C$$

$$\int t \cos bt \, dt = \frac{1}{b^2}\cos bt + \frac{t}{b}\sin bt + C$$

$$\int_0^{2\pi/\omega} \sin(\omega t + \alpha) \, dt = 0$$

$$\int_0^{2\pi/\omega} \cos(\omega t + \alpha) \, dt = 0$$

$$\int_0^{2\pi/\omega} \sin(n\omega t + \alpha) \, dt = 0; \ n \text{ an integer}$$

$$\int_0^{2\pi/\omega} \cos(n\omega t + \alpha) \, dt = 0; \ n \text{ an integer}$$

$$\int_0^{2\pi/\omega} \sin(m\omega t + \alpha) \cos(n\omega t + \alpha) \, dt = 0; \ m, n \text{ integers}$$

$$\int_0^{2\pi/\omega} \sin^2(\omega t + \alpha) \, dt = \pi/\omega$$

$$\int_0^{2\pi/\omega} \cos^2(\omega t + \alpha) \, dt = \pi/\omega$$

$$\int_0^{2\pi/\omega} \cos(m\omega t + \alpha) \cos(n\omega t + \beta) \, dt = 0, \ m \neq n; \ m, n \text{ integers}$$
$$= \pi \cos(\alpha - \beta)/\omega, \ m = n$$

ELECTRIC CIRCUIT ANALYSIS

David E. Johnson
Birmingham-Southern College

Johnny R. Johnson
University of North Alabama

John L. Hilburn
President, Microcomputer Systems Inc.

 Prentice Hall, Englewood Cliffs, New Jersey 07632

Library of Congress Cataloging-in-Publication Data

Johnson, David E.
    Electric circuit analysis / David E. Johnson, Johnny R. Johnson,
John L. Hilburn.
        p.   cm.
    Expanded version of: Basic electric circuit analysis. 3rd ed.
© 1986.

    Bibliography: p.
    Includes index.

    ISBN 0-13-247776-9

    1. Electric circuit analysis.   I. Johnson, Johnny Ray.
II. Hilburn, John L.   III. Johnson, David E.   Basic
electric circuit analysis.   IV. Title.
TK454.J565   1989
621.319′2—dc19   88-19942
                                                      CIP

Editorial/production supervision: Colleen Brosnan
Interior design: Jayne Conte
Cover design: Jayne Conte
Manufacturing buyer: Mary Noonan

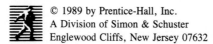 © 1989 by Prentice-Hall, Inc.
A Division of Simon & Schuster
Englewood Cliffs, New Jersey 07632

Printed in the United States of America
10 9 8 7 6 5 4 3 2

ISBN 0-13-247776-9

Prentice-Hall International (UK) Limited, *London*
Prentice-Hall of Australia Pty. Limited, *Sydney*
Prentice-Hall Canada Inc., *Toronto*
Prentice-Hall Hispanoamericana, S.A., *Mexico*
Prentice-Hall of India Private Limited, *New Delhi*
Prentice-Hall of Japan, Inc., *Tokyo*
Simon & Schuster Asia Pte. Ltd., *Singapore*
Editora Prentice-Hall do Brasil, Ltda., *Rio de Janeiro*

Contents

Contents

8 Simple *RC* and *RL* Circuits 195

9 Second-Order Circuits 244

10 Sinusoidal Excitation and Phasors 291

11 AC Steady-State Analysis 328

12 AC Steady-State Power 357

13 Three-Phase Circuits 387

14 Complex Frequency and Network Functions 419

15 Frequency Response 469

16 Transformers 502

17 Fourier Series 537

18 Fourier Transforms 579

19 Laplace Transforms 602

20 Laplace Transform Applications 645

Preface

This book was written for a two-semester or three-quarter course in linear circuit analysis. This is a basic course in electrical engineering and is usually the student's first encounter with his or her chosen field of study. It is important, therefore, for the textbook to cover thoroughly the fundamentals of the subject and at the same time be as easy to understand as possible. These were our objectives throughout the writing of the book.

The book is an expanded treatment of our earlier work, *Basic Electric Circuit Analysis*, and is designed primarily for readers desiring a more complete treatment of Laplace transforms and Fourier analysis. We hope that the earlier book will continue to serve the needs of students who are primarily interested in nontransform methods, and that this book will be attractive to those interested in a broader coverage of operational mathematics methods.

Most students in basic circuit theory will already have studied electricity and magnetism in a physics course. This background is helpful, of course, but is not a prerequisite for reading the book. The material presented here may be easily understood by a student who has had standard courses in differential and integral calculus. The differential equations theory required in circuit analysis is fully developed in the book and integrated with the appropriate circuit theory topics. Even determinants, Gaussian elimination, and complex number theory are presented in appendices.

The operational amplifier is introduced immediately after the discussion of the resistor, and appears, as a matter of course, along with resistors, capacitors, and inductors, as a basic element throughout the book. Likewise, dependent sources and their construction using operational amplifiers are discussed early and encountered routinely in almost every chapter.

To help the reader understand the textual material, examples are liberally supplied and numerous exercises, with answers, are given at the end of virtually every section. Most of the worked-out examples are clearly marked and numbered for easy reference. Problems, some more difficult and some less difficult than the end-of-section exercises, are given at the end of every chapter, and answers to the odd-numbered problems are given in an appendix. Most of the exercises and problems have been designed to yield easier answers, such as 6 volts rather than 6.127 volts, without sacrificing their practical aspects. We hope in this way to minimize unnecessarily tedious arithmetic and make the subject of circuit theory more interesting and rewarding.

A special effort has been made to include a number of problems and exercises with realistic element values. Network scaling, also presented, can be used to make all the remaining problems practical. In the chapter on amplitude and phase responses, problems are given on such practical circuits as electric filters. Active filters, using operational amplifiers, as well as passive filters, are used as examples. Finally, a select few exercises and problems are used to extend the theory discussed in the chapters. In this way, optional material is included without adding to the text.

The first nine chapters of the book are devoted to terminology and time-domain analysis and the last eleven chapters deal with frequency-domain analysis. Some material may be omitted without any loss in continuity. The chapter on network topology, an interesting subject, could be covered entirely, in part, or not at all. Chapters 17–20 on Fourier and Laplace methods constitute a detailed treatment of these subjects and may be omitted if these topics are to be covered in a separate course. Indeed, one could omit the classical differential equations approach to circuit theory and go directly to the Laplace and Fourier transform methods.

For the reader interested in computer solutions of circuits, the computer-aided circuit analysis program SPICE is described in an appendix, and examples using SPICE are given in the last sections of selected chapters. A separate section of computer-aided problems follow problem sets at the end of the chapter. In this way the computer-aided material can be easily omitted if so desired. Color is used throughout the book to highlight the more important equations, to help clarify the figures, and generally to make the book more readable. Finally, to make the subject of electric circuits more real and enjoyable to the student, we have opened each chapter with a picture and a short biography of a famous electrical pioneer, noted engineer, or inventor whose work has contributed importantly to circuit theory.

For the chapter-opening illustrations we are grateful to the Print Collection, Miriam and Ira D. Wallach Division of Art, Prints and Photographs, The New York Public Library, Astor, Lenox, and Tilden Foundations for Chapters 1, 5, 6, 7, 19, and 20; the Library of Congress for Chapters 2, 3, 8, 11, and 17; the Smithsonian Institution for Chapters 4, 9, 12, 14, and 18; and the marvelous book, *Dictionary of American Portraits* (Dover Publications, Inc., 1967, edited by Haywood and Blanche Cirker), for Chapters 15 and 16. The photographs for Chapters 10 and 13 are courtesy of the General Electric Company, whom we also gratefully acknowledge.

There are many people who have provided invaluable assistance and advice concerning this book. We are indebted to our colleagues and our students for the form the book has taken, and to Professors M. E. Van Valkenburg, A. P. Sage, S. R. Laxpati, and S. K. Mitra, who reviewed the first-edition manuscript of *Basic Electric Circuit Analysis* and made many helpful comments and suggestions. We are also grateful for the invaluable reviews of this current edition by

Artice M. Davis, San Jose State University
John A. Fleming, Texas A & M University
Deverl Humphreys, Brigham Young University
Tim Jordanides, California State University, Long Beach
K.S.P. Kumar, University of Minnesota
Terry W. Martin, University of Arkansas
Michael P. Smyth, Widener University
J. Eldon Steelman, New Mexico State University
James Svoboda, Clarkson University
Beth L. Koester, Concord, MA

David E. Johnson
Johnny R. Johnson
John L. Hilburn

Alessandro Volta
1745–1827

1

Introduction

This endless circulation of the electric fluid may appear paradoxical, but it is no less true and real, and you may feel it with your hands.

Alessandro Volta

Electric circuit theory had its real beginning on March 20, 1800, when the Italian physicist Alessandro Volta announced his invention of the electric battery. This magnificent device allowed Volta to produce *current* electricity, a steady, continuous flow of electricity, as opposed to *static* electricity, produced in bursts by previous electrical machines such as the Leyden jar and Volta's own *electrophorus.*

Volta was born in the Italian city of Como, then a part of the Austrian Empire, and at age 18 he was performing electrical experiments and corresponding with well-known European electrical investigators. In 1782 he became professor of physics at the University of Padua, where he became involved in controversy with another well-known electrical pioneer, Luigi Galvani, professor of anatomy at Bologna. Galvani's experiments with frogs had led him to believe that current electricity was *animal electricity* caused by the organisms themselves. Volta, on the other hand, maintained that current electricity was *metallic electricity,* the source of which was the dissimilar metal probes attached to the frogs' legs. Both men were right. There is an animal electricity, and Galvani became famous as a founder of nerve physiology. Volta's great invention, however, revolutionized the use of electricity and gave the world one of its greatest benefits, the electric current. Volta was showered with honors during his lifetime. Napoleon made him a senator and later a count in the French Empire. After Napoleon's defeat, the Austrians allowed Volta to return to his Italian estate as a citizen in good stead. Volta was rewarded 54 years after his death when the unit of electromotive force was officially named the *volt.* ∎

Electric circuit analysis, in nearly every electrical engineering curriculum, is the first course taken in the major area by an electrical engineering student. Virtually all branches of electrical engineering, such as electronics, power systems, communication systems, rotating machinery, and control theory, are based on circuit theory. The only topic in electrical engineering more basic than circuits is electromagnetic field theory, and even there many problems are solved by means of equivalent electric circuits. Thus it is no exaggeration to say that the basic circuit theory course a student first encounters in electrical engineering is the most important course in his or her curriculum.

To begin our study of electric circuits we need to know what an electric circuit is, what we mean by its analysis, what quantities are associated with it, in what units these quantities are measured, and the basic definitions and conventions used in circuit theory. These are the topics we shall consider in this chapter.

1.1

DEFINITIONS AND UNITS

An electric *circuit,* or electric *network,* is a collection of electrical elements interconnected in some specified way. Later we shall define the electrical elements in a formal manner, but for the present we shall be content to represent a general *two-terminal* element as shown in Fig. 1.1. The terminals *a* and *b* are accessible for connections with other elements. Examples with which we are all familiar, and which we shall formally consider in later sections, are resistors, inductors, capacitors, batteries, generators, etc.

FIGURE 1.1 General two-terminal electrical element

More complicated circuit elements may have more than two terminals. Transistors and operational amplifiers are common examples. Also a number of simple elements may be combined by interconnecting their terminals to form a single package having any number of accessible terminals. We shall consider some multiterminal elements later, but our main concern will be simple two-terminal devices.

An example of an electric circuit with six elements is shown in Fig. 1.2. Some authors distinguish a circuit from a network by requiring a circuit to contain at least one closed path such as path *abca*. We shall use the terms interchangeably, but we may note that without at least one closed path the circuit is of little or no practical interest.

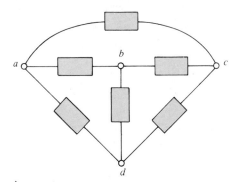

FIGURE 1.2 Electric circuit

To be more specific in defining a circuit element we shall need to consider certain quantities associated with it, such as *voltage* and *current*. These quantities and others, when they arise, must be carefully defined. This can be done only if we have a standard system of units so that when a quantity is described by measuring it, we can all agree on what the measurement means. Fortunately, there is such a standard system of units that is used today by virtually all the professional engineering societies and the authors of most modern engineering textbooks. This system, which we shall use throughout the book, is the *International System of Units* (abbreviated SI), adopted in 1960 by the General Conference on Weights and Measures.

There are six basic units in the SI, and all other units are derived from them. Four of the basic units, the meter, kilogram, second, and coulomb, are important to circuit theorists, and we shall consider them in some detail. The remaining two basic units are the degree Kelvin and the candela, which are important to such people as the electron device physicist and the illumination engineer.

The SI units are very precisely defined in terms of permanent and reproducible quantities. However, the definitions are highly esoteric and in some cases are comprehensible only to atomic scientists.[1] Therefore, we shall be content to name the basic units and relate them to the very familiar *British System of Units*, which includes inches, feet, pounds, etc.

---

[1]Complete definitions of the basic units may be found in a number of sources, such as, for example, "IEEE Recommended Practice for Units in Published Scientific and Technical Work," by C. H. Page et al. (*IEEE Spectrum*, vol. 3, no. 3, pp. 169–173, March 1966).

The basic unit of length in the SI is the *meter,* abbreviated m, which is related to the British system by the fact that 1 inch is 0.0254 m. The basic unit of mass is the *kilogram* (kg), and the basic unit of time is the *second* (s). In terms of the British units, 1 pound-mass is exactly 0.45359237 kg, and the second is the same in both systems.

The fourth unit in the SI is the *coulomb* (C), which is the basic unit used to measure electric charge. We shall defer the definition of this unit until the next section when we consider charge and current. The name coulomb was chosen to honor the French scientist, inventor, and army engineer Charles Augustin de Coulomb (1736–1806), who was an early pioneer in the fields of friction, electricity, and magnetism.

We might note at this point that all SI units named for famous people have abbreviations that are capitalized; otherwise, lowercase abbreviations are most often used. It is also worth mentioning that we could choose units other than the ones we have selected to form the basic units. For example, instead of the coulomb we could take the *ampere* (A), the unit of electric current to be considered later. In this case the coulomb could then be obtained as a derived unit.

There are three derived units in addition to the ampere that we shall find useful in circuit theory. They are the units used to measure force, work or energy, and power. The fundamental unit of force is the *newton* (N), which is the force required to accelerate a 1-kg mass by 1 meter per second per second (1 m/s^2). Thus 1 N = 1 kg-m/s^2. The newton is named, of course, for the great English scientist, astronomer, and mathematician Sir Isaac Newton (1642–1727). Newton's accomplishments are too numerous to be listed in a mere chapter.

The fundamental unit of work or energy is the *joule* (J), named for the British physicist James P. Joule (1818–1889), who shared in the discovery of the law of conservation of energy and helped establish the idea that heat is a form of energy. A joule is the work done by a constant 1-N force applied through a 1-m distance. Thus 1 J = 1 N-m.

The last derived unit we shall consider is the *watt* (W), which is the fundamental unit of power, the rate at which work is done or energy is expended. The watt is defined to be 1 J/s and is named in honor of James Watt (1736–1819), the Scottish engineer whose engine design first made steam power practicable.

Before we leave the subject of units we should point out that one of the greatest advantages the SI has over the British system is its incorporation of the decimal system to relate larger and smaller units to the basic unit. The various powers of 10 are denoted by standard prefixes, some of which are given, along with their abbreviations, in Table 1.1.

TABLE 1.1 Prefixes in the SI

Multiple	Prefix	Symbol
10^9	Giga	G
10^6	Mega	M
10^3	Kilo	k
10^{-3}	Milli	m
10^{-6}	Micro	μ
10^{-9}	Nano	n
10^{-12}	Pico	p

As an example, at one time a second was thought to be a short time, and fractions such as 0.1 or 0.01 of a second were unimaginably short. Nowadays in some applications, such as digital computers, the second is an impractically large unit. As a result, times such as 1 nanosecond (1 ns or 10^{-9} s) are in common use. Another common example is 1 gram (g) $= 10^{-3}$ kg.

EXERCISES

1.1.1 Find the number of nanoseconds in 0.4 s.

Answer 4×10^8

1.1.2 Find the number of kilometers in 2 miles if a mile equals 5280 ft.

Answer 3.22

1.1.3 Find the work done by a constant force of 300 μN applied to a mass of 8 g for a distance of 50 m.

Answer 15 mJ

1.2

CHARGE AND CURRENT

We are familiar with gravitational forces of attraction between bodies, which are responsible for holding us on the earth and which cause an apple dislodged from a tree to fall to the ground rather than to soar upward into the sky. There are bodies, however, that attract each other by forces far out of proportion to their masses. Also, such forces are observed to be repulsive as well as attractive and are clearly not gravitational forces.

We explain these forces by saying that they are electrical in nature and caused by the presence of *electrical charges*. We explain the existence of forces of both attraction and repulsion by postulating that there are two kinds of charges, positive and negative, and that unlike charges attract and like charges repel.

As we know, according to modern theory, matter is made up of atoms, which are composed of a number of fundamental particles. The most important of these particles are protons (positive charges) and neutrons (neutral, with no charge) found in the nucleus of the atom and electrons (negative charges) moving in orbit about the nucleus. Normally the atom is electrically neutral, the negative charge of the electrons balancing the positive charge of the protons. Particles may become positively charged by losing electrons to other particles and become negatively charged by gaining electrons from other particles.

As an example, we may produce a negative charge on a balloon by rubbing it against our hair. The balloon will then stick to a wall or the ceiling, which are uncharged. Relative to the negatively charged balloon, the neutral wall and ceiling are oppositely charged.

We now define the *coulomb* (C), discussed in the previous section, by stating that the charge of an electron is a negative one of 1.6021×10^{-19} coulombs. Putting it another way, a coulomb is the charge of about 6.24×10^{18} electrons. These are, of course, mind-boggling numbers, but their sizes enable us to use more manageable numbers, such as 2 C, in the circuit theory to follow.

The symbol for charge will be taken as Q or q, the capital letter usually denoting constant charges such as $Q = 4$ C, and the lowercase letter indicating a time-varying charge. In the latter case we may emphasize the time dependency by writing $q(t)$. This practice involving capital and lowercase letters will be carried over to the other electrical quantities as well.

The primary purpose of an electric circuit is to move or transfer charges along specified paths. This motion of charges constitutes an *electric current,* denoted by the letters i or I, taken from the French word "intensité." Formally, current is the time rate of change of charge, given by

$$i = \frac{dq}{dt} \tag{1.1}$$

The basic unit of current is the *ampere* (A), named for André Marie Ampère (1775–1836), a French mathematician and physicist who formulated laws of electro-magnetics in the 1820s. An ampere is 1 coulomb per second.

In circuit theory current is generally thought of as the movement of positive charges. This convention stems from Benjamin Franklin (1706–1790), who guessed that electricity traveled from positive to negative. We now know that in metal conductors the current is the movement of electrons that have been pulled loose from the orbits of the atoms of the metal. Thus we should distinguish *conventional* current (the movement of positive charges), which is used in electric network theory, and *electron* current. Unless otherwise stated, our concern will be with conventional current.

As an example, suppose the current in the wire of Fig. 1.3(a) is $I = 3$ A. That is, 3 C/s pass some specific point in the wire. This is symbolized by the arrow labeled 3 A, whose direction indicates that the motion is from left to right. This situation is equivalent to that depicted by Fig. 1.3(b), which indicates -3 C/s or -3 A in the direction from right to left.

FIGURE 1.3 Two representations of the same current

Figure 1.4 represents a general circuit element with a current i flowing from the left toward the right terminal. The total charge entering the element between time t_0 and t is found by integrating (1.1). The result is

$$q_T = q(t) - q(t_0) = \int_{t_0}^{t} i \, dt \tag{1.2}$$

We should note at this point that we are considering the network elements to be *electrically neutral.* That is, no net positive or negative charge can accumulate in the

FIGURE 1.4 Current flowing in a general element

element. A positive charge entering must be accompanied by an equal positive charge leaving (or, equivalently, an equal negative charge entering). Thus the current shown entering the left terminal in Fig. 1.4 must leave the right terminal.

There are several types of current in common use, some of which are shown in Fig. 1.5. A constant current, as shown in Fig. 1.5(a), will be termed a *direct current,* or dc. An *alternating current,* or ac, is a sinusoidal current, such as that of Fig. 1.5(b). Figures 1.5(c) and (d) illustrate, respectively, an *exponential* current and a *sawtooth* current.

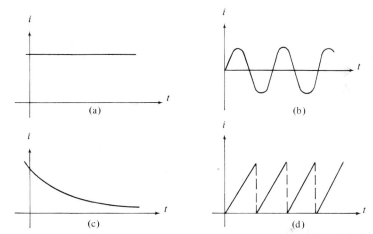

FIGURE 1.5 (a) Dc; (b) ac; (c) exponential current; (d) sawtooth current

There are many commercial uses for dc, such as in flashlights and power supplies for electronic circuits, and, of course, ac is the common household current found all over the world. Exponential currents appear quite often (whether we want them or not!) when a switch is actuated to close a path in an energized circuit. Sawtooth waves are useful in equipment, such as oscilloscopes, used for displaying electrical characteristics on a screen.

EXERCISES

1.2.1 Find the charge in picocoulombs represented by 1,000,000 electrons.
Answer 0.16021

1.2.2 The total charge entering a terminal of an element is given by

$$q = 6t^2 - 12t \text{ mC}$$

Find the current i at $t = 0$ and at $t = 3$ s.
Answer −12, 24 mA

1.2.3 The current entering a terminal is given by

$$i = 6t^2 - 2t \text{ A}$$

Find the total charge entering the terminal between $t = 1$ s and $t = 3$ s.
Answer 44 C

1.3

VOLTAGE, ENERGY, AND POWER

Charges in a conductor, exemplified by free electrons, may move in a random manner. However, if we want some concerted motion on their part, such as is the case with an electric current, we must apply an external or so-called *electromotive force* (EMF). Thus work is done on the charges. We shall define *voltage* "across" an element as the work done in moving a unit charge (+1 C) through the element from one terminal to the other. The unit of voltage, or *potential difference,* as it is sometimes called, is the *volt* (V), named in honor of the Italian physicist Alessandro Giuseppe Antonio Anastasio Volta (1745–1827), who invented the voltaic battery.

Since voltage is the number of joules of work performed on 1 coulomb, we may say that 1 V is 1 J/C. Thus the volt is a derived SI unit, expressible in terms of other units.

We shall represent a voltage by v or V and use the $+$, $-$ polarity convention shown in Fig. 1.6. That is, terminal A is v volts positive with respect to terminal B. Putting it another way in terms of potential difference, terminal A is at a *potential of* v volts higher than terminal B. In terms of work, it is clear that moving a unit charge from B to A requires v joules of work.

FIGURE 1.6 Voltage polarity convention

Some authors prefer to describe the voltage across an element in terms of voltage *drops* and *rises*. Referring to Fig. 1.6, a voltage drop of v volts occurs in moving from A to B. In contrast, a voltage rise of v volts occurs in moving from B to A.

As examples, Figs. 1.7(a) and (b) are two versions of exactly the same voltage. In (a), terminal A is $+5$ V above terminal B, and in (b), terminal B is -5 V above A (or $+5$ V below A).

We may also use a *double-subscript* notation v_{ab} to denote the potential of point a with respect to point b. In this case we have in general, $v_{ab} = -v_{ba}$. Thus in Fig. 1.7(a), $v_{AB} = 5$ V and $v_{BA} = -5$ V.

In transferring charge through an element work is being done, as we have said. Or, putting it another way, energy is being supplied. To know whether energy is being supplied *to* the element or *by* the element to the rest of the circuit, we must know not only the polarity of the voltage across the element, but also the direction of the current

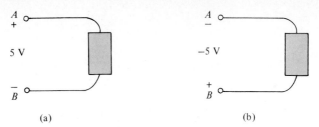

FIGURE 1.7 Two equivalent voltage representations

through the element. If a positive current enters the positive terminal, then an external force must be driving the current and is thus supplying or *delivering* energy to the element. The element is *absorbing* energy in this case. If, on the other hand, a positive current leaves the positive terminal (enters the negative terminal), then the element is delivering energy to the external circuit.

As examples, in Fig. 1.8(a) the element is absorbing energy. A positive current enters the positive terminal. This is also the case in Fig. 1.8(b). In Figs. 1.8(c) and (d) a positive current enters the negative terminal, and therefore the element is delivering energy in both cases.

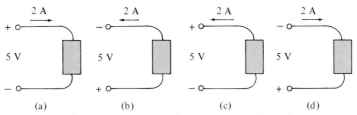

FIGURE 1.8 Various voltage–current relationships.

Let us consider now the *rate* at which energy is being delivered to or by a circuit element. If the voltage across the element is v and a small charge Δq is moved through the element from the positive to the negative terminal, then the energy absorbed by the element, say Δw, is given by

$$\Delta w = v\,\Delta q$$

If the time involved is Δt, then the rate at which the work is being done, or the energy w is being expended, is given by

$$\lim_{\Delta t \to 0} \frac{\Delta w}{\Delta t} = \lim_{\Delta t \to 0} v\frac{\Delta q}{\Delta t}$$

or

$$\frac{dw}{dt} = v\frac{dq}{dt} = vi \qquad (1.3)$$

Since by definition the rate at which energy is expended is power, denoted by p, we have

$$p = \frac{dw}{dt} = vi \qquad (1.4)$$

We might observe that (1.4) is dimensionally correct since the units of vi are (J/C)(C/s) or J/s, which is watts (W), defined earlier.

The quantities v and i are generally functions of time, which we may also denote by $v(t)$ and $i(t)$. Therefore p given by (1.4) is a time-varying quantity. It is sometimes called the *instantaneous* power because its value is the power at the instant of time at which v and i are measured.

Summarizing, the typical element of Fig. 1.9 is absorbing power, given by $p = vi$. If either the polarity of v or i (but not both) is reversed, then the element is delivering power, $p = vi$, to the external circuit. Of course, to say that an element delivers a negative power, say -10 W, is equivalent to saying that it absorbs a positive power, in this case $+10$ W.

FIGURE 1.9 Typical element with voltage and current

EXAMPLE 1.1   As examples, in Fig. 1.8(a) and (b) the element is absorbing power of $p = (5)(2) = 10$ W. [In Fig. 1.8(b) the 2 A leave the negative terminal, and thus 2 A enter the positive terminal.] In Figs. 1.8(c) and (d) it is delivering 10 W to the external circuit, since the 2 A leave the positive terminal, or, equivalently, -2 A enter the positive terminal.

Before ending our discussion of power and energy, let us solve (1.4) for the energy w delivered to an element between time t_0 and t. We have, upon integrating both sides between t_0 and t,

$$w(t) - w(t_0) = \int_{t_0}^{t} vi \, dt \qquad (1.5)$$

EXAMPLE 1.2   For example, the energy delivered to the element of Fig. 1.8(a) between $t = 0$ and $t = 2$ s is given by

$$w(2) - w(0) = \int_{0}^{2} (5)(2) \, dt = 20 \text{ J}$$

Since the left member of (1.5) represents the energy delivered to the element between t_0 and t, we may interpret $w(t)$ as the energy delivered to the element between the beginning of time and t and $w(t_0)$ as the energy between the beginning of time and t_0. In the beginning of time, which let us say is $t = -\infty$, the energy delivered to the element was zero; that is,

$$w(-\infty) = 0$$

If $t_0 = -\infty$ in (1.5), then we shall have the energy delivered to the element from the beginning up to t, given by

$$w(t) = \int_{-\infty}^{t} vi \, dt \qquad (1.6)$$

This is consistent with (1.5) since

$$w(t) = \int_{-\infty}^{t} vi \, dt$$

$$= \int_{-\infty}^{t_0} vi \, dt + \int_{t_0}^{t} vi \, dt$$

By (1.6) this may be written

$$w(t) = w(t_0) + \int_{t_0}^{t} vi \, dt$$

which is (1.5).

EXERCISES

1.3.1 Find i if the element shown is (a) absorbing power of $p = 18$ mW and (b) delivering to the external circuit a power $p = 12$ W.
Answer (a) 3 mA; (b) −2 A

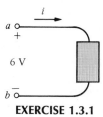

EXERCISE 1.3.1

1.3.2 If the current in Ex. 1.3.1 is $i = 3$ A, find (a) the power absorbed by the element and (b) the energy delivered to the element between 2 and 4 s.
Answer (a) 18 W; (b) 36 J

1.3.3 A two-terminal element absorbs an energy w as shown. If the current entering the positive terminal is

$$i = 100 \cos 1000\pi t \text{ mA}$$

find the element voltage at $t = 1$ ms and at $t = 4$ ms.
Answer −50, 5 V

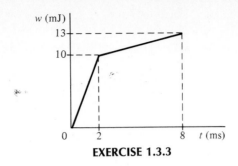

EXERCISE 1.3.3

1.4

PASSIVE AND ACTIVE ELEMENTS

We may classify circuit elements into two broad categories, *passive* elements and *active* elements, by considering the energy delivered to or by them.

A circuit element is said to be passive if the total energy delivered to it from the rest of the circuit is always nonnegative. That is, referring to (1.6), for all t we have

$$w(t) = \int_{-\infty}^{t} p(t) \, dt = \int_{-\infty}^{t} vi \, dt \geq 0 \tag{1.7}$$

The polarities of v and i are as shown in Fig. 1.9. As we shall see later, examples of passive elements are resistors, capacitors, and inductors.

An active element is one that is not passive, of course. That is, (1.7) does not hold for all time. Examples of active elements are generators, batteries, and electronic devices that require power supplies.

We are not ready at this stage to begin a formal discussion of the various passive elements. This will be done in later chapters. In this section we shall give a brief discussion of two very important active elements, the independent voltage source and the independent current source.

An *independent voltage source* is a two-terminal element, such as a battery or a generator, that maintains a specified voltage between its terminals. The voltage is completely independent of the current through the element. The symbol for a voltage source having v volts across its terminals is shown in Fig. 1.10. The polarity is as shown, indicating that terminal a is v volts above terminal b. Thus if $v > 0$, then terminal a is at a higher potential than terminal b. The opposite is true, of course, if $v < 0$.

In Fig. 1.10, the voltage v may be time varying, or it may be constant, in which case we would probably label it V. Another symbol that is often used for a constant voltage source, such as a battery with V volts across its terminals, is shown in Fig. 1.11. In the case of constant sources we shall use Figs. 1.10 and 1.11 interchangeably.

We might observe at this point that the polarity marks on Fig. 1.11 are redundant since the polarity could be defined by the positions of the longer and shorter lines. We shall leave the polarity marks off in most cases in the future. There are times, however, in analyzing circuits when it is convenient to use the polarity marks.

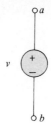

FIGURE 1.10 Independent voltage source

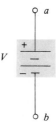

FIGURE 1.11 Constant voltage source

An *independent current source* is a two-terminal element through which a specified current flows. The current is completely independent of the voltage across the element. The symbol for an independent current source is shown in Fig. 1.12, where i is the specified current. The direction of the current is indicated by the arrow.

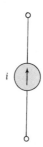

FIGURE 1.12 Independent current source

Independent sources are usually meant to deliver power to the external circuit and not to absorb it. Thus if v is the voltage across the source and its current i is directed out of the positive terminal, then the source is delivering power, given by $p = vi$, to the external circuit. Otherwise it is absorbing power. For example, in Fig. 1.13(a) the battery is delivering 24 W to the external circuit. In Fig. 1.13(b) the battery is absorbing 24 W, as would be the case when it is being charged.

The sources that we have discussed here, as well as the circuit elements to be considered later, are *ideal elements*. That is, they are *mathematical models* that approximate the actual or physical elements only under certain conditions. For example, an ideal automobile battery supplies a constant 12 V, no matter what external circuit is connected to it. Since its current is completely arbitrary, it could theoretically deliver an infinite amount of power. This, of course, is not possible in the case of an

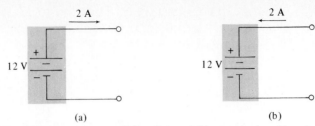

(a)   (b)

FIGURE 1.13 (a) Source-delivering and (b) source-absorbing power

actual device. A real 12-V automobile battery supplies approximately constant voltage only as long as the current it delivers is low. When the current exceeds a few hundred amperes, the voltage drops appreciably from 12 V.

We shall consider practical sources in a later chapter and see under what conditions they may be approximated by ideal sources. Also later, we shall consider *dependent* sources whose voltage (or current) is controlled by another voltage or current somewhere else in the circuit.

EXERCISES

1.4.1 Find the power being supplied by the sources shown.
Answer (a) 36; (b) 20; (c) −24; (d) −45 W

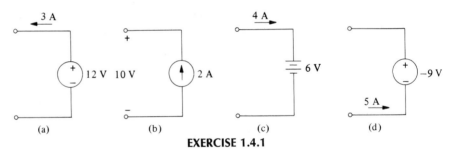

(a)   (b)   (c)   (d)

EXERCISE 1.4.1

1.4.2 The terminal voltage of a voltage source is $v = 6 \sin 2t$ V. If the charge leaving the positive terminal is $q = -2 \cos 2t$ mC, find the power supplied by the source at any time and the energy supplied by the source between 0 and t seconds.
Answer $24 \sin^2 2t$ mW, $12t - 3 \sin 2t$ mJ

1.5

CIRCUIT ANALYSIS

Let us now look at the words *circuit analysis,* which are contained in the title of the book, and see what they mean. Generally, if an electric circuit is subjected to an *input* or *excitation* in the form of, say, a voltage or a current provided by an independent source, then an *output* or *response* is produced. The output or response may also be a voltage or a current associated with some element in the circuit. There may be, of course, more than one input and more than one output.

There are two main branches of circuit theory, and they are derived from the following three key words: input, output, and circuit. The first branch is *circuit analysis,* which, given the circuit and the input, is concerned with finding the output. The other branch is *circuit synthesis,* which, given the input and output, is concerned with finding the circuit itself.

Network synthesis is much more complex in general than analysis and will probably be encountered by the student in a later course. Circuit analysis will be our concern in this book. We may be interested only in finding one or more outputs, such as a voltage or current existing somewhere in the circuit, or in determining the energy or power delivered to one element or another. Or we may wish to perform a complete analysis, finding every unknown current and voltage in the circuit. In any case, in the chapters to come we shall develop systematic methods of analysis that can be applied generally to any circuit of the type we consider. The methods not only will be systematic and general, but they also will be simple and straightforward to apply.

PROBLEMS

1.1 Let $f(t)$ in the graph shown be the current $i(t)$ in milliamperes entering an element terminal as a function of time. Find (a) the charge that enters the terminal between 0 and 10 s and (b) the rate at which the charge is entering at $t = 2$ s and $t = 4$ s.

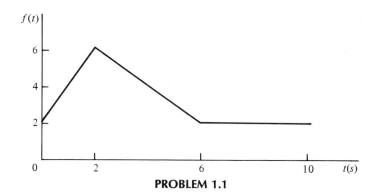

PROBLEM 1.1

1.2 Let $f(t)$ in Prob. 1.1 be the charge $q(t)$ in coulombs that has entered the positive terminal of an element as a function of time. Find (a) the total charge that has entered the terminal between 1 and 2 s, (b) the total charge that has entered between 6 s and 10 s, and (c) the current entering the terminal at 1 s, 4 s, and 8 s.

1.3 Find the power delivered to an element at $t = 1$ ms if the charge entering the positive terminal is

$$q = 10 \sin 250\pi t \text{ mC}$$

and the voltage is

$$v = 4 \cos 250\pi t \text{ V}$$

1.4 The power delivered to an element is $p = 18e^{-6t}$ mW and the charge entering the positive terminal of the element is $q = 3 - 3e^{-3t}$ mC. Find (a) the voltage across the element, and (b) the energy delivered to the element between 0 and 0.5 s.

1.5 The current entering the positive terminal of an element is $i = -2e^{-t}$ A. Find the power delivered to the element as a function of time and the energy delivered to the element between 0 and 1 s if (a) $v = 6i$, (b) $v = 3\, di/dt$, and (c) $v = 2 \int_0^t i\, dt + 4$. (*Note:* The element voltage v is in volts if the current i is in amperes.)

1.6 If the current entering the positive terminal of an element is $i = -2e^{-t}$ A and the voltage is $v = 6i$ V, find the energy delivered to the element between $t = 0$ and $t = 3$ s.

1.7 If a current $i = 2$ A is entering the positive terminal of a battery with voltage $v = 6$ V, then the battery is in the process of being charged. (It is absorbing rather than delivering power.) Find (a) the energy supplied to the battery, and (b) the charge delivered to the battery in 2 h (hours). Note the consistency of the units, 1 V = 1 J/C.

1.8 Find the current needed in Prob. 1.7 to deliver the same charge as in part (b) in 30 min.

1.9 Suppose the voltage v in Prob. 1.7 varies linearly from 6 to 18 V as t varies from 0 to 10 min. If $i = 2$ A during this time, find (a) the total energy supplied and (b) the total charge delivered to the battery.

1.10 Let the current entering the positive terminal of an element be $i = 0$ for $t < 0$, and $i = 6 \sin 2t$ A for $t > 0$. (a) If the voltage is $v = 4\, di/dt$ V, show that the energy delivered to the element is nonnegative for all time. (The element is passive.) (b) Repeat part (a) if

$$v = 2 \int_0^t i\, dt \text{ V}.$$

1.11 In Prob. 1.10(b) find the total charge delivered to the element at $t = \pi/4$ s and the power absorbed at $\pi/8$ s.

1.12 The voltage across an element is 4 V and the charge q entering the positive terminal is as shown. Find the power delivered to the element at $t = 6$ ms and the total charge and total energy delivered to the element between 1 and 10 ms.

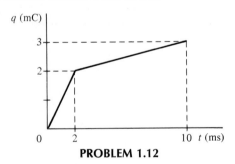

PROBLEM 1.12

1.13 If the graph of Prob. 1.12 is the current i (A) versus t (s), and $v = 2(di/dt)$ (V), find p at 1 s and at 6 s.

1.14 If the graph of Prob. 1.12 is the current i (A) versus t (s), and $p = 4i^2$, find p in terms of time on the intervals $0 < t < 2$ s and $2 < t < 10$ s.

1.15 If the graph of Prob. 1.12 is the power p (mW) versus t (ms), find the energy delivered to the element between 0 and 10 ms.

André Marie Ampère
1775–1836

2

Resistive Circuits

I shall call the first "electric tension" [voltage] and the second "electric current."

André Marie Ampère

On September 11, 1820, the exciting announcement was read to the French Academy of Sciences of the discovery by the Danish physicist Hans Christian Oersted that an electric current produces a magnetic effect. One member of the Academy, André Marie Ampère, a French mathematics professor, was highly impressed and within 1 week had repeated Oersted's experiment, given a mathematical explanation of it, and—in addition—discovered that electric currents in parallel wires exert a magnetic force on each other.

Ampère was born in Lyon, France, and at an early age had read all the great works in his father's library. At age 12 he was introduced to the Lyon library and because many of its best mathematical works were in Latin, he mastered that language in a few weeks. In spite of two crushing personal tragedies—at age 18 he witnessed his father's execution on the guillotine by the French Revolutionaries and later his young, beloved wife died suddenly after only 4 years of marriage—Ampère was a brilliant and prolific scientist. He formulated many of the laws of electricity and magnetism and was the father of electrodynamics. The unit of electric current, the *ampere*, was chosen in his honor in 1881. ■

The simplest and most commonly used circuit element is the resistor. All electrical conductors exhibit properties which are characteristic of a resistor. When currents flow in conductors, electrons which make up the current collide with the lattice of atoms in the conductor. This, of course, on the average, impedes or *resists* the motion of the electrons. The larger the number of collisions, the greater the *resistance* of the conductor. We shall consider a *resistor* to be any device which exhibits solely a resistance. Materials which are commonly used in fabricating resistors include metallic alloys and carbon compounds.

In this chapter, we shall first introduce the terminal relations for a resistor based on Ohm's law. Two laws necessary for systematic solutions of networks, known as Kirchhoff's laws, are then examined. With these laws, we shall begin our study of circuit analysis by finding solutions for "single-loop" and "single-node-pair" resistive networks having independent sources as inputs. We shall conclude the chapter with a discussion of simple measuring instruments followed by a discussion of practical resistors.

2.1

OHM'S LAW

Georg Simon Ohm (1787–1854), a German physicist, is credited with formulating the current-voltage relationship for a resistor based on experiments performed in 1826. In 1827 he published the results in a paper titled "The Galvanic Chain, Mathematically Treated." As a result of this work, the unit of resistance is called the *ohm*. It is ironic, however, that Henry Cavendish (1731–1810), a British chemist, discovered the same results 46 years earlier. Had he not failed to publish his findings, the unit of resistance might well be known as the *caven*.

Ohm's law states that the voltage across a resistor is directly proportional to the current flowing through the resistor. The constant of proportionality is the resistance value of the resistor in ohms. The circuit symbol for the resistor is shown in Fig. 2.1. For the current and voltage shown, Ohm's law is

$$v = Ri \qquad (2.1)$$

where $R \geq 0$ is the resistance in ohms.

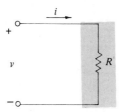

FIGURE 2.1 Circuit symbol for the resistor

The symbol used to represent the ohm is the capital Greek letter *omega* (Ω). Thus by (2.1) we have $R = v/i$, so that

$$1\ \Omega = 1\ \text{V/A}$$

In some applications, such as electronics circuits, the ohm is an inconveniently small unit and units such as *kilo-ohms* or simply *kilohms* (kΩ) and *mega-ohms* or *megohms* (MΩ) are common.

EXAMPLE 2.1   As an example, if $R = 3\ \Omega$ and $v = 6$ V in Fig. 2.1, the current is

$$i = \frac{v}{R} = \frac{6\ \text{V}}{3\ \Omega} = 2\ \text{A}$$

If R is changed to 1 kΩ, the current is

$$i = \frac{6\ \text{V}}{1\ \text{k}\Omega} = \frac{6\ \text{V}}{10^3\ \Omega} = 6 \times 10^{-3}\ \text{A} = 6\ \text{mA}$$

The process is obviously shortened by noting that 1 V/kΩ = 1 mA, 1 V/MΩ = 1 μA, and so forth.

Since R is constant, (2.1) is the equation of a straight line. For this reason, the resistor is called a *linear resistor*. A graph of v versus i is shown in Fig. 2.2, which is a line passing through the origin with a slope of R. Obviously, a straight line is the only graph possible for which the ratio of v to i is constant for all i.

Resistors whose resistances do not remain constant for different terminal currents are known as *nonlinear resistors*. For such a resistor, the resistance is a function of the current flowing in the device. A simple example of a nonlinear resistor is an incandescent lamp. A typical voltage-current characteristic for this device is shown in Fig. 2.3, where we see that the graph is no longer a straight line. Since R is not a constant, the analysis of a circuit containing nonlinear resistors is more difficult.

In reality, all practical resistors are nonlinear because the electrical characteristics of all conductors are affected by environmental factors such as temperature.

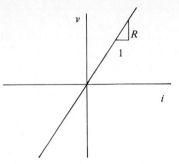

FIGURE 2.2 Voltage-current characteristic for a linear resistor

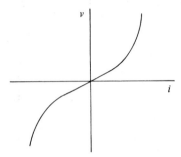

FIGURE 2.3 Typical voltage-current characteristic for a nonlinear resistor

Many materials, however, closely approximate an ideal linear resistor over a desired operating region. We shall concentrate on these types of elements and simply refer to them as resistors.

An examination of (2.1), in conjunction with Fig. 2.1, shows if $i > 0$ (current entering the upper terminal), then $v > 0$. Thus the current enters the terminal of higher potential and exits from that of the lower potential. Next, suppose that $i < 0$ (current entering the lower terminal). Then $v < 0$, and the lower terminal is higher in potential than the upper one. Once again, the current enters the terminal of higher potential. Since charges are transported from a higher to a lower potential in passing through the resistor, the energy lost by a charge q (energy $= qv$) is absorbed by the resistor in the form of heat. The rate at which energy is dissipated is, by definition, the instantaneous power

$$
p(t) = v(t)i(t) = Ri^2(t) = \frac{v^2(t)}{R} \tag{2.2}
$$

A graph of (2.2), shown in Fig. 2.4, reveals that $p(t)$ is a parabolic (and thus nonlinear) function of $i(t)$ or $v(t)$ which is always positive. (The horizontal scales are, of course, different in the two cases.) Thus, for a linear resistor, the instantaneous power is nonlinear even though the voltage-current relationship is linear.

The condition for passivity, given in (1.7), is

$$w(t) = \int_{-\infty}^{t} p(t)\, dt \geq 0$$

Therefore, since $p(t)$ is always positive, we see that the above integral is positive and that the resistor is, indeed, a passive element.

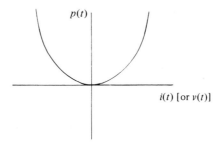

FIGURE 2.4 Graph of the instantaneous power for a resistor

The beginning student often encounters difficulty in determining the proper algebraic sign in applying Ohm's law when the voltage assignment differs from that of Fig. 2.1. If the current enters the negative voltage terminal, as in Fig. 2.5, Ohm's law is

$$v = -Ri$$

since in this case the sign of the voltage v has been changed from its value in Fig. 2.1.

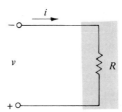

FIGURE 2.5 Resistor with a reversed voltage assignment

The voltage assignment has no effect on the power, however, since

$$p = \frac{v^2}{R} = \frac{(-v)^2}{R}$$

In addition to its resistance, a resistor is also characterized by its *power rating,* or *wattage rating,* which is the maximum power the resistor can dissipate without being damaged by overheating. Thus if a resistor is to dissipate a power p its power rating should be at least p and preferably higher. (The power used in the power rating is *average power,* to be discussed in Chapter 12, but for direct currents the average and instantaneous powers are the same.)

Another important quantity which is very useful in circuit analysis is known as *conductance,* defined by

$$G = \frac{1}{R} \tag{2.3}$$

The unit of conductance is the *mho* (A/V), which is ohm spelled backwards. The symbol representing the mho is an inverted omega $(\mho)$. (The SI unit for conductance is *siemens*, symbolized by S and named for the brothers Werner and William Siemens, two noted German engineers of the late nineteenth century. The mho, however, is still widely used in the United States.) Combining (2.1)–(2.3), we see that alternative expressions for Ohm's law and instantaneous power are

$$i = Gv \tag{2.4}$$

and

$$p(t) = \frac{i^2(t)}{G} = Gv^2(t) \tag{2.5}$$

As a final note, the concept of resistance may be used to define two very common circuit theory terms, *short circuit* and *open circuit*. A short circuit is an ideal conductor between two points and thus may be thought of as a resistance of zero ohms. It can carry any current, depending on the rest of the circuit, but the voltage across it is zero. Analogously, an open circuit is a break in the circuit through which no current can flow. Thus it may be considered to be an infinite resistance, and it may have any voltage, again depending on the rest of the circuit.

EXAMPLE 2.2

As an example, let us find the current i and the power absorbed by the 1-kΩ resistor of Fig. 2.6. From (2.3) and (2.4), $G = \frac{1}{1000} = 10^{-3}$ $\mho$ and $i = 10^{-3} \times 12$ A = 12 mA. Also, (2.5) yields $p(t) = 10^{-3} \times 12^2$ W = 144 mW, which is the minimum power rating required for the resistor.

The current in this example is a direct current since its value does not change with time. Suppose we now replace the 12-V source by the time-varying voltage $v = 10 \cos t$ V and repeat the above procedure. The current is

$$i = \frac{10 \cos t \text{ V}}{1 \text{ k}\Omega} = 10 \cos t \text{ mA}$$

FIGURE 2.6 Current-voltage example

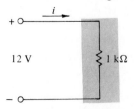

and the instantaneous power is

$$p = 0.1 \cos^2 t \text{ W}$$

which is always nonnegative. The current, in this case, is an alternating current.

EXERCISES

2.1.1 The terminal current of a 10-kΩ resistor is 5 mA. Find (a) the conductance, (b) the terminal voltage, and (c) the minimum wattage of the resistor.
Answer (a) 0.1 m℧; (b) 50 V; (c) 0.25 W

2.1.2 The instantaneous power absorbed by a 5-kΩ resistor is 2 $\sin^2$ 377t W. Find v and i.
Answer 100 sin 377t V, 20 sin 377t mA

2.1.3 Find i and the power delivered to the resistor.
Answer −10 μA, 0.5 mW

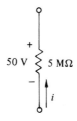

EXERCISE 2.1.3

2.2

KIRCHHOFF'S LAWS

Thus far we have considered Ohm's law and how it may be used to find the current, voltage, and power associated with a resistor. However, Ohm's law by itself cannot be used to analyze even the simplest circuit. In addition we must have two laws first stated by the German physicist Gustav Kirchhoff (1824–1887) in 1847. The two laws are formally known as Kirchhoff's current law and Kirchhoff's voltage law. These laws, together with the terminal characteristics for the various circuit elements, permit systematic methods of solution for any electrical network. We shall not attempt to prove Kirchhoff's laws here since the concepts necessary for the proof are developed in a later, interesting study of electromagnetic field theory.

A circuit consists of two or more circuit elements connected by means of perfect conductors. Perfect conductors are zero-resistance wires which allow current to flow freely but accumulate no charge and no energy. In this case, the energy can be considered to reside, or be lumped, entirely within each circuit element, and thus the network is called a *lumped-parameter circuit*.

A point of connection of two or more circuit elements is called a *node*. An example of a circuit with three nodes is shown in Fig. 2.7(a). Node 1 consists of the entire connection at the top of the circuit. The beginner quite often mistakes points *a*

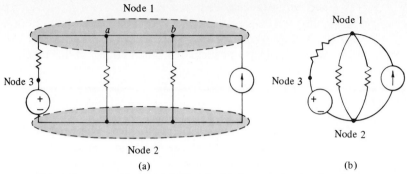

FIGURE 2.7 (a) Three-node circuit; (b) three-node circuit redrawn

and b for nodes. It should be noticed, however, that a and b are connected by a perfect conductor and can be considered electrically as being identical points. This is readily demonstrated by redrawing the circuit in the form of Fig. 2.7(b), where at node 1 all connections are shown at a single point. Similar comments apply for node 2. Node 3 is required for the interconnection of the independent voltage source and the resistor. With these concepts, we are now ready to discuss the all-important laws of Kirchhoff.

Kirchhoff's current law (KCL) states that

The algebraic sum of the currents entering any node is zero.

To demonstrate the use of this law, consider currents flowing into a node, as shown in Fig. 2.8. KCL states that

$$i_1 + i_2 + (-i_3) + i_4 = 0$$

where we recall that i_3 flowing out of the node is equivalent to $-i_3$ entering the node.

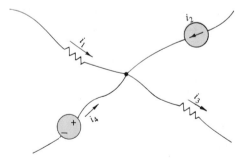

FIGURE 2.8 Currents flowing into a node

For the sake of argument, let us suppose that the sum is not zero. In such a case, we would have

$$i_1 + i_2 - i_3 + i_4 = \Psi \neq 0$$

where Ψ has units of C/s and hence must be the rate at which charges are accumulated in the node. However, a node consists of perfect conductors and cannot accumulate charges. In addition, a basic principle of physics states that charges can neither be created nor destroyed (conservation of charge). Therefore, our assumption is not valid, and Ψ must be zero, demonstrating the plausibility of KCL.

Suppose in our example we now consider the sum of the currents leaving the node. From Fig. 2.8, the sum is

$$-i_1 - i_2 + i_3 - i_4 = 0$$

or, multiplying both sides by -1, we see that

$$i_1 + i_2 - i_3 + i_4 = 0$$

which is identical to our previous result. This demonstrates an equivalent statement for KCL, which states that

The algebraic sum of the currents leaving any node is zero.

Let us now rearrange the above equation in the form

$$i_1 + i_2 + i_4 = i_3$$

where i_1, i_2, and i_4 are entering the node and i_3 is leaving. This form of the equation illustrates another statement for KCL, stated as

The sum of the currents entering any node equals the sum of the currents leaving the node.

In general, a mathematical expression of KCL is

$$\sum_{n=1}^{N} i_n = 0 \qquad\qquad (2.6)$$

where i_n is the nth current entering (or leaving) the node and N is the number of node currents.

EXAMPLE 2.3   As an example of KCL, let us find the current i in Fig. 2.9. Summing the currents entering the node, we have

$$5 + i - (-3) - 2 = 0$$

or

$$i = -6 \text{ A}$$

FIGURE 2.9 Example of KCL

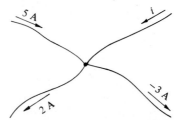

We note that i is -6 A entering the node is equivalent to 6 A leaving the node. Therefore it is not necessary to guess the correct current direction prior to solving the problem. We still arrive at the correct answer in the end.

We may find the current i more directly by considering it as entering the node and thus equating it to the other three currents leaving the node. The result is

$$i = -3 + 2 + (-5) = -6 \text{ A}$$

which agrees with the previous answer.

We now move on to Kirchhoff's voltage law (KVL), which states that

The algebraic sum of the voltages around any closed path is zero.

As an illustration, application of this statement to the closed path $abcda$ of Fig. 2.10 gives

$$-v_1 + v_2 - v_3 = 0 \tag{2.7}$$

where the algebraic sign for each voltage has been taken as positive when going from $+$ to $-$ (higher to lower potential) and negative when going from $-$ to $+$ (lower to higher potential) in traversing the element. Using this convention we are equating the sum of the voltage drops around the loop to zero. We could as well use the opposite convention, in which case the sum of the voltage rises is zero.

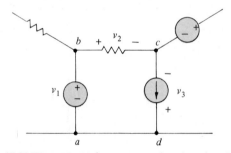

FIGURE 2.10 Voltages around a closed path

As in the case of KCL, we shall not attempt a proof of KVL. However, to illustrate the plausibility of (2.7), let us assume that its right member is not zero. That is,

$$-v_1 + v_2 - v_3 = \Phi \neq 0$$

The left member of this equation is by definition the work required to move a unit charge around the path $adcba$. A lumped-parameter circuit is a *conservative* system, which means that the work required to move a charge around any closed path is zero. (Proof of this will come in a later course involving an interesting study of electromagnetic theory.) Thus, our assumption is not valid, and Φ is indeed zero.

We should point out, however, that all electrical systems are not conservative. In fact, electrical power generation, radio waves, and sunlight, to mention only a few, are consequences of nonconservative systems.

The application of KVL is independent of the direction in which the path is traversed. Consider, for example, the path *adcba* in Fig. 2.10. Summing the voltages, we find

$$v_3 - v_2 + v_1 = 0$$

which is equivalent to (2.7).

In general, a mathematical representation for KVL is

$$\sum_{n=1}^{N} v_n = 0 \qquad (2.8)$$

where v_n is the nth voltage in a loop of N voltages. The sign of each voltage is chosen as described earlier for (2.7).

EXAMPLE 2.4

As an example of the use of KVL, let us find v in Fig. 2.11. Traversing the circuit in a clockwise direction, we have

$$-5 + v + 10 - 2 = 0$$

or $v = -3$ V. Suppose we now perform a counterclockwise traversal. In such a case,

$$5 + 2 - 10 - v = 0$$

or $v = -3$ V, which, of course, is the same result obtained for the clockwise traversal.

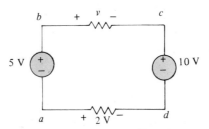

FIGURE 2.11 Circuit to illustrate KVL

Still another version of KVL for Fig. 2.11 yields

$$10 + v = 2 + 5$$

where the sum of the voltages with one polarity is equated to the sum of the voltages with the opposite polarity. (Stated another way, the voltage rises equal the voltage drops.)

Finally, we may solve for v directly by noting that it is the voltage v_{bc} and thus is equal to the sum of the voltages from b to c through the other three elements. That is,

$$v = 5 + 2 - 10 = -3 \text{ V}$$

In each of the previous examples, KVL has been applied around conducting paths, such as *abcda* above. The law, however, is valid for *any* closed path. Consider, for instance, the path *acda* of Fig. 2.11. We note that movement directly from *a* to *c* is not along a conducting path. Applying KVL to this closed path yields $v_{ac} + 10 - 2 = 0$, where v_{ac} is the potential of point *a* with respect to *c*. Thus $v_{ac} = -8$ V. We could also have chosen the path *abca*, for which

$$-5 + v - v_{ac} = -5 - 3 - v_{ac} = 0$$

Therefore, $v_{ac} = -8$ V, which demonstrates the use of different closed paths to obtain the same result.

EXAMPLE 2.5

As another example of the application of KCL and KVL, consider finding i_x and v_x in the network of Fig. 2.12. Summing the currents entering node *a* gives $-4 + 1 + i_1 = 0$, or $i_1 = 3$ A. At node *b*, $-i_1 + 2 - i_2 = 0$, or $i_2 = -1$ A. At node *c*, $i_2 + i_3 - 3 = 0$, or $i_3 = 4$ A. Therefore, at node *d*, $-i_x - 1 - i_3 = 0$, or $i_x = -5$ A. Next, KVL about the path *abcda* gives $-10 + v_2 - v_x = 0$. From Ohm's law $v_2 = 5i_2 = -5$ V. Therefore, $v_x = -15$ V.

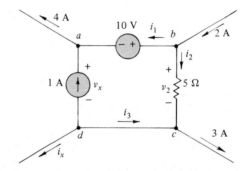

FIGURE 2.12 Network for example of KCL and KVL

Before concluding our discussion of Kirchhoff's laws, consider the network of

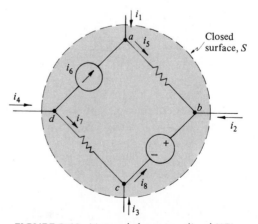

FIGURE 2.13 Network for generalized KCL

Fig. 2.13, in which several elements are shown within a closed surface S. We recall that the current entering each element equals that leaving the device so that each element stores zero net charge. Therefore, the total net charge stored within the surface is zero, requiring that

$$i_1 + i_2 + i_3 + i_4 = 0$$

This result illustrates a generalization of KCL, which states

The algebraic sum of the currents entering any closed surface is zero.[1]

To illustrate the plausibility of the generalized KCL, let us write KCL equations at nodes a, b, c, and d of Fig. 2.13. The results are

$$i_1 = i_5 - i_6$$
$$i_2 = -i_5 - i_8$$
$$i_3 = -i_7 + i_8$$
$$i_4 = i_6 + i_7$$

Adding these equations yields

$$i_1 + i_2 + i_3 + i_4 = 0$$

as previously noted.

From the generalized KCL, we see immediately in Fig. 2.12 for a surface enclosing points a, b, c, and d that $-i_x - 4 + 2 - 3 = 0$, or $i_x = {}^-5$ A.

EXERCISES

2.2.1 Find i and v_{ab}.
Answer 3 A, 19 V

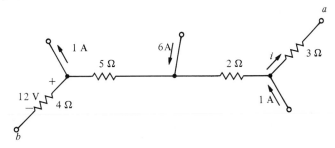

EXERCISE 2.2.1

2.2.2 Find v_x.
Answer 31 V

---

[1] The surface cannot pass through an element, which is considered to be concentrated at a point in lumped-parameter circuits.

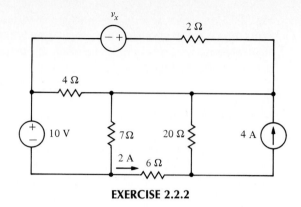

EXERCISE 2.2.2

2.3

SERIES RESISTANCE AND VOLTAGE DIVISION

Now that the laws of Ohm and Kirchhoff have been introduced, we are prepared to analyze resistive circuits. We begin with a *simple circuit,* which we define as one that can be completely described by a single equation. One type, which we shall consider in this section, is a circuit consisting of a single closed path, or loop, of elements. By KCL each element has a common current, say i. Then Ohm's law and KVL applied around the loop yield a single equation in i that completely describes the circuit.

Elements are said to be connected in *series* when they all carry the same current. Clearly, the networks of this section consist entirely of elements connected in series. An important circuit of this type, consisting of two resistors and an independent voltage source, provides an excellent starting point. We shall first analyze this special case and then develop the more general case.

A single-loop circuit having two resistors and an independent voltage source is shown in Fig. 2.14(a). The first step in the analysis procedure is the assignment of currents and voltages to all elements in the network. In this circuit, it is obvious from KCL that all elements carry the same current. We may *arbitrarily* call this current i in the direction shown (clockwise). (The novice often attempts to guess the true current direction in making assignments. The correct assignment is not necessary, as we will see, and usually is not possible, even for the expert.) We next make the voltage

FIGURE 2.14 (a) Single-loop circuit; (b) equivalent circuit

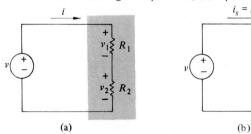

assignments for R_1 and R_2 as v_1 and v_2, respectively. These assignments are also arbitrary but in the figure have been chosen to satisfy Ohm's law for a positive algebraic sign.

The second step in the analysis is the application of KVL, which yields

$$v = v_1 + v_2$$

where, from Ohm's law,

$$v_1 = R_1 i$$

$$v_2 = R_2 i \tag{2.9}$$

Combining these equations, we find

$$v = R_1 i + R_2 i$$

Solving for i yields

$$i = \frac{v}{R_1 + R_2} \tag{2.10}$$

Let us now consider a simple circuit consisting of the voltage source v connected to a resistance R_S, as shown in Fig. 2.14(b). If R_S is selected such that

$$i_S = \frac{v}{R_S} = i \tag{2.11}$$

then the network is called an *equivalent circuit* of Fig. 2.14(a) because an identical current (response) is produced for a voltage v (excitation). In general, two circuits are said to be equivalent when they exhibit identical voltage-current relationships at their terminals. In other words, a source at the terminals sees the same resistance in each circuit.

Comparing (2.10) and (2.11), we see that

$$R_s = R_1 + R_2 \tag{2.12}$$

Interpreting (2.12), we see that if the series combination of R_1 and R_2 is replaced by the resistor R_s, the same current flows from the source v. Therefore R_s is the *equivalent resistance* of the series connection.

Combining (2.9) and (2.10), we see that

$$v_1 = \frac{R_1}{R_1 + R_2} v$$

$$v_2 = \frac{R_2}{R_1 + R_2} v \tag{2.13}$$

The potential of source v divides between resistances R_1 and R_2 in direct proportion to their resistances, demonstrating the *principle of voltage division* for two series resistors. We see in (2.13) that the greater voltage appears across the larger resistor.

The instantaneous powers absorbed by R_1 and R_2 are

$$p_1 = \frac{v_1^2}{R_1} = \frac{R_1}{(R_1 + R_2)^2} v^2$$

and

$$p_2 = \frac{v_2^2}{R_2} = \frac{R_2}{(R_1 + R_2)^2} v^2$$

respectively. The total power absorbed is

$$p_1 + p_2 = \frac{v^2}{R_1 + R_2} = v\left(\frac{v}{R_1 + R_2}\right) = vi$$

The power delivered by the source also equals vi, indicating that the power delivered by the source equals that absorbed by R_1 and R_2. This result is known as *conservation of power* (sometimes also referred to as Tellegen's theorem), a property which is often useful in circuit analysis.

Let us now digress briefly and repeat the analysis for the counterclockwise current i_1 of Fig. 2.15. Application of KVL yields

$$v = v_1 + v_2$$

in which

$$v_1 = -R_1 i_1$$

$$v_2 = -R_2 i_1$$

Combining these equations, we have

$$v = -R_1 i_1 - R_2 i_1$$

from which

$$i_1 = -\frac{v}{R_1 + R_2}$$

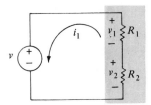

FIGURE 2.15 Single-loop circuit with counterclockwise current

Comparing our result with (2.10), we see that $i_1 = -i$, which is also evident from an inspection of Figs. 2.14(a) and 2.15. Hence, if the proper direction for positive current happens not to be chosen, a negative sign will result. This, of course, means that the positive current flows in the opposite direction. Similar statements apply to voltage assignments. If the polarity for positive voltage happens not to be chosen, a negative sign will occur in its solution, leading, of course, to the correct answer.

EXAMPLE 2.6
As an example of the utility of the preceding analysis, suppose $v = 120 \sin t$ V, $R_1 = 90 \ \Omega$, and $v_1 = 72 \sin t$ V in Fig. 2.14(a). Let us now determine R_2 R_s, i, and the instantaneous power associated with each element. From (2.13)

$$72 \sin t = \frac{90}{90 + R_2} \, 120 \sin t$$

which yields $R_2 = 60 \ \Omega$. Hence, $R_s = R_1 + R_2 = 150 \ \Omega$, and, from (2.10), $i = (120 \sin t)/150 = 0.8 \sin t$ A. The instantaneous power to R_1 and R_2 is $p_1 = R_1 i^2 = 57.6 \sin^2 t$ W and $p_2 = R_2 i^2 = 38.4 \sin^2 t$ W. Thus the power delivered by the source is $96 \sin^2 t$ W. We also observe that the power absorbed by R_1 and R_2 should equal that which is absorbed by R_s in Fig. 2.14(b). The power to R_s is $R_s i^2 = 96 \sin^2 t$ W.

Let us now extend our analysis to include the series connection of N resistors and an independent voltage source, as shown in Fig. 2.16. KVL gives

$$v = v_1 + v_2 + \ldots + v_N$$

in which

$$v_1 = R_1 i$$

$$v_2 = R_2 i$$

$$\vdots$$

$$v_N = R_N i$$

(2.14)

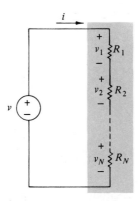

FIGURE 2.16 Single-loop circuit with N series resistors

Therefore,

$$v = R_1 i + R_2 i + \ldots + R_N i$$

Solving this equation for i yields

$$i = \frac{v}{R_1 + R_2 + \ldots + R_N} \qquad (2.15)$$

Let us now select R_s in the circuit of Fig. 2.14(b) so that (2.11) is satisfied. Equivalence of (2.11) and (2.15) requires that

$$R_s = R_1 + R_2 + \ldots + R_N = \sum_{n=1}^{N} R_n \qquad (2.16)$$

Therefore, the equivalent resistance of N series resistors is simply the sum of the individual resistances.

Substituting (2.15) and (2.16) into (2.14), we find

$$
\begin{aligned}
v_1 &= \frac{R_1}{R_s} v \\[2mm]
v_2 &= \frac{R_2}{R_s} v \\[2mm]
&\quad . \\
&\quad . \\
&\quad . \\
v_N &= \frac{R_N}{R_s} v
\end{aligned}
\qquad (2.17)
$$

which are the equations describing the voltage division property for N series resistors. Again, we see that the voltage divides in direct proportion to the resistance.

The instantaneous power delivered to the series combination, from (2.2) and (2.17), is

$$
\begin{aligned}
p &= \frac{v_1^2}{R_1} + \frac{v_2^2}{R_2} + \ldots + \frac{v_N^2}{R_N} \\[2mm]
&= \frac{R_1}{R_s^2} v^2 + \frac{R_2}{R_s^2} v^2 + \ldots + \frac{R_N}{R_s^2} v^2 \\[2mm]
&= \frac{v^2}{R_s} = vi
\end{aligned}
$$

This power is equal to that delivered by the source, verifying conservation of power for the series connection of N resistors.

EXERCISES

2.3.1 Find (a) the equivalent resistance seen by the source, (b) the current i, (c) the power delivered by the source, (d) v_1 (e) v_2, and (f) the minimum wattage required for R_1. *Answer* (a) 12 Ω; (b) 0.5 A; (c) 3 W; (d) 3 V; (e) -2 V; (f) 1 W

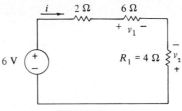

EXERCISE 2.3.1

2.3.2 In Fig. 2.14(a), $v = 16e^{-t}$ V, $v_2 = 4e^{-t}$ V, and $R_1 = 24$ Ω. Find (a) R_2, (b) the instantaneous power delivered to R_2, and (c) the current i.
Answer (a) 8 Ω; (b) $2e^{-2t}$ W; (c) $0.5e^{-t}$ A

2.3.3 A resistive load[2] requires 4 V and dissipates 2 W. A 12-V storage battery is available to operate the load. Referring to Fig. 2.14(a), if R_2 represents the load and v the 12-V battery, find (a) the current i, (b) the necessary resistance R_1, and (c) the minimum wattage of R_1.
Answer (a) 0.5 A; (b) 16 Ω; (c) 4 W

2.3.4 If $v_1 = v/8$ and the power delivered by the source is 8 mW, find R, v, v_1, and i.
Answer 18 kΩ, 16 V, 2 V, 0.5 mA

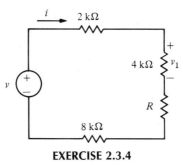

EXERCISE 2.3.4

2.4

PARALLEL RESISTANCE AND CURRENT DIVISION

Another important simple circuit is the single-node-pair resistive circuit. In analyzing these networks, we shall first examine a special case and then develop the more general case, as was performed for the single-loop network.

Elements are connected in *parallel* when the same voltage is common to each of them. A single-node-pair circuit consisting of the parallel connection of two resistors and an independent current source is shown in Fig. 2.17(a). To begin the analysis, we first assign voltages and currents to each circuit element. As in the case of the single-loop circuit, the assignments are completely arbitrary. We have chosen the assignments given by i_1, i_2, and v.

Next, we need to apply either KCL or KVL. Inspection of the circuit shows that

---

[2]A *load* is an element or collection of elements connected between the output terminals. In this case, the load is a resistor.

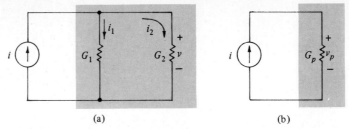

FIGURE 2.17 (a) Single-node-pair circuit; (b) equivalent circuit

a single node (either the upper or the lower) is common to all elements. This suggests that for single-node-pair circuits KCL is the most efficient choice. Applying KCL at the upper node yields

$$i = i_1 + i_2$$

where, from Ohm's law,

$$i_1 = G_1 v$$
$$i_2 = G_2 v$$

(2.18)

Combining these equations gives

$$i = G_1 v + G_2 v$$

and solving for v, we find

$$v = \frac{i}{G_1 + G_2}$$

(2.19)

In the circuit of Fig. 2.17(b), if G_p is selected such that

$$v_p = \frac{i}{G_p} = v$$

(2.20)

then the network is an equivalent circuit to that of Fig. 2.17(a). Comparing (2.19) and (2.20), we see that

$$G_p = G_1 + G_2$$

(2.21)

Clearly, G_p is the equivalent conductance of the two parallel conductances. In terms of resistances, (2.21) becomes

$$G_p = \frac{1}{R_p} = \frac{1}{R_1} + \frac{1}{R_2}$$

or

$$R_p = \frac{R_1 R_2}{R_1 + R_2}$$

(2.22)

Therefore, the equivalent resistance of two resistors connected in parallel is equal to the product of their resistances divided by their sum. It is interesting to note that G_p is greater than either G_1 or G_2; therefore, R_p is less than either R_1 or R_2. From this result, we see that connecting resistors in parallel reduces the overall resistance. In the special case $R_2 = R_1$, we see from (2.22) that $R_p = R_1/2$.

Substituting (2.19) into (2.18) gives

$$i_1 = \frac{G_1}{G_1 + G_2} i$$

$$i_2 = \frac{G_2}{G_1 + G_2} i$$

(2.23)

The current of the source i divides between conductances G_1 and G_2 in direct proportion to their conductances, demonstrating the *principle of current division*. It is common practice to give the resistor values in circuit diagrams in ohms (resistance) and not mhos (conductance). In terms of resistance values, (2.23) becomes

$$i_1 = \frac{R_2}{R_1 + R_2} i$$

$$i_2 = \frac{R_1}{R_1 + R_2} i$$

(2.24)

Therefore, the current divides in inverse proportion to the resistances. We see that the larger current flows through the smaller resistance. The power absorbed by the parallel combination is

$$p_1 + p_2 = R_1 i_1^2 + R_2 i_2^2$$

$$= \frac{R_2^2 i^2}{(R_1 + R_2)^2} R_1 + \frac{R_1^2 i^2}{(R_1 + R_2)^2} R_2$$

$$= \frac{R_1 R_2}{R_1 + R_2} i^2 = vi$$

which equals that delivered to the network by the current source.

EXAMPLE 2.7   Suppose in Fig. 2.17(a) that $R_1 = 3\ \Omega$, $R_2 = 6\ \Omega$, and $i = 3$ A. Then, from (2.22), $R_p = (3)(6)/(3 + 6) = 2\ \Omega$. From (2.24), $i_1 = \frac{6}{9}(3) = 2$ A, and $i_2 = \frac{3}{9}(3) = 1$ A. The voltage $v = R_1 i_1 = R_2 i_2 = (3)(2) = 6$ V. The current of the source flows through an equivalent resistance of R_p. Hence the voltage is also given by $v = R_p i = (2)(3) = 6$ V.

Let us now consider the more general case of N parallel conductances and an independent current source, as shown in Fig. 2.18. KCL gives

FIGURE 2.18 Single-node pair circuit with N parallel conductances

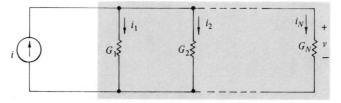

$$i = i_1 + i_2 + \ldots + i_N$$

for which

$$i_1 = G_1 v$$

$$i_2 = G_2 v$$

$$\cdot$$
$$\cdot$$ \hspace{4cm} (2.25)
$$\cdot$$

$$i_N = G_N v$$

Therefore, we have

$$i = G_1 v + G_2 v + \ldots + G_N v$$

from which

$$v = \frac{i}{G_1 + G_2 + \ldots + G_N} \hspace{3cm} (2.26)$$

If we now select G_p in Fig. 2.17(b) such that (2.20) is satisfied, then (2.26) requires that

$$G_p = G_1 + G_2 + \ldots + G_N = \sum_{i=1}^{N} G_i \hspace{2cm} (2.27)$$

In terms of resistances, this equation becomes

$$\frac{1}{R_p} = \frac{1}{R_1} + \frac{1}{R_2} + \ldots + \frac{1}{R_n} = \sum_{i=1}^{N} \frac{1}{R_i} \hspace{2cm} (2.28)$$

Hence the reciprocal of the equivalent resistance is simply the sum of the reciprocals of the resistances.

Combining (2.25)–(2.28), we find

$$i_1 = \frac{G_1}{G_p} i = \frac{R_p}{R_1} i$$

$$i_2 = \frac{G_2}{G_p} i = \frac{R_p}{R_2} i$$

$$\cdot$$
$$\cdot$$ \hspace{4cm} (2.29)
$$\cdot$$

$$i_N = \frac{G_N}{G_p} i = \frac{R_p}{R_N} i$$

Again the currents divide in inverse proportion to the resistances.

EXAMPLE 2.8 We observe in (2.28) that for $N > 2$ an expression for R_p is more complicated than (2.22). Formulas could be obtained, of course, for $N = 3, 4$, etc., but it is usually easier to apply (2.28) directly. As an example, suppose for $N = 3$ that $R_1 = 4\ \Omega$, $R_2 = 12\ \Omega$, and $R_3 = 6\ \Omega$. Then

$$\frac{1}{R_p} = \frac{1}{4} + \frac{1}{12} + \frac{1}{6} = \frac{1}{2}\ \mho$$

and $R_p = 2\ \Omega$.

EXAMPLE 2.9 Let us now find the equivalent resistance R_{eq} of the network of Fig. 2.19(a), as viewed from terminals x-y. Such reductions are very helpful in analyzing many types of circuits, as we shall see in the next section. The process is carried out by successive combinations of parallel and series connected resistors. In Fig. 2.19(a), the student often errs in taking combinations such as the 7- and 12-Ω resistors to be in series. We see, however, that at node a, a current in the 7-Ω resistor would divide between the 1- and 12-Ω resistors; hence they cannot be in series. The 1- and 5-Ω resistors, however, would carry the same current. Therefore they are in series, having an equivalent resistance of 6 Ω as shown in Fig. 2.19(b). We now observe that the same voltage would occur across the 6- and 12-Ω resistors, indicating a parallel connection having an equivalent resistance of $(6)(12)/(6 + 12) = 4\ \Omega$, as shown in Fig. 2.19(c). It is apparent that the 7- and 4-Ω resistors of this network are in series, yielding an equivalent resistance for the entire network of 11 Ω [Fig. 2.19(d)]. Therefore, from terminals x-y, the network could be replaced by a single resistor of 11 Ω. This is useful in determining, for instance, the power delivered by a source connected to terminals x-y. Suppose a 22-V source is applied. Then the current flowing from the source is $i = \frac{22}{11} = 2$ A, which gives an instantaneous power $p(t) = (22)(2) = 44$ W delivered to the resistor network.

FIGURE 2.19 Steps in determining the equivalent resistance of a network

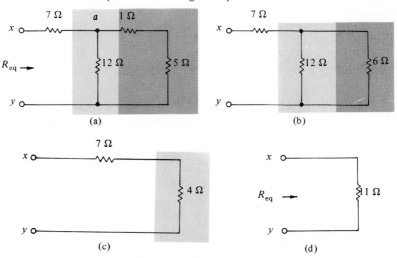

(a)

(b)

(c)

(d)

EXERCISES

2.4.1 Find the equivalent resistance seen by the source and use the result to find i, i_1, and v.

Answer 10 Ω, 9 A, 8 A, 72 V

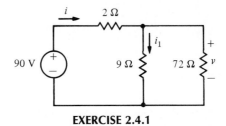

EXERCISE 2.4.1

2.4.2 In Fig. 2.18, if $N = 3$, $R_1 = 9$ Ω, $R_2 = 72$ Ω, and $v = 12 \sin t$ V and the instantaneous power delivered by the current source is $24 \sin^2 t$ W, find (a) R_3, (b) i, and (c) i_3.

Answer (a) 24 Ω; (b) $2 \sin t$ A; (c) $0.5 \sin t$ A

2.4.3 A load requires 4 A and absorbs 24 W. If only a 6-A current source is available, find the required resistance to place in parallel with the load.

Answer 3 Ω

2.4.4 Find the equivalent resistance seen by the source and the current i.

Answer 10 Ω, 0.2 A

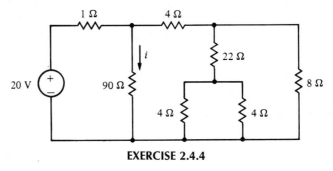

EXERCISE 2.4.4

2.5

ANALYSIS EXAMPLES

EXAMPLE 2.10 For our first example, let us consider a single-loop circuit containing three resistors and two independent voltage sources, as shown in Fig. 2.20(a). KVL and Ohm's law give

$$-20 + 20i + 30i + 30 + 50i = 0$$

Thus the reduced equation is

$$10 + 100i = 0$$

for which $i = -0.1$ A and $v_1 = 30i = -3$ V.

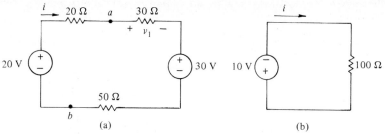

FIGURE 2.20 (a) Single-loop circuit; (b) equivalent circuit

We see that the reduced equation is satisfied by the equivalent circuit of Fig. 2.20(b), where the voltage sources are replaced by their algebraic sum and the resistances by their series equivalent. From this circuit, the power absorbed by all resistors is $p_{100\,\Omega} = 100i^2 = 1$ W. Therefore, the net power delivered by the two voltage sources must be 1 W. We recall that the power absorbed by an element is the product of the voltage and the current, where the current enters the positive voltage terminal. For the 20-V source, we see that $p_{20\,\text{V}} = (20)(-i) = (20)(0.1) = 2$ W absorbed. Physically, this means that current is flowing into the positive terminal and that the source is receiving power, or is being charged. The absorbed power in the 30-V source is $p_{30\,\text{V}} = 30i = (30)(-0.1) = -3$ W. The minus sign indicates that the source is delivering 3 W to the circuit. Hence the net power from the two sources is 1 W, which is the power delivered to the resistors. Finally, let us find the potential of point a with respect to point b, which we denote by v_{ab}. From KVL, $-20 + 20i + v_{ab} = 0$, or $v_{ab} = 20 - 20i = 20 - (20)(-0.1) = 18$ V.

EXAMPLE 2.11   As a second example, let us find i and v for the circuit of Fig. 2.21, in which three conductances and two independent current sources are connected in parallel. Applying KCL to the upper node yields

$$10 \sin \pi t - 0.01v - 0.02v - 5 - 0.07v = 0$$

or

$$(10 \sin \pi t - 5) - 0.1v = 0$$

It is apparent from this result that a current source of $(10 \sin \pi t - 5)$ A connected to a conductance of 0.1 ℧ (a 10-Ω resistor) would be an equivalent circuit as far as v is concerned [see Fig. 2.17(b)].

FIGURE 2.21 Single-node-pair circuit

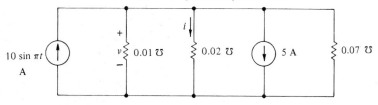

Solving for v and subsequently for i, we have

$$v = 100 \sin \pi t - 50 \text{ V}$$

$$i = 0.02v = 0.02(100 \sin \pi t - 50) = 2 \sin \pi t - 1 \text{ A}$$

Now let us consider conservation of power for the circuit of Fig. 2.21. The total power absorbed by the conductances is

$$p_{abs} = G_p v^2 = 0.1(100 \sin \pi t - 50)^2$$

$$= 1000 \sin^2 \pi t - 1000 \sin \pi t + 250 \text{ W}$$

The power delivered by the leftmost source is

$$p_1 = 10 \sin \pi t (100 \sin \pi t - 50) \text{ W}$$

Similarly, the 5-A source delivers

$$p_2 = -5(100 \sin \pi t - 50) \text{ W}$$

Thus the total power delivered by these sources is

$$p_{del} = p_1 + p_2 = 1000 \sin^2 \pi t - 1000 \sin \pi t + 250 \text{ W}$$

which equals that absorbed by the conductances.

EXAMPLE 2.12 As a third example, let us find i, v, the power delivered by the source, and the power absorbed by the 8-Ω load of Fig. 2.22(a). We begin by obtaining successive combinations of parallel and series resistor connections. The 4- and 8-Ω resistances (in series) add to give 12 Ω. These 12 Ω are in parallel with the 6-Ω resistor, giving an equivalent value of $(12)(6)/(12 + 6) = 4 \Omega$ [Fig. 2.22(b)]. We now add

FIGURE 2.22 Circuit for analysis example using voltage division

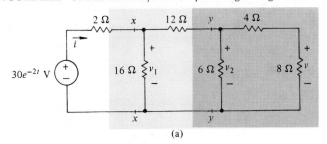

(a)

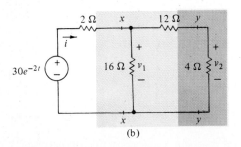

(b)

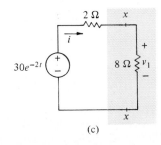

(c)

the 12- and 4-Ω resistances, which are in parallel with the 16 Ω, giving $(16)(16)/(16 + 16) = 8$ Ω, as shown in Fig. 2.22(c). This is the equivalent resistance as seen looking into the circuit at x-x. Applying KVL to the simplified circuit, we have

$$-30e^{-2t} + 2i + 8i = 0$$

from which

$$i = 3e^{-2t} \text{ A}$$

Therefore, the power delivered by the source is

$$p_s = (30e^{-2t})(3e^{-2t}) = 90e^{-4t} \text{ W}$$

By voltage division we see that

$$v_1 = \left(\frac{8}{2 + 8}\right)30e^{-2t} = 24e^{-2t} \text{ V}$$

which is the voltage across points x-x in the circuit. Proceeding to Fig. 2.22(b), we see that v_1 is the voltage across the series combination of the 12- and 4-Ω resistors; hence, again using voltage division, we find

$$v_2 = \left(\frac{4}{12 + 4}\right)v_1 = 6e^{-2t} \text{ V}$$

which is the voltage across points y-y in the circuit. In Fig. 2.22(a), it is obvious that v_2 is the voltage across the series connection of the 4- and 8-Ω resistors. Therefore, voltage division requires that

$$v = \left(\frac{8}{8 + 4}\right)v_2 = 4e^{-2t} \text{ V}$$

The power to the 8-Ω load is

$$p_2 = \frac{v^2}{8} = 2e^{-4t} \text{ W}$$

EXAMPLE 2.13   As a final example, let us find the current i in Fig. 2.23(a). We first combine the two 6-Ω resistors on the right of Fig. 2.23(a) and obtain the equivalent resistance which is in parallel with the 4-Ω resistor. The result, $(4)(12)/(4 + 12) = 3$ Ω, is shown in Fig. 2.23(b). If we now replace the two series resistors to the right of points x-x and the parallel 3- and 6-Ω resistors to the left of x-x by their respective series and parallel equivalents, we obtain the circuit of Fig. 2.23(c). Using current division, we find

$$i_1 = \left(\frac{2}{2 + 6}\right) \times 12 = 3 \text{ A}$$

A second application of current division to Fig. 2.23(a) reveals immediately that

$$i = \left(\frac{4}{4 + 6 + 6}\right)i_1 = \left(\frac{1}{4}\right) \times 3 = \frac{3}{4} \text{ A}$$

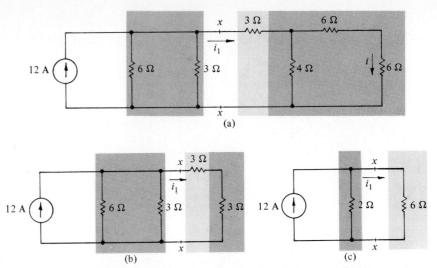

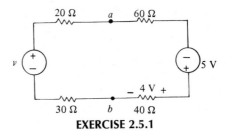

FIGURE 2.23 Circuit for analysis example using current division

---

EXERCISES

2.5.1 Find v, v_{ab}, and the power delivered by the 5-V source.
Answer 10 V, 5 V, 0.5 W

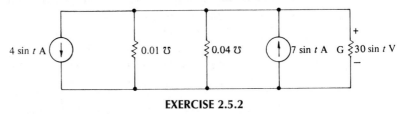

EXERCISE 2.5.1

2.5.2 Find G and construct an equivalent circuit having one current source and a single conductance.
Answer $G = 0.05\, \mho$, $i = 3 \sin t$ A directed upward, 0.1 $\mho$

EXERCISE 2.5.2

2.5.3 The power absorbed by the 15-Ω resistor is 15 W. Find R.
Answer 4 Ω

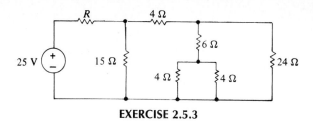

EXERCISE 2.5.3

2.5.4 Find v and i.

Answer 2 V, 9 A

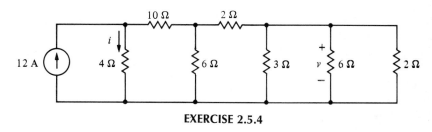

EXERCISE 2.5.4

2.6

AMMETERS, VOLTMETERS, AND OHMMETERS

A good example of the usefulness of current and voltage division is demonstrated in the design of simple two-terminal measuring instruments, such as ammeters, voltmeters, and ohmmeters. An *ideal ammeter* measures the current flowing through its terminals and has zero voltage across its terminals. In contrast, an *ideal voltmeter* measures the voltage across its terminals and has a terminal current of zero. An *ideal ohmmeter* measures the resistance connected between its terminals and delivers zero power to the resistance.

The practical measuring instruments that we shall consider only approximate the ideal devices. The ammeters, for instance, will not have zero terminal voltages. Likewise the voltmeters will not have zero terminal currents, and the ohmmeters will not have zero power delivered from their terminals.

A popular type of ammeter consists of a mechanical movement known as a D'Arsonval meter. This device is constructed by suspending an electrical coil between the poles of a permanent magnet. A dc current passing through the coil causes a rotation of the coil, as a result of magnetic forces, that is proportional to the current. A pointer is attached to the coil so that the rotation, or meter deflection, can be visually observed. D'Arsonval meters are characterized by their *full-scale current*, which is the current that will cause the meter to read its greatest value. Meter movements are common having full-scale currents from 10 μA to 10 mA.

An equivalent circuit for the D'Arsonval meter consists of an ideal ammeter in series with a resistance R_M, as shown in Fig. 2.24. In this circuit, R_M represents the resistance of the electrical coil. Clearly, a voltage appears across the ammeter termi-

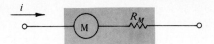

FIGURE 2.24 Equivalent circuit for a D'Arsonval meter

nals as a result of the current i flowing through R_M. R_M is usually a few ohms, and the terminal voltage for a full-scale current is nominally from 20 to 200 mV.

The D'Arsonval meter of Fig. 2.24 is an ammeter which is suitable for measuring dc currents not greater than the full-scale current I_{FS}. Suppose we wish, however, to measure a current which exceeds I_{FS}. It is apparent that we must not allow a current greater than I_{FS} to flow through the device. A circuit to accomplish this is shown in Fig. 2.25, where R_p is a parallel resistance which reduces the current flowing through the meter coil.

From current division, we see that

$$I_{FS} = \frac{R_p}{R_M + R_p} i_{FS}$$

where i_{FS} is the current which produces I_{FS} in the D'Arsonval meter. (Clearly, this is the maximum current the ammeter can measure.) Solving for R_p, we have

$$R_p = \frac{R_M I_{FS}}{i_{FS} - I_{FS}} \tag{2.30}$$

A dc voltmeter can be constructed using the basic D'Arsonval meter by placing a resistance R_s in series with the device, as shown in Fig. 2.26. It is obvious that the full-scale voltage, $v = v_{FS}$, occurs when the meter current is I_{FS}. Therefore, from KVL,

$$-v_{FS} + R_s I_{FS} + R_M I_{FS} = 0$$

from which

$$R_s = \frac{v_{FS}}{I_{FS}} - R_M \tag{2.31}$$

The *current sensitivity* of a voltmeter, expressed in ohms per volt, is the value obtained by dividing the resistance of the voltmeter by its full-scale voltage. Therefore,

$$\Omega/V \text{ rating} = \frac{R_s + R_M}{v_{FS}} \approx \frac{R_s}{v_{FS}} \tag{2.32}$$

(*Note:* "$\approx$" means approximately equal to.)

FIGURE 2.25 Ammeter circuit **FIGURE 2.26** Voltmeter circuit

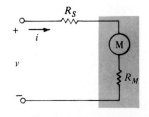

A simple ohmmeter circuit employing a D'Arsonval meter for measuring an unknown resistance R_x is shown in Fig. 2.27. In this circuit, the battery E causes a current i to flow when R_x is connected into the circuit. Applying KVL, we have

$$-E + (R_s + R_M + R_x)i = 0$$

from which

$$R_x = \frac{E}{i} - (R_s + R_M)$$

We select E and R_s such that for $R_x = 0$, $i = I_{FS}$. Therefore,

$$I_{FS} = \frac{E}{R_s + R_M}$$

Combining the last two equations, we find

$$R_x = \left(\frac{I_{FS}}{i} - 1\right)(R_s + R_M) \tag{2.33}$$

A very popular general-purpose meter which combines the three previously described circuits is the *VOM* (voltmeter-ohmmeter-milliammeter). In the VOM, provisions are made for changing R_p and R_s so that a wide dynamic range of operation is provided.

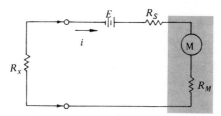

FIGURE 2.27 Ohmmeter circuit

EXERCISES

2.6.1 A D'Arsonval meter has $I_{FS} = 1$ mA and $R_M = 50$ Ω. Determine R_p in Fig. 2.25 so that i_{FS} is (a) 1.0 mA, (b) 10 mA, and (c) 100 mA.
Answer (a) infinite; (b) 5.556 Ω; (c) 0.505 Ω

2.6.2 In Fig. 2.26, determine R_s and the Ω/V rating for a voltmeter to have a full-scale voltage of 100 V using a D'Arsonval meter with (a) $R_M = 100$ Ω and $I_{FS} = 50$ μA and (b) $R_M = 50$ Ω and $I_{FS} = 1$ mA.
Answer (a) 2 MΩ, 20 kΩ/V; (b) 100 kΩ, 1 kΩ/V

2.6.3 What voltage would each meter design of Ex. 2.6.2 measure in the circuit shown? Why are the two measurements different?
Answer 99.5 V, 90.9 V

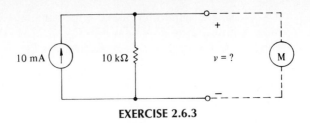

2.6.4 The meter movement of Ex. 2.6.1 is used to form the ohmmeter circuit of Fig. 2.27. Determine R_s and E so that $i = I_{FS}/2$ mA when $R_x = 10$ kΩ.
Answer 9.95 kΩ, 10 V

2.7

PHYSICAL RESISTORS

Resistors are manufactured from a variety of materials and are available in many sizes and values. Their characteristics include a nominal resistance value, an accuracy with which the actual resistance approximates the nominal value (known as *tolerance*), a power dissipation, and a stability as a function of temperature, humidity, and other environmental factors.

The most common type of resistor found in electrical circuits is the carbon composition or carbon film resistor. The composition type is made of hot-pressed carbon granules. The carbon film device consists of carbon powder which is deposited on an insulating substrate. A typical resistor of this type is shown in Fig. 2.28. Multicolored bands, shown as *a*, *b*, *c*, and % tolerance, are painted on the resistor body to indicate the nominal value of the resistance. The color code for the bands is given in Table 2.1.

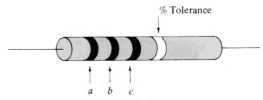

FIGURE 2.28 Carbon resistor

Bands *a*, *b*, and *c* give the nominal resistance of the resistor, and the tolerance band gives the percentage by which the resistance may deviate from its nominal value. Referring to Fig. 2.28, the resistance is

$$R = (10a + b)10^c \pm \% \text{ tolerance} \qquad (2.34)$$

by which we mean that the % tolerance of the nominal resistance is to be added or subtracted to give the range in which the resistance lies.

EXAMPLE 2.14 As an example, suppose we have a resistor with band colors of yellow, violet, red, and silver. The resistor will have a value given by

TABLE 2.1 Color Code for Carbon Resistors

Color	Value	Color	Value
	Bands *a*, *b*, and *c*		
Silver*	−2	Yellow	4
Gold*	−1	Green	5
Black	0	Blue	6
Brown	1	Violet	7
Red	2	Gray	8
Orange	3	White	9
	% Tolerance Band		
Gold	±5%		
Silver	±10%		

*These colors apply to band *c* only.

$$R = (4 \times 10 + 7) \times 10^2 \pm 10\%$$

$$= 4700 \pm 470 \ \Omega$$

Therefore the resistance value lies between 4230 and 5170 Ω.

Values of carbon resistors range from 2.7 to $2.2 \times 10^7 \ \Omega$, with wattages from $\frac{1}{8}$ to 2 W. For resistance values less than 10 Ω, we see from (2.34) that the third band must be gold or silver. Carbon resistors are inexpensive but have the disadvantage of a relatively high variation of resistance with temperature.

Another resistor type which is commonly used in applications requiring a high power dissipation is the wire-wound resistor. This device consists of a metallic wire, usually a nickel-chromium alloy, wound on a ceramic core. Low-temperature-coefficient wire permits the fabrication of resistors that are very precise and stable, having accuracy and stability of the order of ±1% to ±0.001%.

The metal film resistor is another valuable and useful resistor type. These resistors are made by vacuum-deposition of a thin metal layer on a low-thermal-expansion substrate. The resistance is then adjusted by etching or grinding a pattern through the film. Accuracy and stability for these resistors approaches that of wire-wound types, and high resistance values are much easier to attain.

In the future, the reader may be more likely to encounter resistors in *integrated circuits,* which were developed in the late 1950s and came into their own in the 1960s. An integrated circuit is a single monolithic chip of *semiconductor* (material with conducting properties between those of a conductor and an insulator) in which active

FIGURE 2.29 Integrated-circuit resistor

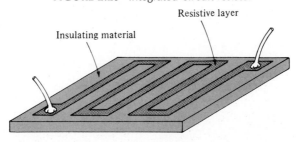

Resistive layer

Insulating material

and passive elements are fabricated by diffusion and deposition processes. Integrated circuits containing hundreds of elements are available on chips about $\frac{1}{8}$ in. square. An integrated-circuit resistor has the typical structure shown in Fig. 2.29.

EXERCISE

2.7.1 Find the resistance range of carbon resistors having color bands of (a) brown, black, red, silver; (b) red, violet, yellow, silver; and (c) blue, gray, gold, gold.
Answer (a) 900–1100 Ω; (b) 243–297 kΩ; (c) 6.46–7.14 Ω

PROBLEMS

2.1 A 4-kΩ resistor is connected to a battery and 3 mA flow. What current will flow if the battery is connected to a 400-Ω resistor?

2.2 A toaster is essentially a resistor that becomes hot when it carries a current. If a toaster is dissipating 600 W at a voltage of 120 V, find its current and its resistance.

2.3 If a toaster with a resistance of 15 Ω is operated at 120 V for 10 s, find the energy it uses.

2.4 Find i_1, i_2, and v_{ba}.

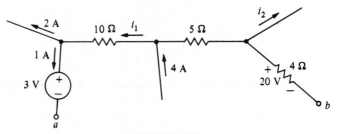

PROBLEM 2.4

2.5 Find i_1, i_2, and v.

PROBLEM 2.5

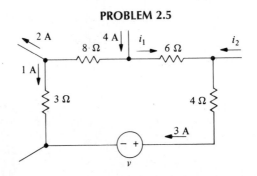

2.6 Find v.

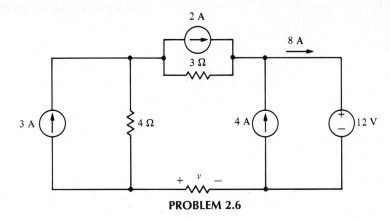

PROBLEM 2.6

2.7 Find i, v_{ab}, and an equivalent circuit for i containing a single source and a single resistor.

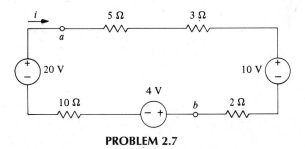

PROBLEM 2.7

2.8 A 10-V source in series with several resistors carries a current of 50 mA. What resistance must be connected in series with the source and the resistors to limit the current to 20 mA?

2.9 A 50-V source and two resistors, R_1 and R_2, are connected in series. If $R_2 = 3R_1$, find the voltages across the two resistors.

2.10 Design a voltage divider to provide 4, 10, 20, and 45 V, all with a common negative terminal, from a 50-V source. The source is to deliver 50 mW of power.

2.11 A voltage divider is to be constructed with a 60-V source and a number of 10-kΩ resistors. Find the minimum number of resistors required if the output voltage is (a) 40 V and (b) 30 V.

2.12 Find i and the power delivered to the 3-Ω resistor.

PROBLEM 2.12

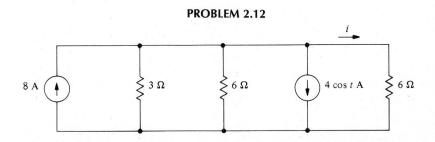

2.13 A 40-Ω resistor, a 60-Ω resistor, and a resistor R are connected in parallel to form an equivalent resistance of 8 Ω. Find R and the current it carries if a 6-A current source is connected to the combination.

2.14 A current divider consists of 10 resistors in parallel. Nine of them have equal resistances of 60 kΩ and the tenth is a 20-kΩ resistor. Find the equivalent resistance of the divider, and, if the total current entering the divider is 40 mA, find the current in the tenth resistor.

2.15 Find all the possible values of equivalent resistance that can be obtained by someone having three 12-Ω resistors.

2.16 Two 1-kΩ resistors are in series. When a resistor R is connected in parallel with one of them, the resistance of the combination is 1750 Ω. Find R.

2.17 Find i_1 and i_2.

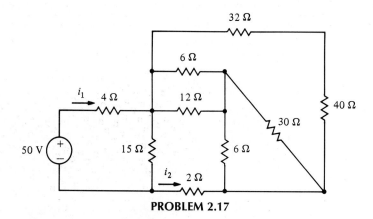

PROBLEM 2.17

2.18 Find v and i.

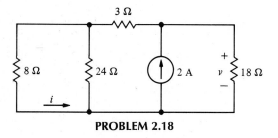

PROBLEM 2.18

2.19 Find i.

PROBLEM 2.19

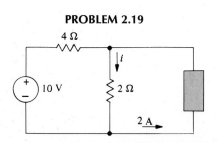

2.20 (a) Find the equivalent resistance looking in terminals *a-b* if terminals *c-d* are open, and if terminals *c-d* are shorted together. (b) Find the equivalent resistance looking in terminals *c-d* if terminals *a-b* are open, and if terminals *a-b* are shorted together.

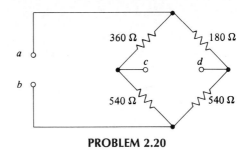

PROBLEM 2.20

2.21 Find i, v_1, and v_2.

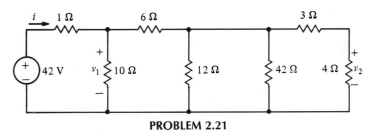

PROBLEM 2.21

2.22 Find R and v using current and voltage division.

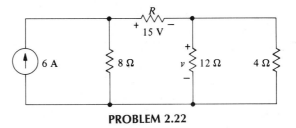

PROBLEM 2.22

2.23 Find i and v.

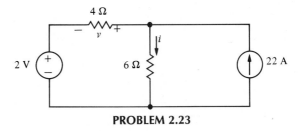

PROBLEM 2.23

2.24 Find i_1 and i_2.

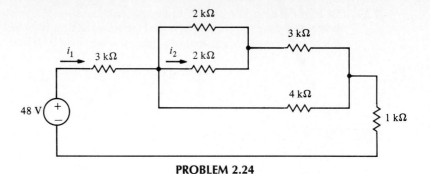

PROBLEM 2.24

2.25 Find i_1, i_2, and v.

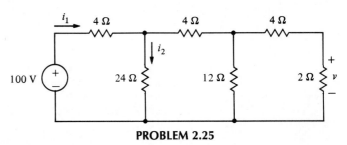

PROBLEM 2.25

2.26 A D'Arsonval meter has a full-scale current of 1 mA and a resistance of 4.9 Ω. If a 0.1-Ω parallel resistor is used in Fig. 2.25, what is i_{FS}? What voltage occurs across the meter?

2.27 A 20,000-Ω/V voltmeter has a full-scale voltage of 120 V. What current flows in the meter when measuring 90 V?

2.28 Two 10-kΩ resistors are connected in series across a 100-V source. What voltage will the voltmeter of Prob. 2.27 measure across one of the 10-kΩ resistors? Repeat for two 1-MΩ resistors in series.

2.29 The D'Arsonval meter of Prob. 2.26 is used for the ohmmeter of Fig. 2.27. What value of series resistance is required if $E = 1.5$ V? What value of unknown resistance will cause a one-quarter full-scale deflection?

2.30 Determine the color codes for resistors having the following resistance ranges: (a) 4.23–5.17 Ω, (b) 6460–7140 Ω, and (c) 3.135–3.465 MΩ.

3

Georg Simon Ohm
1787–1854

Dependent Sources

The most basic and most widely used of all the laws of electricity, Ohm's Law, was published in 1827 by the German physicist Georg Simon Ohm in his great work, *The Galvanic Chain, Mathematically Treated.* Without Ohm's Law we could not analyze the simplest galvanic chain (electric circuit), but at the time of its publication, Ohm's work was denounced by critics as "a web of naked fancies," the "sole effort" of which was "to detract from the dignity of nature."

Ohm was born in Erlangen, Bavaria, the oldest of seven children in a middle-class-to-poor family. He was an early dropout at the University of Erlangen but returned in 1811 and earned his doctorate and the first of several modest, low-paying mathematics teaching positions. To improve his lot, he threw himself into his electrical research at every opportunity allowed by his heavy teaching duties, and his efforts culminated in his famous law. Despite the misplaced criticisms of his work, during his lifetime Ohm received the fame that was due him. The Royal Society of London awarded him the Copley Medal in 1841, and the University of Munich gave him its Professor of Physics chair in 1849. He was also honored after his death when the *ohm* was chosen as the unit of electrical resistance. ∎

The voltage and current sources of Chapters 1 and 2 are independent sources, as defined earlier in Sec. 1.4. We may also have *dependent* sources, which are very important in circuit theory, particularly in electronic circuits. In this chapter we shall define dependent sources and consider an additional circuit element, the operational amplifier, which may be used to obtain dependent sources.

We shall also analyze a few simple circuits containing resistors and sources, both independent and dependent. As we shall see, the analysis is very similar to that performed in Chapter 2, and the results may be used to construct a number of important circuits, such as amplifiers and inverters, which will be defined in the chapter.

3.1

DEFINITIONS

A *dependent* or *controlled voltage source* is one whose terminal voltage depends on, or is controlled by, a voltage or a current existing at some other place in the circuit. A *voltage-controlled voltage source* (VCVS) is a voltage source controlled by a voltage, and a *current-controlled voltage source* (CCVS) is one controlled by a current. The symbol for a dependent voltage source with terminal voltage v is shown in Fig. 3.1(a).

A *dependent,* or *controlled current source,* symbolized by Fig. 3.1(b), is one whose current is dependent on a voltage or a current existing elsewhere in the circuit. A *voltage-controlled current source* (VCCS) is controlled by a voltage, and a *current-controlled current source* (CCCS) is controlled by a current.

Figure 3.2 illustrates the four types of controlled sources and shows the voltage or current on which they are dependent. The quantities μ and β are dimensionless constants, commonly referred to as the voltage and current *gain*, respectively. The constants r and g have units of ohms and mhos, respectively.

As an example, in the circuit of Fig. 3.3 we have an independent source, a

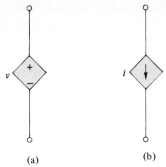

(a)   (b)

FIGURE 3.1 (a) Dependent voltage source; (b) dependent current source

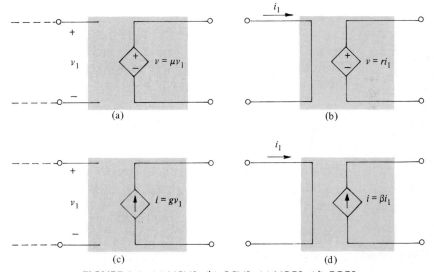

FIGURE 3.2 (a) VCVS; (b) CCVS; (c) VCCS; (d) CCCS

dependent source, and two resistors. The dependent source is a voltage source controlled by the current i_1. The value of r for the dependent source is 0.5 V/A.

Dependent sources are essential components in *amplifier* circuits (those for which the magnitude of the output is greater than that of the input). They also serve

FIGURE 3.3 Circuit containing a dependent source

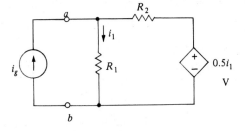

many other functions, such as isolating a desired portion of a circuit from the rest of the circuit, or providing negative resistance. As we know from Chapter 2, the resistor is a passive element with positive resistance. However, by means of dependent sources we may fabricate negative resistance, as we shall see later (Exs. 3.2.2 and 3.2.3, for example).

3.2
CIRCUITS WITH DEPENDENT SOURCES

Circuits containing dependent sources are analyzed in the same manner as those without dependent sources. That is, Ohm's law for resistors and Kirchhoff's voltage and current laws apply, as well as the concepts of equivalent resistance and voltage and current division.

EXAMPLE 3.1

As an illustration of the procedure, let us find the current i in the circuit of Fig. 3.4. The dependent source is a voltage source controlled by the voltage v_1 as shown. Applying Kirchhoff's voltage law around the circuit, we have

$$-v_1 + 3v_1 + 6i = 6 \tag{3.1}$$

and by Ohm's law we have

$$v_1 = -2i \tag{3.2}$$

Using (3.2), we may eliminate v_1 in (3.1), which results in

$$2(-2i) + 6i = 6$$

or $i = 3$ A. Thus the dependent source has complicated matters only to the extent of requiring the extra equation (3.2).

EXAMPLE 3.2

As another example, let us find the voltage v in Fig. 3.5. Applying Kirchhoff's current law to the currents leaving the top node, we have

$$-4 + i_1 - 2i_1 + \frac{v}{2} = 0 \tag{3.3}$$

Also, by Ohm's law we have

$$i_1 = \frac{v}{6} \tag{3.4}$$

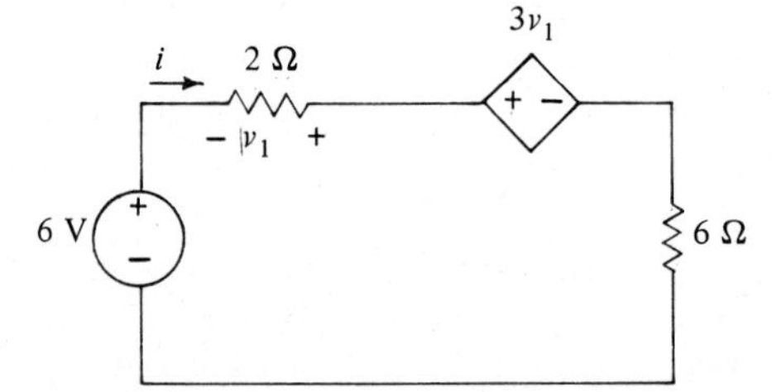

FIGURE 3.4 Dependent source example

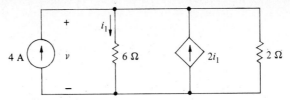

FIGURE 3.5 Another dependent source example

Substituting (3.4) into (3.3) yields

$$-4 - \frac{v}{6} + \frac{v}{2} = 0$$

or $v = 12$ V.

EXERCISES

3.2.1 In the circuit of Fig. 3.4, replace the 2-Ω resistor by a 1-Ω resistor and find i, v_1, and the resistance seen by the source (i.e., $R = 6/i$).
Answer 1.5 A, -1.5 V, 4 Ω

3.2.2 Repeat Ex. 3.2.1 if the 2-Ω resistor is replaced by a 4-Ω resistor. (Note that a negative resistance is possible when a dependent source is present.)
Answer -3 A, 12 V, -2 Ω

3.2.3 In Fig. 3.3, find the resistance seen by the source (that is, looking in terminals *a-b*) if (a) $R_1 = 1.5$ Ω and $R_2 = 2$ Ω and (b) $R_1 = R_2 = 0.1$ Ω.
Answer (a) 1 Ω; (b) $-\frac{1}{30}$ Ω

3.3

OPERATIONAL AMPLIFIERS

A logical question at this point might be, How do we obtain dependent sources? One answer is that they arise as parts of equivalent circuits of electronic devices operating under certain conditions. Another answer is that they can be deliberately constructed by means of certain electronic devices in conjunction with passive elements.

We shall not be interested here in undertaking a study of electronic devices, since the reader will have a detailed encounter with them later in standard electronics courses. However, there is one such device that is extremely useful in the construction of dependent sources and whose ideal mathematical model is both simple and elegant. This is the *operational amplifier,* or op amp, the ideal model of which we shall consider in this section.

The symbol that we shall use for an operational amplifier is shown in Fig. 3.6. The op amp is a multiterminal device, but for simplicity we shall show only the three terminals indicated. Terminal 1 (marked $-$) is the *inverting input terminal,* terminal 2 (marked $+$) is the *noninverting input terminal,* and terminal 3 is the *output terminal.* The purposes of the terminals that are not shown include, in general, dc power supply

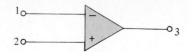

FIGURE 3.6 Operational amplifier symbol

connections, frequency compensation terminals, and offset null terminals. We shall not be interested in discussing these other terminals here, but the interested student may find their purposes and how they are used in any op amp user's manual.

Operational amplifiers are commonly available in integrated circuit form and are normally fabricated in packages having 8 to 14 terminals and containing 1 to 4 op amps. A typical integrated circuit *dual in-line package* (DIP) with eight terminals is shown in Fig. 3.7.

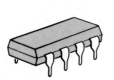

FIGURE 3.7 Dual in-line package op amp

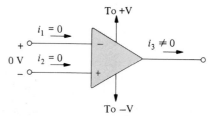

FIGURE 3.8 Op amp with power supply leads shown

The operational amplifier has many characteristics that are important to designers, but the *ideal* model of the op amp has only two properties that the circuit analyst needs to know: that the currents into both input terminals are zero and that the voltage between the input terminals is zero.

It should be pointed out that the generalization of KCL does not hold for the op amp of Fig. 3.6. That is, the input currents being zero does not mean that the output current is zero. This may be seen more clearly in Fig. 3.8, where the power supply leads are shown. Indeed, KCL cannot be applied at terminal 3 of Fig. 3.6 since, because of the unshown terminals, we cannot know the output current.

EXAMPLE 3.3

As an example of a circuit with an operational amplifier, let us consider Fig. 3.9. It is desired to find the current i and the voltage v_3, considering v_g to be a known generator voltage. The symbol attached to node c is the *ground* (to be discussed more thoroughly in Chapter 4), an arbitrary point to which the unshown power supplies are also connected. Thus there should be no temptation to apply KCL at node c because the currents into the ground through the unshown elements are not known. If the ground symbol is not shown, it will generally be understood to be connected to the negative terminal of the voltage source.

Let us write KVL around the loop *abca* through the source. Since the voltage across terminals a and b is zero, we have

$$v_1 - v_g = 0 \tag{3.5}$$

or $v_1 = v_g$. Applying KCL at node b and noting that the current into the negative

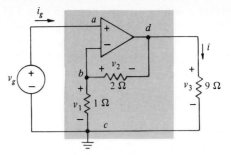

FIGURE 3.9 Circuit containing an op amp

terminal of the op amp is zero, we have

$$\frac{v_1}{1} + \frac{v_2}{2} = 0 \tag{3.6}$$

or $v_2 = -2v_1 = -2v_g$. Next, KVL around loop $cbdc$ through the 9-Ω resistor yields

$$-v_1 + v_2 + v_3 = 0$$

or

$$v_3 = v_1 - v_2 = 3v_g \tag{3.7}$$

Finally, Ohm's law yields

$$i = \frac{v_3}{9} = \frac{3v_g}{9} = \frac{v_g}{3}$$

Thus, if $v_g = 12 \cos 10t$, for example, we have $i = 4 \cos 10t$.

As a by-product of this example, let us note that $i_g = 0$ (the current into an input lead of the op amp) and $v_3 = 3v_g$ from (3.7). Thus we may draw an equivalent circuit, insofar as v_g, i_g, v_3, and i are concerned, as shown in Fig. 3.10. Analysis of the equivalent circuit yields exactly the same v_3 and i for the same v_g and i_g as in the case of Fig. 3.9. Thus the op amp has been used to obtain a controlled source with a gain of 3. (In this case the controlled source is a VCVS, in which v_g controls v_3.)

Before we leave this section, let us observe that the op amp in Fig. 3.9 is being operated in a *feedback* mode. That is, the output v_3 at node d is fed back to the

FIGURE 3.10 Circuit equivalent to that of Fig. 3.9

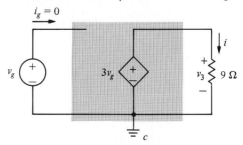

inverting input terminal through the 2-Ω resistor. A practical op amp is a very high-gain device and is generally never used without feedback. In cases when the feedback is to one input terminal rather than to both, it is always to the inverting terminal, for the simple reason that otherwise the op amp will not work. We are not interested here in the reason for this, which is a consequence of the op amp's design. The interested student, in all probability, will have occasion to thoroughly study op amps and their construction in a later course in electronic devices.

3.4

AMPLIFIER CIRCUITS

Equations (3.6) and (3.7), which were the basis for the VCVS circuit of Fig. 3.10, are independent of the current i in the 9-Ω load of Fig. 3.9. This is true in general of VCVS circuits of this type. To see this, let us consider the circuit of Fig. 3.11. The voltage v_2 is the output voltage of the op amp and, as we shall see, is a function only of the input voltage v_1 and the two resistors.

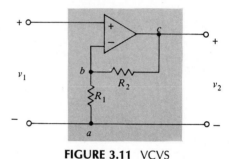

FIGURE 3.11 VCVS

Since there is no voltage across the input terminals of the op amp, we have $v_{ba} = v_1$. Also, KVL around the loop $abca$ containing v_2 yields

$$-v_{ba} + v_{bc} + v_2 = 0$$

or

$$v_{bc} = v_{ba} - v_2 = v_1 - v_2$$

Therefore, applying KCL at node b results in

$$\frac{v_1}{R_1} + \frac{v_1 - v_2}{R_2} = 0$$

Solving for v_2, we have

$$v_2 = \mu v_1 \tag{3.8}$$

where

$$\mu = 1 + \frac{R_2}{R_1} \qquad\qquad (3.9)$$

Figure 3.11 is therefore a VCVS with gain μ. Since it is a three-terminal network (the output and input share a common terminal) and the currents are zero into the input leads of the op amp, an equivalent circuit may be drawn as in Fig. 3.12. We note that since R_1 and R_2 are nonnegative, we have $\mu \geq 1$.

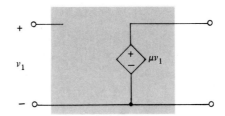

FIGURE 3.12 Circuit equivalent to Fig. 3.11

A special case of Fig. 3.11 is the case $R_2 = 0$ (short circuit) and $R_1 = \infty$ (open circuit), shown in Fig. 3.13. This circuit has $\mu = 1$, or $v_2 = v_1$, and is called a *voltage follower;* that is, v_2 *follows* v_1. It is also called a *buffer amplifier* because it may be used to isolate, or *buffer,* one circuit from another. (The voltages at the two pairs of terminals are the same, but no current can flow from one pair to the other.) An example of buffering is given in Ex. 3.4.3.

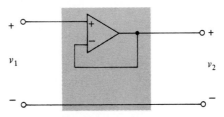

FIGURE 3.13 Voltage follower

Let us consider next the circuit of Fig. 3.14. Applying KCL at the inverting input terminal of the op amp, we have

$$-\frac{v_1}{R_1} - \frac{v_2}{R_2} = 0$$

or

$$v_2 = -\frac{R_2}{R_1} v_1 \qquad\qquad (3.10)$$

(Recall that the voltage across and the currents into the input terminals of the op amp are zero.)

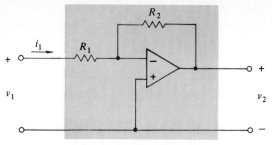

FIGURE 3.14 Inverter

This circuit is called an *inverter* because the polarity of v_2 is opposite that of v_1, as seen in (3.10). It is also a VCVS, but in this case the input current i_1 is not zero, being given by

$$i_1 = \frac{v_1}{R_1} \tag{3.11}$$

An equivalent circuit is shown in Fig. 3.15

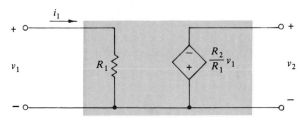

FIGURE 3.15 Equivalent circuit of the inverter

By (3.11) $v_1 = R_1 i_1$, and thus we may eliminate v_1 in Fig. 3.15, resulting in another equivalent circuit, shown in Fig. 3.16. This circuit is evidently a CCVS, since the voltage v_2 is controlled by the current i_1.

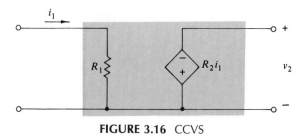

FIGURE 3.16 CCVS

We may also obtain dependent current sources from Fig. 3.14, which we redraw as shown in Fig. 3.17. Since there is no current into the op amp terminals, we have

$$i_2 = i_1 = \frac{v_1}{R_1} \tag{3.12}$$

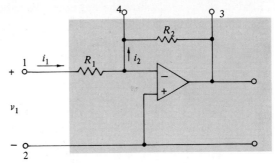

FIGURE 3.17 Inverter circuit redrawn

Insofar as terminals 1, 2, 3, and 4 are concerned, an equivalent circuit is then that of Fig. 3.18, which is a CCCS with a gain of 1.

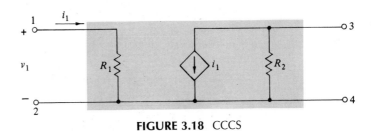

FIGURE 3.18 CCCS

Finally, substituting for i_1 in (3.12), we may redraw Fig. 3.18 as a VCCS, shown in Fig. 3.19. In this case, $g = 1/R_1$.

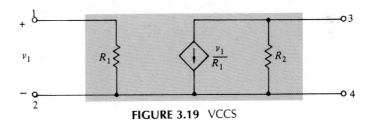

FIGURE 3.19 VCCS

EXERCISES

3.4.1 In Fig. 3.11, let $v_1 = 4$ V and $R_2 = 11$ kΩ. If a load resistor $R = 5$ kΩ is connected across the terminals of v_2, find R_1 so that resistor R carries a current of 3 mA.
Answer 4 kΩ

3.4.2 In the inverter of Fig. 3.14, let $v_1 = 3$ V. Find R_1 and R_2 so that $i_1 = 1.5$ mA and $v_2 = -9$ V.
Answer 2 kΩ, 6 kΩ

3.4.3 Find v_1 and v_2. Note how the buffer amplifier holds v_2 to $v_g/2$ while the 3-kΩ resistor "loads down" the output v_1.

Answer $v_g/4$, $v_g/2$

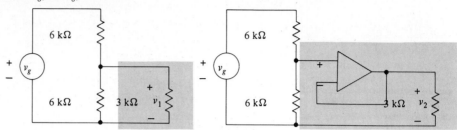

EXERCISE 3.4.3

PROBLEMS

3.1 Find v_1 and the power delivered to the 8-Ω resistor.

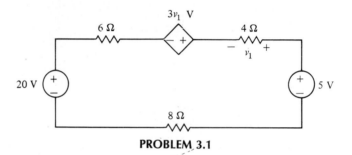

PROBLEM 3.1

3.2 Find i_1 if (a) $R = 1\ \Omega$ and (b) $R = 6\ \Omega$.

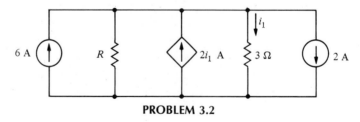

PROBLEM 3.2

3.3 Find i if $R = 10\ \Omega$.

PROBLEM 3.3

3.4 Find R in Prob. 3.3 so that $i = 5$ A.

3.5 Find i_1 and v if (a) $R = 4\ \Omega$ and (b) $R = 12\ \Omega$.

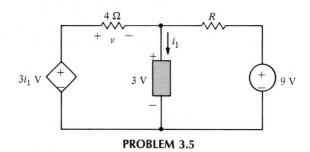

PROBLEM 3.5

3.6 Find i and the resistance seen by the independent current source if (a) $R = 6\ \Omega$ and (b) $R = 1\ \Omega$.

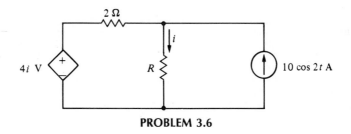

PROBLEM 3.6

3.7 Find v.

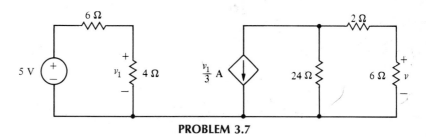

PROBLEM 3.7

3.8 Find v.

PROBLEM 3.8

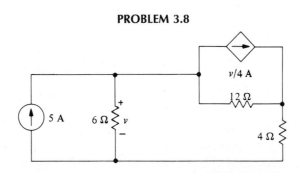

3.9 Show that

$$v_3 = -R_0\left(\frac{v_1}{R_1} + \frac{v_2}{R_2}\right)$$

(This circuit is called a *summer*, since the output voltage is the negative of a weighted sum of the input voltages. Note that the result is independent of the output connections at terminals *a-b*.)

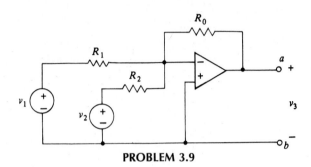

PROBLEM 3.9

3.10 (a) In the figure for Prob. 3.9, let $R_0 = 10\ \text{k}\Omega$ and find R_1 and R_2 so that the magnitude of the output voltage v_3 is the average of the input voltages v_1 and v_2.

(b) Find R_1 if $R_0 = 2\ \text{k}\Omega$, $R_2 = 4\ \text{k}\Omega$, $v_1 = 6\ \text{V}$, $v_2 = 8\ \text{V}$, and the output is to be $v_0 = -16\ \text{V}$.

3.11 Show that regardless of the load at terminals *c-d*, we have

$$v_1 = v_2$$

$$i_1 = \frac{R_2}{R_1}i_2$$

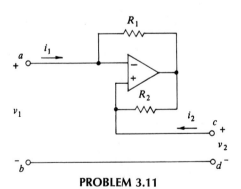

PROBLEM 3.11

3.12 In the figure for Prob. 3.11, let $R_1 = R_2$ and connect a resistor R between terminals *c-d*. Show that the resistance seen at input terminals *a-b* is $R_{ab} = -R$. (The figure for Prob. 3.11 thus converts positive resistance to negative resistance.)

3.13 Find v.

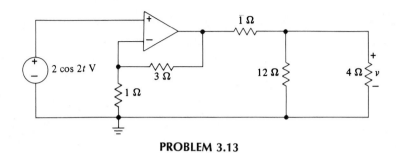

PROBLEM 3.13

3.14 Find i.

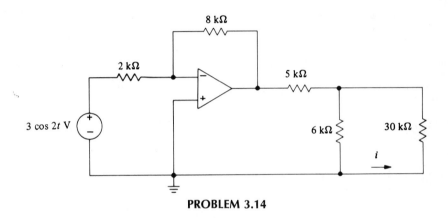

PROBLEM 3.14

3.15 Find v_0.

PROBLEM 3.15

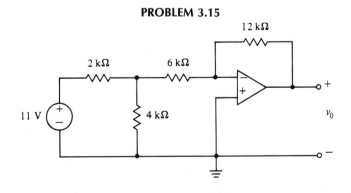

3.16 Find *i*.

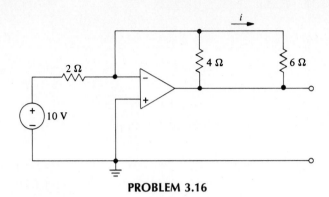

PROBLEM 3.16

3.17 Find *i*.

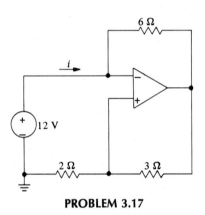

PROBLEM 3.17

3.18 Find *v*.

PROBLEM 3.18

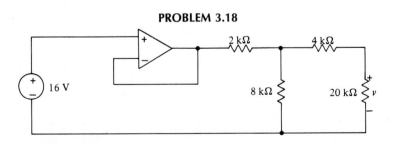

3.19 Find i.

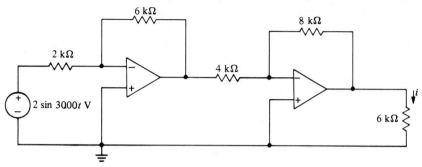

PROBLEM 3.19

3.20 Find v and i.

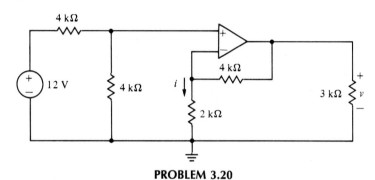

PROBLEM 3.20

3.21 The summer of Prob. 3.9 is an *inverting* summer since the output voltage is the *negative* of the weighted sum of the input voltages. (a) Show that the given circuit is a *noninverting* summer (the output sign is not changed). (b) Use the result in (a) to find v_0 if $v_1 = 3$ V, $v_2 = 2$ V, $R_1 = 4$ kΩ, $R_2 = 3$ kΩ, $R_f = 6$ kΩ, and $R = 1$ kΩ.

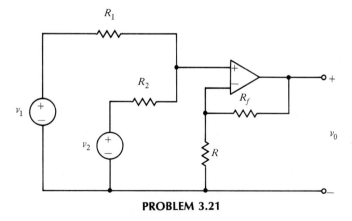

PROBLEM 3.21

3.22 If $R_f = R_1 = 1$ kΩ in the circuit of Prob. 3.21, find the other resistance values so that $v_0 = v_1 + v_2$.

4

Gustav Robert Kirchhoff
1824–1887

Analysis Methods

There must be a fundamental story here [on his research with Bunsen].
Gustav Robert Kirchhoff

Ohm's Law is fundamental to electric circuits, but to analyze even the simplest circuit requires two additional laws formulated in 1847 by the German physicist Gustav Robert Kirchhoff. These laws—Kirchhoff's current law and Kirchhoff's voltage law—are the more remarkable when we consider that Kirchhoff's principal interest was in his pioneering work in spectroscopy with the noted German chemist Robert Bunsen, to whom we owe the Bunsen burner. In that field there is another law of Kirchhoff: Kirchhoff's law of radiation.

Kirchhoff was born in Königsberg, East Prussia, the son of a lawyer. He entered the University of Königsberg at age 18 and graduated with his doctorate five years later. Upon graduation in 1847 he married the daughter of Friedrich Richelot, one of his famous mathematics teachers, and at the same time received a rare travel grant for further study in Paris. The political unrest that led to the 1848 wave of revolutions in Europe forced him to change his plans, and he became a lecturer in Berlin. Two years later he met Bunsen and the two began their famous collaboration. Kirchhoff's great success in spectroscopy drew attention away from his contributions in other branches of physics, but without his electrical laws there would be no circuit theory. ∎

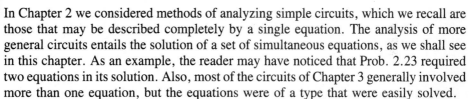

In Chapter 2 we considered methods of analyzing simple circuits, which we recall are those that may be described completely by a single equation. The analysis of more general circuits entails the solution of a set of simultaneous equations, as we shall see in this chapter. As an example, the reader may have noticed that Prob. 2.23 required two equations in its solution. Also, most of the circuits of Chapter 3 generally involved more than one equation, but the equations were of a type that were easily solved.

In this chapter we shall consider systematic ways of formulating and solving the equations that arise in the analysis of more complicated circuits. We shall consider two general methods, one based primarily on Kirchhoff's current law and one on Kirchhoff's voltage law. As we shall see, KCL generally leads to equations in which the unknowns are voltages, whereas KVL leads to equations in which the unknowns are currents.

It should be evident from our work in previous chapters that a complete analysis of a circuit can be performed by finding a relatively few key voltages and/or currents. For example, in a simple circuit consisting of a single loop, a key variable is the current, for if we know the current, we may find every voltage around the loop, and, of course, the current around the loop is the current in every element.

In Sec. 4.1 we shall discuss the case where the selected unknowns are voltages. Quite naturally, our choice of voltages should lead to a set of independent equations. This technique will be referred to as *nodal* analysis. In Sec. 4.5 we shall consider *mesh* analysis, in which the unknowns are currents.

In this chapter we shall discuss the techniques for selecting the voltages or currents to be found and the formulation of the circuit equations. The justification of the methods in the general case will be left until Chapter 6.

4.1

NODAL ANALYSIS

In this section we shall consider methods of circuit analysis in which voltages are the unknowns to be found. A convenient choice of voltages for many networks is the set of *node voltages*. Since a voltage is defined as existing between two nodes, it is convenient to select one node in the network to be a *reference node* or *datum node* and then associate a voltage or a potential with each of the other nodes. The voltage of each of the nonreference nodes with respect to the reference node is defined to be a *node voltage*. It is common practice to select polarities so that the node voltages are positive relative to the reference node. For a circuit containing N nodes, there will be $N - 1$ node voltages, some of which may be known, of course, if voltage sources are present.

Frequently the reference node is chosen to be the node to which the largest number of branches are connected. Many practical circuits are built on a metallic base or chassis, and usually there are a number of elements connected to the chassis, which is a logical choice for the reference node. In many cases, such as in electric power systems, the chassis is the earth itself. For this reason, the reference node is frequently referred to as ground. The reference node is thus at ground potential or zero potential, and the other nodes may be considered to be at some potential above zero.

Since the circuit unknowns are to be voltages, the describing equations are obtained by applying KCL at the nodes. The currents in the elements are proportional to the element voltages, which are themselves either a node voltage (if one element node is ground) or the difference of two node voltages. For example, in Fig. 4.1 the reference node is node 3 with zero or ground potential. The symbol shown attached to node 3 is the standard symbol for ground, as previously noted in Chapter 3. The nonreference nodes 1 and 2 have node voltages v_1 and v_2. Thus the element voltage v_{12} with the polarity shown is

$$v_{12} = v_1 - v_2$$

The other element voltages shown are

$$v_{13} = v_1 - 0 = v_1$$

and

$$v_{23} = v_2 - 0 = v_2$$

These equations may be established by applying KVL around the loops (real or imagined). Evidently, if we know all the node voltages, we may find all the element voltages and thus all the element currents.

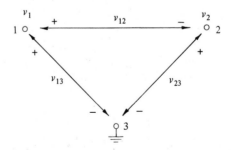

FIGURE 4.1 Reference and nonreference nodes

The application of KCL at a node results in a *node equation,* that is, an equation relating node voltages. Clearly, simplification in writing the resulting equations is possible when the reference node is chosen to be a node with a large number of elements connected to it. As we shall see, however, this is not the only criterion for selecting the reference node, but it is frequently the overriding one.

Since we are going to apply KCL, it seems that the simplest networks to consider are those whose only sources are independent current sources. This is not always true, as we shall see, but we shall begin with an example of this type. In the network shown in Fig. 4.2(a), there are three nodes, dashed and numbered as shown. [This is easier to see in the redrawn version of Fig. 4.2(b).] Since there are four elements connected to node 3, we select it as the reference node, identifying it by the ground connection shown.

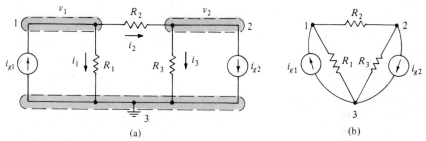

FIGURE 4.2 Circuit containing independent current sources

Before writing the node equations, consider the element shown in Fig. 4.3, where v_1 and v_2 are node voltages. The element voltage v is given by

$$v = v_1 - v_2$$

and thus by Ohm's law we have

$$i = \frac{v}{R} = \frac{v_1 - v_2}{R}$$

or

$$i = G(v_1 - v_2)$$

where $G = 1/R$ is the conductance. That is, the current from node 1 to node 2 is the difference of the *node voltage at node 1* and the *node voltage at node 2* divided by the resistance R, or multiplied by the conductance G. This relation will allow us to rapidly write the node equations by inspection directly in terms of the node voltages.

Now returning to the circuit of Fig. 4.2, the sum of the currents leaving node 1 must be zero, and this results in the equation

$$i_1 + i_2 - i_{g1} = 0$$

In terms of the node voltages, this equation becomes

FIGURE 4.3 Single element

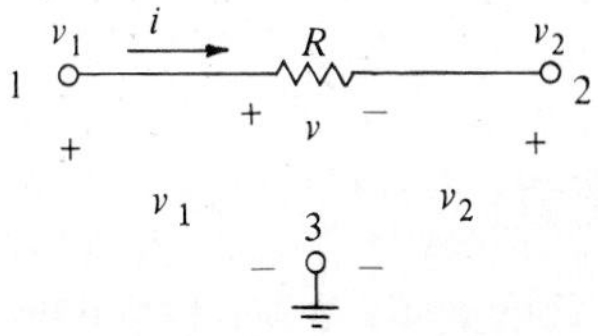

$$G_1v_1 + G_2(v_1 - v_2) - i_{g1} = 0$$

We could have obtained this equation directly using the procedure of the previous paragraph. Applying KCL at node 2 in a similar manner, we obtain

$$-i_2 + i_3 + i_{g2} = 0$$

or

$$G_2(v_2 - v_1) + G_3v_2 + i_{g2} = 0$$

Again, the G's are the conductances (reciprocals of the R's).

We could have equated the sum of currents leaving the node to the sum of currents entering the node. Had we done so, the terms i_{g1} and i_{g2} would have appeared on the right-hand side:

$$G_1v_1 + G_2(v_1 - v_2) = i_{g1}$$

$$G_2(v_2 - v_1) + G_3v_2 = -i_{g2}$$

Rearranging these two equations results in

$$(G_1 + G_2)v_1 - G_2v_2 = i_{g1} \tag{4.1}$$

$$-G_2v_1 + (G_2 + G_3)v_2 = -i_{g2} \tag{4.2}$$

These equations exhibit a symmetry that may be used to write the equations in the rearranged form by inspection. In (4.1) the coefficient of v_1 is the sum of conductances of the elements connected to node 1, while the coefficient of v_2 is the negative of the conductance of the element connecting node 1 to node 2. The same statement holds for (4.2) if the numbers 1 and 2 are interchanged. Thus node 2 plays the role in (4.2) of node 1 in (4.1). That is, it is the node at which KCL is applied. In each equation the right-hand side is the current from a current source which enters the corresponding node.

In general, in networks containing only conductances and current sources, KCL applied at the kth node, with node voltage v_k, may be written as follows. In the left member the coefficient of v_k is the sum of the conductances connected to node k, and the coefficients of the other node voltages are the negatives of the conductances between those nodes and node k. The right member of the equation consists of the net current flowing into node k due to current sources.

To illustrate the process, consider Fig. 4.4, which is a portion of a circuit. The dashed lines indicate connections to nodes other than node 2 (labeled v_2). At node 2 we have, using the shortcut procedure,

$$-G_1v_1 + (G_1 + G_2 + G_3)v_2 - G_2v_3 - G_3v_4 = i_{g1} - i_{g2} \tag{4.3}$$

This may be checked by applying KCL in the usual way, equating currents leaving node 2 to those entering node 2. The result is

$$G_1(v_2 - v_1) + G_2(v_2 - v_3) + G_3(v_2 - v_4) + i_{g2} = i_{g1}$$

Rearranging this result leads to (4.3).

In general, if there are $N - 1$ unknown node voltages, we need $N - 1$ equations, which may be written at any $N - 1$ nodes in the circuit. In a circuit like that of Fig. 4.2 we could even use the datum node for one of the equations. This is because

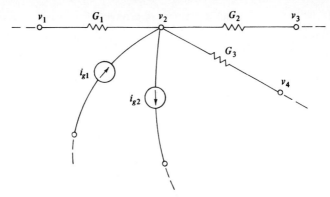

FIGURE 4.4 Part of a circuit

the datum node was chosen arbitrarily and there is no current into the ground. (There is no return path from ground, so there can be no current.)

4.2

AN EXAMPLE

EXAMPLE 4.1   To illustrate the method of nodal analysis, let us consider the circuit of Fig. 4.5. We have taken the reference node as shown and labeled the nonreference nodes as v_1, v_2, and v_3. We note that the conductances are specified for the resistors.

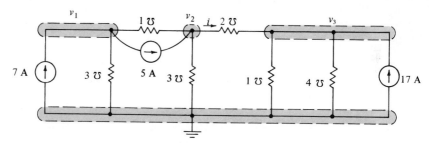

FIGURE 4.5 Example of a circuit

Since there are three nonreference nodes, there will be three equations. At node v_1, using the shortcut method, we have

$$4v_1 - v_2 = 2 \tag{4.4}$$

The sum of the conductances at node v_1 is $3 + 1 = 4$, the conductance between nodes 1 and 2 is 1, the conductance between nodes 1 and 3 is 0, and the net current into the node through the sources is $7 - 5 = 2$. If we had used the conventional KCL method at v_1, we would have obtained

$$1(v_1 - v_2) + 3v_1 + 5 = 7$$

which is equivalent to (4.4).

At nodes v_2 and v_3, we have

$$-v_1 + 6v_2 - 2v_3 = 5$$

$$-2v_2 + 7v_3 = 17$$

(4.5)

We may solve (4.4) and (4.5) for the node voltages using any one of a variety of methods for solving simultaneous equations. Two such methods are Cramer's rule, which employs determinants, and Gaussian elimination. For the reader who is not familiar with these two methods, a complete discussion is given in Appendices A and B at the end of the book.

Using Cramer's rule, we first find the coefficient determinant, given by

$$\Delta = \begin{vmatrix} 4 & -1 & 0 \\ -1 & 6 & -2 \\ 0 & -2 & 7 \end{vmatrix} = 145$$

Then we have

$$v_1 = \frac{\begin{vmatrix} 2 & -1 & 0 \\ 5 & 6 & -2 \\ 17 & -2 & 7 \end{vmatrix}}{145} = \frac{145}{145} = 1 \text{ V}$$

$$v_2 = \frac{\begin{vmatrix} 4 & 2 & 0 \\ -1 & 5 & -2 \\ 0 & 17 & 7 \end{vmatrix}}{145} = \frac{290}{145} = 2 \text{ V}$$

and

$$v_3 = \frac{\begin{vmatrix} 4 & -1 & 2 \\ -1 & 6 & 5 \\ 0 & -2 & 17 \end{vmatrix}}{145} = \frac{435}{145} = 3 \text{ V}$$

Now that we have the node voltages we may completely analyze the circuit. For example, if we want the current i in the 2-$\mho$ element, it is given by

$$i = 2(v_2 - v_3) = 2(2 - 3) = -2 \text{ A}$$

Equations (4.4) and (4.5) are symmetrical in a way that facilitates the writing of the equations. This symmetry also is present in the general case. For example, the coefficient of v_2 in the first equation is the same as that of v_1 in the second equation. Also, the coefficient of v_3 in the first equation is that of v_1 in the third equation. Finally, the coefficient of v_3 in the second equation is that of v_2 in the third equation. These results follow from the fact that the conductance between nodes 1 and 2 is that between 2 and 1, the conductance between nodes 1 and 3 is that between 3 and 1, etc.

This symmetry shows up also in the coefficient determinant Δ. Its diagonal elements, 4, 6, and 7, are the sums of the conductances connected to the three nonreference nodes, and the off-diagonal elements are symmetrical about the diagonal. The latter elements are, of course, the negatives of the conductances between nodes.

EXERCISES

4.2.1 Using nodal analysis, find v_1 and v_2 in Fig. 4.2(a) if $R_1 = 2\ \Omega$, $R_2 = 2\ \Omega$, $R_3 = 4\ \Omega$, $i_{g1} = 2$ A, and $i_{g2} = -4$ A.
Answer 7 V, 10 V

4.2.2 Using nodal analysis, find v_1, v_2, and i.
Answer 2 V, 18 V, 4 A

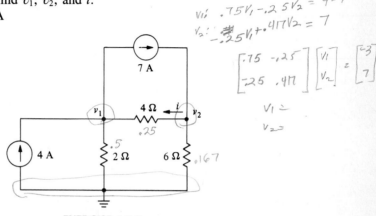

EXERCISE 4.2.2

4.2.3 Using nodal analysis, find v_1, v_2, and v_3.
Answer 24, −4, 20 V

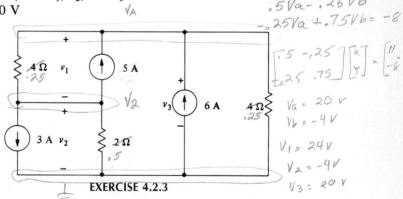

EXERCISE 4.2.3

4.3

CIRCUITS CONTAINING VOLTAGE SOURCES

At first glance it may seem that the presence of voltage sources in a circuit complicates the nodal analysis. We can no longer write the equations using the shortcut method because we do not know the currents through the voltage sources. However, nodal analysis is no more complicated and in many cases is even easier to apply when voltage sources are present, as we shall see.

EXAMPLE 4.2   To illustrate the procedure, let us consider the circuit of Fig. 4.6. We have labeled the nonreference nodes as v_1, v_2, v_3, v_4, and v_5 and have taken the sixth node as reference, as indicated. Again, the resistors are labeled by their conductances.

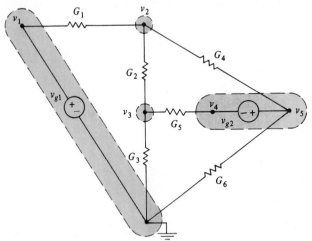

FIGURE 4.6 Circuit containing voltage sources

Since there are five nonreference nodes, we need five equations. Without writing any KCL equations we may note that we have, by inspection,

$$v_1 = v_{g1}$$
$$v_5 - v_4 = v_{g2}$$

(4.6)

We thus need only three KCL equations. To be systematic and at the same time eliminate the need to know the currents in the voltage sources, let us enclose the voltage sources by dashed lines as shown in Fig. 4.6. We may think of these surfaces as *generalized* nodes or, as some authors prefer, *super* nodes. We recall that KCL holds for such a generalized node as well as for an ordinary node. We have, therefore, two generalized nodes and two regular nodes, labeled v_2 and v_3, a total of four nodes. Thus we need only three KCL equations, which together with (4.6) constitute the required set of five equations in the node voltages.

To complete the formulation of the nodal equations, let us apply KCL at nodes v_2 and v_3 and the generalized node containing v_{g2}. The first two are obtained as before, resulting in

$$(G_1 + G_2 + G_4)v_2 - G_1v_1 - G_2v_3 - G_4v_5 = 0$$
$$(G_2 + G_3 + G_5)v_3 - G_2v_2 - G_5v_4 = 0$$

(4.7)

Finally, equating to zero the currents leaving the generalized node, we have

$$G_4(v_5 - v_2) + G_5(v_4 - v_3) + G_6v_5 = 0 \qquad (4.8)$$

The circuit is analyzed by simultaneously solving (4.6), (4.7), and (4.8).

EXAMPLE 4.3

As a second example, let us find v in the circuit of Fig. 4.7(a). The bottom node is taken as the reference and the nonreference nodes are labeled v, v_1, and v_2, as shown in Fig. 4.7(b). By inspection we see that $v_1 = v + 3$ and $v_2 = 20$, as indicated. That is, node v_1 is 3 V above node v and node v_2 is 20 V above the ground. Enclosing the two voltage sources by generalized nodes, shown dashed, we see that there are two "nodes" and thus we need write only one node equation (at either generalized node). This is precisely enough since there is only one unknown—the node voltage v. The node equation for the top generalized node is

$$\frac{v + 3 - 20}{6} + \frac{v + 3}{2} + \frac{v}{4} = 6$$

where if v is in volts, every term has the unit of milliamperes. Solving the equation yields $v = 8$ V.

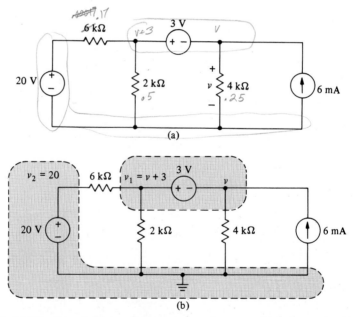

FIGURE 4.7 (a) Circuit containing voltage and current sources; (b) redrawn circuit to show the node voltages

EXAMPLE 4.4

As another example, let us consider the circuit of Fig. 4.8, which contains an independent voltage source and a dependent current source. Anticipating that the presence of the voltage source reduces the number of unknowns by one, we have labeled the node at the top left v_g, the reference node being chosen as indicated. The unknown node voltages are v_1, v_2, and v_3. Applying KCL at these nodes, we have

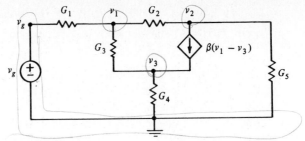

FIGURE 4.8 Circuit with independent and dependent sources

$$(G_1 + G_2 + G_3)v_1 - G_2v_2 - G_3v_3 - G_1v_g = 0$$

$$-G_2v_1 + (G_2 + G_5)v_2 + \beta(v_1 - v_3) = 0$$

$$-G_3v_1 + (G_3 + G_4)v_3 = \beta(v_1 - v_3)$$

These equations may now be solved for the unknown node voltages (v_g is considered to be known). We note that the presence of the dependent source has destroyed the symmetry that was present in the equations of the previous examples. This is true in the general case involving dependent sources.

EXAMPLE 4.5

As a final example, let us consider Fig. 4.9, which contains independent and dependent current and voltage sources. To simplify matters we have chosen the reference node as shown and labeled the nonreference nodes, which in this case may all be expressed in terms of only the two unknowns v_x and i_y. (We have used Ohm's law to obtain v_2, and v_3 was given by $v_2 - 3i_y = -5i_y$.) We therefore need only two equations. Since we have three nodes, the regular node and the two generalized nodes, there are precisely two independent nodal equations. At node v_4 we have

FIGURE 4.9 More complex circuit

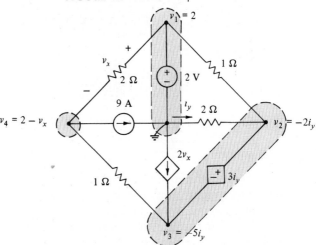

$$-\frac{v_x}{2} + 9 + \frac{2 - v_x - (-5i_y)}{1} = 0$$

and at the generalized node containing the dependent voltage source we have

$$\frac{-2i_y - 2}{1} - i_y - 2v_x + \frac{-5i_y - (2 - v_x)}{1} = 0$$

These equations simplify to

$$-3v_x + 10i_y = -22$$
$$-v_x - 8i_y = 4$$

which have the solution

$$v_x = 4 \text{ V}, \qquad i_y = -1 \text{ A}$$

We may now find all the currents and voltages in the circuit.

EXERCISES

4.3.1 Using nodal analysis, find v if element x is a 4-V independent voltage source with the positive terminal at the top.
Answer 20 V

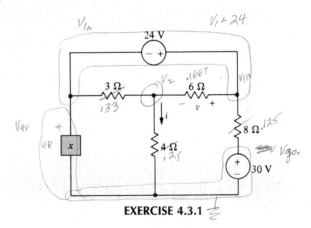

EXERCISE 4.3.1

4.3.2 Find v in Ex. 4.3.1 if element x is a 7-A independent current source directed upward.
Answer 26 V

4.3.3 Find v in Ex. 4.3.1 if element x is a dependent voltage source of $5i$ V with the positive terminal at the top.
Answer 32 V

CIRCUITS CONTAINING OP AMPS

Nodal analysis very often is the best method of analysis when a circuit contains op amps, because in electronic circuits the reference node is usually shown as grounded and all other elements connected to the reference node are often shown individually grounded. Thus the nodes are easily identified for the nodal method, but the loops are not so easily visualized. Therefore a method based on loops, such as we shall consider in the next section, is not so easy to apply. Also, quite often only a relatively few nodal equations are required.

EXAMPLE 4.6 As an example, let us consider the VCVS of Fig. 3.11, redrawn as shown in Fig. 4.10. The reference node is shown as grounded, so that the voltages v_1 and v_2 of Fig. 3.11 are now node voltages, as shown. Since the voltage is zero between the input terminals of the op amp and the currents are zero into the input terminals, we see that $v_3 = v_1$ and that the node equation written at node v_3 is

$$\frac{v_1}{R_1} + \frac{v_1 - v_2}{R_2} = 0 \tag{4.9}$$

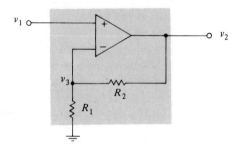

FIGURE 4.10 VCVS

From this we find

$$v_2 = \left(1 + \frac{R_2}{R_1}\right)v_1 = \mu v_1 \tag{4.10}$$

which is the result given earlier in Sec. 3.4.

EXAMPLE 4.7 As another example, let us consider the circuit of Fig. 4.11. We shall take the ground to be the zero voltage reference point so that the inverting input terminal of the op amp is also at zero potential, as labeled. We shall consider the voltage v_1 to be a known input voltage and solve for the output voltage v_2. At node v_3, KCL yields

$$(G_1 + G_2 + G_3 + G_4)v_3 - G_1 v_1 - G_4 v_2 = 0$$

using the shortcut procedure of Sec. 4.1. The sum of the currents entering the inverting input node of the op amp is given by

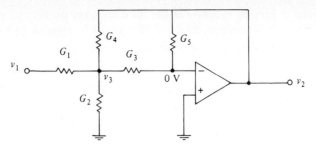

FIGURE 4.11 Circuit containing an op amp

$$G_3 v_3 + G_5 v_2 = 0$$

Eliminating v_3 from these two equations results in

$$(G_1 + G_2 + G_3 + G_4)\left(-\frac{G_5 v_2}{G_3}\right) - G_1 v_1 - G_4 v_2 = 0$$

from which we have

$$v_2 = \frac{-G_1 G_3 v_1}{G_5(G_1 + G_2 + G_3 + G_4) + G_3 G_4}$$

It is always fruitful to write nodal equations at the inverting input nodes of the op amps, as we have done in this example. However, one generally avoids writing nodal equations at output nodes of op amps because it is difficult to find the current out of an op amp. There is no current into the input terminals, but because there are other terminals not shown, the output terminal carries a current. Also, as noted in Chapter 3, when op amps are present the ground node is not arbitrary, and because of the unshown terminals there may be currents into the ground. Thus we should also avoid writing a node equation at the ground node.

EXERCISES

4.4.1   Find i.

Answer $4 \cos 4t$ mA

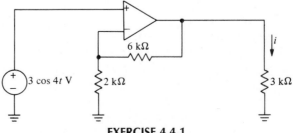

EXERCISE 4.4.1

4.4.2   Find v_0 in terms of the node voltages v_1, v_2, and v_3 and the resistances. (This circuit is a summer, like that of Prob. 3.9, with an additional voltage source.)
Answer $-R_0(v_1/R_1 + v_2/R_2 + v_3/R_3)$

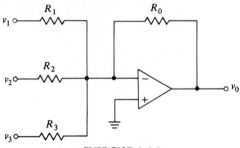

EXERCISE 4.4.2

4.4.3   Find v if $v_g = 6 \cos 2t$ V. (*Suggestion:* Note that the input op amp terminals have the same node voltage v_1, as indicated.)
Answer $2 \cos 2t$ V

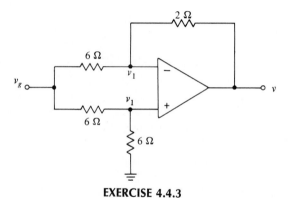

EXERCISE 4.4.3

4.5

MESH ANALYSIS

In the nodal analysis of the previous sections we applied KCL at the nonreference nodes of the circuit. We shall now consider a method, known as *mesh analysis,* or *loop analysis,* in which KVL is applied around certain closed paths in the circuit. As we shall see, in this case the unknowns generally will be currents.

We shall restrict ourselves in this chapter to *planar* circuits, by which we mean circuits that can be drawn on a plane surface in a way such that no element crosses any other element. In this case, the plane is divided by the elements into distinct areas, in the same way that the wooden or metal partitions in a window distinguish the window panes. The closed boundary of each area is called a *mesh* of the circuit. Thus a mesh is a special case of a *loop,* which we consider to be a closed path of elements in the circuit passing through no node or element more than once. In other words, a mesh is a loop that contains no elements within it.

EXAMPLE 4.8   As an example, the circuit of Fig. 4.12 is planar and contains three meshes, identified by the arrows. Mesh 1 contains the elements R_1, R_2, R_3, and v_{g1}; mesh 2 contains R_2, R_4, v_{g2}, and R_5; and mesh 3 contains R_5, v_{g2}, R_6, and R_3.

In the case of nonplanar circuits (that is; those that are not planar), we cannot define meshes. Thus in the analysis using KVL, the closed paths are loops. The procedure is the same, of course, but the equations are not as easily formulated. We shall consider this more general case in detail in Chapter 6.

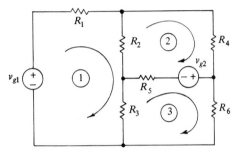

FIGURE 4.12   Planar circuit with three meshes

EXAMPLE 4.9   As an example in which KVL is applied, let us consider the two-mesh circuit of Fig. 4.13. The element currents are I_1, I_2, and I_3. Applying KVL around the first mesh (containing v_{g1}), we have

$$R_1 I_1 + R_3 I_3 = v_{g1} \tag{4.11}$$

Similarly, around the other mesh we have

$$R_2 I_2 - R_3 I_3 = -v_{g2} \tag{4.12}$$

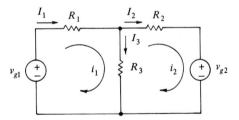

FIGURE 4.13   Circuit with two meshes

We define a *mesh current* as the current which flows around a mesh. The mesh current may constitute the entire current in an element of the mesh, or it may be only a portion of the element current. For example, in Fig. 4.13 the currents i_1 and i_2 are mesh currents, with the directions as shown. The element current is the mesh current in R_1 and R_2, but the element current in R_3 is the composite of two mesh currents.

In general, element currents are algebraic sums of mesh currents. This is illustrated in Fig. 4.13 since the element current in R_1 is

$$I_1 = i_1$$

that in R_2 is

$$I_2 = i_2$$

and that in R_3, by KCL, is

$$I_3 = I_1 - I_2 = i_1 - i_2$$

Using these results, we may rewrite (4.11) and (4.12) as

$$R_1 i_1 + R_3(i_1 - i_2) = v_{g1}$$
$$R_2 i_2 - R_3(i_1 - i_2) = -v_{g2}$$

(4.13)

These are the *mesh equations* of the circuit.

There is also a shortcut method of writing mesh equations which is similar to the shortcut nodal method of Sec. 4.1. Rearranging (4.13) in the form

$$(R_1 + R_3)i_1 - R_3 i_2 = v_{g1}$$
$$-R_3 i_1 + (R_2 + R_3)i_2 = -v_{g2}$$

we note that in the first equation, corresponding to the first mesh, the coefficient of the first current is the sum of the resistances in the first mesh, and the coefficient of any other mesh current is the negative of the resistance common to that mesh and the first mesh. The right member of the first equation is the algebraic sum of the voltage sources driving the first mesh current in its assumed direction. Replacing the word *first* by the word *second* everywhere it appears in these last two sentences will describe the second equation, and so forth. This shortcut procedure is a consequence of selecting all the mesh currents in the same direction (clockwise in Fig. 4.13) and writing KVL as the meshes are traversed in the directions of the currents. Of course, the method applies only when no sources are present except independent voltage sources.

EXAMPLE 4.10   As another example, let us return to Fig. 4.12 and define i_1, i_2, and i_3 as the mesh currents shown in meshes 1, 2, and 3, respectively. Then applying the shortcut method to mesh 1, we have

$$(R_1 + R_2 + R_3)i_1 - R_2 i_2 - R_3 i_3 = v_{g1}$$

This result may be checked by applying KVL to mesh 1, resulting in

$$R_1 i_1 + R_2(i_1 - i_2) + R_3(i_1 - i_3) = v_{g1}$$

The two results are evidently the same.

Applying KVL to meshes 2 and 3 yields, in the same manner,

$$-R_2 i_1 + (R_2 + R_4 + R_5)i_2 - R_5 i_3 = -v_{g2}$$
$$-R_3 i_1 - R_5 i_2 + (R_3 + R_5 + R_6)i_3 = v_{g2}$$

The analysis is completed by solving the three mesh equations for the mesh currents.

The same symmetry is present in the mesh equations as was noted in the nodal equations. If Cramer's rule is to be used for solving for the currents, then the coefficient determinant to be calculated is

$$\Delta = \begin{vmatrix} R_1 + R_2 + R_3 & -R_2 & -R_3 \\ -R_2 & R_2 + R_4 + R_5 & -R_5 \\ -R_3 & -R_5 & R_3 + R_5 + R_6 \end{vmatrix}$$

The diagonal elements are the sums of the resistances in the meshes, and the off-diagonal elements are the negatives of the resistances common to the meshes corresponding to the row and column of the determinant. That is, $-R_2$ in row 1, column 2 or in row 2, column 1 is the negative of the resistance common to meshes 1 and 2, etc. Thus the determinant is symmetric about its diagonal. This symmetry is not preserved, of course, if there are dependent sources present.

EXERCISES

4.5.1 Using mesh analysis, find i_1 and i_2 in Fig. 4.13 if $R_1 = 1\ \Omega$, $R_2 = 2\ \Omega$, $R_3 = 4\ \Omega$. $v_{g1} = 21$ V, and $v_{g2} = 14$ V.
Answer 5 A, 1 A

4.5.2 Using mesh analysis, find i_1 and i_2 if element x is a 6-V independent voltage source with the positive terminal at the top.
Answer 2 A, 1 A

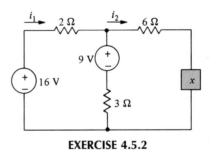

EXERCISE 4.5.2

4.5.3 Repeat Ex. 4.5.2 if element x is a dependent voltage source of $6i_1$ V with the positive terminal at the bottom.
Answer 5 A, 6 A

4.6

CIRCUITS CONTAINING CURRENT SOURCES

As in the case of nodal analysis of circuits with voltage sources, mesh analysis is easier if there are current sources present. To illustrate this point, let us consider the circuit of Fig. 4.14(a), which has two current sources and a voltage source.

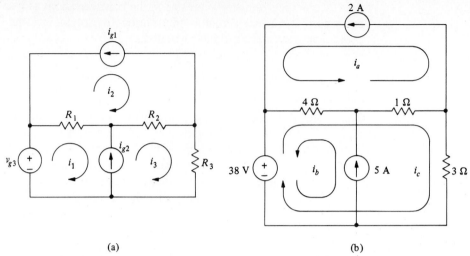

(a)   (b)

FIGURE 4.14 Circuits with two current sources and one voltage source

EXAMPLE 4.11 With the mesh currents i_1, i_2, and i_3 chosen as shown, it is clear that we need three independent equations. Not all of these, however, have to be mesh equations. The presence of the two current sources provides us with two constraints which we may obtain by inspection:

$$i_2 = -i_{g1}$$

$$i_3 - i_1 = i_{g2} \qquad (4.14)$$

We need, therefore, only one more equation. Since it will have to come from KVL, we need to select a closed path in which all the voltages are easily obtained. That is, we need to avoid the current sources since their voltages are not readily obtained.

If we imagine for a moment that the two current sources are removed, that is, opened, then we shall have two less meshes. But we already have two equations so there will still be enough meshes left for the required number of equations. Moreover, the loops left (they may not be meshes) will have only resistors and voltage sources in them, and therefore KVL is easily applied. We must stress that we are not taking the current sources out. We are only imagining them out for a moment in order to *locate* the loops around which KVL is to be applied.

Returning to Fig. 4.14(a) and imagining the current sources open for a moment, we see that we have only one loop left, namely, the loop containing v_{g3}, R_1, R_2, and R_3. Applying KVL to this loop, we have our third equation,

$$R_1(i_1 - i_2) + R_2(i_3 - i_2) + R_3 i_3 = v_{g3} \qquad (4.15)$$

The analysis of the circuit can now be completed by solving (4.14) and (4.15). For example, if $R_1 = 4\ \Omega$, $R_2 = 1\ \Omega$, $R_3 = 3\ \Omega$, $i_{g1} = 2$ A, $i_{g2} = 5$ A, and $v_{g3} = 38$ V, the reader may show that the mesh currents are $i_1 = 1$, $i_2 = -2$, and $i_3 = 6$ A.

In this example we may simplify the process by using loop currents rather than mesh currents. Suppose the problem is to find the current downward in R_3, which is evidently $i_3 = 6$ A. Let us use loop currents i_a, i_b, and i_c, as in Fig. 4.14(b), where i_c is now the desired current downward in $R_3 = 3\ \Omega$, and we have been careful to place only one loop current through a current source. By inspection we have

$$i_a = 2$$
$$i_b = 5$$

and by KVL around the loop of i_c,

$$4(i_a - i_b + i_c) + 1(i_a + i_c) + 3i_c = 38$$

or

$$4(2 - 5 + i_c) + 1(2 + i_c) + 3i_c = 38$$

Solving this equation we have $i_c = 6$ A, as before.

EXAMPLE 4.12   As a final example, let us apply loop analysis to the circuit of Fig. 4.9, which was analyzed previously by the nodal method. The circuit is redrawn in Fig. 4.15. There are four meshes and two constraints to be satisfied by the controlling variables v_x and i_y. We shall need, therefore, six equations.

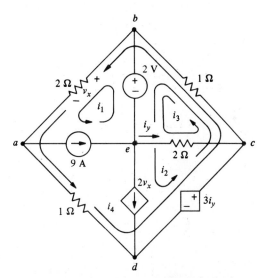

FIGURE 4.15 More complicated circuit

We may choose mesh currents and obtain their relations to the two current sources as we did earlier in Fig. 4.14(a). However, to further illustrate the use of loop currents and also to simplify the resulting equations, we have chosen i_1, i_2, i_3, and i_4 as the unknown currents, as shown in the circuit. This selection results in a single-loop current through the current sources and through the element whose current controls a dependent source. Thus the constraints are simple equations. Before applying KVL, we may write, by inspection of the circuit,

$$i_1 = 9$$

$$i_2 = 2v_x$$

$$i_3 = i_y$$

$$2(i_1 + i_4) = v_x$$

From these results we may express all the loop currents in terms of v_x and i_y. The result is

$$i_1 = 9$$

$$i_2 = 2v_x$$

$$i_3 = i_y$$

$$i_4 = \frac{v_x}{2} - 9$$

(4.16)

As yet we have not written a single loop equation. We need two more equations since there are basically two unknowns, v_x and i_y. By imagining for the moment the current sources as open, we see the two loops to which we shall apply KVL. They are *abecda* and *bceb*. The respective loop equations are

$$-v_x + 2 + 2i_y + 3i_y - 1(i_4) = 0$$

and

$$-1(i_2 + i_3 + i_4) - 2i_y - 2 = 0$$

Substituting (4.16) into these equations, we have

$$-\frac{3v_x}{2} + 5i_y = -11$$

$$\frac{5v_x}{2} + 3i_y = 7$$

The solution is given by

$$v_x = 4 \text{ V}, \qquad i_y = -1 \text{ A}$$

which checks with the result obtained earlier in Sec. 4.3.

The loop currents may now be found from (4.16), completing the analysis. Any other current or voltage in the circuit may be readily obtained also.

In general, before analyzing any circuit, one should note how many equations are required in nodal analysis and in loop analysis, and use the simpler method. Clearly, in the last example the nodal analysis performed earlier was simpler.

EXERCISES

4.6.1 Using mesh analysis, find i.
Answer 3 A

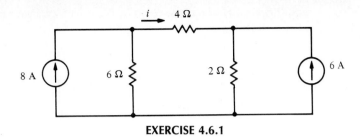

EXERCISE 4.6.1

4.6.2 Using mesh analysis, find v_1.
Answer 3 V

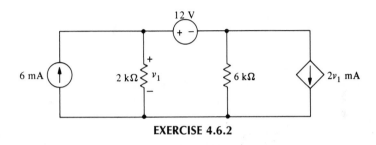

EXERCISE 4.6.2

4.6.3 In Fig. 4.14(a), let $R_1 = 4\ \Omega$, $R_2 = 6\ \Omega$, $R_3 = 2\ \Omega$, $i_{g1} = 4$ A, $i_{g2} = 6$ A, and $v_{g3} = 52$ V. Leave i_2 and i_3 as shown and change i_1 to a loop current clockwise through R_1, R_2, R_3, and v_{g3}, and use loop analysis to find the power delivered to R_3. (Note that in this case the current through R_3 is $i_1 + i_3$.)
Answer 18 W

4.7

DUALITY

The reader may have noticed a similarity in certain pairs of network equations which we have considered so far. For example, Ohm's law may be stated as

$$v = Ri \tag{4.17}$$

or

$$i = Gv \tag{4.18}$$

In the second case we have solved for i, of course, and used the definition $G = 1/R$. Another way of looking at these equations is to note that the second may be obtained from the first by replacing v by i, i by v, and R by G. In like manner, the first may be obtained from the second by replacing i by v, v by i, and G by R.

Similarly, in the case of *series* resistances, R_1, R_2, . . . R_n, the equivalent resistance was shown in Sec. 2.3 to be

$$R_s = R_1 + R_2 + \ldots + R_n \tag{4.19}$$

and for *parallel* conductances, $G_1, G_2, \ldots, G_n$, we saw in Sec. 2.4 that the equivalent conductance was

$$G_p = G_1 + G_2 + \ldots + G_n \tag{4.20}$$

It is clear that one of these equations may be obtained from the other by interchanging resistances and conductances and the subscripts s and p, i.e., series and parallel.

There is thus a definite *duality* between resistance and conductance, current and voltage, and series and parallel. We acknowledge this by defining these quantities as *duals* of each other. That is, R is the dual of G, i is the dual of v, series is the dual of parallel, and vice versa in each case.

Another simple case of dual equations is

$$v = 0 \tag{4.21}$$

and its dual

$$i = 0 \tag{4.22}$$

In the general case, an element described by (4.21) is a *short circuit,* and one described by (4.22) is an *open circuit.* Thus short circuits and open circuits are duals. There are also other dual quantities, as we shall see later.

EXAMPLE 4.13 Every equation in circuit theory has a dual, obtained by replacing each quantity in the equation by its dual quantity. If one equation describes a planar circuit, then the other equation describes the dual of the circuit, or the *dual* circuit. (As the reader may see in a later course, nonplanar circuits do not have duals.) For example, consider the circuit of Fig. 4.16. The mesh equations are given by

$$
\begin{aligned}
(R_1 + R_2)i_1 - R_2i_2 &= v_g \\
-R_2i_1 + (R_2 + R_3)i_2 &= 0
\end{aligned}
\tag{4.23}
$$

To obtain the dual of (4.23), we simply replace the R's by G's, the i's by v's, and v by i. The result is

$$
\begin{aligned}
(G_1 + G_2)v_1 - G_2v_2 &= i_g \\
-G_2v_1 + (G_2 + G_3)v_2 &= 0
\end{aligned}
\tag{4.24}
$$

These are the nodal equations of a circuit having two nonreference node voltages, v_1 and v_2, three conductances, and an independent current source i_g. From our shortcut procedure for nodal equations we see that G_1 and G_2 are connected to the first node, G_2 is common to the two nodes, G_2 and G_3 are connected to the second node, and i_g

FIGURE 4.16 Two-mesh circuit

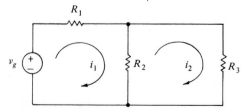

enters the first node. Since G_1 and i_g are not connected to the second node and G_3 is not connected to the first node, these elements are connected to the reference node. Such a circuit is shown in Fig. 4.17 and is a dual of Fig. 4.16.

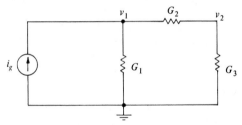

FIGURE 4.17 Dual of Fig. 4.16

Figure 4.16 may be described as v_g in series with R_1 and the parallel combination of R_2 and R_3. Replacing the quantities in this statement by their duals, we see correctly that Fig. 4.17, a dual circuit, may be described as i_g in parallel with G_1 and the series combination of G_2 and G_3.

In this example we note that nodes are duals of meshes and vice versa. This is true in general because of the duality between the mesh currents and the nonreference node voltages. The reference node is the dual of the boundary of the region *outside* the circuit, or what we might call the *outer* mesh.

This last duality (between nodes and meshes) gives us a procedure for finding a dual of a given circuit. We may place one node of the dual network to be obtained inside each mesh, thereby assuring us of the correspondence between each non-reference node and a mesh. Then we may place the reference node outside the circuit (corresponding to the outer mesh). Connecting these nodes together by elements drawn through the elements of the original circuit assures us of correspondence between elements. If these last elements drawn are the duals of those they cross, then the circuit obtained is a dual of the original circuit.

As an example, the circuit of Fig. 4.16 is redrawn using solid lines in Fig. 4.18. The dashed circuit, superimposed in accordance with the dual circuit procedure, is its dual. It should be clear that the latter circuit is the same as that of Fig. 4.17.

FIGURE 4.18 Two dual circuits

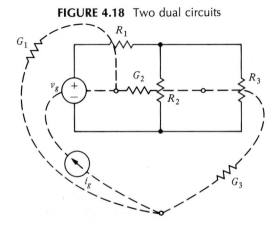

In the case of independent or dependent sources, the polarity of the dual source is specified when the dual circuit is obtained from the dual equations, as was the case in Fig. 4.17. However, in the geometric method of Fig. 4.18, the polarity of i_g cannot be determined since the mesh current directions are not shown and thus are completely arbitrary.

If the mesh currents or node voltages are shown in the original circuit, then the polarity of a source obtained in the dual circuit may be determined from the concept of *driving* a node or a mesh. We shall say that a mesh is *driven* by a source in it if the polarity of the source is such as to drive the current in the direction of the mesh current. A node is driven by a source connected to it if the polarity of the source is such as to apply voltage to the node, i.e., send current toward the node. Thus the dual source will drive or not drive a mesh or a node in accordance with whether the source it crosses in the original circuit drives or does not drive the corresponding node or mesh. This is consistent, as it must be, with the two sets of dual equations of the circuits.

In general, when we have analyzed one circuit, the numerical values of its voltages and currents are the same as those of their duals in the dual circuit. When we solve one circuit, therefore, we have really solved two. As an illustration, the numerical values of i_1 and i_2 in Fig. 4.16 are, respectively, the same as those of v_1 and v_2 in Fig. 4.17.

As a final note in this chapter, we shall not attempt to give a dual of an op amp. We may, of course, obtain duals of circuits containing dependent sources which are equivalent to circuits with op amps.

EXERCISE

4.7.1 Find the dual of the circuit in Fig. 4.14(a).

4.8

COMPUTER-AIDED CIRCUIT ANALYSIS USING SPICE

Numerous computer-aided circuit analysis programs are available for solving electric circuits on virtually all digital computers, from large mainframe to the smaller (in size) personal computers. Although computer programs are very useful and sometimes necessary for solving complicated networks, they are in no way a substitute for a thorough understanding of the rudiments of basic circuit theory. The student should keep in mind that output from these programs is based totally on user input. This requires a thorough knowledge of basic principles for data input and also for valid interpretation of the computed results.

SPICE, developed at the University of California at Berkeley, is a powerful program that has become the de facto standard for analog circuit simulation. PSpice, a version of SPICE executable on personal computers such as the IBM PC family, is used for illustration in this text. A description of the standard commands of SPICE used in our study is given in Appendix E and should be reviewed before continuing in this section. Of particular importance for the dc examples are the E, F, G, H, I, R,

V, and X data statements, as well as the .DC, .END, .ENDS, .LIB, .OP, .PRINT, .SUBCKT, and .TF commands.

EXAMPLE 4.14   As a first example, consider the simple circuit of Fig. 4.19, where the node designations have been encircled for clarity. The procedure for a simulation requires the following steps:

1. Create an input file or *circuit file* using an ASCII text editor (e.g., EDLIN of MS DOS).
2. Run the simulator.
3. Examine the output file.

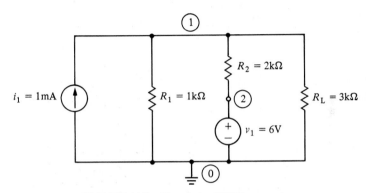

FIGURE 4.19 Circuit for SPICE example

From Appendix E, we see that the creation of a circuit file for our simple dc circuit example, in general, involves (1) title and comment statements, (2) data statements, (3) solution control statements, (4) output control statements, and (5) an end statement. A circuit file, saved as say EX4-19.CIR, that gives a solution is

```
Simple DC Solution for example of Fig. 4.19 (Title statement)
* Data statements for component values
I 1 0 1 DC 1M
V1 2 0 DC 6V
R1 1 0 1KOHM
R2 1 2 2K
RL 1 0 3K
* Solution control statement to print all currents and power
* dissipation for all voltage sources. (Note: .OP is the
* default used if no control statement included.)
.OP
* Output control statement for node voltages 1 and 2. (Optional
* in this case since .OP gives all node voltages and currents
* through voltage sources automatically.)
.PRINT DC V(1) V(2)
* End Statement
.END
```

Running the simulation (step 2) will depend on the hardware configuration being employed. In the case of an IBM PC with a hard disk, the following entry is all that is required:

$$\text{PSPICE EX4-19.CIR}$$

The output is saved on the hard disk as EX4-19.OUT.

Examination of the results (step 3) is easily performed by inspecting the output file using DOS commands TYPE or COPY for viewing on the video monitor or printing on the system printer. A solution in this case appears as follows:

```
NODE   VOLTAGE        NODE   VOLTAGE
(    1)    2.1818     (    2)    6.0000
        VOLTAGE SOURCE CURRENTS
        NAME              CURRENT
        V1                -1.909E-03
        TOTAL POWER DISSIPATION  1.15E-02   WATTS
```

FIGURE 4.20 (a) Circuit containing a CCCS; (b) redrawn for a SPICE simulation with V_d included for use of I_x

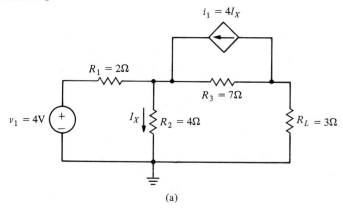

(a)

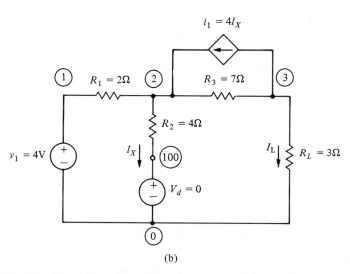

(b)

EXAMPLE 4.15   As a second example, consider the circuit of Fig. 4.20(a), which illustrates an oddity of SPICE. To use the current in a branch, such as I_x of the current-controlled current source (CCCS) i_1, a dummy voltage source such as V_d must be inserted, as shown in Fig. 4.20(b). A circuit file for the node voltages 1 and 2, the voltage between nodes 2 and 3, and the current in R_L is

```
Circuit file for DC solution of Fig. 4.20
* Data statements
V1 1 0 DC 4V
VD 100 0 DC 0
FI1 3 2 VD 4
R1 1 2 2OHM
R2 2 100 4
R3 2 3 7
RL 3 0 3
* Solution control statement for dc solution with V1 = 4 V.
.DC V1 4 4 1
* Output control statement
.PRINT DC V(1) V(2) V(2,3) I(RL)
.END
```

The output file for the solution yields

V1	V(1)	V(2)	V(2,3)	I(RL)
4.000E+00	4.000E+00	1.333E+01	3.733E+01	−8.000E+00

EXAMPLE 4.16   As a final example, let us consider a circuit containing an op amp. A simple VCVS model of a practical op amp is shown in Fig. 4.21. In this model, R_{in} represents the input resistance of the op amp and μ the gain. This model can be used for simulating the op amp in the simple inverter circuit of Fig. 4.22(a). Nominal values of $R_{in} = 10^{10}$ ohms and $\mu = 10^6$ have been used to approximate an ideal device. A circuit file for the solution is

FIGURE 4.21   (a) Op amp symbol; (b) simple VCVS equivalent circuit

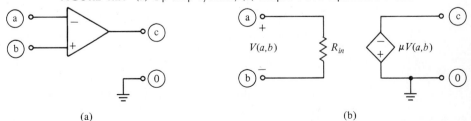

(a)   (b)

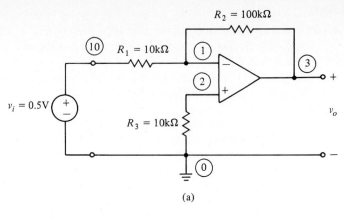

(a)

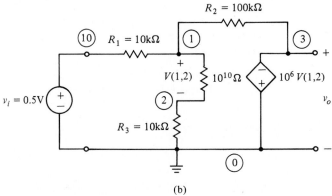

(b)

FIGURE 4.22 (a) Inverter; (b) VCVS equivalent

```
OP AMP INVERTER OF FIG. 4.22.
* Data statements
VI   10 0 DC 0.5V
R1   10 1 10K
R2   1 3   100K
R3   2 0 10K
RIN  1 2 1E+10
EVO 0 3 1 2 1E+6
* Solution control statements for VI = 0.5 V and transfer
* function for V(3)/VI, input and output resistance.
.DC VI 0.5 0.5 1
.TF V(3) VI
* Output control statement
.PRINT DC V(1) V(2) V(3) I(R1)
* End statement
.END
```

The .TF (transfer function) solution control statement added to the circuit file causes the ratio of $V(3)/V_i$, the current in V_i, the input resistance as seen by V_i, and the output resistance of the op amp seen looking into the output terminals to be printed.

A formal discussion of the transfer function is deferred until Chapter 14. The result of the simulation is

```
VI            V(1)         V(2)          V(3)          I(R1)
5.000E−01   5.000E−06   5.000E−12   −5.000E+00   5.000E−05

V(3)/VI = −1.000E+01

INPUT RESISTANCE AT VI =    1.000E+04

OUTPUT RESISTANCE AT V(3) =    0.000E+00
```

A powerful feature of SPICE is the ability to define subcircuits that can be referenced in a data statement. Suppose, for example, we desire to store the equivalent circuit for the op amp used in Example 4.16 and recall it whenever needed in a solution. This can be done by generating a file such as OPAMP.CKT containing the circuit definition desired for the op amp and then calling it using a data statement within a circuit file employing the X command (subcircuit call). Let us begin by using the subcircuit definition feature for file OPAMP.CKT, which can be written as

```
.SUBCKT OPAMP 1 2 3

* 1 DENOTES INVERTING INPUT NODE
* 2 DENOTES NONINVERTING INPUT NODE
* 3 DENOTES OUTPUT NODE
RIN  1 2 1E+10
EVO  0 3 1 2 1E+6
.ENDS
```

The circuit file of Example 4.16, using the subcircuit definition, can now be written as

```
OP AMP INVERTER OF FIG 4.22.

* Data statements
VI  10  0 DC 0.5V
R1  10 1 10K
R2  1 3   100K
R3  2 0 10K
XAMP1 1 2  3 OPAMP

* Define file where OPAMP subcircuit located
.LIB OPAMP.CKT

* Solution control statements
.DC V(3) 0.5 0.5 1

.TF V(3) VI

* Output control statement
.PRINT DC V(1) V(2) V(3) I(R1)

* End statement
.END
```

EXERCISES

4.8.1 Write a circuit file for determining v in Fig. 4.7.

4.8.2 Write a circuit file for finding the node voltages in Fig. 4.9.

4.8.3 Find v_o, v_o/v_i, the input resistance, and the output resistance.
Answer -2.198 V, -10.99, 18.2 kΩ, 0 kΩ

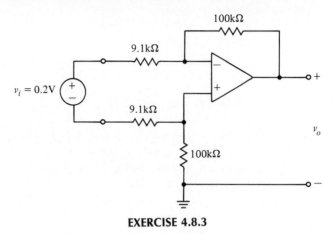

EXERCISE 4.8.3

PROBLEMS

4.1 Using nodal analysis, find v_1 and v_2.

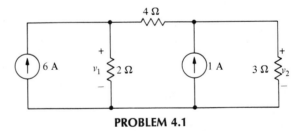

PROBLEM 4.1

4.2 Using nodal analysis, find i.

PROBLEM 4.2

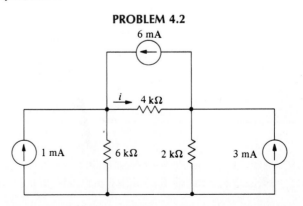

4.3 Using nodal analysis, find i.

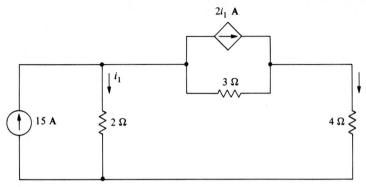

PROBLEM 4.3

4.4 Find the power delivered to the 2-Ω resistor using nodal analysis.

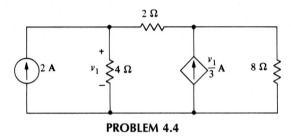

PROBLEM 4.4

4.5 Find v and i using nodal analysis.

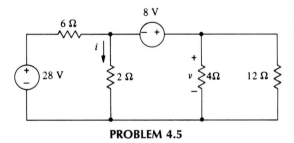

PROBLEM 4.5

4.6 Using nodal analysis, find i.

PROBLEM 4.6

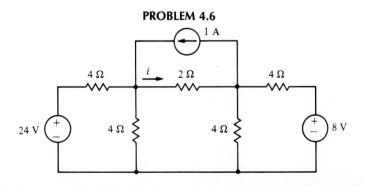

4.7 Using nodal analysis, find v.

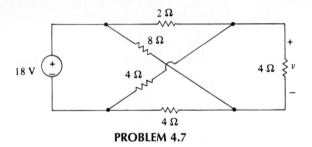

PROBLEM 4.7

4.8 Using nodal analysis, find i.

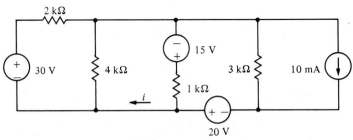

PROBLEM 4.8

4.9 Using nodal analysis, find v and v_1.

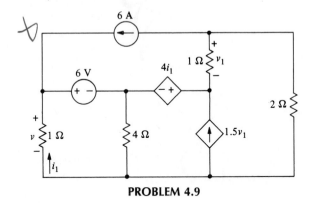

PROBLEM 4.9

4.10 Using nodal analysis, find i_1.

PROBLEM 4.10

4.11   Using nodal analysis, find v.

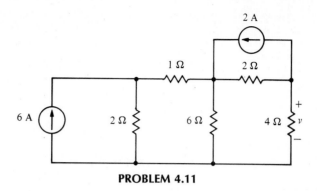

PROBLEM 4.11

4.12   Find v if $v_g = 8 \sin 6t$ V.

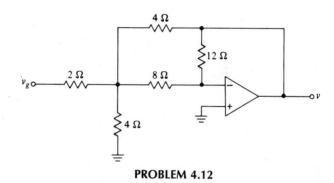

PROBLEM 4.12

4.13   Find v if $v_g = 8 \cos 3t$ V. (*Hint:* Note that $v_1 = v/\mu$, where $\mu = 2$ is the gain of the VCVS.)

PROBLEM 4.13

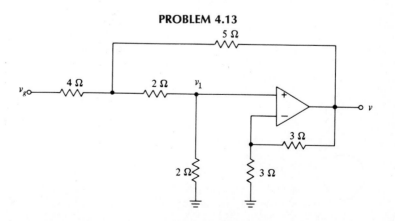

4.14   Find v if $v_g = 4 \cos 6t$ V.

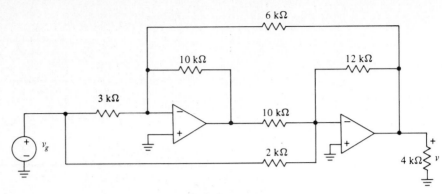

PROBLEM 4.14

4.15   Solve Prob. 4.2 using mesh analysis.

4.16   Solve Prob. 4.3 using mesh analysis.

4.17   Using mesh or loop analysis, find the power delivered to the 4-Ω resistor.

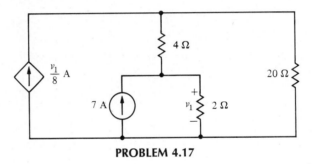

PROBLEM 4.17

4.18   Find v.

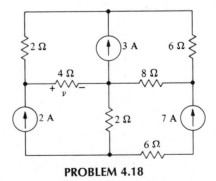

PROBLEM 4.18

4.19   Solve Prob. 4.18 if the 2-A, 3-A, and 7-A current sources are replaced by 17-V, 4-V, and 16-V voltage sources, respectively, with the positive terminal at the top in each case.

4.20 Find i using both nodal and loop analysis.

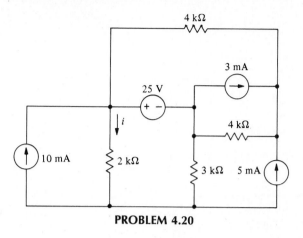

PROBLEM 4.20

4.21 Find v using the method (loop or nodal) that requires the fewer equations.

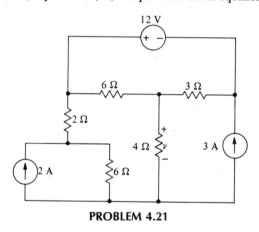

PROBLEM 4.21

4.22 Find the power delivered by the 10-V source, using both nodal and loop analysis.

PROBLEM 4.22

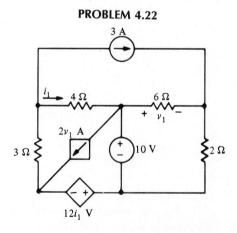

4.23 Find v_2 in terms of R and I_g, and show that if $R = 10 \text{ k}\Omega$, then

$$v_2 \approx -5 \times 10^4 I_g$$

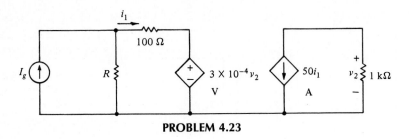

PROBLEM 4.23

4.24 Find v_1.

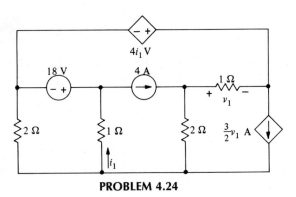

PROBLEM 4.24

4.25 Find i.

PROBLEM 4.25

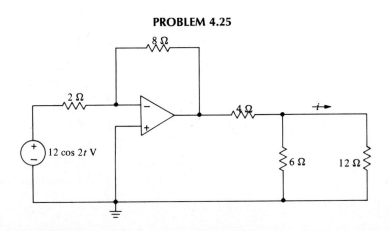

4.26 Find i if $v_g = 6 \cos 1000t$ V.

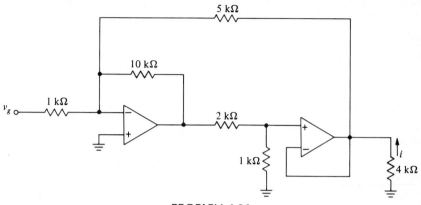

PROBLEM 4.26

4.27 Find the input resistance R_{in}, in terms of the other resistances. Let $R_1 = R_3 = 2$ kΩ and find R_2 so that (a) $R_{in} = 6$ kΩ and (b) $R_{in} = -1$ kΩ.

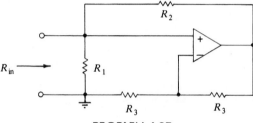

PROBLEM 4.27

4.28 Find v_0 in terms of v_1, v_2, R_1, R_2, R_3, and R_4. Find the relationship satisfied by the resistances so that the output voltage is given by

$$v_0 = \frac{R_2}{R_1}(v_2 - v_1)$$

(Note that this circuit is a *differential amplifier;* it amplifies the difference between the input signals v_1 and v_2.)

PROBLEM 4.28

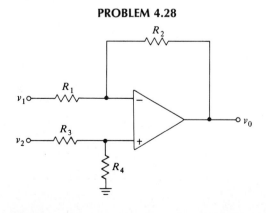

4.29 In the circuit of Prob. 4.28, make $v_2 = v_1$ by connecting the input leads together, make $R_3 = R_4 = R_1$, and show that

$$v_0 = \frac{R_1 - R_2}{2R_1}v_1$$

Thus the difference between R_1 and R_2 is amplified. (The circuit in this case is called a *bridge amplifier*, because it can be used like a *Wheatstone bridge*, a circuit for detecting differences in resistances.)

4.30 Draw a circuit and its dual if the mesh equations of the circuit are given by

$$10i_1 - 2i_2 = 4$$
$$-2i_1 + 8i_2 - i_3 = 0$$
$$-i_2 + 11i_3 = -6$$

4.31 Draw the dual of the circuit of Prob. 4.18. Identify i in the dual circuit corresponding to v in the original circuit and show that they have the same numerical values.

COMPUTER APPLICATION PROBLEMS

4.32 Use SPICE to find (a) i in Prob. 4.8; (b) v in Prob. 4.9; and, (c) i_1 in Prob. 4.22.

4.33 Use SPICE to find v and the input resistance as seen from the terminals of v if $v_g = 1.5$ V in Prob. 4.14.

4.34 Use SPICE to find the current in the 5-kΩ resistor of Prob. 4.26 if $v_g = 2$ V.

4.35 Use the .DC command to find v_1 in Prob. 4.24 for the 18-V source being varied from 10 V to 20 V in 2-V increments.

5

Hermann von Helmholtz
1821–1894

Network Theorems

The force of falling water can only flow down from the hills when rain and snow bring it to them.

Hermann von Helmholtz

The man many think was responsible for Thévenin's theorem, considered in this chapter, was Hermann Ludwig Ferdinand von Helmholtz, a physicist, physician, physiologist, and Germany's greatest scientist of the nineteenth century. He helped prove the law of conservation of energy, invented the ophthalmoscope, constructed a generalized form of electrodynamics, and foresaw the atomic structure of electricity. His anticipation of the existence of radio waves was later proven when they were discovered by one of his students, Heinrich Hertz.

Helmholtz was born in Potsdam, Germany, the oldest of six children of August Helmholtz and Caroline Penne Helmholtz, a descendant of William Penn, the founder of Pennsylvania. He served 8 years as an army doctor to pay his obligations for his medical scholarship during his student years at the Friedrich Wilhelm Institute. His main interest, however, was physics, in which he gained his greatest fame. His 70th birthday was an occasion for nationwide celebrations in Germany. Three years later he died, having raised German science to the great heights to which his famous contemporary, Otto von Bismarck, had raised the German nation. ■

In the previous chapters we have considered fairly straightforward methods of analyzing circuits. However, in many cases the analysis can be shortened considerably by the use of certain network theorems. For example, if we are interested only in what happens to one particular element in a circuit, it may be possible by means of a network theorem to replace the rest of the circuit by an equivalent and simpler circuit.

In this chapter we shall introduce a number of network theorems and illustrate their use in simplifying the analysis of certain circuits. The theorems are applicable, in general, to circuits that are *linear,* a term which we shall also discuss.

Finally, we shall use the network theorems as a motivation for introducing *practical* sources, as distinguished from the ideal sources considered thus far. As we would expect, practical sources are capable of delivering only a finite amount of power, and we shall devote a section to determining the maximum power that can be delivered by a given source.

5.1

LINEAR CIRCUITS

In Chapter 2 we defined a linear resistor as one that satisfied Ohm's law

$$v = Ri$$

and we considered circuits that were made up of linear resistors and independent sources. We defined dependent sources in Chapter 3 and analyzed circuits containing both independent and dependent sources in Chapter 4. The dependent sources that we considered all had describing relations of the form

$$y = kx \tag{5.1}$$

where k is a constant and the variables x and y were either voltages or currents. Clearly, Ohm's law is a special case of (5.1).

In (5.1) the variable y is proportional to the variable x, and the graph of y versus

x is a straight line passing through the origin. For this reason, some authors refer to elements which are characterized by (5.1) as *linear elements*.

For our purposes we shall define a linear element in a more general way, which includes (5.1) as a special case. If x and y are variables, such as voltages and currents, associated with a two-terminal element, then we shall say that the element is *linear* if multiplying x by a constant K results in the multiplication of y by the same constant K. This feature is called the *proportionality property* and evidently holds for (5.1) since

$$Ky = k(Kx)$$

Thus not only is an element described by (5.1) linear, but, in addition, elements described by relations of the forms

$$\frac{dy}{dt} = ax, \quad y = b\frac{dx}{dt} \tag{5.2}$$

are also linear if a and b are nonzero constants.

The ideal op amp is a multiterminal element and is described by more than one equation. However, we shall use the op amp only in a feedback mode, as stated in Chapter 3, and in this case the equivalent circuits which result consist of linear elements with describing relations like (5.1). Therefore we may add the ideal op amp to our list of linear elements.

We shall define a *linear circuit* as one containing only independent sources and/or linear elements. As examples, all the circuits we have considered thus far are linear circuits.

The describing equations of a linear circuit are obtained by applying KVL and KCL, and therefore they contain sums of multiples of voltages or currents. For example, a loop equation is of the form

$$a_1 v_1 + a_2 v_2 + \ldots + a_n v_n = f \tag{5.3}$$

where f is the algebraic sum of the voltages of the independent sources in the loop, the v's are the voltages of the remaining loop elements, and the a's are 0 or ± 1.

EXAMPLE 5.1

To illustrate, KVL around the loop shown in the circuit of Fig. 5.1 yields

$$v_1 + v_2 - v_3 = v_{g1} - v_{g2}$$

FIGURE 5.1 Loop of a linear circuit

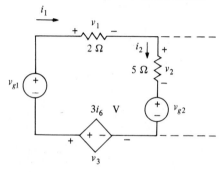

In this case, $a_1 = a_2 = 1$; $a_3 = -1$; the other a's, if any, are all zero; and $f = v_{g1} - v_{g2}$. We also have

$$v_1 = 2i_1, \qquad v_2 = 5i_2, \qquad v_3 = 3i_6$$

where i_6 is a current elsewhere in the circuit.

The proportionality property of a single linear element also holds for a linear circuit in the sense that if all the independent sources of the circuit are multiplied by a constant K, then all the currents and voltages of the remaining elements are multiplied by this same constant K. This is easily seen in (5.3), which multiplied through by K becomes

$$a_1 K v_1 + a_2 K v_2 + \ldots + a_n K v_n = Kf$$

The right member is the consequence of multiplying the independent sources by K, and for equality to still hold all the v's are multiplied by K. Since the elements are linear, multiplying their voltages by K multiplies their currents by K.

EXAMPLE 5.2

To illustrate the proportionality property, let us find the current i in Fig. 5.2. Since by KCL the current to the right in the 2-Ω resistor is $i - i_{g2}$, KVL around the left mesh yields

$$2(i - i_{g2}) + 4i = v_{g1} \tag{5.4}$$

from which

$$i = \frac{v_{g1}}{6} + \frac{i_{g2}}{3} \tag{5.5}$$

If $v_{g1} = 18$ V and $i_{g2} = 3$ A, then $i = 3 + 1 = 4$ A. If we double v_{g1} to 36 V and i_{g2} to 6 A, then i is doubled to 8 A.

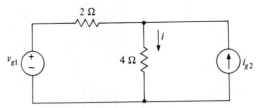

FIGURE 5.2 Linear circuit with two sources

EXAMPLE 5.3

As a final example, let us illustrate another use of the proportionality relation by finding v_1 in the circuit of Fig. 5.3. Such a circuit, because it resembles a ladder, is sometimes called a *ladder* network.

We could write mesh or nodal equations, but to illustrate proportionality we shall introduce an alternative method. Let us simply *assume* a solution,

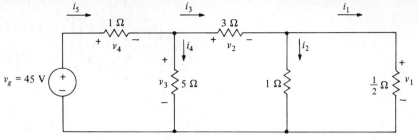

FIGURE 5.3 Ladder network

$$v_1 = 1 \text{ V}$$

and see where it leads us. Referring to the figure, this assumption on v_1 gives $i_1 = 2$ A and $i_2 = 1$ A, and therefore

$$i_3 = i_1 + i_2 = 3 \text{ A}$$

Proceeding down the ladder toward the source, we have, by Ohm's law, KVL, and KCL,

$$v_2 = 3i_3 = 9 \text{ V}$$

$$v_3 = v_1 + v_2 = 10 \text{ V}$$

$$i_4 = \frac{v_3}{5} = 2 \text{ A}$$

$$i_5 = i_3 + i_4 = 5 \text{ A}$$

$$v_4 = 1(i_5) = 5 \text{ V}$$

and, finally, if the guess that $v_1 = 1$ V is correct,

$$v_g = v_3 + v_4 = 15 \text{ V}$$

Our guess was not correct, since actually $v_g = 45$ V, but in view of the laws of probability this should not surprise us. However, by the proportionality relation, if a 15-V source gives an output $v_1 = 1$ V, then our 45-V source will give three times as much, so that the correct answer is

$$v_1 = 3 \text{ V}$$

This method of assuming an answer for the output, working backwards to obtain the corresponding input, and adjusting the assumed output to be consistent with the actual input, by means of the proportionality relation, is particularly easy to apply to the ladder network. There are other circuits to which it also applies, but the ladder network is one of the most-often-encountered circuits, and the method is worthwhile for it alone.

A *nonlinear* circuit is, of course, one that is not linear. That is, it has at least one element whose terminal relation is not of the form (5.1) or (5.2). An example is given in Ex. 5.1.4, for which is it seen that the proportionality property does not apply.

EXERCISES

5.1.1 Find v_1, v_2, and v_3 with (a) the source values as shown, (b) the source values divided by 2, and (c) the source values multiplied by 2. Note how the principle of proportionality applies in (b) and (c).

Answer (a) 4, 8, 28 V; (b) 2, 4, 14 V; (c) 8, 16, 56 V

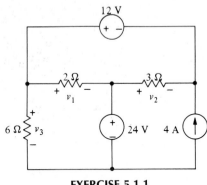

EXERCISE 5.1.1

5.1.2 Find i_1, v_2, i_3, v_3, v_4, and i_5 in Fig. 5.3.

Answer 6 A, 27 V, 9 A, 30 V, 15 V, 15 A

5.1.3 Find v and i using the principle of proportionality.

Answer 4 V, 3 A

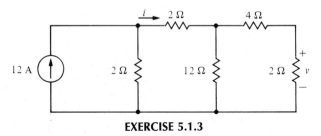

EXERCISE 5.1.3

5.1.4 A circuit is made up of a voltage source v_g, a 2-Ω linear resistor, and a nonlinear resistor in series. The nonlinear resistor is described by

$$v = i^2$$

where v is the voltage across the resistor and i, which is constrained to be nonnegative, is the current flowing into the positive terminal. Find the current flowing out of the positive terminal of the source if (a) $v_g = 8$ V and (b) $v_g = 16$ V. Note that the proportionality property does not apply.

Answer (a) 2 A; (b) 3.123 A

5.2

SUPERPOSITION

In this section we shall consider linear circuits with more than one input. The linearity property makes it possible, as we shall see, to obtain the responses in these circuits by analyzing only single-input circuits.

EXAMPLE 5.4 To see how this may be accomplished, let us first consider the circuit of Fig. 5.2, which was analyzed in the previous section. The output i satisfied the circuit equation (5.4), which we repeat as

$$2(i - i_{g2}) + 4i = v_{g1} \tag{5.6}$$

From the solution,

$$i = \frac{v_{g1}}{6} + \frac{i_{g2}}{3} \tag{5.7}$$

also given earlier, we see that i is made up of two components, one due to each input.

If i_1 is the component of i due to v_{g1} alone, i.e., with $i_{g2} = 0$, then by (5.6) we have

$$2(i_1 - 0) + 4i_1 = v_{g1} \tag{5.8}$$

Similarly, if i_2 is the component of i due to i_{g2} alone, i.e., with $v_{g1} = 0$, then by (5.6) we have

$$2(i_2 - i_{g2}) + 4i_2 = 0 \tag{5.9}$$

Adding (5.8) and (5.9), we have

$$2(i_1 - 0) + 4i_1 + 2(i_2 - i_{g2}) + 4i_2 = v_{g1} + 0$$

or

$$2[(i_1 + i_2) - i_{g2}] + 4(i_1 + i_2) = v_{g1} \tag{5.10}$$

Comparing this result with (5.6), we see that

$$i = i_1 + i_2$$

Also solving (5.8) and (5.9), we have

$$i_1 = \frac{v_{g1}}{6}$$

$$i_2 = \frac{i_{g2}}{3} \tag{5.11}$$

which check with the two components given in (5.7).

Alternatively, we may obtain the components of i directly from the circuit of Fig.

5.2. To find i_1 we need to make the current source i_{g2} zero. Since $i_{g2} = 0$ is the equation of an open circuit, this is accomplished by replacing the current source in the circuit by an open circuit. This operation of making a source zero is sometimes referred to, rather grimly, as "killing" the source, or making the source "dead." The resulting circuit, in this case, is shown in Fig. 5.4(a), from which it is easily seen, using Ohm's law, that i_1, the component of i due to v_{g1} alone, is given by the first equation of (5.11).

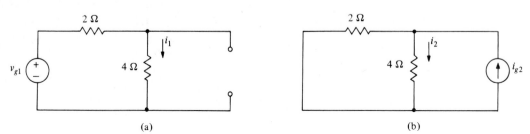

(a)   (b)

FIGURE 5.4   Circuit of Fig. 5.2 with (a) the current source dead; (b) the voltage source dead

To find i_2, the component of i due to i_{g2} alone, we must have $v_{g1} = 0$, the equation of a short circuit. Thus, to kill a voltage source such as v_{g1}, we replace it by a short circuit, as shown in Fig. 5.4(b). Using current division it is easy to see that i_2 is given by the second equation of (5.11).

The method we have illustrated with this example is called *superposition*, because we have *superposed*, or algebraically added, the components due to each independent source acting alone to obtain the total response. The principle of super-position applies to any linear circuit with two or more sources, because the circuit equations are *linear* equations (first-degree equations in the unknowns). This may be seen from (5.3) and the nature of the v–i relations for the linear elements. In particular, in the case of resistive circuits, the response may be found using Cramer's rule and is clearly a sum of components, one due to each independent source alone. (Expanding the numerator determinant by cofactors of the column containing the sources clearly reveals this.)

The linearity of the circuit equations enables us to add (5.8) and (5.9) and to see in (5.10) that the response was the sum of the individual responses. For example, we are able to say that

$$2i_1 + 2i_2 = 2(i_1 + i_2)$$

because these expressions are linear. However, we *cannot* say that

$$2(i_1 + i_2)^2 = 2i_1^2 + 2i_2^2$$

because these are *quadratic* and not linear expressions.

Superposition allows us to analyze linear circuits with more than one independent source by analyzing separately only single-input circuits. This is quite often advantageous since we can use network reduction properties, such as equivalent resistance and voltage division, in analyzing single-input circuits.

Formally, we state the principle of superposition as follows:

In any linear resistive circuit containing two or more independent sources, any circuit voltage (or current) may be calculated as the algebraic sum of all the individual voltages (or currents) caused by each independent source acting alone, i.e., with all other independent sources dead.

EXAMPLE 5.5 As a second example, let us find the voltage v in the circuit with three independent sources, shown in Fig. 5.5. To illustrate the earlier statement that the response is a sum of components, each due solely to an independent source, we shall first solve the circuit by conventional means. Then to contrast the methods we shall use superposition.

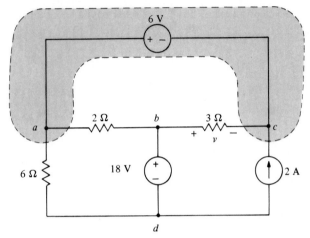

FIGURE 5.5 Circuit with three sources

To illustrate the role each source plays in the response we shall label the sources as

$$v_{g1} = 6 \text{ V}$$
$$i_{g2} = 2 \text{ A} \tag{5.12}$$
$$v_{g3} = 18 \text{ V}$$

Taking d as the reference node, then the node voltage at b is v_{g3}, that at c is $v_{g3} - v$, and that at a is $v_{g3} - v + v_{g1}$. The nodal equation at the generalized node is then

$$\frac{v_{g3} - v + v_{g1}}{6} + \frac{-v + v_{g1}}{2} - \frac{v}{3} - i_{g2} = 0 \tag{5.13}$$

from which we have

$$v = \frac{2}{3}v_{g1} - i_{g2} + \frac{1}{6}v_{g3} \tag{5.14}$$

Thus we see that v is a sum of components due to the individual sources. Finally, substituting (5.12) into (5.14) we have

$$v = 4 - 2 + 3 = 5 \text{ V} \tag{5.15}$$

Solving for v now by the method of superposition, we may write

$$v = v_1 + v_2 + v_3 \tag{5.16}$$

where v_1 is the component due to the 6-V source alone (the 2-A and 18-V sources dead), v_2 is the component due to the 2-A source alone (the two voltage sources dead), and v_3 is the component due to the 18-V source alone (the 6-V and 2-A sources dead). The circuits showing v_1, v_2, and v_3 are given in Figs. 5.6(a), (b), and (c), respectively. In Fig. 5.6(a), killing the 18-V source ties nodes b and d together; in Fig. 5.6(b), killing the two voltages ties nodes a and c together and b and d together; etc.

From Fig. 5.6 it is almost trivial to obtain

$$v_1 = 4 \text{ V}$$

$$v_2 = -2 \text{ V}$$

$$v_3 = 3 \text{ V}$$

which are the values given in (5.15).

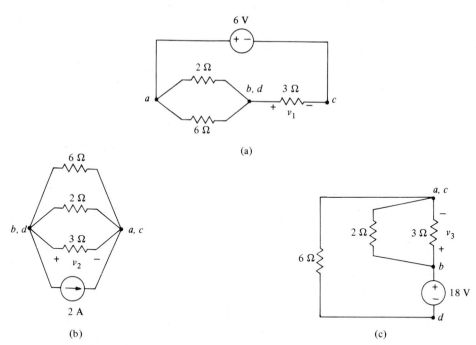

FIGURE 5.6 Circuit of Fig. 5.5 with various sources killed

EXAMPLE 5.6　To illustrate superposition when there is a dependent source present, let us consider the circuit of Fig. 5.7, where it is required to find the power delivered to the 3-Ω resistor. At the outset we should make it clear that power is *not a linear combination* of voltages or currents. In the case we are considering, we have

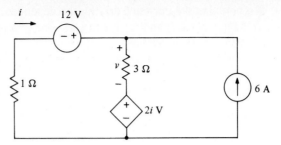

FIGURE 5.7 Circuit with a dependent source

$$p = \frac{v^2}{3}$$

which is a quadratic, and not a *linear,* expression. Therefore superposition will *not* apply to obtaining power directly. That is, we *cannot* find the power due to each independent source acting alone and add the results to obtain the total power. However, we *can* find v by superposition and subsequently obtain p.

Letting v_1 be the component of v when the 12-V source is acting alone [Fig. 5.8(a)] and v_2 be the component of v due to the 6-A source alone [Fig. 5.8(b)], we have, as before,

$$v = v_1 + v_2$$

FIGURE 5.8 Circuit of Fig. 5.7 with (a) the current source killed; (b) the independent voltage source killed

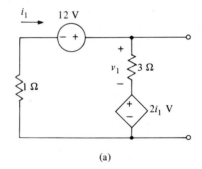

(a)

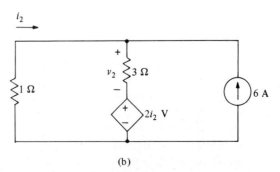

(b)

We note that the current i driving the dependent source also has components, which we have labeled i_1 and i_2, due to each source alone. Solving for v_1 in Fig. 5.8(a) and v_2 in Fig. 5.8(b), we have

$$v_1 = 6 \text{ V}, \qquad v_2 = 9 \text{ V}$$

and, therefore, $v = 15$ V. The power is then given by

$$p = \frac{(15)^2}{3} = 75 \text{ W}$$

There are at least three things that are illustrated by the last example. First, as already mentioned, we cannot use superposition directly to calculate power, but we may use it to get current or voltage, from which power is then found. Second, in applying superposition, *only* the independent sources are killed, *never* the dependent sources. Finally, superposition is often a poor method for solving circuits with dependent sources, because each of the individual single-input circuits frequently is almost as difficult to analyze as the original circuit. For example, there are two meshes in Fig. 5.7, the original circuit, and one mesh current is known. Therefore only one application of KVL is needed, around the left mesh, to solve for the current in the 1-Ω resistor. This is also exactly the case in the circuit of Fig. 5.8(b).

EXERCISES

5.2.1 Solve Ex. 4.2.3 using superposition.

5.2.2 Solve Prob. 3.9 using superposition.

5.2.3 Find the power delivered to the 3-Ω resistors in Fig. 5.8(a) and Fig. 5.8(b) and show that their sum is *not* equal to the total power delivered to the 3-Ω resistor in Fig. 5.7. *Answer* 12, 27 W

5.3

THÉVENIN'S AND NORTON'S THEOREMS

In the previous section we saw that the analysis of some circuits could be greatly simplified by applying the principle of superposition. However, as we saw in the last example of Sec. 5.2, superposition alone may not reduce the complexity of the problems, but its use may lead to additional work. In this section we shall consider *Thévenin's* and *Norton's theorems*, which will in many cases be applicable to and greatly simplify the circuit to be analyzed. As we will see, the use of either of these theorems enables us to replace an entire circuit seen at a pair of terminals by an equivalent circuit made up of a single resistor and a source. Thus, we may determine the voltage or current of a single element of a relatively complex circuit by replacing the rest of the circuit by an equivalent resistor and source and analyzing the resulting, exceedingly simple, circuit.

We shall assume that the circuit to be considered can be separated into two parts, as shown in Fig. 5.9. The part denoted as circuit A is a linear circuit containing resistors, dependent sources, or independent sources. Circuit B may contain nonlinear elements. We also add the constraint that any dependent source in either circuit must have its controlling element in the same circuit. That is, no dependent source in circuit A can be controlled by a voltage or current associated with an element in circuit B and vice versa. The reason for this will become clear in the development which follows.

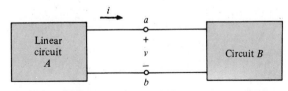

FIGURE 5.9 Partitioned circuit

We shall show that we can replace circuit A by an *equivalent* circuit containing a source and a resistor in such a way that the voltage-current relations at terminals a-b remain the same. Since our aim is to maintain the same terminal relations at a-b, clearly, from the standpoint of circuit A, we may obtain the same effect if we replace circuit B by a voltage source of v volts with the proper polarity, as shown in Fig. 5.10. Insofar as circuit A is concerned, it has the same terminal voltage, and since circuit A itself is unchanged, the same terminal current must flow. We have now obtained a linear circuit and can consequently make use of all the properties we have established for such circuits.

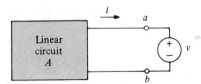

FIGURE 5.10 Replacement of circuit B by a voltage source

In particular, applying superposition to the linear circuit which we have now obtained, we see that the current i will be given by

$$i = i_1 + i_{sc} \tag{5.17}$$

where i_1 is produced by the voltage source v with the network A dead (all its independent sources killed) and i_{sc} is the *short-circuit* current produced by any sources inside circuit A with v killed (replaced by a short circuit). These two cases are shown in Fig. 5.11.

Since the independent sources are dead in circuit A [Fig. 5.11(a)], from the terminals of the source v we see only a resistive circuit, the equivalent resistance of which we shall call R_{th}. By Ohm's law, we therefore have

$$i_1 = -\frac{v}{R_{th}} \tag{5.18}$$

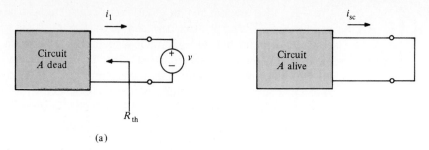

(a)

FIGURE 5.11 Circuits obtained for applying superposition

By (5.17) the expression for the current i is then

$$i = -\frac{v}{R_{th}} + i_{sc} \tag{5.19}$$

Since (5.19) describes network A in the general case, it must hold for any condition at the terminals. Suppose the terminals are open. In this case, $i = 0$, and we shall denote the voltage by $v = v_{oc}$, the *open-circuit* voltage. Substituting these values into (5.19), we have

$$0 = -\frac{v_{oc}}{R_{th}} + i_{sc}$$

or

$$v_{oc} = R_{th}i_{sc} \tag{5.20}$$

Eliminating i_{sc} in (5.19) and (5.20), we have

$$v = -R_{th}i + v_{oc} \tag{5.21}$$

The relations given in (5.19) and (5.21) may be used to obtain two very useful circuits which are equivalent to circuit A. Historically, the first of these was *Thévenin's equivalent circuit,* named in honor of the French telegraph engineer Charles Leon Thévenin (1857–1926), who published his results in 1883. [The circuit might have been more appropriately named for the great German physicist H. L. F. von Helmholtz (1821–1894), who gave a restricted case of it in 1853.]

The Thevenin equivalent circuit is simply one which is described by (5.21) with terminal voltage v and terminal current i, oriented as in Fig. 5.9. (For convenience we will drop the accent mark in Thévenin's name.) To draw the circuit we note that v is the sum of two terms, which therefore must represent two elements in series whose terminal voltages add up to v. The first term evidently corresponds to a resistance R_{th}, called the *Thevenin* resistance, and the second term corresponds to a voltage source with terminal voltage v_{oc}. The result is shown in Fig. 5.12, where the dashed lines represent the connections to the external circuit B of Fig. 5.9. Analysis of the circuit will show that (5.21) is satisfied. The statement that Fig. 5.12 is equivalent at the terminals to Fig. 5.9 is known as *Thevenin's theorem.*

Another circuit that is equivalent to circuit A of Fig. 5.9 is obtained from (5.19). This circuit is the dual of the Thevenin circuit and is called the *Norton equivalent circuit* in honor of the American engineer E. L. Norton (1898–), whose work was

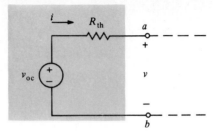

FIGURE 5.12 Thevenin equivalent circuit of circuit *A* of Fig. 5.9

published some 50 years after Thevenin's. From (5.19) we see that *i* is the sum of two terms, which must then represent two parallel elements whose currents add up to *i*. The first term evidently arises from the Thevenin resistance R_{th}, and the second term corresponds to a current source i_{sc}. The result is shown in Fig. 5.13, and the statement of equivalence of Figs. 5.9 and 5.13 is *Norton's theorem*. Again, the dashed lines represent the connections to the external circuit *B*.

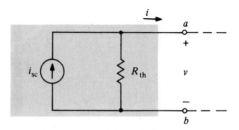

FIGURE 5.13 Norton equivalent circuit of Fig. 5.9

As an example, let us find the Thevenin and Norton equivalent circuits for the network to the left of terminals *a-b* in Fig. 5.14. Then using the results, let us obtain the current *i*, as shown, in terms of the load resistance *R*.

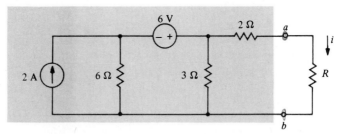

FIGURE 5.14 Circuit with a variable load resistor

EXAMPLE 5.7   To obtain the Thevenin circuit we need to find R_{th} and v_{oc}. The Thevenin resistance R_{th} is found from the dead circuit (the two independent sources made zero), shown in Fig. 5.15(a), from which we have

$$R_{th} = 2 + \frac{(3)(6)}{3 + 6} = 4 \ \Omega$$

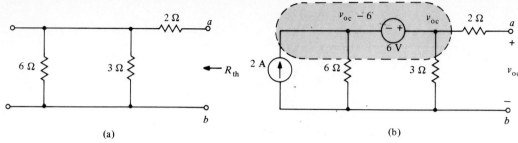

(a)　　　　　　　　　　　　　　　　　　　　　(b)

FIGURE 5.15 Circuits for obtaining the Thevenin circuit of Fig. 5.14

The open-circuit voltage v_{oc} is obtained from Fig. 5.15(b). Since the terminals a-b are open, the voltage v_{oc} is across the 3-Ω resistor. Labeling the nodes as shown, with node b as reference, and writing a nodal equation at the generalized node, shown dashed, we have

$$\frac{v_{oc} - 6}{6} + \frac{v_{oc}}{3} = 2$$

or

$$v_{oc} = 6 \text{ V}$$

The Thevenin equivalent circuit, with the load R connected, is shown in Fig. 5.16. We note that the polarity for $v_{oc} = 6$ V is such that the correct voltage polarity results at terminals a-b when they are opened.

The current i in Fig. 5.14 is the same as that of Fig. 5.16, which in the latter case is readily seen to be

$$i = \frac{6}{R + 4}$$

We may use this result to find the load current for any load R that we choose.

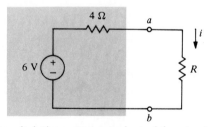

FIGURE 5.16 Loaded Thevenin equivalent of the circuit of Fig. 5.14

EXAMPLE 5.8　To obtain the Norton equivalent circuit we use $R_{th} = 4$ Ω, as before, and calculate i_{sc}. We may short terminals a and b and calculate i_{sc} from the resulting circuit, or we may use the v_{oc} we already have and get i_{sc} from (5.20). In the latter case we have

$$i_{sc} = \frac{6}{4} = 1.5 \text{ A}$$

The Norton equivalent circuit, with the load R connected, is shown in Fig. 5.17. Using current division, we have, as before,

$$i = \left(\frac{4}{R+4}\right)(1.5) = \frac{6}{R+4}$$

The direction of the 1.5-A source is such that when R is replaced by a short, $i_{sc} = 1.5$ A has the correct direction. In this case and in the Thevenin case of Fig. 5.16, it is a simple matter to place the sources correctly, but in a complex example some care needs to be exercised so that the polarities are correct.

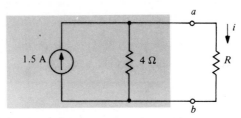

FIGURE 5.17 Loaded Norton equivalent of the circuit of Fig. 5.14

EXAMPLE 5.9 Let us now consider an example containing a dependent source, such as the circuit of Fig. 5.18(a). Suppose we want the Norton equivalent circuit at terminals a-b. We shall need, in this case, R_{th} defined for the dead circuit of Fig. 5.18(b) and i_{sc} shown in Fig. 5.18(c).

We may find i_{sc} from Fig. 5.18(c) by noticing that

$$i_2 = 10 - i_1 - i_{sc}$$

and writing the two mesh equations,

$$-4(10 - i_1 - i_{sc}) - 2i_1 + 6i_1 = 0$$
$$-6i_1 + 3i_{sc} = 0$$

Eliminating i_1, we have

$$i_{sc} = 5 \text{ A} \tag{5.22}$$

We cannot find R_{th} from Fig. 5.18(b) simply by calculating equivalent resistance, because of the dependent source. However, we could excite the circuit with a voltage v (or a current i) at its terminals and calculate the resulting i (or v). Then $R_{th} = v/i$. Alternatively, we could find v_{oc}, as for the Thevenin circuit, and obtain R_{th} from (5.20). This is the method we shall use.

To find v_{oc} we refer to Fig. 5.19, where we have

$$v_{oc} = 6i_1$$

and around the center mesh,

$$-4(10 - i_1) - 2i_1 + 6i_1 = 0$$

From these equations we find $v_{oc} = 30$ V, and thus

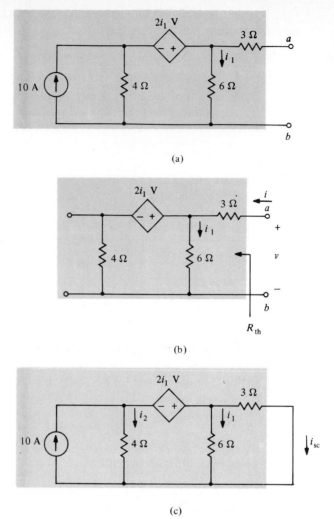

(a)

(b)

(c)

FIGURE 5.18 (a) Circuit to be analyzed, with (b) its source killed; (c) its terminals short-circuited

FIGURE 5.19 Circuit of Fig. 5.18(a) with its terminals opened

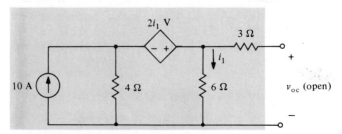

$$R_{\text{th}} = \frac{v_{\text{oc}}}{i_{\text{sc}}} = \frac{30}{5} = 6 \ \Omega$$

EXAMPLE 5.10 As a final example, let us find the Thevenin equivalent of the circuit of Fig. 5.20. To begin with, we see by inspection that since there is no independent source present, we must have

$$v_{\text{oc}} = i_{\text{sc}} = 0 \tag{5.23}$$

Also, the dead circuit is the given circuit itself, so that the R_{th} is simply the resistance seen at the terminals of Fig. 5.20. In view of (5.23), we cannot use $v_{\text{oc}} = R_{\text{th}} i_{\text{sc}}$ to get R_{th}, as we did in the previous example. The only recourse we have is to excite the circuit at its terminals and calculate R_{th} from the results.

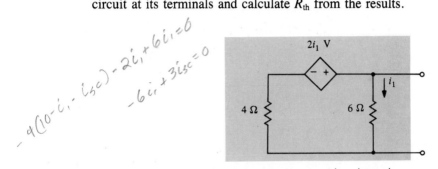

FIGURE 5.20 Circuit with a dependent source

For example, suppose we excite the circuit with a 1-A current source, as shown in Fig. 5.21. Then we have

$$R_{\text{th}} = \frac{v}{1} = v$$

where v is the resulting terminal voltage. Taking the bottom node as reference, the nonreference node voltages are as shown. A nodal analysis yields

$$\frac{v - 2i_1}{4} + \frac{v}{6} = 1$$

FIGURE 5.21 Circuit of Fig. 5.20 excited by a current source

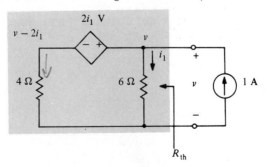

where

$$v = 6i_1$$

From these equations we have $v = 3$ V and thus $R_{th} = 3 \; \Omega$. The Thevenin equivalent (as well as the Norton equivalent) is shown in Fig. 5.22.

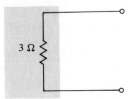

FIGURE 5.22 Thevenin equivalent of Fig. 5.20

EXERCISES

5.3.1 Replace the network to the left of terminals *a-b* by its Thevenin equivalent circuit and use the result to find *i*.
Answer $v_{oc} = 27$ V, $R_{th} = 3 \; \Omega$, $i = 3$ A

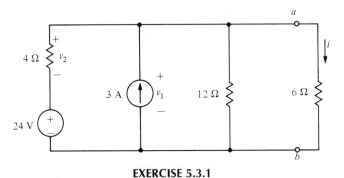

EXERCISE 5.3.1

5.3.2 Replace everything in the circuit of Ex. 5.3.1 except the 3-A source by its Thevenin equivalent circuit and use the result to find v_1.
Answer $v_{oc} = 12$ V, $R_{th} = 2 \; \Omega$, $v_1 = 18$ V

5.3.3 Replace everything in the circuit of Ex. 5.3.1 except the 4-Ω resistor by its Norton equivalent circuit and use the result to find v_2.
Answer $i_{sc} = -3$ A, $R_{th} = 4 \; \Omega$, $v_2 = -6$ V

5.4

PRACTICAL SOURCES

In Chapter 1 we defined independent sources and pointed out that they were *ideal* elements. An ideal 12-V battery, for example, supplies 12 V between its terminals regardless of the load connected to the terminals. However, a real, or *practical*, 12-V battery supplies 12 V when its terminals are open-circuited and supplies less than 12 V as current is drawn through the terminals. A practical voltage source thus appears to have an internal drop in voltage when current flows through its terminals, and this internal drop diminishes the voltage at the terminals.

We may represent a practical source by the mathematical model of Fig. 5.23, consisting of an ideal source v_g in series with an *internal* resistance R_g. The voltage v seen at the terminals of the source now depends on the current i drawn from the source. The relationship is easily seen to be

$$v = v_g - R_g i \tag{5.24}$$

Thus under open-circuit conditions ($i = 0$), we have $v = v_g$, and under short-circuit conditions ($v = 0$), we have $i = v_g/R_g$. If $R_g > 0$, as it is for a practical source, the source can never deliver an infinite current, as an ideal source can.

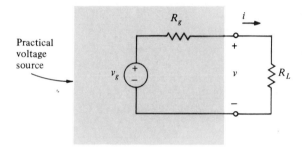

FIGURE 5.23 Practical voltage source connected to a load R_L

For a given practical voltage source (fixed values of v_g and R_g in Fig. 5.23), the load resistance R_L determines the current drawn from the terminals. For example, in Fig. 5.23 the load current is

$$i = \frac{v_g}{R_g + R_L} \tag{5.25}$$

Also, by voltage division we have

$$v = \frac{R_L v_g}{R_g + R_L} \tag{5.26}$$

Therefore as we vary R_L both i and v vary. A sketch of v vs R_L is shown in Fig. 5.24, along with the ideal case, which is dashed. For large values of R_L relative to R_g, v is very nearly equal to the ideal value of v_g. (If R_L is infinite, corresponding to an open circuit, then v is v_g.)

We may replace the practical voltage source of Fig. 5.23 by a practical current source by rewriting (5.24) as

$$i = \frac{v_g}{R_g} - \frac{v}{R_g}$$

which if we define

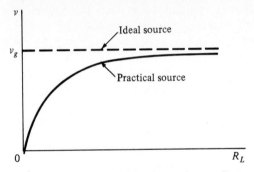

FIGURE 5.24 Practical and ideal voltage source characteristics

$$i_g = \frac{v_g}{R_g} \tag{5.27}$$

becomes

$$i = i_g - \frac{v}{R_g}$$

A circuit described by this equation with voltage v and current i is shown in the shaded rectangle of Fig. 5.25. The circuit is thus a practical current source and is seen to consist of an ideal current source in parallel with an internal resistance.

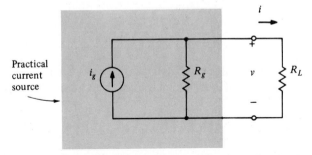

FIGURE 5.25 Practical current source connected to a load R_L

Figures 5.23 and 5.25 are equivalent at the terminals if R_g is the same in both cases and if (5.27) holds. This equivalence is valid, moreover, if the ideal sources are independent or dependent sources. In the case of independent sources the two practical sources are simply the Thevenin and the Norton equivalents of the same circuit.

By current division, we find, in Fig. 5.25,

$$i = \frac{R_g i_g}{R_g + R_L} \tag{5.28}$$

Therefore, for a given current source (fixed values of i_g and R_g), the load current depends on R_L. A sketch of i vs R_L is shown in Fig. 5.26, along with the ideal case, which is dashed.

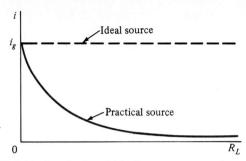

FIGURE 5.26 Practical and ideal current source characteristics

EXAMPLE 5.11   Very often network analysis can be greatly simplified by changing practical voltage sources to practical current sources, and vice versa, by the use of Figs. 5.23 and 5.25, or equivalently, by means of Norton's and Thevenin's theorems. For example, suppose we wish to find the current i shown in Fig. 5.27. We could solve the problem in a number of ways, such as replacing everything except the 4-Ω resistor by its Thevenin equivalent and then finding i. However, we shall illustrate instead the method of successive transformation of sources.

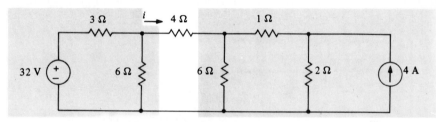

FIGURE 5.27 Circuit with two practical sources

Let us begin by replacing the 32-V source and internal 3-Ω resistance by a practical current source of a 3-Ω internal resistance and a $\frac{32}{3}$-A ideal source. Then let us replace the 4-A source and the internal 2-Ω resistance by a voltage source of $2(4) = 8$ V and a 2-Ω internal resistance. We are applying, respectively, Norton's and Thevenin's theorems, or equivalently, (5.27). The results of these two source transformations are shown in Fig. 5.28.

We may now combine the parallel 3- and 6-Ω resistances and the series 1- and

FIGURE 5.28 Result of two transformations applied to Fig. 5.27

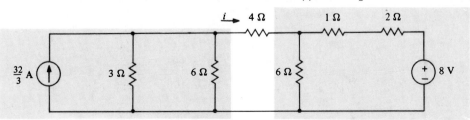

2-Ω resistances, as shown in Fig. 5.29(a), and repeat the source transformation procedure. Continuing the process, as shown in Figs. 5.29(b), (c), and (d), we finally arrive at an equivalent circuit (insofar as i is concerned) which can be analyzed by inspection. In this case, from Fig. 5.29(d), the answer is

$$i = \frac{\dfrac{64}{3} - \dfrac{16}{3}}{2 + 4 + 2} = 2 \text{ A}$$

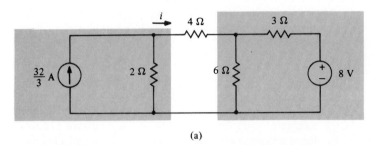

(a)

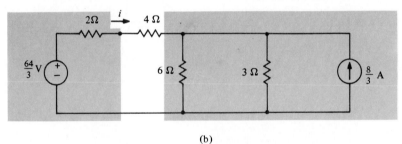

(b)

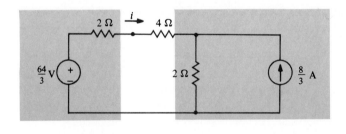

(c)

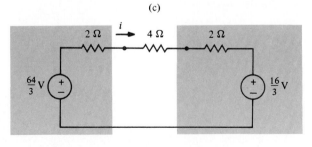

(d)

FIGURE 5.29 Steps in obtaining i in Fig. 5.28

This procedure may seem unduly long, but it should be observed that most of the steps may be carried out mentally.

EXAMPLE 5.12   As a final note in this section, we often may combine sources as we do resistors to obtain equivalent sources. For example, if we are interested only in i in Fig. 5.29(d), we may combine the three series resistors, as we know, but we may also combine the series sources, as we did in Sec. 2.5. They represent a net source of $\frac{64}{3} - \frac{16}{3} = 16$ V with a polarity like that of the larger source. Thus an equivalent circuit insofar as i is concerned is that of Fig. 5.30. Similarly, we may combine parallel current sources to obtain an equivalent source.

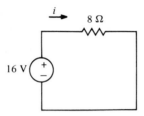

FIGURE 5.30 Circuit equivalent to Fig. 5.29 for finding i

EXERCISES

5.4.1 Solve Ex. 5.3.1 by using source transformations.

5.4.2 By source transformations, replace the entire circuit except for the 8-Ω resistor by an equivalent circuit with a single source and a single resistance R. Using the result, find v.
Answer $R = 12$ Ω, $v = 12$ V

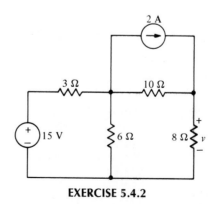

EXERCISE 5.4.2

5.4.3 Convert all the sources in the figure for Ex. 4.2.3 to voltage sources and find v_3.
Answer 20 V

5.5

MAXIMUM POWER TRANSFER

There are many applications in circuit theory where it is desirable to obtain the maximum possible power that a given practical source can deliver. It is very easy, using Thevenin's theorem, to see what maximum power a source is capable of delivering and how to load the source so as to obtain this maximum power. This will be the subject of this section.

Let us begin with the practical voltage source shown earlier in Fig. 5.23 with a load resistance R_L. The power p_L delivered to the resistor R_L is given by

$$p_L = \left(\frac{v_g}{R_g + R_L}\right)^2 R_L \tag{5.29}$$

and it is this quantity which we wish to maximize.

Since the source is assumed to be given, v_g and R_g are fixed, and thus p_L is a function of R_L. To maximize p_L we can make $dp_L/dR_L = 0$ and solve for R_L. From (5.29) we obtain

$$\begin{aligned} \frac{dp_L}{dR_L} &= v_g^2 \left[\frac{(R_g + R_L)^2 - 2(R_g + R_L)R_L}{(R_g + R_L)^4} \right] \\ &= \frac{(R_g - R_L)v_g^2}{(R_g + R_L)^3} = 0 \end{aligned} \tag{5.30}$$

which results in

$$R_L = R_g \tag{5.31}$$

It may be readily shown that

$$\left. \frac{d^2 p_L}{dR_L^2} \right|_{R_L = R_g} = -\frac{v_g^2}{8R_g^3} < 0$$

and therefore (5.31) is the condition which maximizes p_L. We see, therefore, that the maximum power is delivered by a given practical source when the load R_L is equal to the internal resistance of the source. This statement is sometimes called the *maximum power transfer theorem*. We have developed it for a voltage source, but in view of Norton's theorem it also holds for a practical current source.

The maximum power that the practical voltage source is capable of delivering to the load is given by (5.29) and (5.31) to be

$$p_{L_{max}} = \frac{v_g^2}{4R_g} \tag{5.32}$$

In the case of the practical current source, the maximum deliverable power is

$$p_{L_{\max}} = \frac{R_g i_g^2}{4} \tag{5.33}$$

This may be seen from Fig. 5.25 and (5.31) or by (5.32) and Norton's theorem.

We may extend the maximum power transfer theorem to a linear circuit rather than a single source by means of Thevenin's theorem. That is, the maximum power is obtained from a linear circuit at a given pair of terminals when the terminals are loaded by the Thevenin resistance of the circuit. This is obviously true since by Thevenin's theorem the circuit is equivalent to a practical voltage source with internal resistance R_{th}.

EXAMPLE 5.13   As an example, we may draw the maximum power from the circuit of Fig. 5.18(a) if we load terminals a-b with the Thevenin resistance.

$$R_L = R_{\text{th}} = 6 \ \Omega$$

Since $i_{\text{sc}} = 5$ A by (5.22), we may draw the Norton equivalent circuit with the required R_L as shown in Fig. 5.31. The power supplied to the load is given by

$$p = \frac{900 R_L}{(R_L + 6)^2}$$

which for $R_L = 6$ yields

$$p_{\max} = 37.5 \ \text{W}$$

Any other value of R_L will result in a lower value of p. For example, if $R_L = 5 \ \Omega$, then we have $p = 37.19$ W, and for $R_L = 7 \ \Omega$, we have $p = 37.28$ W.

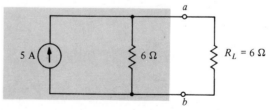

FIGURE 5.31   Equivalent circuit of Fig. 5.18(a) loaded for maximum power transfer

EXERCISES

5.5.1   Find the power delivered to R when (a) $R = 6 \ \Omega$, (b) $R = 2 \ \Omega$, and (c) when R receives the maximum power.
Answer (a) 15.36 W; (b) 14.22 W; (c) 16 W (when $R = 4 \ \Omega$)

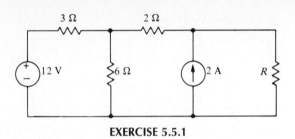

EXERCISE 5.5.1

5.5.2 Show that the two networks are equivalent at terminals *a-b* and find the power dissipated in the 3-Ω resistor in each case.

Answer (a) 12 W; (b) 3 W

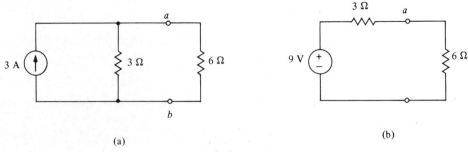

EXERCISE 5.5.2

5.5.3 Find the maximum power delivered to the load R_L in Fig. 5.23 if v_g and $R_L > 0$ are fixed and R_g is variable.

Answer $v_g{}^2/R_L$ when $R_g = 0$

5.6

SPICE AND THEVENIN EQUIVALENT CIRCUITS

SPICE can be used directly for determining the Thevenin equivalent for complex circuits using the .TF command. These equivalent circuits are very useful in determining load conditions for maximum power transfer, as discussed in the previous section.

EXAMPLE 5.14   As an example of the utility of SPICE, consider finding the Thevenin equivalent circuit to the left of R_L in Fig. 5.32(a). Since the open-circuit voltage at terminals *a-b* is equal to the voltage across the 6-ohm resistor (no current flows in the 4-ohm resistor), we need only to apply a SPICE simulation to the circuit of Fig. 5.32(b) and add the 4 ohms to the output resistance found for the open-circuited terminals. It should be noted that

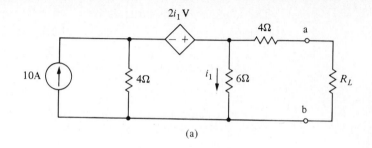

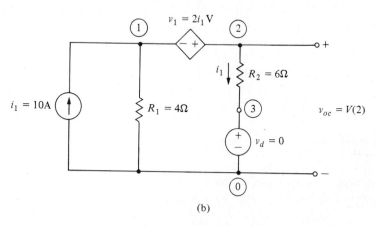

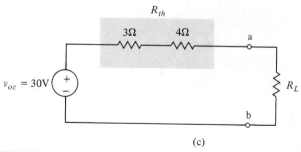

FIGURE 5.32 Circuits for obtaining the Thevenin equivalent using SPICE

a dummy voltage source $v_d = 0$ has been inserted to provide the required current i_1 definition for the CCVS v_1. A circuit file for this circuit is

```
Thevenin equivalent circuit for Fig 5.32 (b)
I1  0 1 DC 10
R1  1 0 4
HV1 2 1 VD 2
R2  2 3 6
VD  3 0 DC 0
.TF V(2) I1
.END
```

The solution for this program is

NODE	VOLTAGE	NODE	VOLTAGE	NODE	VOLTAGE
(1)	20.0000	(2)	30.0000	(3)	0.0000

VOLTAGE SOURCE CURRENTS
NAME   CURRENT

VD   5.000E+00

TOTAL POWER DISSIPATION   0.00E+00 WATTS
****   SMALL-SIGNAL CHARACTERISTICS
V(2)/I1 = 3.000E+00
INPUT RESISTANCE AT I1 = 2.000E+00
OUTPUT RESISTANCE AT V(2) = 3.000E+00

The open-circuit voltage is $V(2) = 30$ V, and the output resistance is 3 ohms. Therefore, the total Thevenin resistance is $3 + 4 = 7$ ohms. The resulting Thevenin equivalent circuit is shown in Fig. 5.32(c). Thus, maximum power transfer occurs when $R = 7$ ohms and the maximum power is $(30/14)^2 \times 7 = 32.14$ W.

EXERCISE

5.6.1 Using SPICE, determine the Thevenin equivalent circuit to the left of terminals *a-b*. Find the value of R for maximum power transfer and the value of the maximum power.
Answer 12 V, 4 kΩ, 4 kΩ, 9 mW

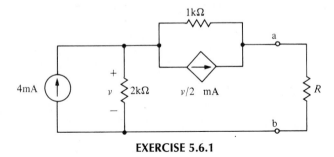

EXERCISE 5.6.1

PROBLEMS

5.1 Solve Prob. 2.21 using the property of proportionality.

5.2 Solve Prob. 2.24 using the property of proportionality. (*Suggestion:* Let $i_2 = 1$ mA, and work toward the source.)

5.3 Solve Ex. 4.4.1 using the property of proportionality.

5.4 Solve Prob. 4.23 using the property of proportionality.

5.5 Solve Ex. 4.6.1 using superposition.

5.6 Find v using superposition.

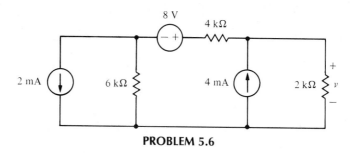

PROBLEM 5.6

5.7 Find v using superposition if $R = 2\ \Omega$.

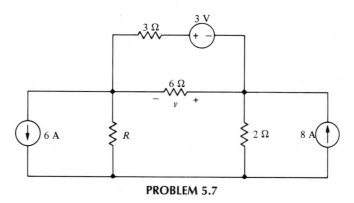

PROBLEM 5.7

5.8 Using superposition, find the power delivered to the 4-Ω resistor in the circuit of Prob. 4.21.

5.9 Find i using superposition.

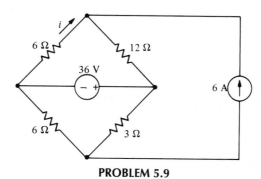

PROBLEM 5.9

5.10 Find v by replacing everything in the circuit except the 4-Ω resistor by its Thevenin equivalent.

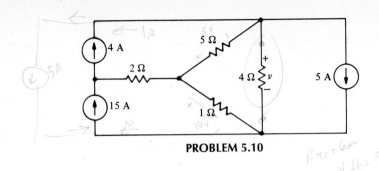

PROBLEM 5.10

5.11 Replace the network to the left of terminals *a-b* by its Thevenin equivalent and use the result to find *i*.

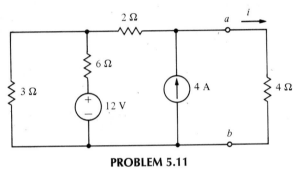

PROBLEM 5.11

5.12 Find the Thevenin equivalent of everything except the 2-Ω resistor in the circuit of Prob. 5.11 and use the result to find the power delivered to the 2-Ω resistor.

5.13 Find *i* by replacing the network to the left of terminals *a-b* by its Norton equivalent.

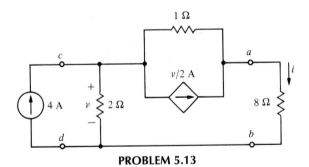

PROBLEM 5.13

5.14 In Prob. 5.13, replace the network to the right of terminals *c-d* by its Thevenin equivalent and use the result to find *v*.

5.15 Replace the circuit to the left of terminals *a-b* by its Thevenin equivalent and use the result to find *v*.

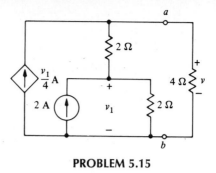

PROBLEM 5.15

5.16 Find the Thevenin equivalent of the circuit external to the 2-Ω resistor and use the result to find i.

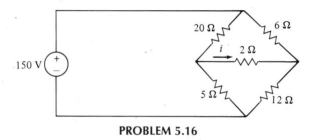

PROBLEM 5.16

5.17 Find the Norton equivalent of the circuit to the left of terminals a-b and use the result to find i.

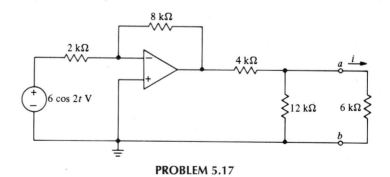

PROBLEM 5.17

5.18 Replace everything except the 2-Ω resistor by its Thevenin equivalent circuit and use the result to find i.

143

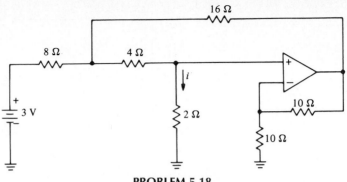

PROBLEM 5.18

5.19 In the circuit of Prob. 5.11, find the power delivered to the 2-Ω resistor by using successive source transformations to obtain the Thevenin equivalent of everything except the 2-Ω resistor.

5.20 Find *i* by using source transformations to obtain an equivalent circuit insofar as the 1-kΩ resistor is concerned, containing only one source and one resistor, in addition to the 1-kΩ resistor.

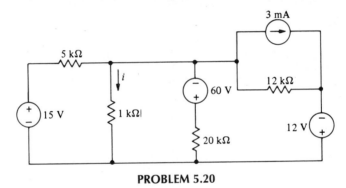

PROBLEM 5.20

5.21 Find the maximum power that can be delivered to resistor *R* in the circuit of Prob. 5.7.

5.22 Find a resistance *R* to be placed between terminals *a-b* to draw the maximum power. Also find the maximum power.

PROBLEM 5.22

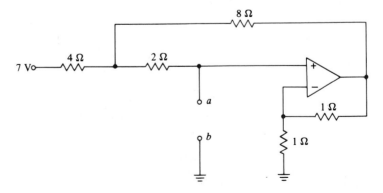

5.23 Find the value of R that will draw the maximum power from the rest of the circuit. Also find the maximum power drawn by R.

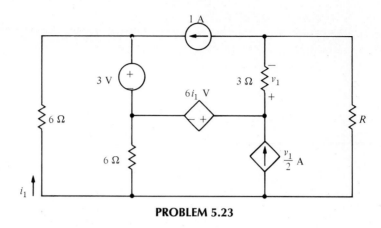

PROBLEM 5.23

COMPUTER APPLICATION PROBLEMS

5.24 Apply SPICE to Prob. 5.15.

5.25 Apply SPICE to Prob. 5.22.

5.26 Apply SPICE to Prob. 5.23.

Leonhard Euler
1707–1783

6

Independence of Equations

Euler calculated without apparent effort, as men breathe, or as eagles sustain themselves in the wind.

Dominique Arago

The application of Kirchhoff's laws to a circuit of many nodes and loops can be extremely difficult, unless we use a branch of mathematics known as *graph theory*, which we introduce in this chapter. (A circuit with only 10 nodes and no parallel elements, for example, could have as many as 10^8 loops.) The father of graph theory was the great Swiss mathematician Leonhard Euler, whose famous 1736 paper, "The Seven Bridges of Königsberg," was the first treatise on the subject. He also made original important contributions to every branch of the mathematics of his day, and Euler's formula is the basis of the phasor method of solving ac circuits discussed in Chapter 10.

Euler was born in Basel, Switzerland, the son of a clergyman. He graduated from the University of Basel in 1724 and joined the Russian Academy of Sciences in Saint Petersburg in 1727 on the invitation of Catherine I. He served in a similar capacity at the German Academy of Sciences at the request of Frederick the Great in 1741. He was perhaps the most prolific mathematician of all time, even continuing to dictate books and papers after he became blind in 1766. He still found time for 13 children and 2 wives, the second of whom he took when he was 69 years old. Swiss mathematicians are still publishing his papers, and it is estimated that his works will eventually fill 60 to 80 large volumes. ■

An electric network is determined by the type of elements it contains and the manner in which the elements are connected. We have spent considerable time in the previous chapters considering the elements themselves and their volt-ampere characteristics. In this chapter we shall consider the manner in which the network elements are connected, or, as it is sometimes called, the network *topology*. As we shall see, a study of the topology of the network provides us with a systematic way of determining how many equations are required in the analysis, which ones are independent, and the best set of equations to select for the most straightforward analysis.

6.1

GRAPH OF A NETWORK

To illustrate the problems involved in the analysis of more complicated networks, let us consider the circuit of Fig. 6.1. The resistors are numbered 1, 2, . . . , 9, with

FIGURE 6.1 Nonplanar circuit

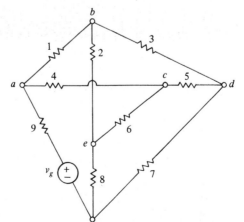

values of resistance, say $R_1, R_2, \ldots, R_9$. Suppose we are required to perform a loop analysis, in which case we need to write a set of independent KVL equations. (We note that the circuit is nonplanar, and thus we cannot perform a mesh analysis. Anyone doubting this is welcome to try redrawing the circuit in a planar fashion.)

There are 15 loops in the circuit, as may be verified by several means, one of which is trial and error (much trial and more error). For the curious reader the 15 loops are (1, 3, 4, 5), (1, 3, 7, 9), (2, 3, 5, 6), (1, 2, 8, 9), (1, 2, 4, 6), (4, 5, 7, 9), (2, 3, 4, 5, 8, 9), (1, 2, 5, 6, 7, 9), (1, 3, 5, 6, 8, 9), (2, 3, 7, 8), (2, 3, 4, 6, 7, 9), (4, 6, 8, 9), (5, 6, 7, 8), (1, 3, 4, 6, 7, 8), and (1, 2, 4, 5, 7, 8). That is, resistors 1, 3, 4, and 5 form a loop; 1, 3, 7, and 9 (with the source v_g) form a loop; etc.

To undertake a loop analysis of Fig. 6.1 we need to know which of these loops are independent and how many are required. To answer these questions we need consider only how the elements are connected; it is unimportant what kind of elements are involved. To facilitate matters, then, we may retain the nodes of the network and replace its elements by lines. The network topology thus is preserved in a much simpler configuration.

The configuration of lines and nodes obtained by replacing the elements of a network by lines is called the *graph* of the network. The lines of the graph are called its *branches,* and the *nodes* of the graph are, of course, the nodes of the network. As an example, the graph of the network of Fig. 6.1 is shown in Fig. 6.2. It has nine branches and six nodes. (We could consider a node between resistor 9 and v_g, but inasmuch as this will not affect the number of loops, we have chosen to consider these two series elements as one element, namely, a nonideal voltage source.)

We shall say that a graph is *connected* if there is a path of one or more branches between any two nodes. The graph of Fig. 6.2 is evidently connected. An example of a graph that is not connected is shown in Fig. 6.3. There is, for instance, no path between nodes a and d. For the present we shall consider only connected graphs.

FIGURE 6.2 Graph of a network

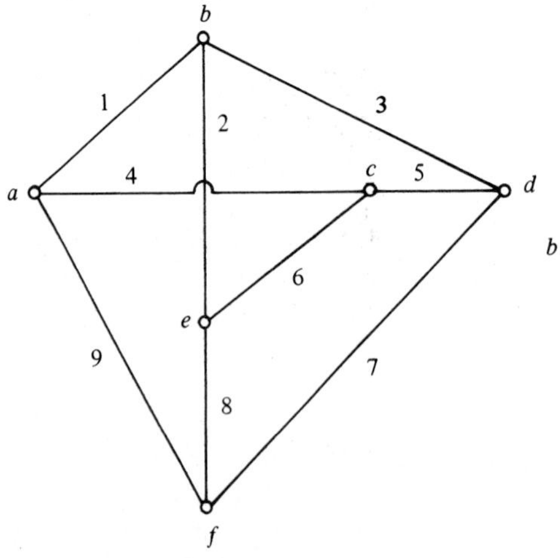

FIGURE 6.3 Unconnected graph

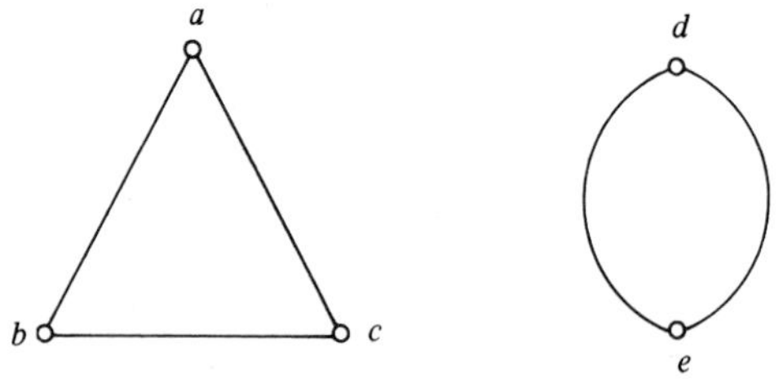

EXERCISES

6.1.1 Show that the graph is planar.

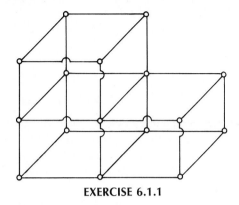

EXERCISE 6.1.1

6.1.2 Show that the graph is planar.

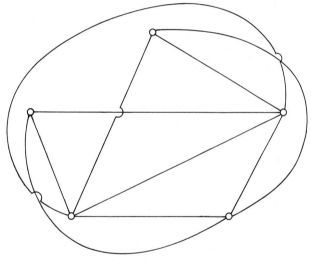

EXERCISE 6.1.2

We define a *tree* of a graph as a connected portion of the graph that contains all the nodes but no loops. As an example, Fig. 6.4(b) is a tree of the graph of Fig. 6.4(a). The tree is connected, has no loops, and contains all the nodes of the graph.

Generally, a graph has many trees. The configuration of Fig. 6.4(c) is evidently another tree of the graph of Fig. 6.4(a), since it satisfies all the requirements. This

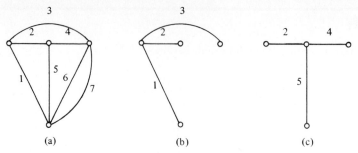

(a)   (b)   (c)

FIGURE 6.4 Graph and two of its trees

particular graph has 24 trees, which the reader may wish to try to discover. It will help in enumerating the trees to notice that each one has exactly three branches, since it takes at least three lines to connect four nodes and more than three lines will form a loop. There are 35 ways to select seven branches three at a time, but 11 of these combinations are not trees.

In the general case, let B be the number of branches and N be the number of nodes in a given graph. Then any tree of the graph contains N nodes and $N - 1$ branches. The number of nodes follows from the definition of a tree, and the number of tree branches may be established by the following construction argument. Let us build the tree starting with one branch and the two nodes to which it is connected. Each additional branch connected to build the tree adds one additional node. The number of nodes is, therefore, one more than the number of branches, and since there are N nodes, there must be $N - 1$ branches.

EXAMPLE 6.1   As examples, the graph of Fig. 6.4(a) has $N = 4$, and thus the number of tree branches in both Figs. 6.4(b) and (c) is $N - 1 = 3$.

Once the branches of a tree are designated, the remaining branches of the graph are called *links*. The number of links in a graph is then $B - (N - 1)$ or $B - N + 1$.

EXAMPLE 6.2   As an example, the tree of Fig. 6.4(b) is redrawn in Fig. 6.5, with the tree branches shown as solid lines and the links as dashed lines. The number of links in this case, since $B = 7$, is $7 - 4 + 1 = 4$.

FIGURE 6.5 Tree branches and links of a graph

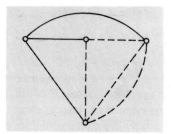

EXERCISES

6.2.1 Find the number of tree branches and links in the graph of Fig. 6.2.

Answer 5, 4

6.2.2 Find all the trees of the graph shown.

Answer (1, 2, 4), (1, 2, 5), (1, 3, 4), (1, 3, 5), (1, 4, 5), (2, 3, 4), (2, 3, 5), (2, 4, 5)

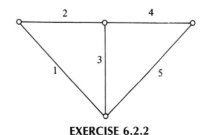

EXERCISE 6.2.2

6.2.3 In the figure for Ex. 6.2.2, let i_2 and i_4 be link currents in elements 2 and 4, directed to the right. Let i_1, i_3, and i_5 be tree currents directed downward. Find the tree currents in terms of the link currents.

Answer $-i_2$, $i_2 - i_4$, i_4

6.3

INDEPENDENT VOLTAGE EQUATIONS

In this section we shall consider the nodal analysis of a circuit whose graph has N nodes and B branches. A tree of the graph then has $N - 1$ branches and, of course, $N - 1$ tree branch voltages. There are also $B - N + 1$ link voltages in the circuit.

Let us imagine that all the branch voltages of any tree are made zero by short-circuiting the branches (i.e., the branches are replaced by short circuits). Then all the nodes of the circuit, being in the tree, are at the same potential, and thus the link voltages are all zero also. Therefore, the link voltages depend on the tree branch voltages, for if a link voltage were independent of the tree voltages, it could not be forced to zero by short-circuiting the tree branches. On the other hand, if one tree branch were not short-circuited, there would be one node at a different potential and thus one tree branch voltage not dependent on the others. We may conclude, therefore, that the $N - 1$ branch voltages of any tree are independent and may be used to find the link voltages.

Another way to see that any link voltage may be expressed in terms of tree voltages is to note that adding the link to the tree completes a circuit whose only other elements are tree branches. Thus by KVL the link voltage is an algebraic sum of tree branch voltages.

EXAMPLE 6.3

As an example, the graph of Fig. 6.6 has tree branch voltages v_1, v_2, and v_3, indicated by the solid lines. The dashed lines are the links whose voltages are v_4, v_5, and v_6. If the link labeled v_4 is added to the tree, the circuit of v_2, v_3, v_4 is formed. By KVL around this circuit, we may obtain

$$v_4 = v_3 - v_2$$

In like manner, adding first link v_5 and then link v_6, we obtain

$$v_5 = v_1 - v_2$$

$$v_6 = v_1 - v_3$$

Thus the link voltages may be found from the tree branch voltages.

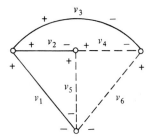

FIGURE 6.6 Tree and link voltages

A systematic way of writing equations involving tree branch voltages is to imagine opening a tree branch, noting that this separates the tree into two parts. Currents flow between the two parts through the tree branch imagined to be open and through links. Thus by KCL the algebraic sum of these currents in a given direction is zero. This procedure may then be repeated for the other tree branches.

EXAMPLE 6.4

To illustrate the use of tree branch voltages, let us consider the circuit in Fig. 6.7(a). A graph of the circuit is shown in Fig. 6.7(b) with the tree branches shown as solid lines and the links as dashed lines. Since the tree branch voltages are independent, we

FIGURE 6.7 Circuit and its graph

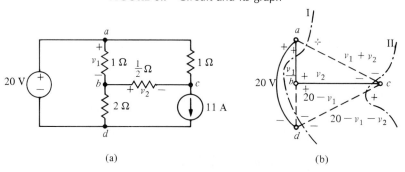

(a)   (b)

have included in the tree the 20-V source. Thus the number of unknowns is reduced by one. By KVL we may find the link voltages in terms of the tree voltages, with the results shown in Fig. 6.7(b).

If we imagine the tree branch (a, b) labeled v_1 as open, then the tree is separated into two parts. These two parts are connected by branch (a, b) and links (a, c), (b, d), and (d, c), as indicated by the line marked I. Summing the currents across the line, we have

$$v_1 + v_2 + v_1 - \frac{20 - v_1}{2} - 11 = 0$$

Repeating the procedure for tree branch (b, c), labeled v_2, leads to line II and the equation

$$-(v_1 + v_2) - 2v_2 + 11 = 0$$

Solving these equations, we have $v_1 = 8$ V and $v_2 = 1$ V. All the link voltages, and consequently all the link and tree currents, may now be found.

In any circuit analysis procedure where the unknowns are voltages, we need to find only the $N - 1$ tree branch voltages which constitute an independent set. This means that only $N - 1$ independent voltage equations are required in the analysis. Since *any* independent set of equations will suffice, then *any* independent set of $N - 1$ voltages constitutes a solution.

Another independent set of $N - 1$ voltages, other than the tree branch voltages, is the set of nondatum node voltages, considered in the nodal method of Chapter 4. To see this, we note that any nondatum node is in the tree and is connected through tree branches to the datum node. Thus every nondatum node voltage is an algebraic sum of tree branch voltages (the tree branches between the nondatum and datum nodes). On the other hand, every tree branch voltage is the difference between its two node voltages. In summary, the node voltages may be determined from the tree voltages and vice versa. Thus the nondatum node voltages are also an independent set.

EXAMPLE 6.5   In the example of Fig. 6.7 we see that if d is the datum node, then the nondatum node voltages v_a, v_b, and v_c are related to the tree voltages v_1, v_2, and 20 by

$$v_a = 20$$
$$v_b = 20 - v_1$$
$$v_c = 20 - v_1 - v_2$$

Conversely, we have

$$v_1 = v_a - v_b$$
$$v_2 = v_b - v_c$$
$$20 = v_a$$

The nodal method, in many cases, is easier to apply than the loop method because the nodes are easy to find. In the loop method, as exemplified by the example of Fig. 6.1, the appropriate loops may be difficult to identify. In the next section we shall consider a method based on graph theory of finding sets of independent loop equations.

EXERCISES

6.3.1 In the graph of the circuit of Prob. 4.5, select the tree of the voltage sources and the 4-Ω resistor. Using the method of this section, write one KCL equation and determine v.
Answer 10 V

6.3.2 Select the tree of the voltage source, and the 3- and 6-Ω resistors, and use the method of this section to find v.
Answer 3 V

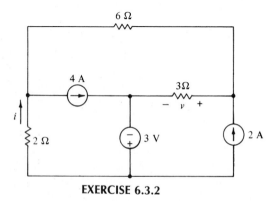

EXERCISE 6.3.2

6.3.3 Using an appropriate tree and the methods of this section, find v in Prob. 4.19. (*Note:* The tree should contain the voltage v and the three sources as well as one other branch.)
Answer 8 V

6.4

INDEPENDENT CURRENT EQUATIONS

As we have seen in the example of Fig. 6.1, it is not always easy to identify the independent loops for a loop analysis of a circuit. To develop a systematic means of writing loop equations, let us consider a general network with B branches and N nodes. Corresponding to a given tree there are $B - N + 1$ links.

Suppose that all the link currents are made zero by open-circuiting the links (replacing the links by open circuits). Since the tree contains no loops, then all the tree branch currents are zero also. The tree currents thus depend on the link currents, i.e., they may be expressed in terms of the link currents, for if a tree current were independent of the link currents it could not be forced to zero by open-circuiting the links. Moreover, if one link is not open-circuited, a loop is left in the graph, and a current will flow in the link. A link current thus is not dependent on the other link currents. In summary, the $B - N + 1$ link currents are an independent set, and the loop analysis of the circuit requires $B - N + 1$ independent equations.

One systematic way to find $B - N + 1$ independent loops is to start with the tree and add one of the links. This determines the loop containing that link, since adding the link to the tree closes a loop. Remove this link and add another link to the tree, determining a second loop. Continue the process until the $B - N + 1$ loops are found. The set is independent because each loop contains a different link.

EXAMPLE 6.6   As an example, let us reconsider the circuit of Fig. 6.1. One tree consists of branches 1, 5, 7, 8, and 9, with corresponding links 2, 3, 4, and 6, as shown in Fig. 6.8. Closing the links one at a time results in the four independent loops I, II, III, and IV shown. Loop I contains link 2 and tree branches 8, 9, and 1; loop II contains 3, 7, 9, and 1; loop III contains 4, 5, 7, and 9; and loop IV contains 6, 5, 7, and 8. These four loops are sufficient for performing a loop analysis.

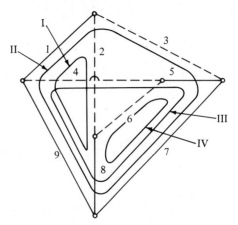

FIGURE 6.8 Loops of Fig. 6.1

EXAMPLE 6.7   To illustrate the use of link currents in circuit analysis, let us return to the example of Fig. 6.7(a). The graph is redrawn in Fig. 6.9, showing the link currents i_1, i_2, and 11 A. We have chosen the current source as a link because the link currents are an independent set. This reduces the number of unknowns by one. Generally, for this reason, one should place voltage sources in the tree and current sources in the links.

The tree branch currents, as in the general case, may be found from the link currents, as shown in Fig. 6.9. Closing the links labeled i_1 and i_2 forms loops 1 and 2, as indicated. Applying KVL to these loops yields, from Figs. 6.7(a) and 6.9,

$$2i_1 - 20 + i_1 - i_2 + 11 = 0$$

$$i_2 - \frac{11 - i_2}{2} - i_1 + i_2 - 11 = 0$$

the solution of which is $i_1 = 6$ A and $i_2 = 9$ A.

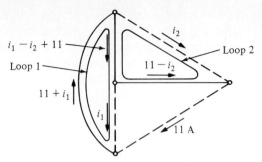

FIGURE 6.9 Graph of Fig. 6.7(a)

Since the link current 11 A is known, we needed only two loops involving link currents i_1 and i_2. Incidentally, in this simple example the links were chosen so that the link currents are also mesh currents. This is, of course, not the case in general.

The results obtained thus far in this chapter are valid for general networks, which may be either planar or nonplanar. In the special case of planar networks, as we saw in Chapter 4, a mesh analysis is possible. In the circuits of that chapter the mesh currents were independent and were sufficient in number to perform the analysis. We shall now show that this is the case in general for planar networks.

Let us begin by taking apart the planar circuit with M meshes and reconstructing it one mesh at a time. The first mesh in the reconstruction has the same number, say k_1, of nodes and branches, for the first branch has two nodes, each additional branch adds one new node, and the last branch adds no nodes since it is tied back to a node of the first branch. This is illustrated by the graph of four meshes in Fig. 6.10(a). The first mesh constructed, shown in Fig. 6.10(b), has the same number of branches and nodes, namely four in this case.

After the first mesh, each subsequent mesh is formed by connecting branches and nodes to previous meshes. Each time the number of nodes added is one less than the number of branches, because each added branch adds one node, except for the last branch, which is connected to a node of a previously added mesh. This process is illustrated in Figs. 6.10(c), (d), and (e).

Thus, if the second mesh adds k_2 branches, then it adds only $k_2 - 1$ nodes. Similarly, the third mesh adds k_3 branches and $k_3 - 1$ nodes, and so forth. The last mesh, the Mth, adds k_M branches and $k_M - 1$ nodes. If in the completed graph the number of branches is B and the number of nodes is N, we have

$$k_1 + k_2 + \ldots + k_M = B \tag{6.1}$$

and

$$k_1 + (k_2 - 1) + \ldots + (k_M - 1) = N \tag{6.2}$$

The latter equation may be written

$$k_1 + k_2 + \ldots + k_M - (M - 1) = N$$

which by (6.1) becomes

$$B - (M - 1) = N$$

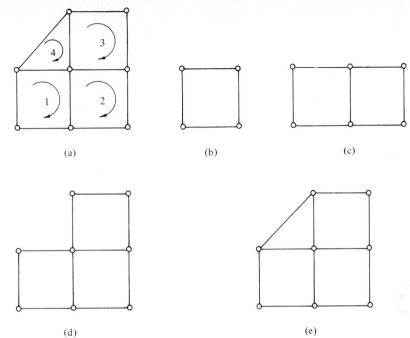

(a)   (b)   (c)

(d)   (e)

FIGURE 6.10 Planar circuit and its meshes

Solving for the number of meshes, we have

$$M = B - N + 1 \tag{6.3}$$

which is also the number of links in the graph.

The mesh currents, therefore, constitute an appropriate set of currents to completely describe a planar network. They are the same in number as the independent set of link currents, and they are independent since each new mesh contains at least one branch not in the previous meshes.

EXERCISE

6.4.1 Show that (6.3) holds for the circuits of Probs. 4.18, 4.20, and 4.24.

6.5

A CIRCUIT APPLICATION

EXAMPLE 6.8 As a final illustration in this chapter we shall analyze a moderately complicated circuit, shown in Fig. 6.11(a). Its graph is shown in Fig. 6.11(b), where we have selected a tree shown by the solid lines.

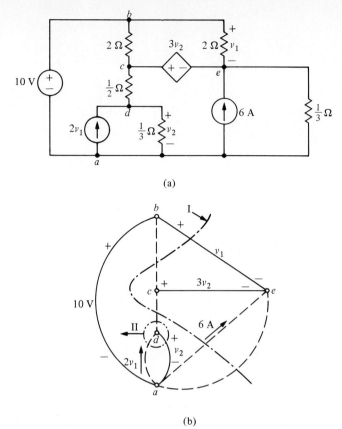

(a)

(b)

FIGURE 6.11 Network and its graph

From the graph we see that there are four tree branches labeled v_1, v_2, $3v_2$, and 10. Thus if the tree branch voltage method is used, there will be only two unknowns, v_1 and v_2, requiring two equations. There are five link currents in the graph, one of which is known (the 6-A source) and another, the $2v_1$ source, which may be expressed in terms of the tree branch current i_{be}, and subsequently in terms of other link currents. Thus if the link current method is used, we must have three equations. Accordingly, we shall analyze the circuit using three branch voltages.

To write the two necessary equations we first imagine tree branch (b, e) opened, which separates the tree into two parts. These parts are connected by branch (b, e) and links (b, c), (c, d), and (a, e) of the 6-A source and link (a, e) of the $\frac{1}{3}$-Ω resistor. These elements are identified by line I. A similar procedure using branch (a, d) leads to line II. Applying KCL to the currents crossing lines I and II in the direction indicated yields

$$\frac{v_1}{2} + \frac{v_{bc}}{2} + 2v_{dc} + 6 + 3v_{ae} = 0 \tag{6.4}$$

and

$$2v_{dc} - 2v_1 + 3v_2 = 0 \tag{6.5}$$

Inspection of the graph shows that

$$v_{bc} = v_1 - 3v_2$$

$$v_{dc} = v_1 - 2v_2 - 10$$

$$v_{ae} = v_1 - 10$$

Substituting these values into (6.4) and (6.5) and solving for the tree branch voltages, we have

$$v_1 = -11 \ V, \qquad v_2 = -20 \ V$$

We may note that nodal analysis is equally easy to apply. The node voltages may be expressed in terms of the two unknowns v_1 and v_2, and thus only two nodal equations are required. We shall leave the details to the problems (Prob. 6.4).

EXERCISES

6.5.1 Using the graph described for the figure for Ex. 6.3.2, write one KVL equation and find i, using the method of Sec. 6.4.
Answer 3 A

6.5.2 Using an appropriate tree for the graph of Prob. 4.18, find the current i flowing to the right in the 4-Ω resistor. (An appropriate tree should *not* contain the current sources or the current i. Thus, using the methods of Sec. 6.4, only one KVL equation is required.)
Answer 6.5 A

6.5.3 Using the method of Sec. 6.4, find the power delivered to the 4-Ω resistor.
Answer 4 W

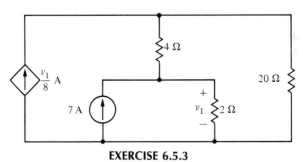

EXERCISE 6.5.3

PROBLEMS

6.1 Find a tree, if possible, that contains all the voltage sources and the branches whose voltages control dependent sources but does not contain current sources or branches whose currents control dependent sources. Use this tree with an appropriate graph theory method to find v_1.

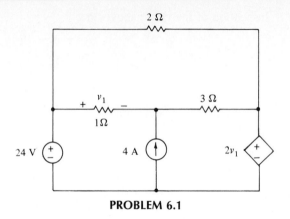

PROBLEM 6.1

6.2 Select a tree as described in Prob. 6.1 and use an appropriate graph theory method to find v_1.

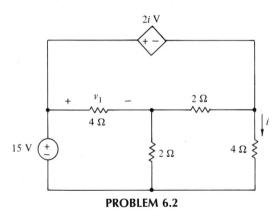

PROBLEM 6.2

6.3 Solve Prob. 4.10 selecting an appropriate tree and using graph theory methods.

6.4 Find v_1 and v_2 in Fig. 6.11 using nodal analysis.

6.5 All the resistances are 1 Ω, element w is also a 1-Ω resistor, elements x and y are independent 1-A current sources directed upward, and element z is an independent 3-A current source directed to the left. Selecting an appropriate tree and using graph theory methods, find i. (Note that this circuit is similar to that of Fig. 6.1 and is thus nonplanar. However, only one loop equation is required in this case.)

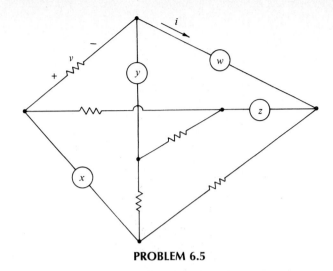

PROBLEM 6.5

6.6 Find v in Prob. 6.5 using graph theory methods if element x is a 4-V source with positive terminal at the top, y is a 2-V source with positive terminal at the bottom, z is a 6-V source with positive terminal at the left, and w is a 4-V source with positive terminal at the left.

6.7 The figure for Ex. 6.2.2 and figures (a) and (b) shown here are examples of graphs of ladder networks. There is a theorem that states that the number of trees in a ladder graph of n branches is the *Fibonacci number* a_n, defined by $a_0 = a_1 = 1$, $a_2 = a_0 + a_1 = 2$, $a_3 = a_1 + a_2 = 3$, $a_4 = a_2 + a_3 = 5$, etc. That is, except for a_0 and a_1, each Fibonacci number is obtained by adding the previous two. Verify that the theorem holds for the ladder graphs shown and also for the graph in the figure for Ex. 6.2.2.

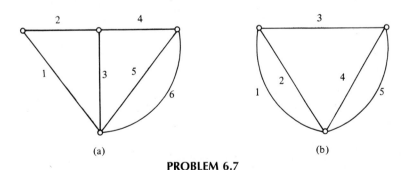

(a)   (b)

PROBLEM 6.7

6.8 The graphs shown in (a) and (b) are two basic nonplanar graphs [the one in (b) is that in the figure for Prob. 6.5 redrawn]. Branch a-b in (a) is an ideal 6-V voltage source with its positive terminal at the top and in (b) is a 4-A ideal current source directed upward. All the other branches in both figures are 1-Ω resistors. Find i shown in each figure using graph theory methods.

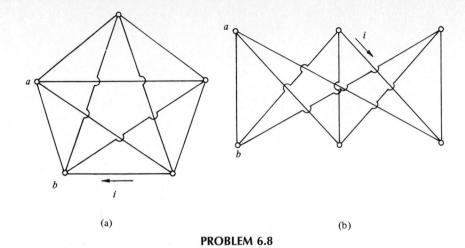

(a)

(b)

PROBLEM 6.8

6.9   Find v_1 and v_2 in Fig. 6.11(a) using link currents as the unknowns.

6.10   Solve Prob. 4.24 using graph theory methods.

Michael Faraday
1791–1867

7

Energy-Storage Elements

My greatest discovery was Michael Faraday.

Sir Humphry Davy

On August 29, 1831, Michael Faraday, the great English chemist and physicist, discovered electromagnetic induction, when he found that moving a magnet through a coil of copper wire caused an electric current to flow in the wire. Since the electric motor and generator are based on this principle, Faraday's discovery profoundly changed the course of world history. When asked by the British prime minister years later what use could be made of his discoveries, Faraday quipped, "Some day it might be possible to tax them."

Faraday, one of 10 children of a blacksmith, was born near London. He was first apprenticed to a bookbinder, but at age 22 he realized his boyhood dream by becoming assistant at the Royal Institution to his idol, the great chemist Sir Humphry Davy. He remained at the Institution for 54 years, taking over Davy's position when Davy retired. Faraday was perhaps the greatest experimentalist who ever lived, with achievements to his credit in nearly all the areas of physical science under investigation in his time. To describe the phenomena he investigated, he and a science-philosopher friend invented new words such as electrolysis, electrolyte, ion, anode, and cathode. To honor him, the unit of capacitance is named the *farad*. ∎

Up to now we have considered only resistive circuits, that is, circuits containing resistors and sources. The terminal characteristics of these elements are simple algebraic equations which lead to circuit equations that are algebraic. In this chapter we shall introduce two important dynamic circuit elements, the capacitor and the inductor, whose terminal equations are differential rather than algebraic equations. These elements are referred to as *dynamic* because, in the ideal case, they store energy which can be retrieved at some later time. Another term which is used, for this reason, is *storage* elements.

We shall first describe the property of capacitance and discuss the mathematical model of an ideal device. The terminal characteristics and energy relations will then be given, followed by derivations for parallel and series connections of two or more capacitors. We shall then repeat this procedure for the inductor.

The chapter will be concluded with a discussion of practical capacitors and inductors and their equivalent circuits.

7.1

CAPACITORS

A *capacitor* is a two-terminal device that consists of two conducting bodies that are separated by a nonconducting material. Such a nonconducting material is known as an insulator or a *dielectric*. Because of the dielectric, charges cannot move from one conducting body to the other within the device. They must therefore be transported between the conducting bodies via external circuitry connected to the terminals of the capacitor. One very simple type, called a parallel-plate capacitor, is shown in Fig. 7.1. The conducting bodies are flat, rectangular conductors that are separated by the dielectric material.

To describe the charge-voltage relationship for the device, let us transfer charge from one plate to the other. Suppose, for instance, by means of some external circuit, that we take a small charge, say Δq, from the lower plate to the upper plate. This, of course, deposits a charge of $+\Delta q$ on the top plate and leaves a charge of $-\Delta q$ on

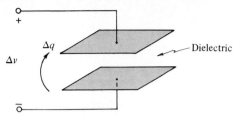

FIGURE 7.1 Parallel-plate capacitor

the bottom plate. Since moving these charges requires the separation of unlike charges (recall that unlike charges attract one another), a small amount of work is performed, and the top plate is raised to a potential of say Δv with respect to the bottom plate.

Each increment of charge Δq that we transfer increases the potential difference between the plates by Δv. Therefore, the potential difference between the plates is proportional to the charge being transferred. This suggests that a change in the terminal voltage by an amount Δv causes a corresponding change in the charge on the upper plate by an amount Δq. Thus the charge is proportional to the potential difference. That is, if a terminal voltage v corresponds to a charge q on the capacitor ($+q$ on the top plate and $-q$ on the bottom plate), then the capacitor has been *charged* to the voltage v, which is proportional to the charge q. We thus may write

$$q = Cv \qquad (7.1)$$

where C is the constant of proportionality, known as the *capacitance* of the device, in coulombs per volt. The unit of capacitance is known as the *farad* (F), named for the famous British physicist Michael Faraday (1791–1867). Capacitors that satisfy (7.1) are called *linear capacitors* since their charge-voltage relationship is the equation of a straight line having a slope of C.

It is interesting to note in the above example that the net charge within the capacitor is always zero. Charges removed from one plate always appear on the other so that the total charge remains zero. We should also observe that charges leaving one terminal enter the other. This fact is consistent with the requirement that current entering one terminal must exit the other in a two-terminal device.

Since the current is defined as the rate of change of charge, differentiating (7.1), we find that

$$i = C\frac{dv}{dt} \qquad (7.2)$$

which is the current-voltage relation for a capacitor.

The circuit symbol for the capacitor and the current-voltage convention which satisfies (7.2) are shown in Fig. 7.2. It is apparent that moving a charge of Δq in Fig. 7.1 from the lower to the upper conductor represents a current flowing into the upper terminal. The movement of this charge causes the upper terminal to become more positive than the lower one by an amount Δv. Hence the current-voltage convention

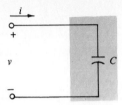

FIGURE 7.2 Circuit symbol for a capacitor

of Fig. 7.2 is satisfied. If v is assigned so that the lower terminal is positive or if the current is assumed to enter the negative terminal, then a minus sign must be used in the right-hand side of (7.2). We recall that this is also necessary in the case of a resistor.

EXAMPLE 7.1   As an example, suppose that the voltage on a 1-μF capacitor is

$$v = 6 \cos 2000t \text{ V}$$

Then the current is

$$i = C\frac{dv}{dt} = 10^{-6}\,(-12{,}000 \sin 2000t) \text{ A}$$

$$= -12 \sin 2000t \text{ mA}$$

In (7.2) we see that if v is constant, then the current i is zero. Therefore, a capacitor acts like an open circuit to a dc voltage. On the other hand, the more rapidly v changes, the larger is the current flowing through its terminals. Consider, for example, a voltage which increases linearly from 0 to 1 V in a^{-1} s, given by

$$v = 0, \qquad t \le 0$$
$$= at, \qquad 0 \le t \le a^{-1}$$
$$= 1, \qquad t \ge a^{-1}$$

If this voltage is applied to the terminals of a 1-F capacitor (an unusually large value which is convenient for illustrative purposes), the resulting current is

$$i = 0, \qquad t < 0$$
$$= a, \qquad 0 < t < a^{-1}$$
$$= 0, \qquad t > a^{-1}$$

Plots of v and i are shown in Fig. 7.3. We see that i is zero when v is constant and that it is equal to a when v increases linearly. If a is made larger, then v changes more rapidly and i increases. It is apparent that if $a^{-1} = 0$ (a is infinite), v changes abruptly (in zero time) from 0 to 1 V.

In general any abrupt or instantaneous changes in voltage, such as in the above example, require that an infinite current flow through the capacitor. An infinite current, however, requires that an infinite power exist at the capacitor terminals, which is a *physical* impossibility. Thus abrupt or instantaneous changes in the voltage

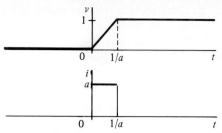

FIGURE 7.3 Voltage and current waveforms for a 1-F capacitor

across a capacitor are *not* possible, and the voltage is continuous even though the current may be discontinuous. (It is possible, of course, to draw on paper circuits that contradict this statement. These circuits are mathematical models that do not sufficiently describe the entire physical picture, as we shall see in Sec. 7.9.) An alternative statement concerning abrupt changes in voltages in circuits containing more than one capacitor is that *the total charge cannot change instantaneously* (conservation of charge).

Let us now find $v(t)$ in terms of $i(t)$ by integrating both sides of (7.2) between times t_0 and t. The result is

$$v(t) = \frac{1}{C} \int_{t_0}^{t} i \, dt + v(t_0) \tag{7.3}$$

where $v(t_0) = q(t_0)/C$ is the voltage on C at time t_0. In this equation, the integral term represents the voltage that accumulates on the capacitor in the interval from t_0 to t, whereas $v(t_0)$ is that which accumulates from $-\infty$ to t_0. The voltage $v(-\infty)$, of course, is taken to be zero. Thus an alternative form of (7.3) is

$$v(t) = \frac{1}{C} \int_{-\infty}^{t} i \, dt$$

EXAMPLE 7.2

In applying this result, we obviously are obtaining the area associated with a plot of i from $-\infty$ to t. In Fig. 7.3, for example, since $v(-\infty) = 0$ and $C = 1$ F, we have

$$v = \frac{1}{1} \int_{-\infty}^{t} (0) \, dt + v(-\infty) = 0, \qquad t \le 0$$

Therefore, $v(0) = 0$, and

$$v = \frac{1}{1} \int_{0}^{t} a \, dt + v(0) = at, \qquad 0 \le t \le a^{-1}$$

Therefore, $v(1/a) = 1$, so that

$$v = \frac{1}{1} \int_{1/a}^{t} (0) \, dt + v\left(\frac{1}{a}\right) = 1, \qquad t \ge a^{-1}$$

which agrees with v in Fig. 7.3.

In this example, we see that v and i do not necessarily have the same shape. Specifically, the maximum and minimum values of v and i do not necessarily occur at the same time, unlike the case for the resistor. In fact, inspection of Fig. 7.3 reveals that the current can be discontinuous even though the voltage is continuous, as stated previously.

EXERCISES

7.1.1 A constant current of 10 mA is charging a 4-μF capacitor (entering its positive voltage terminal). If the capacitor was previously uncharged (zero voltage), find the charge and voltage on it after 20 ms.
Answer 200 μC, 50 V

7.1.2 A 1-μF capacitor has a voltage of 10 sin 2000t V. Find its current.
Answer 20 cos 2000t mA

7.1.3 A 0.3-μF capacitor has a current of $12e^{-4000t}$ mA. Find its voltage $v(t)$ for $t > 0$ if $v(0) = -10$ V.
Answer $-10e^{-4000t}$ V

7.1.4 A 0.4-μF capacitor has a voltage v as shown. Find the current at $t = -4, -1, 2, 5,$ and 9 ms.
Answer 0.5, -2, -0.5, 0, 1 mA

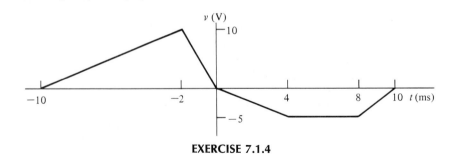

EXERCISE 7.1.4

7.2

ENERGY STORAGE IN CAPACITORS

The terminal voltage across a capacitor is accompanied by a separation of charges between the capacitor plates. These charges have electrical forces acting on them. An *electric field*, a basic quantity in electromagnetic theory, is defined as the force acting on a unit positive charge. Thus the forces acting on the charges within the capacitor can be considered to result from an electric field. It is for this reason that the energy stored or accumulated in a capacitor is said to be stored in the electric field.

The energy stored in a capacitor, from (1.6) and (7.2), is given by

$$w_C(t) = \int_{-\infty}^{t} vi \, dt = \int_{-\infty}^{t} v\left(C\frac{dv}{dt}\right) dt$$

$$= C \int_{-\infty}^{t} v \, dv = \frac{1}{2}Cv^2(t) \, \bigg|_{t=-\infty}^{t}$$

Since $v(-\infty) = 0$, we may write

$$w_C(t) = \frac{1}{2}Cv^2(t) \text{ J} \qquad (7.4)$$

From this result, we see that $w_C(t) \geq 0$. Therefore, from (1.7), the capacitor is a passive circuit element. In terms of the charge on the device, (7.1) and (7.4) yield

$$w_C(t) = \frac{1}{2}\frac{q^2(t)}{C} \text{ J} \qquad (7.5)$$

The ideal capacitor, unlike the resistor, *cannot* dissipate any energy. The energy which is stored in the device can thus be recovered. Consider, for instance, a 1-F capacitor which has a voltage of 10 V. The energy stored is

$$w_C = \frac{1}{2}Cv^2 = 50 \text{ J}$$

Suppose the capacitor is not connected in a circuit; then no current can flow, and the charge, voltage, and energy remain constant. If we now connect a resistor across the capacitor, a current flows until all the energy (50 J) is absorbed as heat by the resistor and the voltage across the combination is zero. The analysis of such a network is found in the next chapter.

As has been pointed out earlier, the voltage on a capacitor is a continuous function. Thus by (7.4) we see that the energy stored in a capacitor is also continuous. This is not surprising since otherwise energy would have to be transported from one place to another in zero time, which is an impossibility.

To illustrate continuity of capacitor voltage, let us consider Fig. 7.4, which contains a switch that is opened at $t = 0$, as indicated. (Ideally, a switch transforms a pair of terminals from an open circuit to a short circuit, or vice versa, in zero time.) To discuss the effect of the switching action we first need to consider two different

FIGURE 7.4 Circuit illustrating continuity of capacitor voltage

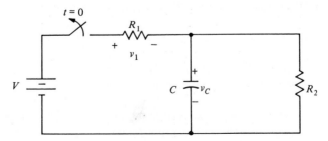

types of time $t = 0$. We shall denote $t = 0^-$ as the time just before the switching action and $t = 0^+$ as the time just after the switching action. Theoretically, of course, no time has elapsed between 0^- and 0^+, but the two times represent radically different states of the circuit. Thus $v_C(0^-)$ is the voltage on the capacitor just before the switch is moved and $v_C(0^+)$ is the voltage immediately after switching. Mathematically, $v_C(0^-)$ is the limit of $v_C(t)$ as t approaches zero through negative $(t < 0)$ values and $v_C(0^+)$ is the limit as t approaches zero through positive $(t > 0)$ values. The same notation applies to v_1 across the resistor R_1.

EXAMPLE 7.3 Suppose that in Fig. 7.4 we have $V = 6$ V and $v_C(0^-) = 4$ V. Just prior to the switching action $(t = 0^-)$ we have $v_1(0^-) = V - v_C(0^-) = 2$ V. Immediately after the switch is opened we have $v_1(0^+) = 0$, since no current is flowing in R_1. However, since v_C is continuous we have

$$v_C(0^+) = v_C(0^-) = 4 \ V$$

Thus the voltage on the resistor has changed abruptly, but that on the capacitor has not.

Obviously, one could consider circuits on paper in which capacitor voltages are forced to change abruptly. For example, if two capacitors having different voltages are suddenly connected in parallel by a switching action, their resulting common voltage cannot be the same as both their previous, different voltages. We shall consider such *singular* circuits in Sec. 7.9, where we shall see that stored energy has *appeared* to change abruptly. The apparent change cannot be accounted for in the lumped circuit models we are using, but it is a remarkable fact that the lumped models are valid before and after (though not during) the switching action. Physical circuits, however, have resistance associated with the capacitor (such as in the leads and the dielectric) which precludes the infinite currents that must accompany discontinuous capacitor voltages. These are the type circuits we shall be concerned with, in general.

EXERCISES

7.2.1 A 0.2-μF capacitor is charged to 10 V. Find the charge and energy.
Answer 2 μC, 10 μJ

7.2.2 In Fig. 7.4, let $C = \frac{1}{4}$F, $R_1 = R_2 = 4 \ \Omega$, and $V = 20$ V. If the current in R_2 at $t = 0^-$ is 2 A directed downward, find at $t = 0^-$ and at $t = 0^+$ (a) the charge on the capacitor, (b) the current in R_1 directed to the right, and (c) the current in C directed downward.
Answer (a) 2,2 C; (b) 3, 0 A; (c) 1, -2 A

7.3

SERIES AND PARALLEL CAPACITORS

In this section we shall determine the equivalent capacitance for series and parallel connections of capacitors. As we shall see, equivalent capacitance is in direct analogy with equivalent conductance.

170

Let us first consider the series connection of N capacitors, as shown in Fig. 7.5(a).

Applying KVL, we find

$$v = v_1 + v_2 + \ldots + v_N \qquad (7.6)$$

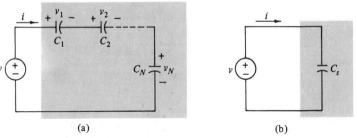

FIGURE 7.5 (a) Series connection of N capacitors; (b) equivalent circuit

From (7.3), this equation can be written

$$v(t) = \frac{1}{C_1} \int_{t_0}^{t} i\, dt + v_1(t_0) + \frac{1}{C_2} \int_{t_0}^{t} i\, dt + v_2(t_0) + \ldots + \frac{1}{C_N} \int_{t_0}^{t} i\, dt + v_N(t_0)$$

$$= \left(\frac{1}{C_1} + \frac{1}{C_2} + \ldots + \frac{1}{C_N} \right) \int_{t_0}^{t} i\, dt + v_1(t_0) + v_2(t_0) + \ldots + v_N(t_0)$$

or, by (7.6),

$$v(t) = \left(\sum_{n=1}^{N} \frac{1}{C_n} \right) \int_{t_0}^{t} i\, dt + v(t_0)$$

In Fig. 7.5(b), we see that

$$v(t) = \frac{1}{C_s} \int_{t_0}^{t} i\, dt + v(t_0)$$

where $v(t_0)$ is the voltage on C_s at $t = t_0$.

Suppose we require that the circuit of Fig. 7.5(b) be an equivalent circuit for that of Fig. 7.5(a). Comparing the last two equations, we see that

$$\frac{1}{C_s} = \frac{1}{C_1} + \frac{1}{C_2} + \ldots + \frac{1}{C_N} = \sum_{n=1}^{N} \frac{1}{C_n} \qquad (7.7)$$

from which we may find the equivalent capacitance C_s.

In the case of two series capacitors C_1 and C_2, (7.7) may be simplified to

$$C_s = \frac{C_1 C_2}{C_1 + C_2}$$

In other words, the equivalent capacitance is the product over the sum of the two individual capacitances. This is directly analogous to the equivalence of two *parallel* resistances.

EXAMPLE 7.4

As an example of the utility of (7.6) and (7.7), consider the series connection of 1- and $\frac{1}{3}$-F capacitors having initial voltages of 4 and 6 V, respectively. Then

$$\frac{1}{C_s} = 1 + 3$$

or

$$C_s = 0.25 \text{ F}$$

and

$$v(t_0) = 4 + 6 = 10 \text{ V}$$

Let us now consider the parallel connection of N capacitors, as shown in Fig. 7.6(a). Application of KCL gives

$$i = i_1 + i_2 + \ldots + i_N$$

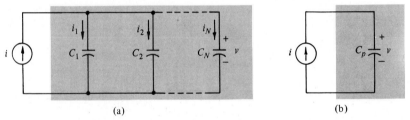

(a)   (b)

FIGURE 7.6 (a) Parallel connection of N capacitors; (b) equivalent circuit

Substituting from (7.2), we have

$$i = C_1 \frac{dv}{dt} + C_2 \frac{dv}{dt} + \ldots + C_N \frac{dv}{dt}$$

$$= \left(C_1 + C_2 + \ldots + C_N \right) \frac{dv}{dt} = \left(\sum_{n=1}^{N} C_n \right) \frac{dv}{dt}$$

In the circuit of Fig. 7.6(b), the current is

$$i = C_p \frac{dv}{dt}$$

If we now require that this circuit be an equivalent circuit for that of Fig. 7.6(a), the above equations give

$$C_p = C_1 + C_2 + \ldots + C_N = \sum_{n=1}^{N} C_n \qquad (7.8)$$

Thus the equivalent capacitance of N parallel capacitors is simply the sum of the individual capacitances. An initial voltage, of course, would be equal to that which is present across the parallel combination.

It is interesting to notice that the equivalent capacitance of series and parallel capacitors is analogous to the equivalent conductance of series and parallel conductances.

EXERCISES

7.3.1 Find the equivalent capacitance.
Answer 10 μF

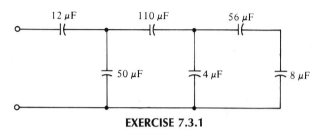

EXERCISE 7.3.1

7.3.2 Derive an equation for current division between two parallel capacitors by finding i_1 and i_2.
Answer $\dfrac{C_1}{C_1 + C_2} i, \dfrac{C_2}{C_1 + C_2} i$

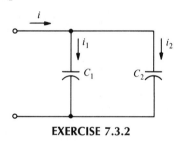

EXERCISE 7.3.2

7.3.3 Derive an equation for voltage division between two uncharged series capacitors by finding v_1 and v_2.
Answer $\dfrac{C_2}{C_1 + C_2} v, \dfrac{C_1}{C_1 + C_2} v$

EXERCISE 7.3.3

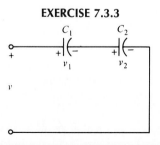

INDUCTORS

In the previous sections, we found that the electrical characteristics of the capacitor are the result of forces that exist between electric charges. Just as static charges exert forces upon one another, it is found that moving charges, or currents, also influence one another. The force which is experienced by two neighboring current-carrying wires was experimentally determined by Ampere in the early nineteenth century. These forces can be characterized by the existence of a *magnetic field*. The magnetic field, in turn, can be thought of in terms of a *magnetic flux* that forms closed loops about electric currents. The origin of the flux, of course, is the electric currents. The study of magnetic fields, like that of electric fields, comes in a later course on electromagnetic theory.

An inductor is a two-terminal device that consists of a coiled conducting wire. A current flowing through the device produces a magnetic flux ϕ which forms closed loops encircling the coils making up the inductor, as shown by the simple model of Fig. 7.7. Suppose that the coil contains N turns and that the flux ϕ passes through each turn. In this case, the total flux linked by the N turns of the coil is

$$\lambda = N\phi$$

This total flux is commonly referred to as the *flux linkage*. The unit of magnetic flux is the *weber* (Wb), named for the German physicist Wilhelm Weber (1804–1891).

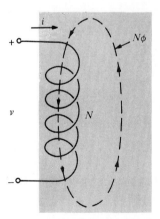

FIGURE 7.7 Simple model of an inductor

In a linear inductor, the flux linkage is directly proportional to the current flowing through the device. Therefore, we may write

$$\lambda = Li \qquad (7.9)$$

where L, the constant of proportionality, is the *inductance* in webers per ampere. The unit of 1 Wb/A is known as the *henry* (H), named for the American physicist Joseph Henry (1797–1878).

In (7.9) we see that an increase in i produces a corresponding increase in λ. This increase in λ produces a voltage in the N-turn coil. The fact that voltages occur with changing magnetic flux was first discovered by Henry. Henry, however, repeating the mistake of Cavendish with the resistor, failed to publish his findings. As a result, Faraday is credited with discovering the law of electromagnetic induction. This law states that the voltage is equal to the time rate of change of the total magnetic flux. In mathematical form, the law is

$$v = \frac{d\lambda}{dt}$$

which with (7.9) yields

$$v = L\frac{di}{dt} \qquad (7.10)$$

Clearly, as i increases, a voltage is developed across the terminals of the inductor, the polarity of which is shown in Fig. 7.7. This voltage opposes an increase in i, for if this were not the case, that is, if the polarity were reversed, the induced voltage would "aid" the current. This physically cannot be true because the current would increase indefinitely.

The circuit symbol and the current-voltage convention for the inductor are shown in Fig. 7.8. Just as in the cases of the resistor and the capacitor, if either the current direction or the voltage assignment, but not both, are reversed, then a negative sign must be employed in the right-hand side of (7.10).

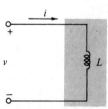

FIGURE 7.8 Circuit symbol for an inductor

An examination of (7.10) shows that if i is constant, then the voltage v is zero. Therefore, an inductor acts like a short circuit to a dc current. On the other hand, the more rapidly i changes, the greater is the voltage that appears across its terminals.

EXAMPLE 7.5 Consider, for instance, a current that decreases linearly from 1 to 0 A in b^{-1} s, defined by

$$i = 1, \qquad t \le 0$$
$$= 1-bt, \qquad 0 \le t \le b^{-1}$$
$$= 0, \qquad t \ge b^{-1}$$

A 1-H inductor having this terminal current has a terminal voltage given by

$$v = 0, \qquad t < 0$$
$$= -b, \qquad 0 < t < b^{-1}$$
$$= 0, \qquad t > b^{-1}$$

Plots of i and v for this case are shown in Fig. 7.9. We see that v is zero when i is constant and is equal to $-b$ when i decreases linearly. If b is made larger, i changes more rapidly and v becomes more negative. Clearly, if $b^{-1} = 0$ (b infinite), then i changes abruptly from 1 to 0 A, and v becomes infinite.

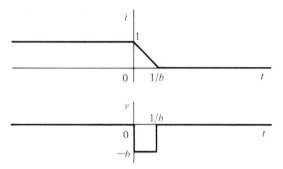

FIGURE 7.9 Current and voltage waveforms for a 1-H inductor

In general, abrupt changes in the current, such as in the above example, require that an infinite voltage appear across the terminals of the inductor. As described in the case of the capacitor, this requires that an infinite power exist at the terminals of the inductor, a physical impossibility. Thus instantaneous changes in the current through an inductor are *not* possible. We observe that the current is continuous even though the voltage may be discontinuous.

An alternative statement concerning abrupt changes in the currents flowing in circuits containing more than one inductor is that *the total flux linkage cannot change instantaneously*. That is, for a circuit containing inductors $L_1, L_2, \ldots, L_N$, the sum $\lambda_1 + \lambda_2 + \ldots + \lambda_N$ cannot change instantaneously. If we compare (7.1) and (7.9), we see that the flux linkage in an inductor is analogous to the charge on a capacitor. Thus the sum of the flux linkages given above (conservation of flux linkage) is analogous to conservation of charge. An example employing the conservation of flux linkage is given in Sec. 7.9.

Let us now find the current $i(t)$ in terms of the voltage $v(t)$. Integrating (7.10) from time t_0 to t and solving for $i(t)$, we have

$$i(t) = \frac{1}{L} \int_{t_0}^{t} v(t)\, dt + i(t_0) \qquad (7.11)$$

In this equation, the integral term represents the current buildup from time t_0 to t, whereas $i(t_0)$ is the current at t_0. Obviously, $i(t_0)$ is the current which accumulates from $t = -\infty$ to t_0, where $i(-\infty) = 0$. Thus an alternative expression is

$$i(t) = \frac{1}{L} \int_{-\infty}^{t} v(t)\, dt$$

EXAMPLE 7.6   In the application of (7.11), we are obtaining the net area under the graph of v from $-\infty$ to t, since $i(t_0)$ represents the area from $-\infty$ to t_0. In Fig. 7.9, for instance, since $i(0) = 1$, we have, for $L = 1$ H,

$$i(t) = \frac{1}{L} \int_0^t (-b) \, dt + i(0) = -bt + 1, \qquad 0 \leq t \leq b^{-1}$$

Thus $i(1/b) = 0$, and

$$i(t) = \frac{1}{L} \int_{1/b}^t (0) \, dt + i\left(\frac{1}{b}\right) = 0, \qquad b^{-1} \leq t$$

In the above example, we see that v and i, just as in the case of the capacitor, do not necessarily have the same variation in time. Inspection of Fig. 7.9, for example, shows that the voltage can be discontinuous even though the current is continuous, as previously mentioned.

EXERCISES

7.4.1 A 10-mH inductor has a current of $50e^{-1000t}$ mA. Find its voltage.
Answer $-0.5e^{-1000t}$ V

7.4.2 Find the current $i(t)$ for $t > 0$ in a 200-mH inductor having a voltage of $10 \sin 500t$ V if $i(0) = -0.1$ A.
Answer $-0.1 \cos 500t$ A.

7.4.3 If the figure of Ex. 7.1.4 represents the voltage on a 0.5-H inductor, find its current at $t = 0$, 4, and 10 ms. Take the current at -10 ms to be 20 mA.
Answer 120, 100, 50 mA

7.5

ENERGY STORAGE IN INDUCTORS

A current i flowing through an inductor causes a total flux linkage λ to be produced that passes through the turns of the coils making up the device. Just as work was performed in moving charges between the plates of a capacitor, a similar work is necessary to establish the flux ϕ in the inductor. The work or energy required in this case is said to be stored in the magnetic field.

The energy stored in an inductor, employing (1.6) and (7.10), is given by

$$w_L(t) = \int_{-\infty}^t vi \, dt = \int_{-\infty}^t \left(L \frac{di}{dt}\right) i \, dt$$

$$= L \int_{-\infty}^t i \, di = \frac{1}{2} L i^2(t) \Big|_{t=-\infty}^t$$

Recalling that $i(-\infty) = 0$, we have

$$w_L(t) = \frac{1}{2}Li^2(t) \text{ J} \qquad\qquad (7.12)$$

Inspection of this equation reveals that $w_L(t) \geq 0$. Therefore, from (1.7), we see that the inductor is a passive circuit element.

The ideal inductor, like the ideal capacitor, does not dissipate any power. Therefore the energy stored in the inductor can be recovered. Consider, for example, a 2-H inductor that is carrying a current of 5 A. The energy stored is

$$w_L = \frac{1}{2}Li^2 = 25 \text{ J}$$

Suppose the inductor, by means of an external circuit, is connected in parallel with a resistor. In this case, a current flows through the inductor-resistor combination until all the energy previously stored in the inductor (25 J) is absorbed by the resistor and the current is zero. Solutions of circuits of this type are found in the next chapter.

EXAMPLE 7.7   Since inductor currents are continuous, it follows that the energy stored in an inductor, like that stored in a capacitor, is also continuous. To illustrate this let us consider the circuit of Fig. 7.10, which contains a switch that is closed at $t = 0$, as indicated. Suppose that $i_L(0^-) = 2$ A and $I = 3$ A. Then by KCL, $i_1(0^-) = 3 - 2 = 1$ A. After the switch is closed ($t = 0^+$), we have $i_1(0^+) = 0$ since a short circuit is placed across R_1. However, we have

$$i_L(0^+) = i_L(0^-) = 2 \text{ A}$$

Thus the resistor current has changed abruptly but the inductor current has not.

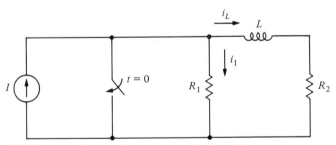

FIGURE 7.10   Circuit illustrating continuity of inductor current

An example of a *singular* circuit for which inductor currents *appear* to be discontinuous is given in Sec. 7.9. As in the case of singular capacitive circuits, the apparent discontinuity in the energy stored in the inductors cannot be accounted for in the lumped circuit model. However, lumped circuit theory is valid before and after (though not during) the switching action. Physical circuits having inductors contain associated resistance which does not permit the infinite inductor voltages that must accompany abrupt changes in inductor currents. We shall be concerned primarily with circuits of this type.

EXERCISES

7.5.1 Derive an expression for the energy stored in an inductor in terms of the flux linkage λ and the inductance L.
Answer $\lambda^2/2L$

7.5.2 A 20-mH inductor has a current of 100 mA. Find the flux linkage and the energy.
Answer 2 mWb, 100 μJ

7.5.3 In Fig. 7.10, let $I = 5$ A, $R_1 = 6\ \Omega$, $R_2 = 4\ \Omega$, $L = 2$ H, and $i_1(0^-) = 2$ A. If the switch is open at $t = 0^-$, find $i_L(0^-)$, $i_L(0^+)$, $i_1(0^+)$, and $di_L(0^+)/dt$.
Answer 3 A, 3 A, 0, -6 A/s

7.6

SERIES AND PARALLEL INDUCTORS

In this section we shall determine the equivalent inductance for series and parallel connections of inductors. Let us first consider a series connection of N inductors, as shown in Fig. 7.11(a). Applying KVL, we see that

$$v = v_1 + v_2 + \ldots + v_N$$

from which we may write

$$v = L_1 \frac{di}{dt} + L_2 \frac{di}{dt} + \ldots + L_N \frac{di}{dt}$$

$$= (L_1 + L_2 + \ldots + L_N)\frac{di}{dt}$$

$$= \left(\sum_{n=1}^{N} L_n\right) \frac{di}{dt}$$

For the circuit of Fig. 7.11(b), the voltage is

$$v = L_s \frac{di}{dt}$$

FIGURE 7.11 (a) Series connection of N inductors; (b) equivalent circuit

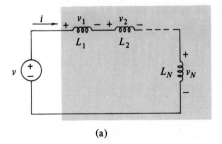

(a)

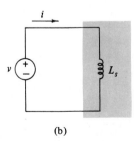

(b)

If we now require that this circuit be an equivalent circuit for the series connection, then the above equations yield

$$L_s = L_1 + L_2 + \ldots + L_N = \sum_{n=1}^{N} L_n \qquad (7.13)$$

Therefore the equivalent inductance of N series inductors is simply the sum of the individual inductances. In addition, an initial current would clearly be equal to that flowing in the series connection.

Let us now consider the parallel connection of N inductors, as shown in Fig. 7.12(a). Application of KCL gives

$$i = i_1 + i_2 + \ldots + i_N \qquad (7.14)$$

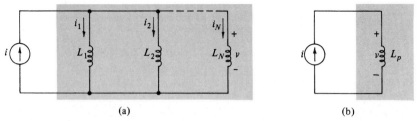

(a)   (b)

FIGURE 7.12 (a) Parallel connection of N inductors; (b) equivalent circuit

Substituting from (7.11), we have

$$i(t) = \frac{1}{L_1} \int_{t_0}^{t} v \, dt + i_1(t_0) + \frac{1}{L_2} \int_{t_0}^{t} v \, dt + i_2(t_0) + \ldots + \frac{1}{L_N} \int_{t_0}^{t} v \, dt + i_N(t_0)$$

$$= \left(\frac{1}{L_1} + \frac{1}{L_2} + \ldots + \frac{1}{L_N} \right) \int_{t_0}^{t} v \, dt + i_1(t_0) + i_2(t_0) + \ldots + i_N(t_0)$$

or, by (7.14),

$$i(t) = \left(\sum_{n=1}^{N} \frac{1}{L_n} \right) \int_{t_0}^{t} v \, dt + i(t_0)$$

In Fig. 7.12(b), we see that

$$i(t) = \frac{1}{L_p} \int_{t_0}^{t} v \, dt + i(t_0)$$

where $i(t_0)$ is the current in L_p at $t = t_0$. If this circuit is an equivalent network for the parallel connection, then the above equations require that the equivalent parallel inductance be given by

$$\frac{1}{L_p} = \frac{1}{L_1} + \frac{1}{L_2} + \ldots + \frac{1}{L_N} = \sum_{n=1}^{N} \frac{1}{L_n} \qquad (7.15)$$

In the case of two parallel inductors L_1 and L_2, (7.15) may be simplified to

$$L_p = \frac{L_1 L_2}{L_1 + L_2}$$

which is directly analogous to the equivalence of two parallel resistances.

EXAMPLE 7.8 As an example, suppose we have two parallel inductors of 6 and 3 H carrying initial currents of 2 and 1 A, respectively. The parallel combination could be replaced by a single inductance,

$$L_p = \frac{6 \times 3}{6 + 3} = 2 \text{ H}$$

carrying an initial current of

$$i(t_0) = 2 + 1 = 3 \text{ A}$$

In the case of inductors, it is interesting to observe that the equivalent inductance for series and parallel connections is analogous to the equivalent resistance of series and parallel resistors.

EXERCISES

7.6.1 Find the equivalent inductance.
Answer 10 mH

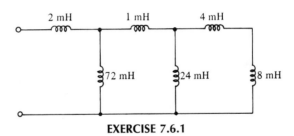

EXERCISE 7.6.1

7.6.2 Derive an equation for voltage division between two series inductors by finding v_1 and v_2.

Answer $\dfrac{L_1}{L_1 + L_2} v, \dfrac{L_2}{L_1 + L_2} v$

EXERCISE 7.6.2

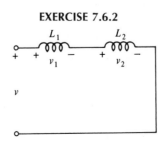

7.6.3 Derive an equation for current division between two parallel inductors with no initial current by finding i_1 and i_2.

Answer $\dfrac{L_2}{L_1 + L_2} i, \dfrac{L_1}{L_1 + L_2} i$

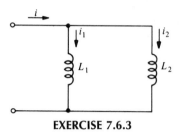

EXERCISE 7.6.3

7.7

PRACTICAL CAPACITORS AND INDUCTORS

Commercially available capacitors are manufactured in a wide variety of types, values, and voltage ratings. The capacitor type is generally classified by the kind of dielectric used, and its capacitance is determined by the type of dielectric and the physical geometry of the device. The voltage rating, or *working voltage,* is the maximum voltage that can safely be applied to the capacitor. Voltages exceeding this value may permanently damage the device by destroying or breaking down the dielectric.

Simple capacitors are often constructed employing two sheets of metal foil which are separated by a dielectric material. The foil and dielectric are pressed together into a laminar form and are then rolled or folded into a compact package. Electrical conductors attached to each metal-foil sheet constitute the terminals of the capacitor.

Practical capacitors, unlike ideal capacitors, generally dissipate a small amount of power. This is due primarily to *leakage* currents that occur within the dielectric material in the device. Practical dielectrics have a nonzero conductance which allows an *ohmic* current to flow between the capacitor plates. This current is easily included in an equivalent circuit for the device by placing a resistance in parallel with an ideal capacitance, as shown in Fig. 7.13. In this figure, R_c represents the ohmic losses of the dielectric and C the capacitance. The leakage resistance R_c is inversely proportional to the capacitance C. Therefore the product of the leakage resistance and

FIGURE 7.13 Simple equivalent circuit for a practical capacitor

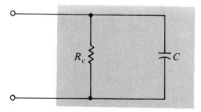

capacitance R_cC, a quantity often given by manufacturers, is useful in specifying the capacitor loss.

Common types of capacitors include ceramic (barium titanate), Mylar, Teflon, and polystyrene. These types are available in capacitance values ranging typically from 100 pF to 1 μF having tolerances of 3, 10, and 20%. Resistance-capacitance products for these types range from 10^3 Ω-F (ceramic) to 2×10^6 Ω-F (Teflon).

Another type of capacitor which gives larger values of C is the electrolytic capacitor. This capacitor is constructed of polarized layers of aluminum oxide or tantalum oxide and has values of 1 to 100,000 μF. Resistance-capacitance products, however, range from 10 to 10^3 Ω-F, which indicates that electrolytics are more *lossy* than nonelectrolytic types. Also, since electrolytic capacitors are polarized, they must be connected into a circuit with the proper voltage polarity. If the incorrect polarity is used, the oxide will be reduced, and heavy conduction will occur between the plates.

Practical inductors, like practical capacitors, usually dissipate a small amount of power. This dissipation results from ohmic losses associated with the wire making up the inductor coil and *core* losses due to induced currents arising in the core on which the coil is wound. An equivalent circuit for an inductor can be realized by placing a resistance in series with an ideal inductor, as shown in Fig. 7.14, where R_L represents the ohmic losses and L the inductance.

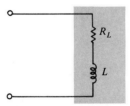

FIGURE 7.14 Simple equivalent circuit for a practical inductor

Inductors are available with values ranging from less than 1 μH to 100 H. Large inductance values are obtained by employing many turns and ferrous (iron) core materials; hence as the inductance increases, the series resistance generally increases.

Like the resistor and the operational amplifier, the capacitor can be fabricated in integrated-circuit form. However, attempts at integrating the inductor have not been very successful because of geometry constraints and because semiconductors do not exhibit the necessary magnetic properties. For this reason, in many applications, circuits are designed using only resistors, capacitors, and electronic devices, such as op amps.

EXERCISE

7.7.1 Mylar capacitors have a resistance-capacitance product of 10^5 Ω-F. Find the equivalent parallel resistor in Fig. 7.13 for the following capacitors; (a) 100 pF, (b) 0.1 μF, and (c) 1 μF.
Answer (a) 10^{15} Ω; (b) 10^{12} Ω; (c) 10^{11} Ω

7.8

DUALITY AND LINEARITY

Let us now determine the dual relationships for the capacitor and the inductor. This is easily done by considering the current-voltage relations of (7.2) and (7.10) for the elements. Repeating these equations for convenience, we have

$$i = C\frac{dv}{dt} \tag{7.16}$$

and

$$v = L\frac{di}{dt} \tag{7.17}$$

Comparing the equations, we see that replacing i by v, v by i, and C by L in the first equation yields the second equation. Therefore it is clear that the capacitor and the inductor are dual elements and that C and L are dual quantities. A similar comparison of the equations for charge and flux, given by (7.1) and (7.9), shows that these are also dual quantities. A summary of the dual quantities that we shall consider in the book is given in Table 7.1.

TABLE 7.1 Dual Quantities

Voltage	Current
Charge	Flux
Resistance	Conductance
Inductance	Capacitance
Short circuit	Open circuit
Impedance	Admittance*
Nonreference node	Mesh
Reference node	Outer mesh
Tree branch	Link*
Series	Parallel
KCL	KVL

*Tree branch and link were considered in Chapter 6, which the reader may have omitted, and impedance and admittance will be considered in Chapter 11.

EXAMPLE 7.9   We are now able to construct dual circuits for networks containing the dual quantities listed in Table 7.1. Consider, for example, the two-mesh network of Fig. 7.15(a). Using the geometric method of Sec. 4.7, the dual circuit is shown dashed in Fig. 7.15(a) and is redrawn in Fig. 7.15(b). The solutions for the currents and voltages in networks including combinations of resistors, capacitors, and inductors will be considered in the coming chapters.

Let us now consider the property of linearity for capacitors and inductors, which are defined by (7.16) and (7.17), respectively. Comparing these equations to that of (5.2), we see that the elements satisfy the proportionality property and that they are, therefore, linear elements. Thus circuits containing any combination of independent

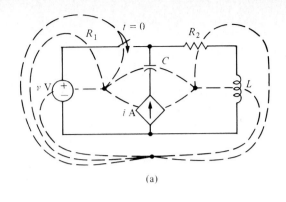

(a)

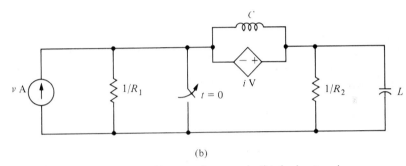

(b)

FIGURE 7.15 (a) Two-mesh network; (b) dual network

sources, and linear dependent sources, resistors, capacitors, and inductors are linear, and superposition and Thevenin's or Norton's theorems are applicable. These topics will be considered in later chapters.

EXERCISE

7.8.1 Construct dual circuits for the networks of (a) Fig. 7.4, (b) Fig. 7.10, (c) Fig. 7.12, and (d) Fig. 7.14.

7.9

SINGULAR CIRCUITS

A circuit in which a switching action takes place that *appears* to produce discontinuities in capacitor voltages or inductor currents is sometimes called a *singular* circuit. In this section we shall consider two such circuits, one containing capacitors and one containing inductors.

EXAMPLE 7.10 Let us consider first Fig. 7.16, where the 1-F capacitors C_1 and C_2 have voltages of 1 and 0 V, respectively, prior to the closing of the switch. That is, $v_1(0^-) = 1$ V and

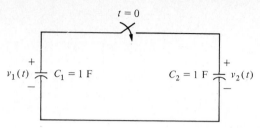

FIGURE 7.16 Circuit containing two capacitors which are switched at time $t = 0$

$v_2(0^-) = 0$ V. We shall now determine $w_1(0^+)$ and $w_2(0^+)$, the energies stored in C_1 and C_2 at $t = 0^+$.

The energy stored prior to closing the switch is

$$w_1(0^-) = \frac{1}{2}C_1 v_1^2(0^-) = \frac{1}{2} \text{ J}$$

and

$$w_2(0^-) = \frac{1}{2}C_2 v_2^2(0^-) = 0 \text{ J}$$

so that the total stored energy in the circuit is

$$w(0^-) = w_1(0^-) + w_2(0^-) = \frac{1}{2} \text{ J} \tag{7.18}$$

The current out of a generalized node at the top enclosing the switch is given by

$$C_1 \frac{dv_1}{dt} + C_2 \frac{dv_2}{dt} = 0$$

which integrated from $t = 0^-$ to $t = 0^+$ yields

$$\int_{0-}^{0+} (C_1 dv_1 + C_2 dv_2) = C_1[v_1(0^+) - v_1(0^-)] + C_2[v_2(0^+) - v_2(0^-)] = 0 \tag{7.19}$$

Substituting for C_1, C_2, $v_1(0^-)$, and $v_2(0^-)$, we have

$$v_1(0^+) + v_2(0^+) = 1 \tag{7.20}$$

For $t > 0$ we see that $v_1 = v_2$, and thus

$$v_1(0^+) = v_2(0^+)$$

which with (7.20) gives

$$v_1(0^+) = v_2(0^+) = \frac{1}{2} \text{V}$$

Therefore, the total energy stored at $t = 0^+$ is

$$w(0^+) = \frac{1}{2}C_1 v_1^2(0^+) + \frac{1}{2}C_2 v_2^2(0^+) = \frac{1}{8} + \frac{1}{8} = \frac{1}{4} \text{ J}$$

which by (7.18) compares with $\frac{1}{2}$ J stored at $t = 0^-$.

We know that capacitors do not dissipate power. What, then, has happened to the $\frac{1}{4}$ J from $t = 0^-$ to $t = 0^+$? Looking back over our work, we see that v_1 changes abruptly from 1 to $\frac{1}{2}$ V at $t = 0$. As pointed out in Sec. 7.1, instantaneous changes in the voltage are not possible. Therefore, during the infinitesimal time from $t = 0^-$ to $t = 0^+$, our mathematical model is not valid. In reality, what has happened is the following. When the switch closes at time $t = 0$, a large current is produced as charges are transferred from C_1 to C_2. This rapidly changing current gives rise to an electromagnetic wave which *radiates* $\frac{1}{4}$ J of energy. The voltage v_1 changes in a short, but nonzero, time from 1 to $\frac{1}{2}$ V. Our network during this interval does not behave as a lumped-parameter circuit, and concepts from electromagnetic theory (a later course) are required for the solution we have described.

Although our circuit model is not valid at the instant the switch closes, the solutions for the voltages and energies before and after the closing of the switch *are correct*. This is due entirely to the fact that the total charge did not change during this time interval. This may be seen from (7.19), written in the form

$$C_1 v_1(0^-) + C_2 v_2(0^-) = C_1 v_1(0^+) + C_2 v_2(0^+)$$

or, equivalently,

$$q_1(0^-) + q_2(0^-) = q_1(0^+) + q_2(0^+)$$

This, of course, is the statement of conservation of charge. (The total charge remains constant during the switching instant.)

As pointed out earlier, most circuit models do not permit an infinite current in a capacitor. Physical circuits normally have a finite value of resistance and inductance which limit such currents. As a result, the capacitor voltages and energies are continuous functions. If, for example, a series resistor is included in the circuit of Fig. 7.16, the voltage on each capacitor is continuous. That is,

$$v_1(0^-) = v_1(0^+)$$

and

$$v_2(0^-) = v_2(0^+)$$

Analyses for circuits of this type are given in Chapter 8.

EXAMPLE 7.11   For a second example, consider the circuit of Fig. 7.17. The 2-H and 1-H inductors L_1 and L_2 have currents of 1 and 0 A, respectively, before the switch is closed at time $t = 0$. Therefore $i_1(0^-) = 1$ A and $i_2(0^-) = 0$ A. We shall now determine $w_1(0^+)$ and $w_2(0^+)$, the energy stored in the inductors at $t = 0^+$.

FIGURE 7.17   Circuit containing two inductors which are switched at time $t = 0$

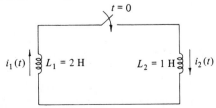

The energy stored prior to the closing of the switch is

$$w_1(0^-) = \frac{1}{2}L_1 i_1^2(0^-) = 1 \text{ J}$$

and

$$w_2(0^-) = \frac{1}{2}L_2 i_2^2(0^-) = 0 \text{ J}$$

so that the total energy is

$$w(0^-) = w_1(0^-) + w_2(0^-) = 1 \text{ J}$$

The flux linkage of each inductor at this time is

$$\lambda_1(0^-) = L_1 i_1(0^-) = 2 \text{ Wb}$$

and

$$\lambda_2(0^-) = L_2 i_2(0^-) = 0 \text{ Wb}$$

so that the total flux linkage is

$$\lambda(0^-) = L_1 i_1(0^-) + L_2 i_2(0^-) = 2 \text{ Wb}$$

[Since flux linkage Li is the integral of the voltage $L(di/dt)$, the signs on the terms in the total flux linkage are the same as those on voltages used in KVL around the loops.]

After the switch is closed, by duality with conservation of charge for a capacitor we see from KCL that

$$i_1(0^+) = i_2(0^+)$$

Also, when the switch is closed, conservation of flux linkage requires that the total flux remain constant; hence

$$L_1 i_1(0^-) + L_2 i_2(0^-) = L_1 i_1(0^+) + L_2 i_2(0^+)$$
$$= (L_1 + L_2)i_1(0^+)$$

or

$$2(1) + 1(0) = 3i_1(0^+)$$

Therefore,

$$i_1(0^+) = i_2(0^+) = \frac{2}{3} \text{ A}$$

Thus the energy stored in each inductor at $t = 0^+$ is

$$w_1(0^+) = \frac{1}{2}L_1 i_1^2(0^+) = \frac{4}{9} \text{ J}$$

and

$$w_2(0^+) = \frac{1}{2}L_2 i_2^2(0^+) = \frac{2}{9} \text{ J}$$

If we now compare the total energy stored in the network, we see at $t = 0^-$ that

$$w_1(0^-) + w_2(0^-) = 1 \text{ J}$$

and at $t = 0^+$ that

$$w_1(0^+) + w_2(0^+) = \frac{2}{3} \text{ J}$$

which indicates that $\frac{1}{3}$ J has been lost by the circuit even though ideal inductors can dissipate no power.

Looking back over the problem, we see that $i_1(t)$ changes abruptly from 1 to $\frac{2}{3}$ A at $t = 0$. We know, however, that abrupt changes in the current are not possible. Therefore during the infinitesimal time from $t = 0^-$ to $t = 0^+$, our mathematical model is once again not valid. As pointed out previously, a rapidly changing current gives rise to an electromagnetic wave which radiates energy. In this case $\frac{1}{3}$ J is radiated, and the current changes from 1 to $\frac{2}{3}$ A in a short, but nonzero, time. Our circuit, of course, does not behave as a lumped-parameter network during this interval of time.

As in the case of a capacitive circuit having infinite currents, most inductive circuit models do not permit infinite voltages to occur across an inductor as a result of abrupt currents. As discussed in the case of the capacitor, physical circuits having inductors contain a finite value of resistance and capacitance which limit such voltages. The currents and energies in these circuits are continuous functions. If, for instance, a parallel resistance is included in Fig. 7.17, the currents are continuous at $t = 0$. Circuits of this type are studied in Chapter 8.

EXERCISES

7.9.1 In Fig. 7.16 let $C_1 = \frac{1}{2}$ F, $C_2 = 1$ F, $v_1(0^-) = 10$ V, $v_2(0^-) = 4$ V, and find $v_1(0^+)$, $v_2(0^+)$, and the total energy stored in the circuit at $t = 0^-$ and at $t = 0^+$.
Answer 6 V, 6 V, 33 J, 27 J

7.9.2 In Fig. 7.17 let $L_1 = 4$ H, $L_2 = 2$ H, $i_1(0^-) = 3$ A, and $i_2(0^-) = 6$ A, and find $i_1(0^+)$, $i_2(0^+)$, and the total energy stored in the circuit at $t = 0^-$ and at $t = 0^+$.
Answer 4 A, 4 A, 54 J, 48 J

PROBLEMS

7.1 How long will it take for a constant 25-mA current to deliver a charge of 100 μC to a 1-μF capacitor? What will then be the capacitor voltage?

7.2 Find the charge residing on each plate of a 1-μF capacitor that is charged to 10 V. If the same charge resides on a 2-μF capacitor, what is the voltage?

7.3 Find the current flowing in a 2-μF capacitor having terminal voltage (a) 100 V, (b) $10t$ V, (c) $5(1 - e^{-2t})$ V, and (d) 15 sin $100t$ V.

7.4 The voltage across a 10-μF capacitor is as shown. Find the current and the power at $t = 10$ ms and at $t = 35$ ms.

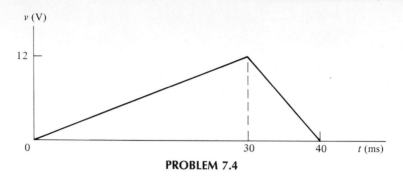

PROBLEM 7.4

7.5 The initial voltage (at $t = 0$) on a 0.5-F capacitor is 4 V. Find the capacitor voltage for $t > 0$ if the current is (a) 5 A, (b) $10t$ A, (c) $5e^{-2t}$ A, and (d) $3 \cos 2t$ A.

7.6 Find the voltage for $t > 0$ across a 2-Ω resistor and a 0.5-F capacitor in series if the current entering their positive terminals is $4 \cos 2t$ A. The capacitor is uncharged at $t = 0$.

7.7 A voltage of $4e^{-t}$ V appears across a parallel combination of a 1-Ω resistor and a 0.25-F capacitor. Find the power absorbed by the combination.

7.8 The current entering the positive terminal of a 0.25-F capacitor is $i = 4t$ A and the initial voltage is $v(0) = 8$ V. Find the energy stored in the capacitor at $t = 1$ s.

7.9 The current entering the positive terminal of a 0.25-F capacitor is $i = 2t - 2$ A and initially the voltage is $v(0) = 8$ V. Find the minimum stored energy and the time at which it occurs.

7.10 Find $i_1(0^-)$, $i_1(0^+)$, $i_2(0^-)$, and $i_2(0^+)$ if the switch is opened at $t = 0$, and $v_1(0^-) = 18$ V and $v_2(0^-) = 6$ V.

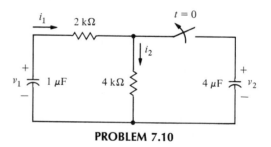

PROBLEM 7.10

7.11 Find $i(0^-)$ and $i(0^+)$ if the switch is opened at $t = 0$ and $v_c(0^-) = 12$ V.

PROBLEM 7.11

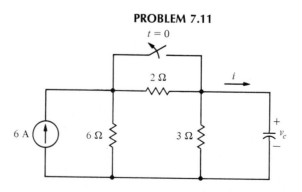

Chapter 7   Energy-Storage Elements

7.12 Find $i_1(0^-)$, $i_2(0^-)$, $i_3(0^-)$, $i_1(0^+)$, $i_2(0^+)$, and $i_3(0^+)$ if $v_c(0^-) = 6$ V and the switch is closed at $t = 0$.

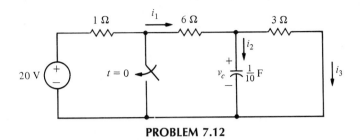

PROBLEM 7.12

7.13 (a) Find the maximum and minimum values of capacitance that can be obtained from ten 1-μF capacitors. (b) Find a connection using all ten capacitors that yields an equivalent capacitance of 2.5 μF. (c) Repeat part (b) for an equivalent capacitance of 0.4 μF.

7.14 The capacitances shown are all in μF. Find C_{eq} at terminals $a - b$.

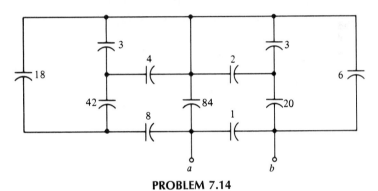

PROBLEM 7.14

7.15 Find the voltage across a 0.1-H inductor at $t = 10$ ms, 70 ms, and 90 ms if its current is as shown.

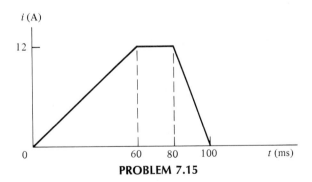

PROBLEM 7.15

7.16 Find the terminal voltage of a 10-mH inductor if the current is (a) 4 A, (b) 10t A, (c) 10 cos 100t A, and (d) 10(1 $- e^{-10t}$) A.

7.17 Find v if $i = 2e^{-10t}$ A.

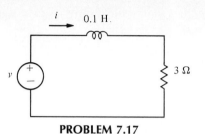

PROBLEM 7.17

7.18 A voltage source of 20 sin 1000t V is connected across a parallel combination of a 10-Ω resistor and a 10-mH inductor. If the initial inductor current is 2 A entering the positive terminal (at $t = 0$), find the current delivered by the source for $t > 0$.

7.19 The voltage of a $\frac{1}{4}$-H inductor is $v = 12 \cos 4t$ V and the current at $t = 0$ is zero. Find the energy stored in the inductor at (a) $\pi/8$ s and (b) $\pi/16$ s.

7.20 The voltage of a 0.5-H inductor is $v = 4 \cos 4t$ V and the initial current is $i(0) = 4$ A. Find for $t > 0$ the minimum and maximum values of the stored energy.

7.21 Find $v(0^-)$, $v(0^+)$, and $di(0^+)/dt$ if $i(0^-) = 2$ A and the switch is opened at $t = 0$.

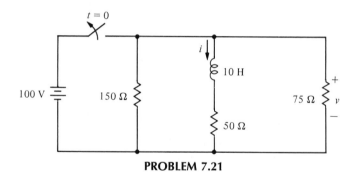

PROBLEM 7.21

7.22 Find $v(0^+)$ and $di(0^+)/dt$ if $v(0^-) = 4$ V and the switch is closed at $t = 0$.

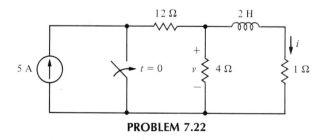

PROBLEM 7.22

7.23 If $v(0^-) = 9$ V, $i(0^-) = 1$ A, and the switch is opened at $t = 0$, find $i(0^+)$ and $di(0^+)/dt$.

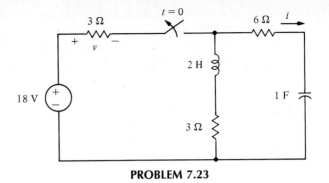

PROBLEM 7.23

7.24 If $v_1(0^-) = 5$ V and $v_2(0^-) = 25$ V, find at $t = 0^+$ the values of v_1 and v_2. The switch is opened at $t = 0$.

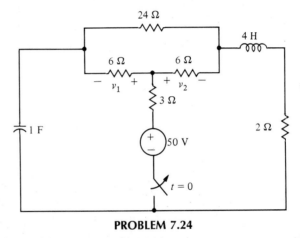

PROBLEM 7.24

7.25 (a) Find the maximum and minimum values of inductance that can be obtained from ten 5-mH inductors. (b) Find a connection using all ten inductors that yields an equivalent inductance of 20 mH.

7.26 Determine L_{eq} (values shown are in henrys).

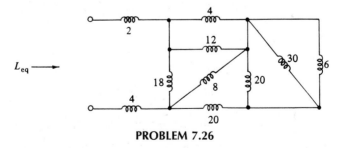

PROBLEM 7.26

7.27 A 400- and a 600-pF ceramic capacitor are connected in parallel. The resistance-capacitance product for a ceramic capacitor is 10^3 Ω-F. What is the equivalent capacitance and parallel resistance for the combination?

7.28 Determine a dual circuit for the network of (a) Fig. 7.13, (b) the figure for Prob. 7.14, and (c) the figure for Prob. 7.26.

7.29 In the circuit of Fig. 7.16, if $v_1(0^-) = 6$ V and 4 J of energy is radiated by the switching action, find $v_2(0^-)$.

7.30 If in Fig. 7.17, i_2 is reversed in direction, find $i_1(0^+)$ and $i_2(0^+)$ if $i_1(0^-) = 9$ A, $i_2(0^-) = 3$ A, $L_1 = 2$ H, and $L_2 = 4$ H. Find also the energy radiated by the switching action.

8

Simple *RC* and *RL* Circuits

Joseph Henry
1797–1878

Blot out these two names [Joseph Henry and Michael Faraday] and the civilization of the present world would become impossible.

H. S. Carhart

Michael Faraday's great discovery in 1831 of electromagnetic induction was being independently duplicated at about the same time by an American physicist Joseph Henry, but Faraday was credited with the discovery because his results were published first. Henry became famous, however, as the discoverer of the inductance (called self-inductance) of a coil and as the developer of powerful electromagnets capable of lifting thousands of pounds of weight. He was also America's foremost nineteenth-century physicist and the first secretary of the newly-formed Smithsonian Institution.

Henry was born near Albany, New York, and his early years were spent in poverty. His ambition was to become an actor until by chance at age 16 he happened upon a book of science, which caused him to devote his life to the acquisition of knowledge. He enrolled in the Albany Academy and upon graduation became a teacher there. In 1832 he joined the faculty of the College of New Jersey, now Princeton, and in 1846 joined the Smithsonian Institution. In his honor the unit of inductance was given the name *henry* 12 years after his death. ∎

In this chapter we shall consider simple circuits containing resistors and capacitors or resistors and inductors, which we shall refer to for brevity as *RC* or *RL* circuits, respectively. The application of Kirchhoff's laws to these networks gives rise to *differential equations* that, in general, are more difficult to solve than the algebraic equations encountered in the previous chapters. Several methods of solving these equations will be presented.

We shall first concern ourselves with source-free *RC* and *RL* circuits, so-called because they contain no independent sources. As we shall see, the source-free responses result from energies stored in the dynamic circuit elements and are characterized by the nature of the circuit itself. For this reason the response is known as the natural response of the circuit.

Following our study of source-free circuits, we shall consider driven *RC* and *RL* circuits in which the forcing or driving functions are constant independent sources that are suddenly applied to the networks. We shall find that the responses in these networks consist of two parts, a natural response, similar in form to that of the source-free case, and a forced response, characterized by the forcing function.

8.1

SOURCE-FREE *RC* CIRCUIT

We shall begin our study of a source-free network by considering the series connection of a capacitor and a resistor, as shown in Fig. 8.1. We shall assume that the capacitor is charged to a voltage of V_0 at an initial time, which we shall take as $t = 0$. Since there are no current or voltage sources in the network, the circuit response (v or i) is due entirely to the energy which is stored initially in the capacitor. The energy in this case, at time $t = 0$, is

$$w(0) = \frac{1}{2}CV_0^2 \tag{8.1}$$

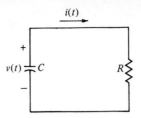

FIGURE 8.1 Source-free *RC* circuit

Let us now determine $v(t)$ and $i(t)$ for $t \geq 0$. Applying KCL at the top node, we find that

$$C\frac{dv}{dt} + \frac{v}{R} = 0$$

or

$$\frac{dv}{dt} + \frac{1}{RC}v = 0 \tag{8.2}$$

which is a *first-order* differential equation. (The *order* of a differential equation is the order of the highest-order derivative in the equation.)

Numerous methods are available for solving differential equations of the form of (8.2). One straightforward method is to rearrange the terms in the equation so as to *separate* the variables v and t. Then simply integrating the result leads to the solution. In (8.2) the variables may be separated by first writing

$$\frac{dv}{dt} = -\frac{1}{RC}v$$

from which we obtain

$$\frac{dv}{v} = -\frac{1}{RC}\,dt \tag{8.3}$$

Taking the indefinite integral of each side, we have

$$\int \frac{dv}{v} = -\frac{1}{RC}\int dt$$

or

$$\ln v = -\frac{t}{RC} + K$$

where K is a constant of integration.

For the solution to be valid for $t \geq 0$, we see that K must be selected such that the initial condition of $v(0) = V_0$ is satisfied. Therefore, at $t = 0$, we have

$$\ln v(0) = \ln V_0 = K$$

Substituting this value of K into the solution yields

$$\ln v - \ln V_0 = \ln \frac{v}{V_0} = -\frac{t}{RC}$$

If we now recall the relationship

$$e^{\ln x} = x$$

we see that

$$v(t) = V_0 e^{-t/RC} \tag{8.4}$$

In Fig. 8.1, we see that this is the voltage across R; therefore, the current is

$$i(t) = \frac{v(t)}{R} = \frac{V_0}{R} e^{-t/RC}$$

Another method of solving the separated equation (8.3) is to integrate each side of the equation between appropriate limits. In our case v has a value of V_0 at time 0, and thus

$$\int_{V_0}^{v} \frac{dv}{v} = -\frac{1}{RC} \int_0^t dt \tag{8.5}$$

where the integrals in this equation are *definite* integrals. Performing the integrations, we have

$$\ln v - \ln V_0 = -\frac{t}{RC}$$

which is equivalent to (8.4).

A graph of (8.4) is shown in Fig. 8.2. We see that the voltage is initially V_0 and that it decays exponentially toward zero as t becomes large. The rate at which the voltage decays is determined solely by the product of the resistance and the capacitance of the network. Since the response is characterized by the circuit elements and not by an external voltage or current source, the response is called the *natural response* of the circuit.

The energy stored in the network at $t = 0$ is given in (8.1). As time increases, the voltage across the capacitor, and hence the energy stored in the capacitor, decreases. From a physical standpoint, it is apparent that all the energy stored in the capacitor at $t = 0$ must be dissipated by the resistor as time becomes infinite. The instantaneous power absorbed by the resistor is

FIGURE 8.2 Graph of the voltage response in the simple *RC* circuit of Fig. 8.1

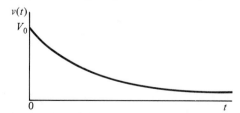

$$p_R(t) = \frac{v^2(t)}{R} = \frac{V_0^2}{R}e^{-2t/RC} \text{ W}$$

Therefore, the energy absorbed by the resistor as time becomes infinite is

$$w_R(\infty) = \int_0^\infty p_R(t)\, dt$$

$$= \int_0^\infty \frac{V_0^2}{R}e^{-2t/RC}\, dt$$

$$= -\frac{1}{2}CV_0^2 e^{-2t/RC}\Big|_0^\infty$$

$$= \frac{1}{2}CV_0^2$$

which is, indeed, equal to the energy initially stored in the network.

EXAMPLE 8.1

As an example, let us find the voltage v for $t > 0$ in the circuit of Fig. 8.1 if $R = 100$ kΩ, $C = 0.01$ μF, and $v(0) = 6$ V. We note in this case that $RC = (10^5)(10^{-8}) = 10^{-3}$. Therefore, we have

$$v = 6e^{-t/10^{-3}} = 6e^{-1000t} \text{ V}$$

As a final note in this section on the RC circuit, we observe that if the initial time is some $t = t_0$ rather than $t = 0$—i.e., $v(t_0) = V_0$—then the lower limit in the right member of (8.5) is t_0. As the reader may verify, this yields the more general result,

$$v(t) = V_0 e^{-(t-t_0)/RC}, \qquad t \ge t_0 \tag{8.6}$$

EXERCISES

8.1.1 In Fig. 8.1, let $t_0 = 0$, $V_0 = 10$ V, $R = 1$ kΩ, and $C = 1$ μF. Find v, i, and w_c at $t = 1$ ms.
Answer 3.68 V, 3.68 mA, 6.8 μJ

8.1.2 If the switch is opened at $t = 0$ and $i(0^-) = 2$ A, find v for $t > 0$.
Answer $14e^{-5t}$ V

EXERCISE 8.1.2

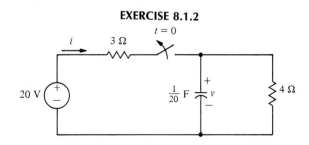

8.1.3 If $i(0^+) = 1$ A, find v for $t > 0$. (*Suggestion:* Find the equivalent resistance seen by the capacitor.)

Answer $24e^{-2t}$ V

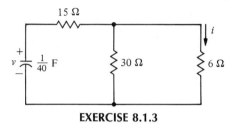

EXERCISE 8.1.3

8.2

TIME CONSTANTS AND DC STEADY STATE

In networks that contain energy-storage elements it is very useful to characterize with a single number the rapidity with which the natural response decreases. To describe such a number, let us consider the network of Fig. 8.1 and the voltage response,

$$v = V_0 e^{-t/RC}$$

where V_0 is the voltage at $t = 0$.

Graphs of v for $RC = k = $ a constant, $RC = 2k$, and $RC = 3k$ are shown in Fig. 8.3. We see that the smaller the RC product, the more rapidly the exponential function $v(t)$ decreases. In fact, the voltage for $RC = k$ decays to a specific value in one-half the time of that required for $RC = 2k$ and in one-third the time of that required for $RC = 3k$. It is also clear that the voltage response remains unchanged if R is increased and C is decreased, or vice versa, such that the product RC is the same. For instance, if we double R and halve C, the voltage response is unchanged.

The current in the network of Fig. 8.1 is

FIGURE 8.3 Graphs of v for various values of RC

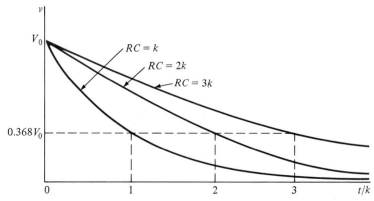

$$i = \frac{V_0}{R} e^{-t/RC}$$

Clearly, the current decreases in the same manner as the voltage. It should be noticed that changing R and C such that the product of RC remains constant causes a change in the initial current V_0/R. The current response, however, still decreases in the same fashion because $e^{-t/RC}$ is unchanged.

The time required for the natural response to decay by a factor of $1/e$ is defined as the *time constant* of a circuit, which we shall denote by τ. In our case, this requires that

$$V_0 e^{-(t+\tau)/RC} = \frac{V_0}{e} e^{-t/RC} = V_0 e^{-(t+RC)/RC}$$

which yields

$$\tau = RC$$

The units of τ are $\Omega\text{-F} = (\text{V/A})(\text{C/V}) = (\text{C/A}) = \text{s}$. In terms of the time constant, the voltage response is

$$v = V_0 e^{-t/\tau} \tag{8.7}$$

The response at the end of one time constant is reduced to $e^{-1} = 0.368$ of its initial value. At the end of two time constants it is equal to $e^{-2} = 0.135$ of its initial value, and at the end of five time constants it has become $e^{-5} = 0.0067$ of its initial value. Therefore, after four or five time constants, the response is essentially zero.

An interesting property of exponential functions is shown in Fig. 8.4. A tangent to the curve at $t = 0$ intersects the time axis at $t = \tau$. This is easily verified by considering the equation of a straight line tangent to the curve at $t = 0$, given by

$$v_1 = mt + V_0$$

where m is the slope of the line. Differentiating v, we have

FIGURE 8.4 Graph illustrating the relation between a line tangent to v at $t = 0$ and τ

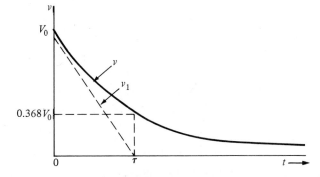

$$\frac{dv}{dt} = -\frac{V_0}{\tau} e^{-t/\tau}$$

Therefore,

$$m = \frac{dv}{dt}\bigg|_{t=0} = -\frac{V_0}{\tau}$$

and

$$v_1 = -\frac{V_0}{\tau} t + V_0$$

The line intersects the time axis at $v_1 = 0$, which requires that $t = \tau$. It can also be shown that a tangent to the curve at a time t_1 intersects the time axis at $t_1 + \tau$ (see Prob. 8.6). This fact is often useful in sketching the exponential function.

From Fig. 8.4, we see that an alternative definition for the time constant is the time required for the natural response to become zero if it decreases at a constant rate equal to the initial rate of decay.

Knowledge of the time constant allows us to predict the general form of the response (8.7), but to complete the solution we must find the initial voltage $v(0^+) = V_0$. For a capacitor, $v(0^+) = v(0^-)$, so we may often find V_0 from the circuit at $t = 0^-$, just prior to the actuating of the switch at $t = 0$. This is relatively easy if at $t = 0^-$ the circuit is in a *dc steady-state* condition, by which we mean that all the voltages and currents have reached constant values. As we shall see in Sec. 8.4 on driven circuits, a dc steady-state condition is established when a switch has been open or closed for a long period of time in the presence of a dc source.

EXAMPLE 8.2

To illustrate the procedure, let us find the capacitor voltage $v(t)$ in Fig. 8.5(a), given that the circuit was in dc steady state just before the opening of the switch. Thus at $t = 0^-$, the switch is still closed, the capacitor is an open circuit to steady-state dc, and so the circuit is as shown in Fig. 8.5(b). The resistance seen by the capacitor to the left of its terminals is given by

$$R_{eq} = 8 + \frac{3(2 + 4)}{3 + (2 + 4)} = 10 \ \Omega$$

and by voltage division we have, from Fig. 8.5(b),

$$v(0^-) = \left(\frac{10}{10 + 15}\right) \times 100 = 40 \text{ V}$$

Therefore, $V_0 = v(0^+) = v(0^-) = 40$ V.

For $t > 0$ the battery is switched out of the circuit, as shown in Fig. 8.5(c), where resistors to the left of the capacitor have been replaced by their equivalent R_{eq}. The time constant for the network is simply the product of the capacitance and the equivalent resistance, given by

$$\tau = R_{eq}C = 10 \text{ s}$$

Therefore by (8.7) the voltage is

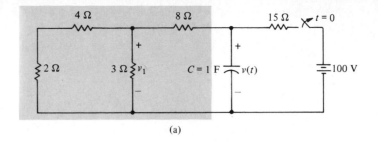

(a)

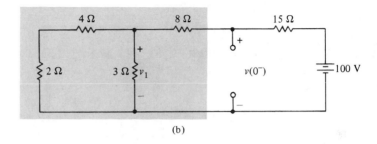

(b)

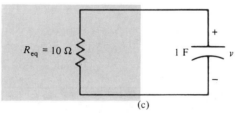

(c)

FIGURE 8.5 (a) More general *RC* circuit; (b) its equivalent at $t = 0^-$; (c) its equivalent for $t > 0$

$$v(t) = 40e^{-t/10} \text{ V}$$

If we want v_1 in Fig. 8.5(a), we may find it from v using voltage division. Since v_1 is across the equivalent resistance of $6(3)/(6 + 3) = 2 \, \Omega$, then

$$v_1 = \frac{2}{2 + 8} v = 8e^{-t/10} \text{ V}$$

EXERCISES

8.2.1 In a series *RC* circuit, determine (a) τ for $R = 10 \text{ k}\Omega$ and $C = 10 \, \mu\text{F}$, (b) C for $R = 100 \text{ k}\Omega$ and $\tau = 10 \, \mu\text{s}$, and (c) R for $v(t)$ to halve every 10 ms on a 1-μF capacitor.

Answer (a) 0.1 s; (b) 100 pF; (c) 14.43 kΩ

8.2.2 A series *RC* circuit consists of a 50-kΩ resistor and a 0.02-μF capacitor. It is desired to increase the current in the network by a factor of 5 without changing the capacitor voltage. Find the necessary values of R and C.

Answer 10 kΩ, 0.1 μF

8.2.3 The circuit is in steady state at $t = 0^-$ and the switch is moved from position 1 to position 2 at $t = 0$. Find v for $t > 0$.

Answer $16e^{-4t}$ V

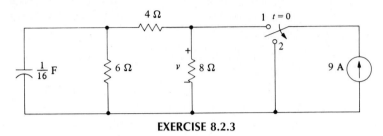

EXERCISE 8.2.3

8.2.4 Find i for $t > 0$ if the circuit is in steady state at $t = 0^-$.

Answer $2e^{-5t}$ A

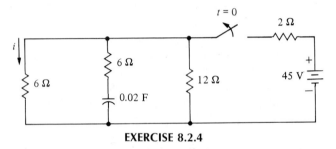

EXERCISE 8.2.4

8.3

SOURCE-FREE *RL* CIRCUIT

In this section we shall study the series connection of an inductor and a resistor, as shown in Fig. 8.6. We assume that the inductor is carrying a current I_0 at time $t = 0$. As in the case of the source-free *RC* circuit, there are no current or voltage sources in the network, and the current and voltage responses are due entirely to the energy stored in the inductor. The stored energy at $t = 0$ is given by

$$w_L(0) = \frac{1}{2}LI_0^2 \tag{8.8}$$

FIGURE 8.6 Source-free *RL* circuit

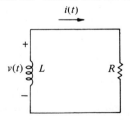

Summing the voltages around the circuit, we have

$$L\frac{di}{dt} + Ri = 0$$

or

$$\frac{di}{dt} + \frac{R}{L}i = 0 \tag{8.9}$$

This equation is of the same form as that of (8.2) for the RC circuit. We may therefore solve it by separating the variables. Instead, however, let us introduce a second, very powerful method, which we shall generalize in the next chapter. The method consists of assuming or guessing (a perfectly legitimate mathematical technique) a general form of the solution based on an inspection of the equation to be solved. In guessing a solution, we shall include several unknown constants and determine their values so that our assumed solution satisfies the differential equation and the initial conditions for the network.

A close inspection of (8.9) reveals that i must be a function that does not change its form upon differentiation; that is, di/dt is a multiple of i. The only function which satisfies this requirement is an exponential function of t. Realizing also that the function must be multiplied by a constant to allow for a proper amplitude in the response, we make an obvious guess that the solution is of the form

$$i(t) = Ae^{st} \tag{8.10}$$

where A and s are constants to be determined. Substituting the solution into (8.9), we obtain

$$\left(s + \frac{R}{L}\right)Ae^{st} = 0$$

From this result, we see that our solution is valid if $Ae^{st} = 0$ or if $s = -R/L$. The first case is disregarded since, by (8.10), it results in $i = 0$ for all t and cannot satisfy $i(0) = I_0$. Thus we take the case $s = -R/L$, and (8.10) becomes

$$i(t) = Ae^{-Rt/L}$$

The constant A can now be determined from the initial condition $i(0) = I_0$. This condition requires that

$$i(0) = I_0 = A$$

Therefore, the solution becomes

$$i(t) = I_0e^{-Rt/L} \tag{8.11}$$

Since the solution is an exponential function, as in the RC case, it also has a time constant τ. In terms of τ we may write the current in the general form

$$i(t) = I_0e^{-t/\tau}$$

where by comparison with (8.11) we see that $\tau = L/R$. Evidently, the units of τ are $H/\Omega = (V\text{-}s/A)/(V/A) = s$. Increasing L, like increasing C in the RC circuit, increases the time constant. However, an increase in R, in contrast to the RC circuit, lowers the value of the time constant. A graph of a typical current response is shown in Fig. 8.7.

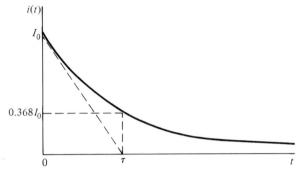

FIGURE 8.7 Current response of a simple RL circuit

The instantaneous power delivered to the resistor in Fig. 8.6 is

$$p(t) = Ri^2(t) = RI_0^2 e^{-2Rt/L}$$

Therefore, the energy absorbed by the resistor as time becomes infinite is given by

$$w(\infty) = \int_0^\infty p(t)\, dt$$

$$= \int_0^\infty RI_0^2 e^{-2Rt/L}\, dt$$

$$= \tfrac{1}{2}LI_0^2$$

Comparing this result with (8.8), we see that the energy initially stored in the inductor is dissipated by the resistor, as expected.

Suppose we had chosen to find the inductor voltage v in the circuit instead of the current i. Applying KCL, we find

$$\frac{v}{R} + \frac{1}{L}\int_0^t v\, dt + i(0) = 0$$

which is an *integral* equation. Differentiating this equation with respect to time, we see that

$$\frac{1}{R}\frac{dv}{dt} + \frac{1}{L}v = 0$$

or

$$\frac{dv}{dt} + \frac{R}{L}v = 0$$

This equation is a differential equation that we can solve using one of the methods discussed previously. It is also interesting to note that if we replace v by iR, we have

$$\frac{di}{dt} + \frac{R}{L}i = 0$$

which is (8.9), obtained using KVL. From these last two results we see that v and i satisfy the same equation and thus have the same form.

EXAMPLE 8.3 Let us now determine i and v in the more general RL circuit of Fig. 8.8, which we assume is in a dc steady-state condition at $t = 0^-$. Therefore, recalling that an inductor is a short circuit to dc, we have

$$i(0^-) = \frac{100}{50} = 2 \text{ A}$$

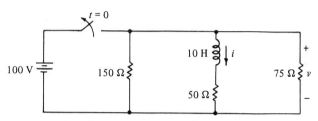

FIGURE 8.8 More general RL circuit

Since the current in the inductor is continuous at $t = 0$, we have

$$i(0^+) = i(0^-) = 2 \text{ A}$$

The time constant for the network for $t > 0$ is clearly the ratio of the inductance and the equivalent resistance as seen from the terminals of the inductor. The equivalent resistance is

$$R_{eq} = 50 + \frac{(75)(150)}{75 + 150} = 100 \ \Omega$$

and hence the time constant is

$$\tau = \frac{L}{R_{eq}} = 0.1 \text{ s}$$

Therefore, since $I_0 = i(0^+) = 2$ A, we have

$$i(t) = 2e^{-10t} \text{ A}$$

Summing the voltages of the inductor and the 50-Ω resistor, the voltage $v(t)$ is given by

$$v(t) = 10\frac{di}{dt} + 50i$$

$$= -100e^{-10t} \text{ V}$$

EXAMPLE 8.4

As a final example, consider the network of Fig. 8.9, which contains a dependent voltage source. The initial current is $i(0) = I_0$. Summing the voltages around the loop, we find that

$$L \frac{di}{dt} + Ri + ki = 0$$

or

$$\frac{di}{dt} + \left(\frac{R + k}{L} \right) i = 0$$

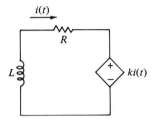

FIGURE 8.9 *RL* circuit containing a dependent voltage source

Comparing this equation to (8.9), we see that the equations are identical if R in (8.9) is replaced by $R + k$. Thus from (8.11) we have

$$i(t) = I_0 e^{-(R+k)t/L}$$

The time constant in this case, which is modified by the presence of the dependent source, is given by

$$\tau = \frac{L}{R + k}$$

EXERCISES

8.3.1 In a series *RL* circuit, determine (a) τ for $R = 200\ \Omega$ and $L = 10$ mH, (b) L for $R = 1\ k\Omega$ and $\tau = 100$ ms, and (c) R for the stored energy in a 0.1-H inductor to halve every 10 ms.
Answer (a) 50 μs; (b) 100 H; (c) 3.47 Ω

8.3.2 A series *RL* circuit has a time constant of 1 ms with an inductance of 2 H. It is desired to halve the inductor voltage without changing the current response. Find the new values of inductance and resistance required.
Answer 1 H, 1 kΩ

8.3.3 The circuit is in steady state at $t = 0^-$. Find i and v for $t > 0$.
Answer $2e^{-3t}$ A, $-6e^{-3t}$ V

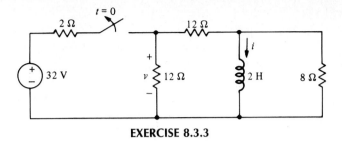

EXERCISE 8.3.3

8.3.4 The circuit shown is in a dc steady-state condition at $t = 0^-$. Find v for $t > 0$.
Answer $-16e^{-4t}$ V

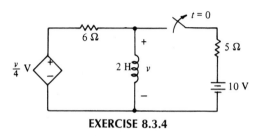

EXERCISE 8.3.4

8.4

RESPONSE TO A CONSTANT FORCING FUNCTION

In the preceding sections we have considered source-free circuits whose responses have been the result of initial energies stored in capacitors and inductors. All independent current or voltage sources were removed or switched out of the circuits prior to finding the natural responses. It was shown that these responses, when arising in circuits containing a single capacitor or inductor and an equivalent resistor, die out with increasing time.

In this section we shall examine circuits which, in addition to having initial stored energies, are *driven* by constant independent current or voltage sources, or *forcing functions*. For these circuits we shall obtain solutions which are the result of inserting or switching sources into the networks. We shall find that the responses in these cases, unlike those of source-free circuits, consist of two parts, one of which is always a constant.

Let us begin by considering the circuit of Fig. 8.10. The network consists of the parallel connection of a constant current source and a resistor which is switched at time $t = 0$ across a capacitor having a voltage $v(0^-) = V_0$ V. For $t > 0$, the switch is closed and a nodal equation at the upper node is given by

$$C\frac{dv}{dt} + \frac{v}{R} = I_0$$

or

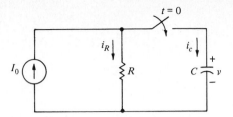

FIGURE 8.10 Driven RC network

$$\frac{dv}{dt} + \frac{1}{RC}v = \frac{I_0}{C} \tag{8.12}$$

Equations of this type that have constant forcing functions (I_0 in this case) can be solved by the method of separation of variables. We may first write (8.12) in the form

$$\frac{dv}{dt} = -\frac{v - RI_0}{RC}$$

Multiplying both sides by $dt/(v - RI_0)$ and forming indefinite integrals, we have

$$\int \frac{dv}{v - RI_0} = -\frac{1}{RC}\int dt + K$$

where K is a constant of integration. Thus, performing the integrations, we obtain

$$\ln(v - RI_0) = -\frac{t}{RC} + K$$

This result can be written as

$$v - RI_0 = e^{-(t/RC)+K}$$

or, solving for v, we find

$$v = Ae^{-t/RC} + RI_0 \tag{8.13}$$

where we have taken $A = e^K$, a constant to be determined by the initial condition of the circuit.

From (8.13) we see that the general solution for the voltage response consists of two parts, an exponential function and a constant function. The exponential function is of the identical form as that of the natural response in a source-free circuit composed of R and C. Since this part of the solution is characterized entirely by the RC time constant, we shall refer to it as the *natural response* v_n of the driven circuit. As in the case of the source-free circuit, this response approaches zero as time increases.

The second part of the solution, given by RI_0, bears a close resemblance to the forcing function I_0. In fact, as time increases, the natural response disappears, and the solution is simply RI_0. This component is due entirely to the forcing function, and we shall call it the *forced response* v_f of the driven circuit.

A reader who has had a course in differential equations will, of course, recognize the natural response v_n and the forced response v_f as the homogeneous response v_h and the particular response v_p.

Let us now evaluate the constant A in (8.13). As in the case of the source-free circuit, its value must be selected so that the initial voltage in the circuit is satisfied. At $t = 0^+$, we see that

$$v(0^+) = v(0^-) = V_0$$

Therefore, at $t = 0^+$, (8.13) requires that

$$V_0 = A + RI_0$$

or

$$A = V_0 - RI_0$$

Substituting this value of A back into our solution yields

$$v(t) = RI_0 + (V_0 - RI_0)e^{-t/RC} \tag{8.14}$$

We should observe in this solution that the constant A is now determined not only by the initial voltage (or energy) on the capacitor but also by the forcing function I_0.

Graphs of v_n, v_f, and v are shown in Figs. 8.11(a) and (b). In (a) the natural response v_n for $V_0 - RI_0 > 0$ and the forced response v_f are shown. In (b) the complete response is shown.

FIGURE 8.11 Graphs of voltage response for the driven RC network of Fig. 8.10. (a) Natural and forced responses; (b) complete response

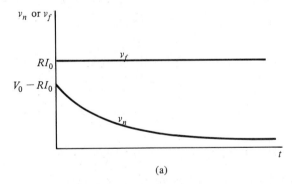

(a)

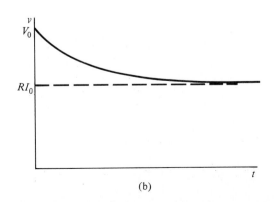

(b)

The current in the capacitor for $t > 0$ is

$$i_C = C\frac{dv}{dt} = -\frac{V_0 - RI_0}{R}e^{-t/RC}$$

whereas the current in the resistor is

$$i_R = I_0 - i_C = I_0 + \frac{V_0 - RI_0}{R}e^{-t/RC}$$

It is interesting to note that the resistor voltage has changed abruptly from RI_0 at $t = 0^-$ to V_0 at $t = 0^+$. The capacitor voltage, as previously pointed out, is continuous.

The solutions that we have encountered so far in this chapter are often referred to in other more descriptive terms. Two such terms that are very popular are the *transient response* and the *steady-state response*. The transient response is the transitory portion of the complete response which approaches zero as time increases. The steady-state response, on the other hand, is that part of the complete response which remains after the transient response has become zero. In the case of dc sources, the steady-state response is constant and is the dc steady state discussed in Sec. 8.2.

In our example, we see that the transient response and the natural response are identical, as are those for the source-free circuits of the previous sections. The steady-state response is therefore identical to the forced response. In our example these responses are $v = RI_0$, $i_C = 0$, and $i_R = I_0$, dc values that constitute the dc steady-state condition.

We should not conclude from the above discussion that the natural and forced responses are always the transient and steady-state responses. If the forcing function is a transitory function, for instance, the steady-state response is zero, as we shall see in the next chapter. In this case the complete response is the transient response.

EXERCISES

8.4.1   Find v for $t > 0$ if $v(0^-) = 6$ V.
Answer $10 - 4e^{-50t}$ V

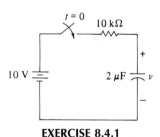

EXERCISE 8.4.1

8.4.2   The circuit is in steady state at $t = 0^-$. Find i for $t > 0$ if the switch is moved from position 1 to position 2 at $t = 0$. (*Suggestion:* Write KCL at the top node and note that $v = 2\,di/dt$.)
Answer $4e^{-2t} + 2$ A

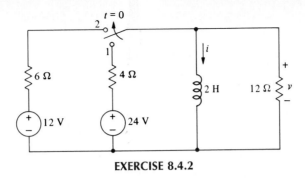

EXERCISE 8.4.2

8.4.3 Find v for $t > 0$ in Ex. 8.4.2 if the 2-H inductor is replaced by a $\frac{1}{16}$-F capacitor.
Answer $10e^{-4t} + 8$ V

8.5

THE GENERAL CASE

The equations describing the networks of the previous sections are all special cases of a general expression given by

$$\frac{dy}{dt} + Py = Q \qquad (8.15)$$

where y is the unknown variable, such as v or i, and P and Q are constants. If, for instance, we compare this equation to that of (8.2) for the source-free circuit of Sec. 8.1, we see that $y = v$, $P = 1/RC$, and $Q = 0$. The same relations are valid for the forced RC circuit of Sec. 8.4 except that $Q = I_0/C$.

A solution of (8.15) can be found by separation of variables. However, let us introduce another method which also is applicable when Q is a function of time, an important case in later chapters. This method, known as the *integrating factor method*, consists of multiplying the equation by a factor which makes its left-hand side a perfect derivative and simply integrating both sides.

Let us begin by considering the derivative of a product, given by

$$\frac{d}{dt}(ye^{Pt}) = \frac{dy}{dt}e^{Pt} + Pye^{Pt}$$

$$= \left(\frac{dy}{dt} + Py\right)e^{Pt}$$

From this result we see that if we multiply both sides of (8.15) by e^{Pt}, we have

$$\frac{d}{dt}(ye^{Pt}) = Qe^{Pt}$$

Integrating both sides of the equation, we find

$$ye^{Pt} = \int Qe^{Pt}\,dt + A$$

where A is a constant of integration. Solving for y, we have

$$y = e^{-Pt} \int Qe^{Pt}\,dt + Ae^{-Pt} \qquad (8.16)$$

which is valid, of course, if Q is a function of time or a constant. If Q is not a constant, then we must carry out the integration to find y. Examples of this type are given in Ex. 8.5.2 and Ex. 8.5.4.

In the important dc case where Q is a constant, (8.16) becomes

$$y = Ae^{-Pt} + \frac{Q}{P}$$
$$= y_n + y_f \qquad (8.17)$$

where $y_n = Ae^{-Pt}$ and $y_f = Q/P$ are the natural and forced responses. We observe that y_n has the same mathematical form as the source-free natural response and that y_f is *always* a constant which is proportional to Q. In addition, $1/P$ is the time constant in the natural response.

EXAMPLE 8.5 To illustrate the method, let us find i_2 for $t > 0$ in the circuit of Fig. 8.12, given that $i_2(0) = 1$ A. Although the circuit is a somewhat complex combination of elements, the solution of (8.17) is valid since the network contains a constant forcing function and a *single* energy-storage element (the inductor). The loop equations for the circuit are

$$8i_1 - 4i_2 = 10$$

$$-4i_1 + 12i_2 + \frac{di_2}{dt} = 0$$

Eliminating i_1 from these equations, we find that

$$\frac{di_2}{dt} + 10i_2 = 5$$

FIGURE 8.12 Driven RL circuit

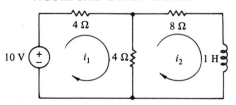

Comparing this equation with (8.15), we see that $P = 10$ and $Q = 5$. Hence (8.17) yields

$$i_2 = Ae^{-10t} + \frac{1}{2}$$

Applying the initial condition, we have

$$i_2(0) = A + \frac{1}{2} = 1$$

Therefore, $A = \frac{1}{2}$ and the solution is given by

$$i_2 = \frac{1}{2}e^{-10t} + \frac{1}{2} \text{ A}$$

We may also obtain (8.17) for the case Q constant by observing (8.15) and (8.16). Making $Q = 0$ in (8.16) yields the natural response $y = y_n = Ae^{-Pt}$, which then must be the solution of (8.15) when $Q = 0$. That is, the natural response satisfies

$$\frac{dy}{dt} + Py = 0$$

and may be found as in Sec. 8.3 by *trying*

$$y_n = Ae^{st}$$

which results in

$$s + P = 0$$

Thus we have $s = -P$, which yields $y_n = Ae^{-Pt}$, as before. The forced response y_f may be found by *trying* in (8.15) a function like Q. Since Q is constant in this case, the *trial solution* is then a constant. That is, we try

$$y_f = K$$

which substituted into (8.15) gives

$$0 + PK = Q$$

or

$$y_f = K = \frac{Q}{P}$$

as before.

EXERCISES

8.5.1 Find i for $t > 0$ if $i(0) = 3$ A.
Answer $e^{-7t} + 2$ A

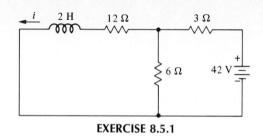

EXERCISE 8.5.1

8.5.2 Solve Ex. 8.5.1 if the 42-V source is replaced by a source of $60e^{-2t}$ V.
Answer $4e^{-2t} - e^{-7t}$ A.

8.5.3 Find v for $t > 0$ if $i(0) = 1$ A.
Answer $6 + 6e^{-5t}$ V

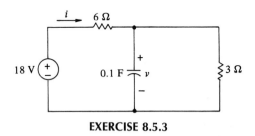

EXERCISE 8.5.3

8.5.4 Solve Ex. 8.5.3 if the 18-V source is replaced by a $(6e^{-4t})$-V source.
Answer $-10e^{-5t} + 10e^{-4t}$ V

8.6

A SHORTCUT PROCEDURE

Let us now introduce a shortcut procedure that is very useful for finding the currents and voltages in many circuits, particularly those with no dependent sources. The technique involves formulating the solution by merely inspecting the circuit.

EXAMPLE 8.6   Consider, for instance, Example 8.5 of the previous section (Fig. 8.12). We know that

$$i_2 = i_{2n} + i_{2f}$$

where i_{2n} and i_{2f} are the natural and forced responses, respectively. Since i_{2n} has the same form as the source-free response, we can look at the network in the absence of the forcing function (i.e., make the 10-V source zero by replacing it by a short circuit), as shown in Fig. 8.13(a). The natural response is then

$$i_{2n} = Ae^{-10t}$$

The forced response is constant; therefore, insofar as the forced response is concerned, it does not matter at what time we look at the circuit. We may choose then

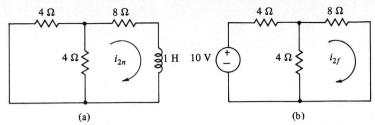

(a)   (b)

FIGURE 8.13 Circuits for finding the response of Fig. 8.12. (a) Circuit for finding i_{2n}; (b) circuit for finding i_{2f}

to look at the circuit in the steady state when i_{2n} is zero. At this time the inductor is a short circuit, as shown in Fig. 8.13(b), and

$$i_{2f} = \frac{1}{2}$$

Therefore,

$$i_2 = Ae^{-10t} + \frac{1}{2}$$

The constant A is now determined as before from the initial condition, $i_2(0) = 1$.

A word of caution is appropriate at this point. When evaluating the constant A, the student should *always* apply the initial condition to the complete response—never to the natural response alone—because the initial condition is always given for the current, not for a part of it.

EXAMPLE 8.7   As a second example, let us find i for $t > 0$ in Fig. 8.14, given $v(0) = 24$ V. The current is given by

$$i = i_n + i_f$$

To obtain i_n we note that it has the same *form* as v_n, the natural response of the capacitor voltage. In fact, the natural response of *every* current or voltage in the circuit has the same form as v_n. This is true because all the other currents and voltages in the source-free circuit may be obtained from v_n by applying one or more of the operations of addition, subtraction (in KCL and KVL), differentiation, and integration, none of

FIGURE 8.14 Driven *RC* circuit

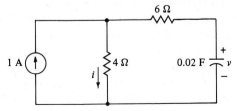

which changes the nature of the exponential $e^{-t/\tau}$. Examining the source-free circuit (the current source open circuited), we see that the time constant for the capacitor voltage is $\tau = 0.2$ s. Therefore,

$$i_n = Ae^{-5t}$$

In the steady state the capacitor is an open circuit, and the forced response is, by inspection,

$$i_f = 1 \text{ A}$$

Therefore,

$$i(t) = Ae^{-5t} + 1$$

To evaluate A, we must find the value of $i(0^+)$. Since $v(0) = v(0^+) = 24$ V, summing the voltages around the right-hand mesh, for $t = 0^+$, we have

$$-4i(0^+) + 6[1 - i(0^+)] + 24 = 0$$

or

$$i(0^+) = 3$$

Substituting this initial current into our solution, we find that

$$3 = A + 1$$

Therefore, $A = 2$ and

$$i = 1 + 2e^{-5t} \text{ A}$$

EXAMPLE 8.8

Before concluding this section let us, as a final example, determine i and v in the circuit of Fig. 8.15(a). The network is in a dc steady-state condition at $t = 0^-$ with the switch open; therefore the inductor and capacitor are a short circuit and an open circuit, respectively, at this time. The capacitor voltage is equal to the voltage that appears across the 20-Ω resistor, and the inductor current is equal to the current in the 15-Ω resistor. By current division, the currents in the 15- and 20-Ω resistors are easily shown to be 2 and 3 A, respectively. Thus

$$i(0^-) = 2 \text{ A}$$

and

$$v(0^-) = 60 \text{ V}$$

When the switch closes at $t = 0$, we observe that nodes a and b are short-circuited together, and we can redraw the network as shown in Fig. 8.15(b). It should be noticed that the 30-Ω resistor need not be included in this circuit because the switch is a short circuit across its terminals. The combination is equivalent to a 30-Ω resistor in parallel with a 0-Ω resistor, which, of course, is 0 Ω or a short circuit.

Let us next consider the current i leaving a in Fig. 8.15(b) through the 15-Ω resistor. From KCL, this same current must enter a through the 1-H inductor; hence no current flowing in the circuit to the left of a can enter the other part of the circuit to the right of a, and vice versa. Thus after the switch is closed, the network reduces to two independent circuits, each of which can be solved individually.

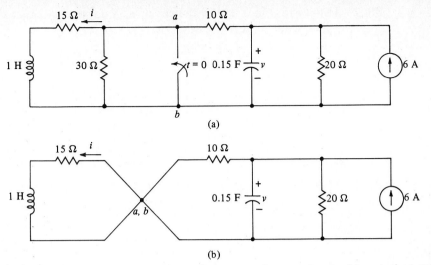

FIGURE 8.15 (a) Circuit containing an inductance and a capacitance; (b) equivalent circuit for $t > 0$

The first circuit, consisting of the 1-H inductor and the 15-Ω resistor, is simply a source-free *RL* network having $i(0^+) = i(0^-) = 2$ A. Therefore,

$$i = 2e^{-15t} \text{ A}$$

The second circuit, composed of all the elements to the right of a, is simply a driven *RC* network with $v(0^+) = v(0^-) = 60$ V. From our shortcut procedure, we find

$$v = 40 + 20e^{-t} \text{ V}$$

The shortcut procedure presented in this section is applicable also to circuits containing dependent sources. However, no savings in time or effort usually result because the circuit equations still have to be written for the source-free and dc steady-state cases.

EXERCISES

8.6.1 Find v and i for $t > 0$ if the circuit is in dc steady state at $t = 0^-$.
Answer $2(1 - e^{-10^5 t})$ V, $4(1 + e^{-10^5 t})$ mA

EXERCISE 8.6.1

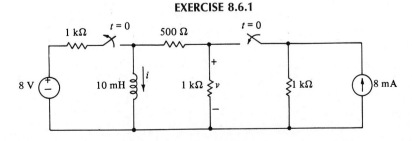

8.6.2 Find i and v for $t > 0$ if the circuit is in dc steady-state at $t = 0^-$.

Answer $-2e^{-10t}$ mA, $-20e^{-10^5t}$ V

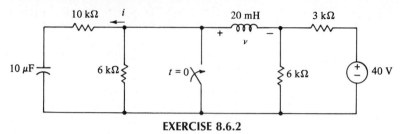

EXERCISE 8.6.2

8.6.3 Find i for $t > 0$ in the circuit of Ex. 8.6.1 if both switches are closed at $t = 0$ and $i(0) = 0$. (*Suggestion:* Find i_n as before and note that i_f is due to two sources.)

Answer $12 - 12e^{-50,000t}$ mA

8.7

THE UNIT STEP FUNCTION

In the previous sections we have analyzed circuits in which energy sources have been suddenly inserted into the networks. At the instant these sources are applied the voltages or currents, at the points of application, change abruptly. Forcing functions whose values change in this manner are called *singularity functions*. There are many singularity functions that are useful in circuit analysis. One of the most important is the unit step function, so named by the English engineer Oliver Heaviside (1850–1925).

The *unit step function* is a dimensionless function which is equal to zero for all negative values of its argument and which is equal to one for all positive values of its argument. If we denote the unit step function by the symbol $u(t)$, then a mathematical description is

$$
\begin{aligned}
u(t) &= 0, && t < 0 \\
&= 1, && t > 0
\end{aligned}
\tag{8.18}
$$

From a graph of (8.18), shown in Fig. 8.16, we see that at $t = 0$, $u(t)$ changes

FIGURE 8.16 Graph of the unit step function $u(t)$

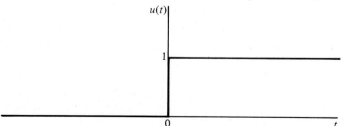

abruptly from 0 to 1. Some authors define $u(0)$ to be 1, but we are leaving $u(t)$ undefined at $t = 0$.

Since the unit step function is a dimensionless quantity, a voltage or current source can be obtained by multiplying $u(t)$ by a voltage or a current. A voltage step of V volts is represented by the product $Vu(t)$. Clearly, this voltage is 0 for $t < 0$ and V volts for $t > 0$. A voltage step source of V volts is shown in Fig. 8.17(a). A circuit which is equivalent to this source is shown in Fig. 8.17(b). A short circuit exists for $t < 0$, and the voltage is, of course, zero. For $t > 0$, a voltage V appears at the terminals. We have assumed in our model that the switching action occurs in zero time.

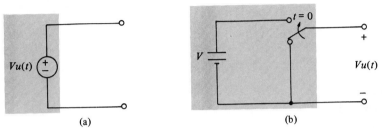

FIGURE 8.17 (a) Voltage step source of V volts; (b) equivalent circuit

Equivalent circuits for a current step source of I amperes are shown in Fig. 8.18. An open circuit exists for $t < 0$, and the current is zero. For $t > 0$, the switching action causes a terminal current of I amperes to flow.

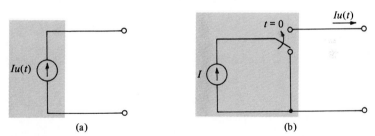

FIGURE 8.18 (a) Current step source of I amperes; (b) equivalent circuit

The switching action shown in Fig. 8.17 can only be approximated in actual circuits. However, in many cases, it is not necessary to require that the voltage source be a short circuit for $t < 0$, as we shall see in the next section. If the terminals of a network to which the source is to be connected remain at 0 V for $t < 0$, then a series connection of a source V and a switch is equivalent to the voltage step generator, as shown in Fig. 8.19.

Equivalent circuits for a current step generator in a network are shown in Fig. 8.20. In each case the current in the network terminals must be zero for $t < 0$.

Let us now return to our definition of the unit step function given in (8.18). We may generalize this definition by replacing t by $t - t_0$ in the three places that it occurs, which results in

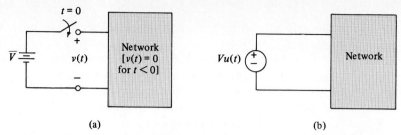

FIGURE 8.19 (a) Network with V applied at $t = 0$; (b) equivalent circuit

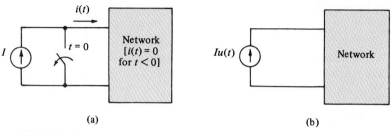

FIGURE 8.20 (a) Network with I applied at $t = 0$; (b) equivalent circuit

$$u(t - t_0) = 0, \qquad t < t_0$$
$$= 1, \qquad t > t_0 \qquad\qquad (8.19)$$

The function $u(t - t_0)$ is the function $u(t)$ *delayed* by t_0 seconds, as shown in Fig. 8.21.

Multiplying (8.19) by V or I gives us a voltage step source or a current step source whose value changes abruptly at time t_0. Equivalent networks for these sources are obtained in Figs. 8.17–8.20 by taking all actions related to switching to occur at $t = t_0$.

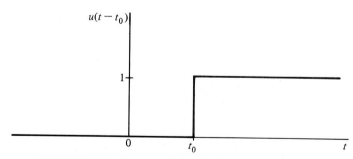

FIGURE 8.21 Graph of the unit step function $u(t - t_0)$

EXAMPLE 8.9 Step functions are very useful in formulating more complex functions. Take, for instance, the rectangular voltage pulse of Fig. 8.22(a). From this figure, we see that

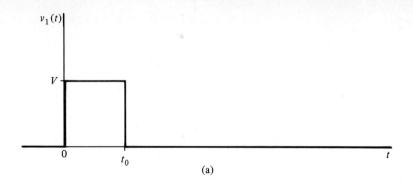

(a)

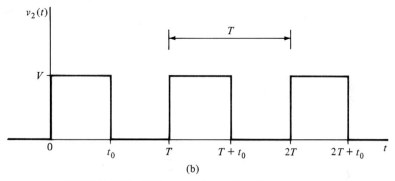

(b)

FIGURE 8.22 (a) Rectangular pulse; (b) square wave

$$v_1(t) = 0, \qquad t < 0$$
$$= V, \qquad 0 < t < t_0$$
$$= 0, \qquad t > t_0$$

Since $u(t)$ becomes 1 for $t > 0$ and $-u(t - t_0)$ becomes -1 for $t > t_0$, we may write

$$v_1(t) = V[u(t) - u(t - t_0)] \qquad (8.20)$$

To check this result, we see that, for $t < 0$,

$$v_1(t) = V(0 - 0) = 0$$

For $0 < t < t_0$,

$$v_1(t) = V(1 - 0) = V$$

and for $t > t_0$,

$$v_1(t) = V(1 - 1) = 0$$

Now suppose that we wish to produce a train of these pulses with one occurring every T seconds, where $T > t_0$, as shown in Fig. 8.22(b). Such a wave is called a *square wave*. The first pulse is given by (8.20). The second pulse is simply the first pulse delayed by T seconds. Therefore, replacing t by $t - T$ in (8.20), we have

$$\text{Pulse } 2 = V\{u(t - T) - u[t - (T + t_0)]\}$$

The $(n + 1)$th pulse in the pulse train is the first delayed by nT, and therefore

$$\text{Pulse } n + 1 = V\{u(t - nT) - u[t - (nT + t_0)]\}$$

To obtain an expression for the square wave for all $t > 0$, we add the above expressions and obtain

$$v_2(t) = V \sum_{n=0}^{\infty} \{u(t - nT) - u[t - (nT + t_0)]\} \qquad (8.21)$$

Waveforms like those of (8.20) and (8.21) are very common in *digital* circuits such as those in the digital computer.

EXERCISES

8.7.1 Using unit step functions, write an expression for the current $i(t)$ that satisfies

(a) $i(t) = 0,$ $\qquad\qquad t < 0$
$\qquad\quad = -4$ mA, $\qquad t > 0.$

(b) $i(t) = 0,$ $\qquad\qquad t < -10$ ms
$\qquad\quad = 2$ A, $\qquad\quad -10$ ms $< t < 100$ ms
$\qquad\quad = -1$ A, $\qquad\quad 100$ ms $< t.$

(c) $i(t) = 3$ mA, $\qquad t < 0$
$\qquad\quad = 0,$ $\qquad\qquad t > 0.$

(d) $i(t) = 4$ μA, $\qquad t < .1$ s
$\qquad\quad = 0,$ $\qquad\qquad t > 1$ s.

Answer (a) $-4u(t)$ mA; (b) $2u(t + 10^{-2}) - 3u(t - 10^{-1})$ A; (c) $3u(-t)$ mA;
(d) $4u(-t + 1)$ μA

8.7.2 Sketch the voltage given by

$$v(t) = (t + 1)u(t - 1) - \frac{3}{2}tu(t) + \left(\frac{1}{2}t - 1\right)u(t - 2)$$

8.7.3 Using unit step functions, write an expression for $v(t)$ for $-\infty < t < \infty$.
Answer $10 \sin 2\pi t[u(t) - u(t - 1)]$

EXERCISE 8.7.3

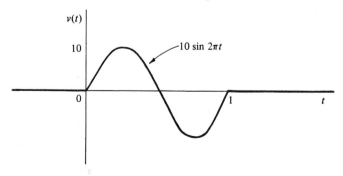

THE STEP RESPONSE

The *step response* is the response of a circuit having only one input which is a unit step function. The response and step input can, of course, be a current or a voltage. The step response is due entirely to the step input since no initial energies are present in the dynamic circuit elements. This is the case because all the currents and voltages in the network are zero at $t = 0^-$ due to the fact that the step function is zero for $-\infty < t < 0$. Thus the step response is the response to a unit step input with no initial energy stored in the circuit.

EXAMPLE 8.10
As an example, let us find the step response v in the simple RC circuit of Fig. 8.23(a) having an input of $v_g = u(t)$ V. Applying KCL, we have

$$C\frac{dv}{dt} + \frac{v - u(t)}{R} = 0$$

or

$$\frac{dv}{dt} + \frac{v}{RC} = \frac{1}{RC}u(t)$$

For $t < 0$, this equation becomes

$$\frac{dv}{dt} + \frac{v}{RC} = 0$$

the solution of which is

$$v = Ae^{-t/RC}$$

Applying the initial condition $v(0^-) = 0$, we see that $A = 0$, and therefore

$$v(t) = 0, \qquad t < 0$$

which confirms our assertion that the response is zero prior to the change in the input. For $t > 0$, our differential equation is

$$\frac{dv}{dt} + \frac{v}{RC} = \frac{1}{RC}$$

FIGURE 8.23 (a) *RC* circuit with a voltage step input; (b) equivalent circuit

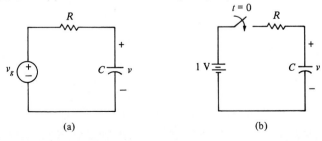

(a)   (b)

225

We know that

$$v = v_n + v_f$$

where

$$v_n = Ae^{-t/RC}$$

and, by inspection,

$$v_f = 1$$

Therefore,

$$v = 1 + Ae^{-t/RC}$$

The initial condition $v(0^+) = v(0^-) = 0$ requires that $A = -1$, and therefore our solution for all t is

$$v(t) = 0, \qquad\qquad t < 0$$
$$= 1 - e^{-t/RC}, \qquad t > 0$$

This may be written more concisely, using the unit step function, as

$$v(t) = (1 - e^{-t/RC})u(t)$$

The voltages across the resistor and the capacitor are zero for $t < 0$. Therefore, an equivalent circuit for our network is satisfied by the circuit of Fig. 8.23(b) provided we specify that $v(0^-) = 0$.

EXAMPLE 8.11 As a second example, let us find $v_2(t)$ in the circuit of Fig. 8.24, consisting of a resistor, a capacitor, and an op amp. A nodal equation at the inverting terminal of the op amp is given by

$$\frac{v_1}{R} + C\frac{dv_2}{dt} = 0$$

since the node voltage and the current of the inverting terminal are both zero. Therefore,

$$\frac{dv_2}{dt} = -\frac{1}{RC}v_1$$

FIGURE 8.24 Integrator

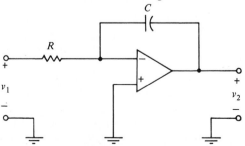

Integrating both sides of this equation between the limits of 0^+ and t, we have

$$v_2(t) = -\frac{1}{RC}\int_{0^+}^{t} v_1 \, dt + v_2(0^+)$$

The response $v_2(t)$ is proportional to the integral of the input voltage v_1 if $v_2(0^+) = 0$. Thus the circuit is called an *integrator*.

EXAMPLE 8.12   Let us now determine the response of the network if $v_1 = Vu(t)$. In this case,

$$v_2(t) = -\frac{V}{RC}\int_{0^+}^{t} u(t) \, dt + v_2(0^+)$$

$$= -\frac{V}{RC}tu(t) + v_2(0^+)$$

The capacitor voltage at $t = 0^+$ is zero, which requires that $v_2(0^+) = 0$. Thus we have

$$v_2 = -\frac{V}{RC}tu(t)$$

A graph of v_2 is shown in Fig. 8.25. This function is called a *ramp* function with a slope of $-V/RC$.

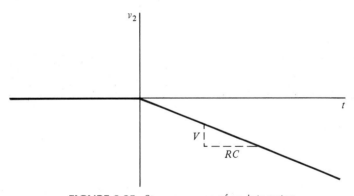

FIGURE 8.25 Step response of an integrator

EXAMPLE 8.13   As a final example, let us find the voltage $v(t)$ in the network of Fig. 8.26, given that $i(0^-) = 0$. The forcing function for the circuit is the current pulse,

$$i_g(t) = 10[u(t) - u(t - 1)] \text{ A}$$

which is shown in Fig. 8.27(a).

Examining the circuit, we see that a zero initial condition $[i(0^-) = 0]$ exists and that at $t = 0$ a current step of 10 A is applied. Thus the network response is identical to the step response until time $t = 1$ s. At this time, the forcing function $i_g(t)$ becomes zero, and the response is simply the source-free response resulting from the energy which the inductor has accumulated during the interval of the current pulse. Therefore,

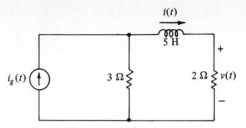

FIGURE 8.26 *RL* circuit driven by $i_g(t)$

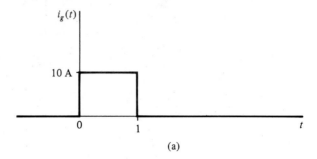

(a)

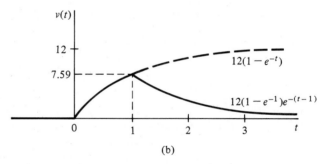

(b)

FIGURE 8.27 (a) Forcing function $i_g(t)$; (b) response of an *RL* circuit to $i_g(t)$

we see that the solution of the problem involves finding the step response of the circuit for $t < 1$ s and then finding the source-free response for $t > 1$ s.

We know that the step response is of the form

$$v = v_n + v_f$$

where

$$v_n = Ae^{-R_{eq}t/L} = Ae^{-5t/5} = Ae^{-t}$$

and, by current division and Ohm's law, the forced response is the steady-state value

$$v_f = 2\left[\frac{(3)(10)}{2 + 3}\right] = 12$$

Combining these equations, we have

$$v = Ae^{-t} + 12$$

Since $i(0^+) = i(0^-) = 0$, then $v(0^+) = v(0^-) = 0$ and $A = -12$. Therefore,

$$v = 0, \qquad\qquad t < 0$$
$$= 12(1 - e^{-t}), \qquad 0 < t < 1$$

For $t > 1$, we know that v is of the form

$$v = Be^{-t}$$

At $t = 1^-$ (the instant just before $t = 1$), our solution for the step response gives

$$v(1^-) = 12(1 - e^{-1})$$

Since the inductor current is continuous, $v(1^-) = v(1^+)$, where 1^+ is the instant just after $t = 1$. Therefore,

$$v(1^+) = Be^{-1} = 12(1 - e^{-1})$$

or

$$B = 12(1 - e^{-1})e$$

and our solution becomes

$$v = 12(1 - e^{-1})e^{-(t-1)}, \qquad t > 1$$

The solution for all t can be written as

$$v(t) = 12(1 - e^{-t})[u(t) - u(t - 1)] + 12(1 - e^{-1})e^{-(t-1)}u(t - 1) \quad (8.22)$$

A graph of this response is shown in Fig. 8.27(b).

EXERCISES

8.8.1 Find the response $i(t)$ to the voltage step $v(t) = 40u(t)$ V.
Answer $(2 - e^{-100t})u(t)$ mA

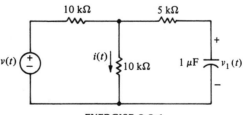

EXERCISE 8.8.1

8.8.2 Find $v_1(t)$ in Ex. 8.8.1 if $v(t) = 20[u(t) - u(t - 0.1)]$ V.
Answer $10[1 - e^{100t}][u(t) - u(t - 0.1)] + 10(e^{10} - 1)e^{-100t}u(t - 0.1)$ V

8.8.3 Find v if $v_g = 5e^{-t}u(t)$ V and there is no initial stored energy. (This circuit is like the integrator of Fig. 8.24, except that the capacitor has a resistor in parallel with it, making it a practical, or *lossy*, capacitor. Thus the circuit is called a *lossy integrator*.)
Answer $5(e^{-2t} - e^{-t})u(t)$ V

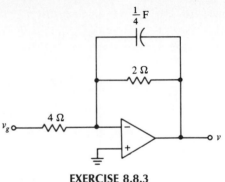

EXERCISE 8.8.3

8.9

APPLICATION OF SUPERPOSITION

In this section we shall consider the use of superposition for obtaining solutions of *RC* and *RL* circuits containing two or more independent sources.

EXAMPLE 8.14 As a first example, let us consider the circuit of Fig. 8.26 of the previous section. The value of the independent current source is given by

$$i_g = 10u(t) - 10u(t - 1)$$

This source is equivalent to a pair of independent current sources connected in parallel. Thus if we let

$$i_g = i_1 + i_2$$

where $i_1 = 10u(t)$ and $i_2 = -10u(t - 1)$, the circuit of Fig. 8.26 can be redrawn as shown in Fig. 8.28.

From the principle of superposition, we may write the output voltage as

$$v = v_1 + v_2$$

where v_1 and v_2 are the responses due to i_1 and i_2 respectively. In our previous solution we found that the response due to the current step i_1 was given by

$$v_1 = 12(1 - e^{-t})u(t)$$

FIGURE 8.28 Equivalent circuit of Fig. 8.26

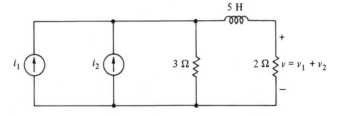

Next, we need the response v_2 due to i_2. We note that i_2 is simply the negative of i_1 delayed 1 s in time. Therefore, v_2 is obtained from v_1 by multiplying v_1 by -1 and replacing t by $t - 1$. The result is given by

$$v_2 = -12(1 - e^{-(t-1)})u(t - 1)$$

Our solution is now given by

$$v = 12(1 - e^{-t})u(t) - 12(1 - e^{-(t-1)})u(t - 1) \qquad (8.23)$$

which is equivalent to (8.22) (see Ex. 8.9.1).

EXAMPLE 8.15   As a second example, let us consider the RC network of Fig. 8.29, which contains two independent sources and an initial capacitor voltage $v(0) = V_0$. Applying KVL around the left mesh, we find that

$$(R_1 + R_2)i + \frac{1}{C} \int_0^t i\, dt + V_0 = V_1 - R_2 I_1$$

Multiplying each term in this equation by a constant K, we have

$$(R_1 + R_2)(Ki) + \frac{1}{C} \int_0^t (Ki)\, dt + KV_0 = KV_1 - R_2(KI_1)$$

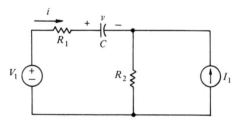

FIGURE 8.29 RC network

Clearly, the current response becomes Ki when the independent sources *and* the initial capacitor voltage are multiplied by the factor K, which demonstrates the proportionality property for a linear network. This result is easily extended to any linear circuit containing one or more capacitors. Thus initial capacitor voltages can be treated as independent voltage sources. In a similar manner, it is easily shown that initial inductor currents can be treated as independent current sources.

EXAMPLE 8.16   We shall now employ superposition to determine the voltage v by finding v_1, v_2, and v_3, the responses due to V_1, I_1, and V_0, respectively. The circuit for finding v_1 is shown in Fig. 8.30(a). This is a simple driven RC circuit having a zero initial capacitor voltage. The solution is given by

$$v_1 = V_1(1 - e^{-t/(R_1+R_2)C})$$

The circuit for finding v_2 is shown in Fig. 8.30(b). This again is a simple driven circuit having a zero initial capacitor voltage, for which we find

$$v_2 = -R_2 I_1(1 - e^{-t/(R_1+R_2)C})$$

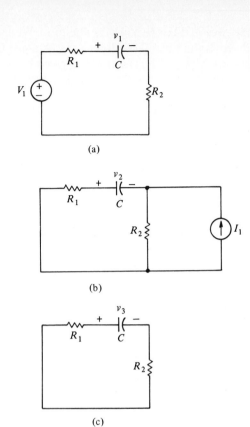

(a)

(b)

(c)

FIGURE 8.30 Circuits for finding (a) v_1; (b) v_2; (c) v_3 for an *RC* network

In the circuit of Fig. 8.30(c), the voltage v_3 is simply the source-free response resulting from the initial capacitor voltage. Since $v_3(0) = V_0$, we find that

$$v_3 = V_0 e^{-t/(R_1+R_2)C}$$

Therefore, the entire response is given by

$$v = v_1 + v_2 + v_3$$
$$= V_1 - R_2 I_1 + (R_2 I_1 - V_1 + V_0)e^{-t/(R_1+R_2)C} \qquad (8.24)$$

Inspecting our solution, we see that it consists of a forced response v_f and a natural response v_n, which we could have anticipated. An alternative method of finding the solution is to obtain v_f and v_n, as described previously. Superposition can, of course, be used in finding v_f. In our case, we know that

$$v_f = v_{1f} + v_{2f}$$

From Fig. 8.30(a) and (b) we see, by inspection, that

$$v_{1f} = V_1$$

and

$$v_{2f} = -R_2 I_1$$

Therefore, the forced response is

$$v_f = V_1 - R_2 I_1$$

The natural response, obtained from the source-free circuit [Fig. 8.30(c)], is given by

$$v_n = Ae^{-t/(R_1+R_2)C}$$

Therefore,

$$v = V_1 - R_2 I_1 + Ae^{-t/(R_1+R_2)C} \tag{8.25}$$

Since $v(0) = V_0$, we have

$$A = R_2 I_1 - V_1 + V_0$$

which substituted into (8.25) gives (8.24).

EXERCISES

8.9.1 Reduce (8.22) to the form of (8.23).

8.9.2 Use superposition to find v in Fig. 8.29 if $R_1 = 2\ \Omega$, $R_2 = 3\ \Omega$, $C = 0.1$ F, $v(0) = 8$ V, $V_1 = 12$ V, and $I_1 = 2$ A.
Answer $6 + 2e^{-2t}$ V

8.9.3 Use superposition to find i for $t > 0$. Assume the circuit is in a steady-state condition at $t = 0^-$.
Answer $5 + 10(1 - e^{-500t})$ mA

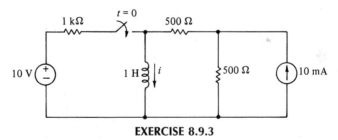

EXERCISE 8.9.3

8.10

SPICE AND THE TRANSIENT RESPONSE

SPICE, as in the dc case, is a very useful tool for obtaining transient responses for networks containing energy-storage elements, such as capacitors and inductors, and excitation sources that are dc, exponential, pulse, sinusoidal, and piecewise linear. As we have seen in previous chapters for simple RC and RL circuits, a solution first requires the determination of the initial conditions of all energy-storage devices at the time the transient response is to begin. The dc analysis of SPICE described in previous chapters can be used for determining initial conditions if the circuit is in a steady-state condition prior to the start of the transient response, which occurs, for instance, as a result of a switch operation. SPICE can then be used to simulate the response just after the switching action using these initial values. Before proceeding to the following discussion, it is recommended that the C and L statements and the .PLOT and .TRAN commands of Appendix E be reviewed.

EXAMPLE 8.17   As an example, consider finding and plotting the capacitor voltage for the circuit of Fig. 8.31(a) for the first 75-ms interval following the switching action at $t = 0$. At the instant just prior to activation of the switches, let us assume that the circuit is in a dc steady-state condition, as shown in Fig. 8.31(b). A SPICE circuit file for determining the initial capacitor voltage is

FIGURE 8.31 (a) *RC* circuit; (b) redrawn at $t = 0^-$ in dc steady condition; (c) redrawn at $t = 0^+$ for transient response

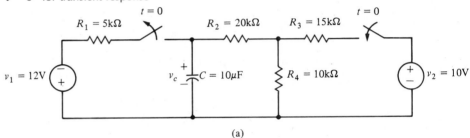

(a)

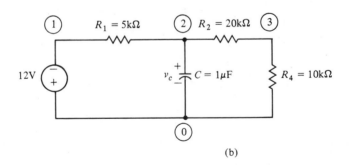

(b)

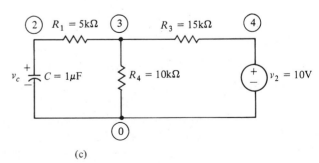

(c)

INITIAL CONDITIONS FOR FIG. 8.31(b)
* Data Statements
V1 1 0 DC -12
R1 1 2 5K
R2 2 3 20K
R4 3 0 10K
C 2 0 1E-6
* Solution control statement
.DC V1 -12 -12 1
* Output control statement
.PRINT DC V(C)
.END

The output file gives the initial capacitor voltage of -10.29 V. A circuit file can now be written for the transient response using this initial value for the circuit of Fig. 8.31(c). Suppose we desire a response in 10-ms intervals for the first 75 ms. A circuit file is

TRANSIENT RESPONSE FOR FIG. 8.31(c)
* Data Statements
R2 2 3 20K
R3 3 4 20K
R4 3 0 10K
C 2 0 1E-6 IC=-10.29
V2 4 0 DC 10
* Solution control statement for transient response
.TRAN 5MS 75MS UIC
* Output control statement for printing and plotting response
.PRINT TRAN V(C)
.PLOT TRAN V(C)
.END

The plot generated on the printer for this program is shown in Fig. 8.32.

FIGURE 8.32 Transient response for circuit of Fig. 8.31

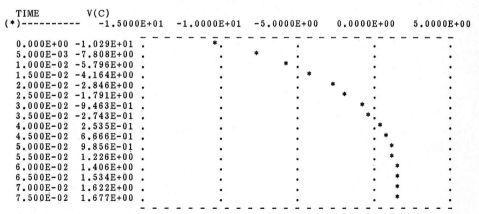

```
  TIME         V(C)
(*)----------     -1.5000E+01   -1.0000E+01   -5.0000E+00    0.0000E+00    5.0000E+00
                  - - - - - - - - - - - - - - - - - - - - - - - - - - - - -
 0.000E+00  -1.029E+01 .             * .            .             .             .
 5.000E-03  -7.808E+00 .             .      *        .             .             .
 1.000E-02  -5.796E+00 .             .             .*             .             .
 1.500E-02  -4.164E+00 .             .             .   *          .             .
 2.000E-02  -2.846E+00 .             .             .      *       .             .
 2.500E-02  -1.791E+00 .             .             .        *     .             .
 3.000E-02  -9.463E-01 .             .             .          * . .             .
 3.500E-02  -2.743E-01 .             .             .           *.             .
 4.000E-02   2.535E-01 .             .             .            .*             .
 4.500E-02   6.666E-01 .             .             .            . *            .
 5.000E-02   9.856E-01 .             .             .            .  *           .
 5.500E-02   1.226E+00 .             .             .            .   *          .
 6.000E-02   1.406E+00 .             .             .            .    *         .
 6.500E-02   1.534E+00 .             .             .            .    *         .
 7.000E-02   1.622E+00 .             .             .            .     *        .
 7.500E-02   1.677E+00 .             .             .            .     *        .
                  - - - - - - - - - - - - - - - - - - - - - - - - - - - - -
```

In Example 8.17, the voltage source v_2 has a dc value of 10 V. SPICE provides a variety of other excitations for transient responses, such as the examples given in Table 8.1.

TABLE 8.1 Examples of excitations for v_2 of Fig. 8.31 in the interval $0 < t < 0.1$ s

v_2	SPICE Statement
$10e^{-10t}$ V	V2 4 0 EXP(10V 0V 0S 0.1S 1S)
$10[u(t) - u(t - 0.02)]$ V	V2 4 0 PULSE(10V 0V 0.02S 0S 0.1S)
$100t$ V	V2 4 0 PWL(0S 0V 0.1S 10V)
$10 \sin[2\pi(2.5t)]$ V	V2 4 0 SIN(0V 10V 2.5HZ)

EXERCISES

8.10.1 Using SPICE, find (a) $i_L(0^-)$; (b) a plot of $i_L(t)$ for $0 < t < 30 \ \mu s$.
Answer -2.475 mA

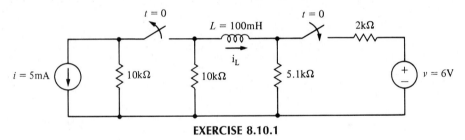

EXERCISE 8.10.1

8.10.2 Repeat Ex. 8.10.1 if $v = 5 \times 10^5 t$ V.

8.10.3 Plot v for $0 < t < 10$ ms if $v(0) = 10$ V.

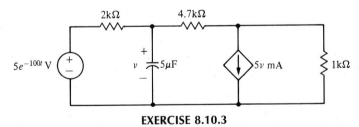

EXERCISE 8.10.3

PROBLEMS

8.1 The current $i(10^{-2}) = 3.68$ mA. Determine (a) $v(0)$, (b) $v(t)$ for $t > 0$, and (c) the energy dissipated by the 10-kΩ resistor as t becomes large. (d) Sketch $v(t)$ for $t > 0$.

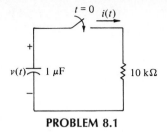

PROBLEM 8.1

8.2 Find v for $t > 0$ if $i(0) = 2$ A.

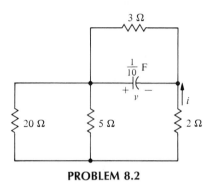

PROBLEM 8.2

8.3 Find v for $t > 0$ if the circuit is in steady state at $t = 0^-$.

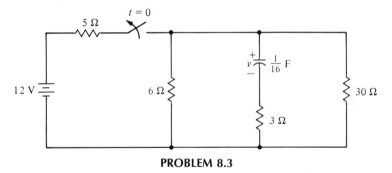

PROBLEM 8.3

8.4 Find i for $t > 0$ if the circuit is in steady state at $t = 0^-$.

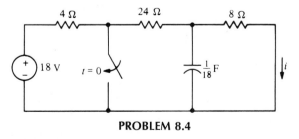

PROBLEM 8.4

8.5 Find i for $t > 0$ if $v(0) = 6$ V.

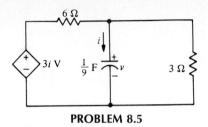

PROBLEM 8.5

8.6 Consider a source-free circuit that has a response $v(t) = V_0 e^{-t/\tau}$. Show that a straight line that is tangent to a graph of this response at time t_1 intersects the time axis (abscissa) at time $t_1 + \tau$.

8.7 (a) Find v for $t > 0$ if the circuit is in steady state at $t = 0^-$. (b) Repeat part (a) if a resistor R_1 is placed in series with the source V_0.

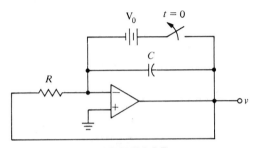

PROBLEM 8.7

8.8 Find v and i for $t > 0$ if the circuit is in steady state at $t = 0^-$.

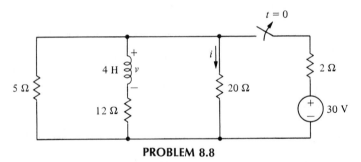

PROBLEM 8.8

8.9 Find v for $t > 0$ if the circuit is in steady state at $t = 0^-$.

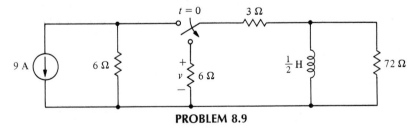

PROBLEM 8.9

8.10 Find i for $t > 0$ if $i(0) = 2$ A.

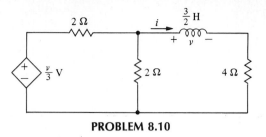

PROBLEM 8.10

8.11 Find i and v for $t > 0$ if $v(0) = -6$ V.

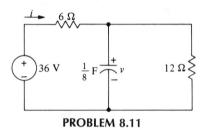

PROBLEM 8.11

8.12 Find v_1 for $t > 0$ if there is no initial stored energy.

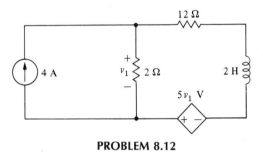

PROBLEM 8.12

8.13 Repeat Prob. 8.11 if the 36-V source is replaced by a $36e^{-3t}$-V source with the same polarity.

8.14 Find i for $t > 0$ if the circuit is in steady state at $t = 0^-$.

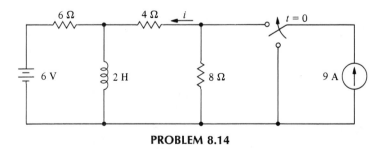

PROBLEM 8.14

8.15 Find v for $t > 0$ if the circuit is in steady state at $t = 0^-$.

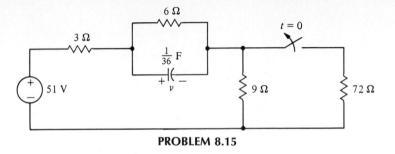

PROBLEM 8.15

8.16 Find i for $t > 0$ if the circuit is in steady state at $t = 0^-$.

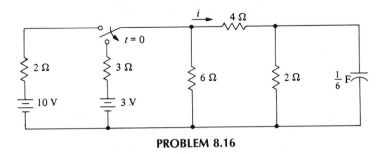

PROBLEM 8.16

8.17 Find i for $t > 0$ if the circuit is in steady state at $t = 0^-$.

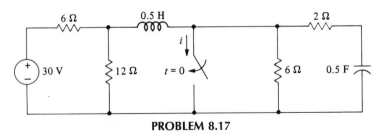

PROBLEM 8.17

8.18 Find v and i for $t > 0$ if the circuit is in steady state at $t = 0^-$.

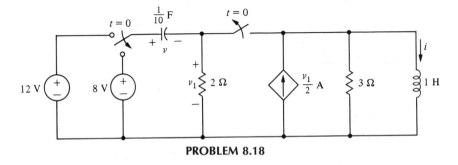

PROBLEM 8.18

8.19 Find v for $t > 0$ if $v_g = 2e^{-3t}$ V and $v_C(0) = 0$.

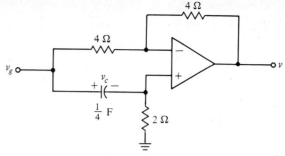

PROBLEM 8.19

8.20   Express $i(t)$ in terms of unit step functions.

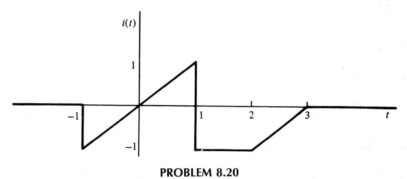

PROBLEM 8.20

8.21   Find the step response i. [Note that this means $v_g = u(t)$ V.]

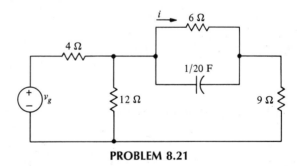

PROBLEM 8.21

8.22   Find v for $-\infty < t < \infty$ if $v_g = 12u(t)$ V.

PROBLEM 8.22

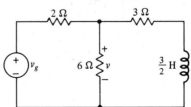

8.23   Repeat Prob. 8.22 if $v_g = 12[u(t) - u(t-1)]$ V.

8.24   Find v for $t > 0$ if $v_1(0) = 0$ and $v_g = 4u(t)$ V.

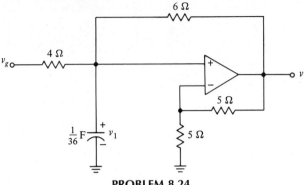

PROBLEM 8.24

8.25   Find v for $t > 0$ if $v_1(0) = 0$ and $v_g = 12u(t)$ V.

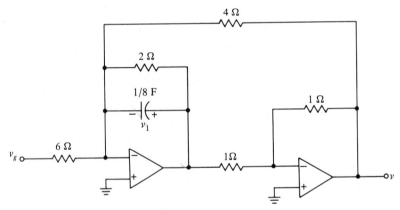

PROBLEM 8.25

8.26   Find v for $t > 0$ using (a) the shortcut procedure of Sec. 8.6 and (b) superposition. Assume the circuit is in steady state at $t = 0^-$.

PROBLEM 8.26

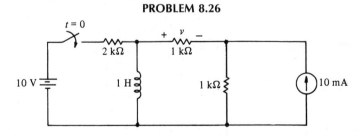

COMPUTER APPLICATION PROBLEMS

8.27 Use SPICE to solve (a) Prob. 8.8; (b) Prob. 8.15; (c) Prob. 8.18.

8.28 Use SPICE to plot the output voltage of the circuit of Prob. 8.24 for $v_g = 10 \sin(1000t)u(t)$ V, with resistor values changed to kilohms (multiplied by 1000), capacitor changed to 1 μF, and $v_1(0) = 4$ V in the interval $0 < t < 0.01$ s.

8.29 Use SPICE to plot the output voltage of the circuit of Prob. 8.25 for v_g being equal to the waveform of Prob. 8.20 with $i(t)$ replaced by $v_g(t)$ in volts. Plot for the interval $0 < t < 4$ s. (*Hint:* Replace t by $t + 1$ in v_g for application of SPICE.)

Hans Christian Oersted
1777–1851

9

Second-Order Circuits

An attempt should be made to see whether electricity, in its most latent stage, has any effect on the magnet as such.

Hans Christian Oersted

Electromagnetism was discovered in the spring of 1820 by the Danish physicist Hans Christian Oersted when, during a lecture to an advanced group of students, he demonstrated that the needle of a compass moved when placed near a current-carrying wire. By July of that year, he was certain that an electric current produced about it a circular magnetic field, and he published his results in a short paper, written in Latin, and carried by the major scientific journals of Europe.

Oersted was born in the town of Rudkobing, on the Danish island of Langeland, the elder son of an apothecary, Soren Christian Oersted. Because of family problems, Hans and his younger brother were placed with a German wigmaker while they were still young boys. The brothers' intellectual abilities and extraordinary thirst for knowledge were soon apparent to the townspeople, who did what they could to educate them. In 1794 the brothers, with no prior formal education, were accepted by the University of Copenhagen, where Hans studied astronomy, chemistry, mathematics, physics, and pharmacy. He completed his training in pharmacy in 1797 and two years later received his doctorate in philosophy. After a brief stint as a pharmacist, he was attracted to the world of science, which was in ferment at the time over Volta's discovery of the electric battery. Between 1800 and 1820, he was a university teacher, researcher, publisher, and one of the most sought-after lecturers of his day.

Oersted's great discovery had an enormous impact on the scientific world. More than a hundred scientists published their comments and researches on electromagnetism within seven years of the discovery, and Oersted was showered with honors and awards. The Royal Society of London gave him the Copley Medal, and the French Academy awarded him a prize of 3000 gold francs. For the rest of his life, he was a recognized leader in science, and in his honor the *oersted* was chosen as the standard cgs unit of magnetic field intensity. ∎

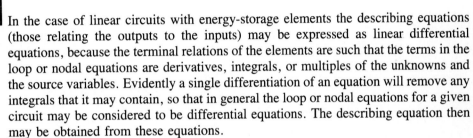

In the case of linear circuits with energy-storage elements the describing equations (those relating the outputs to the inputs) may be expressed as linear differential equations, because the terminal relations of the elements are such that the terms in the loop or nodal equations are derivatives, integrals, or multiples of the unknowns and the source variables. Evidently a single differentiation of an equation will remove any integrals that it may contain, so that in general the loop or nodal equations for a given circuit may be considered to be differential equations. The describing equation then may be obtained from these equations.

The circuits containing storage elements that we have considered so far were first-order circuits. That is, they were described by first-order differential equations. This is always the case when there is only one storage element present or when a switching action converts the circuit into two or more independent circuits each having no more than one storage element.

In this chapter we shall consider second-order circuits, which, as we shall see, contain two storage elements and have describing equations that are second-order differential equations. In general, nth-order circuits, containing n storage elements, are described by nth-order differential equations. The results for first- and second-order circuits ($n = 1$ and $n = 2$) may be readily extended to the general case, but we shall not do so here. However, a solution of a third-order differential equation is outlined in Prob. 9.25, which may be used to solve a third-order circuit given in Prob. 9.26. Higher-order circuits will be treated in more detail in Chapter 14.

Another, very elegant, method of solving higher-order circuits, as well as first- and second-order ones, is the Laplace transform method given in Chapter 19. An interested reader may go directly to this chapter without the need for reading the intervening chapters.

9.1

CIRCUITS WITH TWO STORAGE ELEMENTS

To introduce the subject of *second-order* circuits, let us begin with the circuit of Fig. 9.1, where the output to be found is the mesh current i_2. The circuit contains two storage elements, the inductors, and as we shall see, i_2 satisfies a second-order

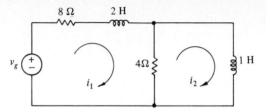

FIGURE 9.1 Circuit with two inductors

differential equation. Methods of solving such equations will be considered in later sections of this chapter.

The mesh equations of Fig. 9.1 are given by

$$2\frac{di_1}{dt} + 12i_1 - 4i_2 = v_g$$

$$-4i_1 + \frac{di_2}{dt} + 4i_2 = 0 \tag{9.1}$$

From the second of these we have

$$i_1 = \frac{1}{4}\left(\frac{di_2}{dt} + 4i_2\right) \tag{9.2}$$

which differentiated results in

$$\frac{di_1}{dt} = \frac{1}{4}\left(\frac{d^2i_2}{dt^2} + 4\frac{di_2}{dt}\right) \tag{9.3}$$

Substituting (9.2) and (9.3) into the first equation of (9.1) to eliminate i_1, we have, after multiplying the resulting equation through by 2,

$$\frac{d^2i_2}{dt^2} + 10\frac{di_2}{dt} + 16i_2 = 2v_g \tag{9.4}$$

The describing equation for the output i_2 is thus a *second-order* differential equation. That is, it is a differential equation in which the highest derivative is second-order. For this reason we shall refer to Fig. 9.1 as a *second-order* circuit and note that, typically, second-order circuits contain two storage elements.

There are exceptions, however, to the rule that two-storage-element circuits have second-order describing equations. For example, let us consider the circuit of Fig. 9.2, which has two capacitors. With the reference node taken as indicated, nodal equations at the nodes labeled v_1 and v_2 are given by

$$\frac{dv_1}{dt} + v_1 = v_g$$

$$\frac{dv_2}{dt} + 2v_2 = 2v_g \tag{9.5}$$

The choice of the node voltages v_1 and v_2 as the unknowns has resulted in two

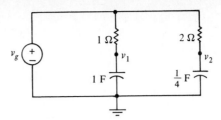

FIGURE 9.2 Circuit with two capacitors

first-order differential equations, each containing only one of the unknowns. When this happens, we say that the equations are *uncoupled,* and thus no elimination procedure is required to separate the variables. It was the elimination procedure which, applied to (9.1), gave the second-order equation of (9.4). The equations of (9.5) may be solved separately by the methods of the previous chapter.

Evidently Fig. 9.2, although it contains two storage elements, is not a second-order circuit. The same voltage v_g is across each RC combination, and thus the circuit may be redrawn as two first-order circuits. If the source were a practical source rather than an ideal source, then the circuit would be a second-order circuit. (See Prob. 9.1.)

EXERCISES

9.1.1 Find the equation satisfied by the mesh current i_2.

Answer $\dfrac{d^2 i_2}{dt^2} + 7\dfrac{di_2}{dt} + 6i_2 = \dfrac{dv_g}{dt}$

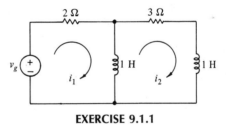

EXERCISE 9.1.1

9.1.2 Let $v_g = 14e^{-2t}$ V, $i_1(0^+) = 6$ A and $i_2(0^+) = 2$ A in Ex. 9.1.1, and find $di_2(0^+)/dt$ (the value of di_2/dt at $t = 0^+$).
Answer -4 A/s

9.1.3 For the values of $i_2(0^+)$, $di_2(0^+)/dt$, and v_g given in Ex. 9.1.2, show that i_2 in Ex. 9.1.1 is given by

$$i_2 = -4e^{-t} + 7e^{-2t} - e^{-6t} \text{ A}$$

(*Suggestion:* Substitute the answer into the differential equation, etc.)

SECOND-ORDER EQUATIONS

In Chapter 8 we considered first-order circuits in some detail and saw that their describing equations were first-order differential equations of the form

$$\frac{dx}{dt} + a_0 x = f(t) \tag{9.6}$$

In Sec. 9.1 we defined second-order circuits as those having two storage elements with describing equations that were second-order differential equations, given generally by

$$\frac{d^2 x}{dt^2} + a_1 \frac{dx}{dt} + a_0 x = f(t) \tag{9.7}$$

In (9.6) and (9.7) the a's are real constants, x may be either a voltage or a current, and $f(t)$ is a known function of the independent sources.

As an example, for the circuit of Fig. 9.1, the describing equation was (9.4). Comparing this equation with (9.7), we see that $a_1 = 10$, $a_0 = 16$, $f(t) = 2v_g$, and $x = i_2$.

From Chapter 8 we know that the complete response satisfying (9.6) is given by

$$x = x_n + x_f \tag{9.8}$$

where x_n is the natural response obtained when $f(t) = 0$ and x_f is the forced response, which satisfies (9.6). The forced response, in contrast to the natural response, contains no arbitrary constants.

Let us see if this same procedure will apply to the second-order equation (9.7). By a solution to (9.7) we shall mean a function x which satisfies (9.7) identically. That is, when x is substituted into (9.7), the left member becomes identically $f(t)$. We shall also require that x contain *two* arbitrary constants since we must be able to satisfy the two conditions imposed by the initial energy stored in the two storage elements.

If x_n is the natural response, i.e., the response when $f(t) = 0$, then it must satisfy the equation

$$\frac{d^2 x_n}{dt^2} + a_1 \frac{dx_n}{dt} + a_0 x_n = 0 \tag{9.9}$$

Since each term contains x_n to the same degree, namely 1 (the right member may be thought of as $0 = 0x_n$), this equation is sometimes called the *homogeneous* equation.

If x_f is to satisfy the original equation, as it did in the first-order case, then by (9.7) we must have

$$\frac{d^2 x_f}{dt^2} + a_1 \frac{dx_f}{dt} + a_0 x_f = f(t) \tag{9.10}$$

Adding (9.9) and (9.10) and rearranging the terms, we may write

$$\frac{d^2}{dt^2}(x_n + x_f) + a_1 \frac{d}{dt}(x_n + x_f) + a_0(x_n + x_f) = f(t) \tag{9.11}$$

The rearrangement is possible, of course, because the equations involved are linear.

Comparing (9.7) and (9.11) we see that (9.8) is our solution, as it was in the first-order case. That is, x satisfying (9.7) is made up of two components, a natural response x_n satisfying the homogeneous equation (9.9) and a forced response x_f satisfying the original equation (9.10) or (9.7). As we shall see, the natural response will contain two arbitrary constants and, as in the first-order case, the forced response will have no arbitrary constants. We shall consider methods of finding the natural and forced responses in the next three sections.

Of course, if the driving, or forcing, functions are such that $f(t) = 0$ in (9.7), then the forced response is zero, and the solution of the differential equation is simply the natural response.

EXERCISES

9.2.1 Show that

$$x_1 = A_1 e^{-t}$$

and

$$x_2 = A_2 e^{-3t}$$

are each solutions of

$$\frac{d^2 x}{dt^2} + 4\frac{dx}{dt} + 3x = 0$$

regardless of the values of the constants, A_1 and A_2.

9.2.2 Show that

$$x = x_1 + x_2 = A_1 e^{-t} + A_2 e^{-3t}$$

is also a solution of the differential equation of Ex. 9.2.1.

9.2.3 Show that if the right member of the differential equation of Ex. 9.2.1 is changed from 0 to 6, then

$$x = A_1 e^{-t} + A_2 e^{-3t} + 2$$

is a solution.

9.3

THE NATURAL RESPONSE

The natural response x_n of the general solution

$$x = x_n + x_f$$

of (9.7) must satisfy the homogeneous equation, which we repeat as

$$\frac{d^2 x}{dt^2} + a_1\frac{dx}{dt} + a_0 x = 0 \tag{9.12}$$

Evidently the solution $x = x_n$ must be a function which does not change its form when it is differentiated. That is, the function, its first derivative, and its second derivative must all have the same form, for otherwise the combination in the left member of the equation could not become identically zero for all t.

We are therefore led to *try*

$$x_n = Ae^{st} \qquad (9.13)$$

since this is the only function which retains its form when it is repeatedly differentiated. This is, of course, the same function that worked so well for us in the first-order case of Chapter 8. Also, as in the first-order case, A and s are constants to be determined.

Substituting (9.13) for x in (9.12), we have

$$As^2 e^{st} + Asa_1 e^{st} + Aa_0 e^{st} = 0$$

or

$$Ae^{st}(s^2 + a_1 s + a_0) = 0$$

Since Ae^{st} cannot be zero [for then by (9.13) $x_n = 0$, and we cannot satisfy any initial energy-storage conditions], we have

$$s^2 + a_1 s + a_0 = 0 \qquad (9.14)$$

This equation is called the *characteristic equation* and is simply the result of replacing derivatives in (9.12) by powers of s. That is, x, the zeroth derivative, is replaced by s^0, the first derivative by s^1, and the second derivative by s^2.

Since (9.14) is a quadratic equation, we have not one solution, as in the first-order case, but two solutions, say s_1 and s_2, given by the quadratic formula as

$$s_{1,2} = \frac{-a_1 \pm \sqrt{a_1^2 - 4a_0}}{2} \qquad (9.15)$$

Therefore, we have two natural components of the form (9.13), which we denote by

$$x_{n1} = A_1 e^{s_1 t}$$
$$\qquad \qquad \qquad \qquad \qquad (9.16)$$
$$x_{n2} = A_2 e^{s_2 t}$$

The coefficients A_1 and A_2 are, of course, arbitrary. Either of the two solutions (9.16) will satisfy the homogeneous equation, because substituting either into (9.12) reduces it to (9.14).

As a matter of fact, because (9.12) is a linear equation, the *sum* of the solutions (9.16) is also a solution. That is,

$$x_n = x_{n1} + x_{n2} \qquad (9.17)$$

is a solution of (9.12). To see this, we have only to substitute the expression for x_n into (9.12). This results in

$$\frac{d^2}{dt^2}(x_{n1} + x_{n2}) + a_1 \frac{d}{dt}(x_{n1} + x_{n2}) + a_0(x_{n1} + x_{n2})$$

(equation continued)

$$= \left(\frac{d^2x_{n1}}{dt^2} + a_1\frac{dx_{n1}}{dt} + a_0x_{n1}\right) + \left(\frac{d^2x_{n2}}{dt^2} + a_1\frac{dx_{n2}}{dt} + a_0x_{n2}\right)$$

$$= 0 + 0 = 0$$

since both x_{n1} and x_{n2} satisfy (9.12).

By (9.16) and (9.17) we have

$$x_n = A_1e^{s_1t} + A_2e^{s_2t} \tag{9.18}$$

which is a more general solution (unless $s_1 = s_2$) than either equation of (9.16). In fact, (9.18) is called the *general solution* of the homogeneous equation if s_1 and s_2 are *distinct* (i.e., not equal) roots of the characteristic equation (9.14).

EXAMPLE 9.1 As an example, the homogeneous equation corresponding to (9.4) in the previous section is given by

$$\frac{d^2i_2}{dt^2} + 10\frac{di_2}{dt} + 16i_2 = 0 \tag{9.19}$$

and thus the characteristic equation is

$$s^2 + 10s + 16 = 0$$

The roots are $s = -2$ and $s = -8$, so that the general solution is given by

$$i_2 = A_1e^{-2t} + A_2e^{-8t} \tag{9.20}$$

The reader may verify by direct substitution that (9.20) satisfies (9.19), regardless of the value of the arbitrary constants.

Because (9.18) is the natural response, the numbers s_1 and s_2 are sometimes called the *natural frequencies* of the circuit. Evidently they play the same role as the negative reciprocal of the time constants considered in Chapter 8. There are, of course, two time constants in the second-order case as compared to one in the first-order case. For example, the natural frequencies of the circuit of Fig. 9.1 are $s = -2, -8$, as displayed in (9.20); the time constants of the two terms are then $\frac{1}{2}$ and $\frac{1}{8}$.

The unit of natural frequency, which is the inverse of that of the time constant, is the reciprocal of seconds. That is, it is a dimensionless quantity divided by seconds. Therefore st is dimensionless, as it must be in e^{st}.

EXERCISES

9.3.1 Given the *linear* differential equation

$$(t - 1)\frac{d^2x}{dt^2} + (t - 2)\frac{dx}{dt} = 0$$

show that $x_1 = te^{-t}$, $x_2 = 1$, and $x_1 + x_2$ are all solutions.

9.3.2 Given the *nonlinear* differential equation

$$x\frac{d^2x}{dt^2} - t\frac{dx}{dt} = 0$$

show that $x_1 = t^2$ and $x_2 = 1$ are both solutions but that $x_1 + x_2$ is *not* a solution.

9.3.3 Given

(a) $\dfrac{d^2x}{dt^2} + 7\dfrac{dx}{dt} + 10x = 0$

(b) $\dfrac{d^2x}{dt^2} + 4\dfrac{dx}{dt} + 4x = 0$

find the characteristic equation and the natural frequencies in each case.
Answer (a) $-2, -5$; (b) $-2, -2$

9.4

TYPES OF NATURAL FREQUENCIES

Since the natural frequencies of a second-order circuit are the roots of a quadratic characteristic equation, they may be real, imaginary, or complex numbers. The nature of the roots is determined by the discriminant $a_1^2 - 4a_0$ of (9.15), which may be positive (corresponding to real, distinct roots), negative (complex roots), or zero (real, equal roots).

EXAMPLE 9.2 For example, consider the circuit of Fig. 9.3, where the response to be found is the voltage v. The nodal equation at node a is

$$\frac{v - v_g}{4} + i + \frac{1}{4}\frac{dv}{dt} = 0$$

and the right mesh equation is

$$Ri + \frac{di}{dt} = v$$

Substituting for i from the first equation into the second, we have

FIGURE 9.3 Second-order circuit

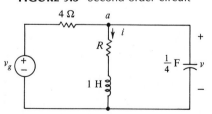

$$-R\left[\frac{1}{4}\left(\frac{dv}{dt} + v - v_g\right)\right] + \frac{d}{dt}\left[-\frac{1}{4}\left(\frac{dv}{dt} + v - v_g\right)\right] = v$$

Differentiating and simplifying the result, we have

$$\frac{d^2v}{dt^2} + (R + 1)\frac{dv}{dt} + (R + 4)v = Rv_g + \frac{dv_g}{dt}$$

The natural component v_n satisfies the homogeneous equation

$$\frac{d^2v_n}{dt^2} + (R + 1)\frac{dv_n}{dt} + (R + 4)v_n = 0$$

from which the characteristic equation is

$$s^2 + (R + 1)s + R + 4 = 0$$

Using the quadratic formula, we have the natural frequencies,

$$s_{1,2} = \frac{-(R + 1) \pm \sqrt{R^2 - 2R - 15}}{2} \tag{9.21}$$

If $R = 6\ \Omega$ in (9.21), the natural frequencies are real and distinct, given by

$$s_{1,2} = -2, -5 \tag{9.22}$$

If $R = 5\ \Omega$, the natural frequencies are real and equal, given by

$$s_{1,2} = -3, -3 \tag{9.23}$$

Finally, if $R = 1\ \Omega$, the natural frequencies are complex numbers, given by

$$s_{1,2} = -1 \pm j2 \tag{9.24}$$

where $j = \sqrt{-1}$. (In electrical engineering we cannot use i, as the mathematicians do, for the imaginary number unit, since this would result in confusion with the current. Complex numbers are considered in Appendix C for the reader who needs to review the subject.)

Distinct Real Roots: Overdamped Case

If the natural frequencies $s_{1,2}$ are real and distinct, then the natural response is given by (9.18). This case is called the *overdamped case,* because for a real circuit s_1 and s_2 are negative so that the response decays, or is *damped* out, with time. As an example, in the case of (9.22) we have

$$v_n = A_1e^{-2t} + A_2e^{-5t}$$

Complex Roots: Underdamped Case

If the natural frequencies are complex, then in general we have

$$s_{1,2} = \alpha \pm j\beta$$

where α and β are real numbers. By (9.18) the natural response in the general case is

$$x_n = A_1 e^{(\alpha+j\beta)t} + A_2 e^{(\alpha-j\beta)t} \qquad (9.25)$$

This appears to be a complex quantity and not a suitable answer for a real current or voltage. However, because A_1 and A_2 are complex numbers, it is mathematically correct, although somewhat inconvenient.

To put the natural response (9.25) in a better form, let us consider *Euler's formula*, given by

$$e^{j\theta} = \cos\theta + j\sin\theta \qquad (9.26)$$

and its alternative form, obtained by replacing θ by $-\theta$

$$e^{-j\theta} = \cos\theta - j\sin\theta \qquad (9.27)$$

These results are derived in Appendix D. They are named for the great Swiss mathematician Leonhard Euler (pronounced "oiler"), who lived from 1707 to 1783. Euler's greatness is attested to by the fact that the symbol e for the base of the natural logarithmic system was chosen in his honor.

Using (9.26) and (9.27), we may write (9.25) as

$$\begin{aligned} x_n &= e^{\alpha t}(A_1 e^{j\beta t} + A_2 e^{-j\beta t}) \\ &= e^{\alpha t}[A_1(\cos\beta t + j\sin\beta t) + A_2(\cos\beta t - j\sin\beta t)] \\ &= e^{\alpha t}[(A_1 + A_2)\cos\beta t + (jA_1 - jA_2)\sin\beta t] \end{aligned}$$

Since A_1 and A_2 are arbitrary, let us rename the constants as

$$A_1 + A_2 = B_1$$
$$jA_1 - jA_2 = B_2$$

so that

$$x_n = e^{\alpha t}(B_1 \cos\beta t + B_2 \sin\beta t) \qquad (9.28)$$

The case of complex roots is called the *underdamped* case. For a real circuit, α is negative so that the response (9.28) is damped out with time. Because of the sinusoidal terms, however, the damping is accompanied by oscillations, which distinguishes this case from the overdamped case.

EXAMPLE 9.3　As an example, in the case of (9.24) we have $\alpha = -1$ and $\beta = 2$, so that

$$v_n = e^{-t}(B_1 \cos 2t + B_2 \sin 2t)$$

where B_1 and B_2 are, of course, arbitrary.

Real Equal Roots: Critically Damped Case

The last type of natural frequencies we may have are those that are real and equal, say

$$s_1 = s_2 = k \qquad (9.29)$$

These characterize the *critically damped* case, which is the dividing line between the overdamped and underdamped cases. In the critically damped case, (9.18) is not the general solution since both x_{n1} and x_{n2} are of the form Ae^{kt}, and thus there is only one independent arbitrary constant. For (9.29) to be the natural frequencies, the characteristic equation must be

$$(s - k)^2 = s^2 - 2ks + k^2 = 0$$

and therefore the homogeneous equation must be

$$\frac{d^2x_n}{dt^2} - 2k\frac{dx_n}{dt} + k^2x_n = 0 \tag{9.30}$$

Since we know that Ae^{kt} is a solution for A arbitrary, let us *try*

$$x_n = h(t)e^{kt}$$

Substituting this expression into (9.30) and simplifying, we have

$$\frac{d^2h}{dt^2}e^{kt} = 0$$

Therefore, $h(t)$ must be such that its second derivative is zero for all t. This is true if $h(t)$ is a polynomial of degree 1, or

$$h(t) = A_1 + A_2t$$

where A_1 and A_2 are arbitrary constants. The general solution in the repeated-root case, $s_{1,2} = k$, is thus

$$x_n = (A_1 + A_2t)e^{kt} \tag{9.31}$$

which may be verified by direct substitution into the homogeneous equation (9.30).

EXAMPLE 9.4 As an example, in the case of (9.23) we have $s_{1,2} = -3, -3$, and thus

$$v_n = (A_1 + A_2t)e^{-3t}$$

EXERCISES

9.4.1 Find the natural frequencies of a circuit described by

$$\frac{d^2x}{dt^2} + a_1\frac{dx}{dt} + a_0x = 0$$

if (a) $a_1 = 5$, $a_0 = 6$; (b) $a_1 = 4$, $a_0 = 13$; and (c) $a_1 = 8$, $a_0 = 16$.
Answer (a) -2, -3; (b) $-2 \pm j3$; (c) -4, -4

9.4.2 Find x in Ex. 9.4.1 with the arbitrary constants determined so that $x(0) = 1$ and $dx(0)/dt = 4$.
Answer (a) $7e^{-2t} - 6e^{-3t}$; (b) $e^{-2t}(\cos 3t + 2 \sin 3t)$; (c) $(1 + 8t)e^{-4t}$

9.4.3 Find x if

$$\frac{d^2x}{dt^2} + 16x = 0$$

Answer $x = A_1 \cos 4t + A_2 \sin 4t$

9.5

THE FORCED RESPONSE

The forced response x_f of the general second-order circuit must satisfy (9.10) and contain no arbitrary constants. There are a number of methods for finding x_f, but for our purposes we shall use the procedure of guessing the solution, which has worked so well for us in the past. We know from our experience with first-order circuits that the forced response has the form of the driving function. A constant source results in a constant forced response, etc. However, the response must satisfy (9.10) identically, which means that first and second derivatives of x_f, as well as x_f itself, will appear in the left member of (9.10). Thus we are led to *try* as x_f a combination of the right member of (9.10) and its derivatives.

EXAMPLE 9.5

As an example, let us consider the case $v_g = 16$ V in Fig. 9.1. Then by (9.4), for $i_2 = x$, we have

$$\frac{d^2x}{dt^2} + 10\frac{dx}{dt} + 16x = 32 \tag{9.32}$$

The natural response was given earlier in (9.20) by

$$x_n = A_1 e^{-2t} + A_2 e^{-8t} \tag{9.33}$$

Since the right member of (9.32) is a constant and all its derivatives are constant (namely zero), let us try

$$x_f = A$$

where A is a constant to be determined. We note that A is not arbitrary but is a particular value that hopefully makes x_f a solution of (9.32). Substituting x_f into (9.32) yields

$$16A = 32$$

or

$$x_f = A = 2$$

Therefore, the general solution of (9.32) is

$$x(t) = A_1 e^{-2t} + A_2 e^{-8t} + 2$$

A knowledge of the initial energy stored in the inductors can now be used to evaluate A_1 and A_2.

In the case of constant forcing functions we may often obtain x_f from the circuit itself. In the example just considered, x_f is the steady-state value of i_2 in Fig. 9.1 when $v_g = 16$ V. At steady state the inductors look like short circuits, as shown in Fig. 9.4, so that, from the figure, we have

$$x_f = i_2 = 2 \text{ A}$$

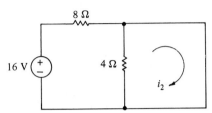

FIGURE 9.4 Circuit of Fig. 9.1 in the steady state

EXAMPLE 9.6   As another example, suppose in Fig. 9.1 we have

$$v_g = 20 \cos 4t \text{ V}$$

Then by (9.4), again for $i_2 = x$, we have

$$\frac{d^2x}{dt^2} + 10\frac{dx}{dt} + 16x = 40 \cos 4t \qquad (9.34)$$

The natural response x_n is given by (9.33), as before. To find the forced response x_f we need to seek a solution which contains all the terms, and their possible derivatives, in the right member of (9.34). The coefficients of these terms will then be determined by requiring x_f to satisfy the differential equation. In the case under consideration, the only term is a $\cos 4t$ term and the trial,

$$x_f = A \cos 4t + B \sin 4t \qquad (9.35)$$

contains this term and all its possible derivatives (which are $\cos 4t$ and $\sin 4t$).

From (9.35) we have

$$\frac{dx_f}{dt} = -4A \sin 4t + 4B \cos 4t$$

$$\frac{d^2x_f}{dt^2} = -16A \cos 4t - 16B \sin 4t$$

Substituting these values and (9.35) into (9.34) and collecting terms, we have

$$40B \cos 4t - 40A \sin 4t = 40 \cos 4t$$

Since this must be an identity, the coefficients of like terms must be the same on both sides of the equation. In the case of the $\cos 4t$ terms we have

$$40B = 40$$

and for the $\sin 4t$ terms we have

$$-40A = 0$$

Thus $A = 0$ and $B = 1$, so that

$$x_f = \sin 4t \qquad (9.36)$$

The general solution of (9.34), from (9.33) and (9.36), is given by

$$i_2 = x = A_1 e^{-2t} + A_2 e^{-8t} + \sin 4t \qquad (9.37)$$

This may be readily verified by direct substitution.

Some of the more common forcing functions $f(t)$ which occur in (9.7) are listed in the first column of Table 9.1. The general form of the corresponding forced response is given in the second column, which may be useful for formulating the trial solution x_f.

TABLE 9.1 Trial Forced Responses

$f(t)$	x_f
k	A
t	$At + B$
t^2	$At^2 + Bt + C$
e^{at}	Ae^{at}
$\sin bt,\ \cos bt$	$A \sin bt + B \cos bt$
$e^{at} \sin bt,\ e^{at} \cos bt$	$e^{at}(A \sin bt + B \cos bt)$

EXERCISES

9.5.1 Find the forced response if

$$\frac{d^2x}{dt^2} + 4\frac{dx}{dt} + 3x = f(t)$$

where $f(t)$ is given by (a) 6, (b) $8e^{-2t}$, and (c) $6t + 14$.
Answer (a) 2; (b) $-8e^{-2t}$; (c) $2t + 2$

9.5.2 If $x(0) = 4$ and $dx(0)/dt = -2$, find the complete solution in Ex. 9.5.1.
Answer (a) $2e^{-t} + 2$; (b) $9e^{-t} - 8e^{-2t} + 3e^{-3t}$; (c) $e^{-t} + e^{-3t} + 2t + 2$

9.6

EXCITATION AT A NATURAL FREQUENCY

Suppose the circuit equation to be solved is given by

$$\frac{d^2x}{dt^2} - (a + b)\frac{dx}{dt} + abx = f(t) \qquad (9.38)$$

where a and $b \neq a$ are known constants. In this case the characteristic equation is

$$s^2 - (a + b)s + ab = 0$$

from which the natural frequencies are

$$s_1 = a, \qquad s_2 = b$$

Therefore, we have the natural response

$$x_n = A_1 e^{at} + A_2 e^{bt} \tag{9.39}$$

where A_1 and A_2 are arbitrary.

Let us suppose now that the excitation function contains a natural frequency, say

$$f(t) = e^{at} \tag{9.40}$$

The usual procedure is to seek a forced response,

$$x_f = A e^{at} \tag{9.41}$$

and determine A so that x_f satisfies (9.38), which in this case is

$$\frac{d^2x}{dt^2} - (a + b)\frac{dx}{dt} + abx = e^{at} \tag{9.42}$$

However, substituting x_f into (9.42) yields

$$0 = e^{at}$$

which is an impossible situation.

This difficulty could have been foreseen by observing that x_f in (9.41) has the form of one of the components of x_n in (9.39), and thus x_f will satisfy the homogeneous equation corresponding to (9.38). That is, x_f substituted into (9.38) makes its left member identically zero. There is no point then in trying such a forced response as (9.41).

Let us consider what happens if we multiply by t the part of x_f that is duplicated in x_n. That is, let us try

$$x_f = A t e^{at} \tag{9.43}$$

instead of (9.41). We then have

$$\frac{dx_f}{dt} = A(at + 1)e^{at}$$

$$\frac{d^2x_f}{dt^2} = A(a^2t + 2a)e^{at}$$

Substituting these values, along with (9.43), into (9.42), we have

$$A e^{at}[a^2t + 2a - (a + b)(at + 1) + abt] = e^{at}$$

which, upon simplification, becomes

$$A(a - b)e^{at} = e^{at}$$

Since this must be an identity for all t, we must have

$$A = \frac{1}{a - b}$$

The general solution of (9.38), using (9.39) and (9.43), is then

$$x = A_1 e^{at} + A_2 e^{bt} + \frac{te^{at}}{a - b}$$

EXAMPLE 9.7 As an example, suppose the excitation in Fig. 9.1 is given by

$$v_g = 6e^{-2t} + 32$$

Then if $i_2 = x$, we have, by (9.4),

$$\frac{d^2x}{dt^2} + 10\frac{dx}{dt} + 16x = 12e^{-2t} + 64 \qquad (9.44)$$

The natural response, as before, is

$$x_n = A_1 e^{-2t} + A_2 e^{-8t}$$

Noting that the right member of the differential equation has the term e^{-2t} in common with x_n, we try

$$x_f = Ate^{-2t} + B$$

The factor t has been inserted into the natural trial solution of x_f to remove the duplication of the term e^{-2t}. Substituting x_f into (9.44) and simplifying we have

$$6Ae^{-2t} + 16B = 12e^{-2t} + 64$$

Therefore, we have $A = 2$ and $B = 4$, so that

$$x_f = 2te^{-2t} + 4$$

The general solution is now

$$i_2 = x = x_n + x_f$$

As a final example, let us consider the case of (9.38) where $b = a$ and $f(t)$ is given by (9.40). That is,

$$\frac{d^2x}{dt^2} - 2a\frac{dx}{dt} + a^2x = e^{at} \qquad (9.45)$$

The characteristic equation is

$$s^2 - 2as + a^2 = 0$$

and thus the natural frequencies are

$$s_1 = s_2 = a$$

The natural response is then

$$x_n = (A_1 + A_2t)e^{at}$$

We know it is fruitless to try as the forced response x_f given in (9.41) because it is duplicated in the natural response. In this case, (9.43) will not work either because it, too, is duplicated. The lowest power of t that is not duplicated is 2; thus we are led to try

$$x_f = At^2 e^{at}$$

Substituting this expression into (9.45) we have

$$2Ae^{at} = e^{at}$$

so that $A = \frac{1}{2}$. The forced and complete responses follow as before.

EXERCISES

9.6.1 Find the forced response if

$$\frac{d^2x}{dt^2} + 4\frac{dx}{dt} + 3x = f(t)$$

where $f(t)$ is given by (a) $2e^{-3t} + 6e^{-4t}$ and (b) $4e^{-t} + 2e^{-3t}$.
Answer (a) $2e^{-4t} - te^{-3t}$; (b) $t(2e^{-t} - e^{-3t})$

9.6.2 Find the forced response if

$$\frac{d^2x}{dt^2} + 4\frac{dx}{dt} + 4x = f(t)$$

where $f(t)$ is given by (a) $6e^{-2t}$ and (b) $6te^{-2t}$. [*Suggestion:* In (b), try $x_f = At^3 e^{-2t}$.]
Answer (a) $3t^2 e^{-2t}$; (b) $t^3 e^{-2t}$

9.6.3 Find the complete response if

$$\frac{d^2x}{dt^2} + 4x = 8 \sin 2t$$

and $x(0) = dx(0)/dt = 0$. [*Suggestion:* Try $x_f = t(A \cos 2t + B \sin 2t)$.]
Answer $\sin 2t - 2t \cos 2t$

9.7

THE COMPLETE RESPONSE

In the previous sections we have noted that the complete response of a circuit is the sum of a natural and a forced response and that the natural response, and thus the complete response, contains arbitrary constants. These constants, as in the first-order cases of Chapter 8, are determined so that the complete response satisfies specified initial energy-storage conditions.

EXAMPLE 9.8 To illustrate the procedure, let us find $x(t)$, for $t > 0$, which satisfies the system of equations

$$\frac{dx}{dt} + 2x + 5 \int_0^t x \, dt = 16e^{-3t}$$

$$x(0) = 2 \tag{9.46}$$

To begin, let us differentiate the first of these equations to eliminate the integral; this results in

$$\frac{d^2x}{dt^2} + 2\frac{dx}{dt} + 5x = -48e^{-3t}$$

The characteristic equation is

$$s^2 + 2s + 5 = 0$$

with roots

$$s_{1,2} = -1 \pm j2$$

Therefore, the natural response is

$$x_n = e^{-t}(A_1 \cos 2t + A_2 \sin 2t)$$

Trying as the forced response

$$x_f = Ae^{-3t}$$

we see that

$$8Ae^{-3t} = -48e^{-3t}$$

so that $A = -6$. The complete response is therefore

$$x(t) = e^{-t}(A_1 \cos 2t + A_2 \sin 2t) - 6e^{-3t} \tag{9.47}$$

To determine the arbitrary constants we need two initial conditions. One, $x(0) = 2$, is given in (9.46). To obtain the other we may evaluate the first equation of (9.46) at $t = 0$, resulting in

$$\frac{dx(0)}{dt} + 2x(0) + 5\int_0^0 x\, dt = 16$$

Noting the value of $x(0)$ and that the integral term is zero, we have

$$\frac{dx(0)}{dt} = 12 \tag{9.48}$$

Applying the second equation of (9.46) to (9.47), we have

$$x(0) = A_1 - 6 = 2$$

or $A_1 = 8$. To apply (9.48) we may differentiate (9.47), obtaining

$$\frac{dx}{dt} = e^{-t}(-2A_1 \sin 2t + 2A_2 \cos 2t) - e^{-t}(A_1 \cos 2t + A_2 \sin 2t) + 18e^{-3t}$$

from which

$$\frac{dx(0)}{dt} = 2A_2 - A_1 + 18 = 12 \tag{9.49}$$

From this, knowing A_1, we find $A_2 = 1$.

At this point let us digress for a moment to note a very easy way to get (9.49).

We may differentiate $x(t)$ and immediately replace t by 0 before we write down the result. That is, in (9.47), the derivative of x at $t = 0$ is e^{-t} at $t = 0$ (which is 1) times the derivative of $(A_1 \cos 2t + A_2 \sin 2t)$ at $t = 0$ (which is $2A_2$) plus $(A_1 \cos 2t + A_2 \sin 2t)$ at $t = 0$ (which is A_1) times the derivative of e^{-t} at $t = 0$ (which is -1) plus the derivative of $-6e^{-3t}$ at $t = 0$ (which is 18). These steps are written down in (9.49) and can be done mentally, avoiding the intermediate prior step.

Returning to our problem, we now have the arbitrary constants, so that by (9.47), the final answer is

$$x = e^{-t}(8 \cos 2t + \sin 2t) - 6e^{-3t}$$

EXAMPLE 9.9   As a last example, let us find v, $t > 0$, in the circuit of Fig. 9.5 if $v_1(0) = v(0) = 0$ and $v_g = 5 \cos 2000t$ V. The nodal equation at node v_1 is

$$2 \times 10^{-3}v_1 - 10^{-3}v_g - \frac{1}{2} \times 10^{-3}v + 10^{-6}\frac{dv_1}{dt} = 0$$

or

$$4v_1 - v + 2 \times 10^{-3}\frac{dv_1}{dt} = 2v_g = 10 \cos 2000t \qquad (9.50)$$

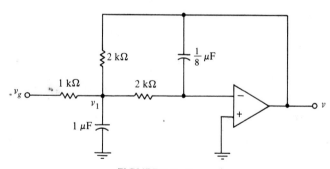

FIGURE 9.5 Example

and the nodal equation at the inverting input of the op amp is

$$\frac{1}{2} \times 10^{-3}v_1 + \frac{1}{8} \times 10^{-6}\frac{dv}{dt} = 0$$

or

$$v_1 = -\frac{1}{4} \times 10^{-3}\frac{dv}{dt} \qquad (9.51)$$

Substituting (9.51) into (9.50) and simplifying, we have

$$\frac{d^2v}{dt^2} + 2 \times 10^3\frac{dv}{dt} + 2 \times 10^6v = -2 \times 10^7 \cos 2000t$$

The characteristic equation is

$$s^2 + 2 \times 10^3 s + 2 \times 10^6 = 0$$

so that the natural frequencies are $s_{1,2} = 1000(-1 \pm j1)$. The natural response is therefore

$$v_n = e^{-1000t}(A_1 \cos 1000t + A_2 \sin 1000t)$$

For the forced response we shall try

$$v_f = A \cos 2000t + B \sin 2000t$$

which substituted into the differential equation yields

$$(-2A + 4B) \cos 2000t + (-4A - 2B) \sin 2000t = -20 \cos 2000t$$

Therefore, equating coefficients of like terms, we have

$$-2A + 4B = -20$$

$$-4A - 2B = 0$$

from which $A = 2$ and $B = -4$. The complete response is then

$$v = e^{-1000t}(A_1 \cos 1000t + A_2 \sin 1000t) + 2 \cos 2000t - 4 \sin 2000t \quad (9.52)$$

From (9.51), for $t = 0^+$, we see that

$$v_1(0^+) = -\frac{1}{4} \times 10^{-3} \frac{dv(0^+)}{dt}$$

and since $v_1(0^+) = v_1(0^-) = 0$ we have

$$\frac{dv(0^+)}{dt} = 0 \quad (9.53)$$

Like v_1, v is also a capacitor voltage (across the $\frac{1}{8}$-μF capacitor), so that

$$v(0^+) = v(0^-) = 0 \quad (9.54)$$

From (9.52) and (9.54) we have

$$A_1 + 2 = 0$$

or $A_1 = -2$, and from (9.52) and (9.53) we have

$$1000A_2 - 1000A_1 - 8000 = 0$$

from which $A_2 = 6$. Thus the complete response is

$$v = e^{-1000t}(-2 \cos 1000t + 6 \sin 1000t) + 2 \cos 2000t - 4 \sin 2000t \text{ V}$$

EXERCISES

9.7.1 Find x, $t > 0$, where

$$\frac{dx}{dt} + 2x + \int_0^t x \, dt = f(t)$$

$$x(0) = -1$$

and (a) $f(t) = 1$ and (b) $f(t) = t^2$.
Answer (a) $(-1 + 2t)e^{-t}$; (b) $3(1 + t)e^{-t} + 2t - 4$

9.7.2 Find i, v, di/dt, and dv/dt at $t = 0^+$.
Answer 0, 0, 2 A/s, 40 V/s

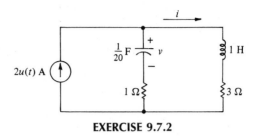

EXERCISE 9.7.2

9.7.3 For $t > 0$ in Ex. 9.7.2, find (a) v and (b) i. (*Suggestion:* Because of Kirchhoff's laws and the terminal relationships of the elements, i has the same natural frequencies as v. Thus v_n is easily obtained after i_n is found; its forced response is evident by inspection of the circuit.)
Answer (a) $e^{-2t}(-6 \cos 4t + 7 \sin 4t) + 6$;
   (b) $e^{-2t}(-2 \cos 4t - \frac{1}{2} \sin 4t) + 2$

9.8

THE PARALLEL *RLC* CIRCUIT

One of the most important second-order circuits is the parallel *RLC* circuit of Fig. 9.6(a). We shall assume that at $t = 0$ there is an initial inductor current.

$$i(0) = I_0 \tag{9.55}$$

and an initial capacitor voltage,

FIGURE 9.6 (a) Parallel *RLC* circuit, with (b) the source killed

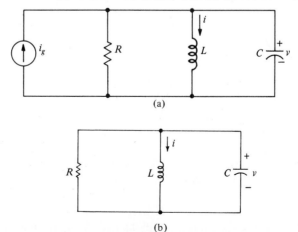

$$v(0) = V_0 \tag{9.56}$$

and analyze the circuit by finding v for $t > 0$.

The single nodal equation that is necessary is given by

$$\frac{v}{R} + \frac{1}{L}\int_0^t v\, dt + I_0 + C\frac{dv}{dt} = i_g \tag{9.57}$$

which is an integrodifferential equation that becomes, upon differentiation,

$$C\frac{d^2v}{dt^2} + \frac{1}{R}\frac{dv}{dt} + \frac{1}{L}v = \frac{di_g}{dt}$$

To find the natural response we make the right member zero, resulting in

$$C\frac{d^2v}{dt^2} + \frac{1}{R}\frac{dv}{dt} + \frac{1}{L}v = 0 \tag{9.58}$$

This result follows also from killing the current source, as in Fig. 9.6(b), and writing the nodal equation. From (9.58) the characteristic equation is

$$Cs^2 + \frac{1}{R}s + \frac{1}{L} = 0$$

from which the natural frequencies are

$$s_{1,2} = \frac{-\dfrac{1}{R} \pm \sqrt{\dfrac{1}{R^2} - \dfrac{4C}{L}}}{2C} = -\frac{1}{2RC} \pm \sqrt{\left(\frac{1}{2RC}\right)^2 - \frac{1}{LC}} \tag{9.59}$$

As in the general second-order case already discussed, there are three types of responses, depending on the nature of the discriminant, $1/R^2 - 4C/L$, in (9.59). We shall now look briefly at these three cases. For simplicity we will take $i_g = 0$ and consider the source-free case of Fig. 9.6(b). The forced response is then zero and the natural response is the complete response.

Overdamped Case

If the discriminant is positive, i.e.,

$$\frac{1}{R^2} - \frac{4C}{L} > 0$$

or, equivalently,

$$L > 4R^2C \tag{9.60}$$

then the natural frequencies of (9.59) are real and distinct negative numbers, and we have the overdamped case,

$$v = A_1 e^{s_1 t} + A_2 e^{s_2 t} \tag{9.61}$$

From the initial conditions and (9.57) evaluated at $t = 0^+$, we obtain

$$\frac{dv(0^+)}{dt} = -\frac{V_0 + RI_0}{RC} \qquad (9.62)$$

which together with (9.56) can be used to determine the arbitrary constants.

EXAMPLE 9.10   As an example, suppose $R = 1\,\Omega$, $L = \frac{4}{3}$ H, $C = \frac{1}{4}$ F, $V_0 = 2$ V, and $I_0 = -3$ A. Then by (9.59) we have $s_{1,2} = -1, -3$, and hence

$$v = A_1 e^{-t} + A_2 e^{-3t}$$

Also, by (9.56) and (9.62) we have

$$v(0) = 2 \text{ V}$$

$$\frac{dv(0^+)}{dt} = 4 \text{ V/s}$$

which may be used to obtain $A_1 = 5$ and $A_2 = -3$, and thus

$$v = 5e^{-t} - 3e^{-3t}$$

This overdamped case is easily sketched, as shown by the solid line of Fig. 9.7, by sketching the two components and adding them graphically.

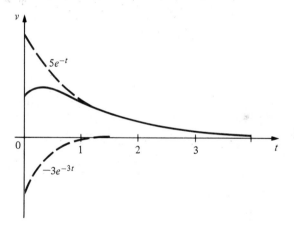

FIGURE 9.7   Sketch of an overdamped response

The reason for the term *overdamped* may be seen from the absence of oscillations (fluctuations in sign). The element values are such as to "damp out" any oscillatory tendencies. It is, of course, possible for the response to change signs *once*, depending on the initial conditions.

Underdamped Case

If the discriminant in (9.59) is negative, i.e.,

$$L < 4R^2 C \qquad (9.63)$$

then we have the underdamped case, where the natural frequencies are complex, and the response contains sines and cosines, which of course are oscillatory-type functions. In this case it is convenient to define a *resonant frequency*,

$$\omega_0 = \frac{1}{\sqrt{LC}} \qquad (9.64)$$

a *damping coefficient*,

$$\alpha = \frac{1}{2RC} \qquad (9.65)$$

and a *damped frequency*,

$$\omega_d = \sqrt{\omega_0^2 - \alpha^2} \qquad (9.66)$$

Each of these is a dimensionless quantity "per second." The resonant and damped frequencies are defined to be radians per second (rad/s) and the damping coefficient is nepers per second (Np/s).

Using these definitions, the natural frequencies, by (9.59), are

$$s_{1,2} = -\alpha \pm j\omega_d$$

and therefore the response is

$$v = e^{-\alpha t}(A_1 \cos \omega_d t + A_2 \sin \omega_d t) \qquad (9.67)$$

which is oscillatory in nature, as expected.

EXAMPLE 9.11 As an example, suppose $R = 5\ \Omega$, $L = 1$ H, $C = \frac{1}{10}$ F, $V_0 = 0$, and $I_0 = -\frac{3}{2}$ A. Then we have

$$v = e^{-t}(A_1 \cos 3t + A_2 \sin 3t)$$

From the initial conditions we have $v(0) = 0$ and $dv(0^+)/dt = 15$ V/s, from which $A_1 = 0$ and $A_2 = 5$. Therefore, the underdamped response is

$$v = 5e^{-t} \sin 3t$$

This response is readily sketched if it is observed that since $\sin 3t$ varies between $+1$ and -1, v must be a sinusoid that varies between $5e^{-t}$ and $-5e^{-t}$. The response is shown in Fig. 9.8, where it may be seen that it is oscillatory in nature. The response goes through zero at the points where the sinusoid is zero, which is determined, in general, by the damped frequency ω_d.

Critically Damped Case

When the discriminant of (9.59) is zero, we have the critically damped case, for which

$$L = 4R^2C \qquad (9.68)$$

In this case the natural frequencies are real and equal, given by

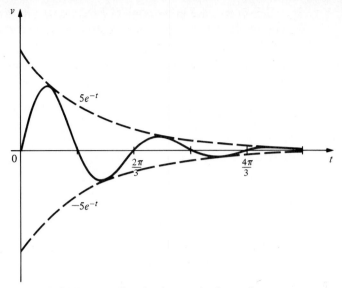

FIGURE 9.8 Sketch of an underdamped response

$$s_{1,2} = -\alpha, -\alpha$$

where α is given by (9.65). The response is then

$$v = (A_1 + A_2 t)e^{-\alpha t} \qquad (9.69)$$

EXAMPLE 9.12 As an example, consider the case $R = 1\ \Omega$, $L = 1$ H, $C = \frac{1}{4}$ F, $V_0 = 0$, and $I_0 = -1$ A. In this case we have $\alpha = 2$, $A_1 = 0$, and $A_2 = 4$. Thus the response is

$$v = 4te^{-2t}$$

This is easily sketched by plotting $4t$ and e^{-2t} and multiplying the two together. The result is shown in Fig. 9.9.

FIGURE 9.9 Sketch of a critically damped response

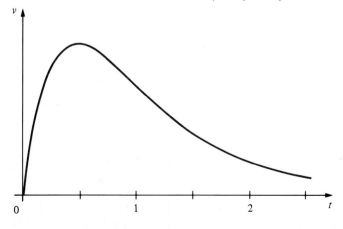

For every case in the parallel *RLC* circuit, the steady-state value of the natural response is zero, because each term in the response contains a factor e^{at}, where $a < 0$.

EXERCISES

9.8.1 In a source-free parallel *RLC* circuit, $R = 1$ kΩ and $C = 1$ μF. Find L so that the circuit is (a) overdamped with $s_{1,2} = -250, -750$ s⁻¹, (b) underdamped with $\omega_d = 250$ rad/s, and (c) critically damped.
Answer (a) $\frac{16}{3}$ H; (b) $\frac{16}{5}$ H; (c) 4 H

9.8.2 Find the differential equation satisfied by i in Fig. 9.6(b). Use this result to find i, for $t > 0$, if $R = 5$ Ω, $L = 1$ H, $C = 0.1$ F, $v(0) = 0$, and $i(0) = 3$ A.

Answer $\dfrac{d^2 i}{dt^2} + \dfrac{1}{RC}\dfrac{di}{dt} + \dfrac{1}{LC} i = 0$; $e^{-t}(3 \cos 3t + \sin 3t)$

9.8.3 The larger the value of R, the less damping there is in the underdamped case of the parallel *RLC* circuit (because $\alpha = 1/2RC$). Let $R = \infty$ (open circuit) and show that, in the source-free case,

$$\frac{d^2 v}{dt^2} + \omega_0^2 v = 0$$

For this case, find the general solution for v.
Answer $A_1 \cos \omega_0 t + A_2 \sin \omega_0 t$

9.9

THE SERIES *RLC* CIRCUIT

The series *RLC* circuit, shown in Fig. 9.10, is the dual of the parallel circuit, considered in the previous section. Therefore all the results for the parallel circuit have dual counterparts for the series circuit, which may be written down by inspection. In this section we shall simply list these results using duality and leave to the reader their verification by conventional means.

Referring to Fig. 9.10, the initial conditions will be taken as

$$v(0) = V_0$$
$$i(0) = I_0$$

FIGURE 9.10 Series *RLC* circuit

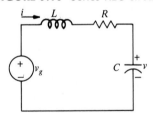

The single-loop equation necessary in the analysis is

$$L\frac{di}{dt} + Ri + \frac{1}{C}\int_0^t i\,dt + V_0 = v_g \qquad (9.70)$$

which is valid for $t > 0$. The resulting characteristic equation is

$$Ls^2 + Rs + \frac{1}{C} = 0 \qquad (9.71)$$

and the natural frequencies are

$$s_{1,2} = -\frac{R}{2L} \pm \sqrt{\left(\frac{R}{2L}\right)^2 - \frac{1}{LC}} \qquad (9.72)$$

The series RLC circuit is overdamped if

$$C > \frac{4L}{R^2} \qquad (9.73)$$

and the response is

$$i = A_1 e^{s_1 t} + A_2 e^{s_2 t} \qquad (9.74)$$

The circuit is critically damped if

$$C = \frac{4L}{R^2} \qquad (9.75)$$

in which case $s_1 = s_2 = -R/2L$, and the response is

$$i = (A_1 + A_2 t)e^{s_1 t} \qquad (9.76)$$

Finally, the circuit is underdamped if

$$C < \frac{4L}{R^2} \qquad (9.77)$$

in which case the resonant frequency is

$$\omega_0 = \frac{1}{\sqrt{LC}} \qquad (9.78)$$

the damping coefficient is

$$\alpha = \frac{R}{2L} \qquad (9.79)$$

and the damped frequency is

$$\omega_d = \sqrt{\omega_0^2 - \alpha^2} \qquad (9.80)$$

The underdamped response is

$$i = e^{-\alpha t}(A_1 \cos \omega_d t + A_2 \sin \omega_d t) \qquad (9.81)$$

EXAMPLE 9.13 As an example, suppose it is required to find v, for $t > 0$, in Fig. 9.11, given that

$$v(0) = 6 \text{ V}, \qquad i(0) = 2 \text{ A}$$

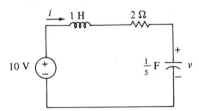

FIGURE 9.11 Driven series *RLC* circuit

We know that

$$v = v_n + v_f$$

where the natural response v_n contains the natural frequencies. The natural frequencies of the current i are the same as those of v because obtaining one from the other, in general, requires only Kirchhoff's laws and the operations of addition, subtraction, multiplication by constants, integration, and differentiation. None of these operations changes the natural frequencies. Therefore, since the natural frequencies of i are easier to get (only one loop equation is required), let us obtain them. Around the loop we have

$$\frac{di}{dt} + 2i + 5 \int_0^t i \, dt + 6 = 10 \tag{9.82}$$

The characteristic equation, following differentiation, is

$$s^2 + 2s + 5 = 0$$

with roots

$$s_{1,2} = -1 \pm j2$$

Thus we have

$$v_n = e^{-t}(A_1 \cos 2t + A_2 \sin 2t)$$

[We could have obtained $s_{1,2}$ from (9.72) directly, but it is probably easier to write the characteristic equation, since it can be done by inspection.]

The forced response v_f is a constant in this case and may be obtained by inspection of the circuit in the steady state. Since in the steady state the capacitor is an open circuit and the inductor is a short circuit, $i_f = 0$ and $v_f = 10$. Therefore, the complete response is

$$v = e^{-t}(A_1 \cos 2t + A_2 \sin 2t) + 10$$

From the initial voltage, we have

$$v(0) = 6 = A_1 + 10$$

or $A_1 = -4$. Also, we have

$$\frac{1}{5}\frac{dv(0^+)}{dt} = i(0) = 2$$

$$\frac{dv(0^+)}{dt} = 10 = 2A_2 - A_1$$

Therefore, $A_2 = 3$, and we have

$$v = e^{-t}(-4 \cos 2t + 3 \sin 2t) + 10 \text{ V}$$

EXAMPLE 9.14   As a final example, let us find v in Fig. 9.11 if the source is

$$v_g = 4 \cos t \text{ V}$$

In this case we shall need the differential equation for v, which we may get from (9.82) and

$$i = \frac{1}{5}\frac{dv}{dt} \qquad (9.83)$$

We may also obtain it directly from the figure since the voltages across the inductor and resistor are

$$\frac{di}{dt} = \frac{1}{5}\frac{d^2v}{dt^2}$$

and

$$2i = \frac{2}{5}\frac{dv}{dt}$$

and that across the capacitor is v, of course. In either case, the result by KVL is

$$\frac{d^2v}{dt^2} + 2\frac{dv}{dt} + 5v = 20 \cos t$$

The natural response is the same as in the previous example. To get the forced response we shall try

$$v_f = A \cos t + B \sin t$$

which substituted into the differential equation results in

$$(4A + 2B) \cos t + (4B - 2A) \sin t = 20 \cos t$$

Equating like coefficients and solving for A and B, we have

$$A = 4, \qquad B = 2$$

Therefore, the general solution is

$$v = e^{-t}(A_1 \cos 2t + A_2 \sin 2t) + 4 \cos t + 2 \sin t$$

From the initial voltage, we have

$$v(0) = 6 = A_1 + 4$$

or $A_1 = 2$. From the initial current and (9.83), we have

$$\frac{dv(0^+)}{dt} = 10 = 2A_2 - A_1 + 2$$

or $A_2 = 5$.

The complete response is therefore

$$v = e^{-t}(2 \cos 2t + 5 \sin 2t) + 4 \cos t + 2 \sin t$$

EXERCISES

9.9.1 Let $R = 4\ \Omega$, $L = 1$ H, $v_g = 0$, $v(0) = 4$ V, and $i(0) = 2$ A in Fig. 9.10. Find i for $t > 0$ if C is (a) $\frac{1}{20}$ F, (b) $\frac{1}{4}$ F, and (c) $\frac{1}{3}$ F.

Answer (a) $2e^{-2t}(\cos 4t - \sin 4t)$ A; (b) $2(1 - 4t)e^{-2t}$ A; (c) $-3e^{-t} + 5e^{-3t}$ A

9.9.2 Find v for $t > 0$ if $R = 40\ \Omega$, $L = 10$ mH, and $C = 5\ \mu$F.

Answer $e^{-2000t}(4 \cos 4000t - 3 \sin 4000t)$ V

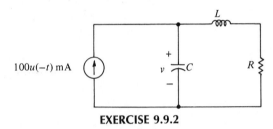

EXERCISE 9.9.2

9.9.3 Find v for $t > 0$ if the circuit is in steady state at $t = 0$.

Answer $12 - 12(1 + t)e^{-2t}$ V

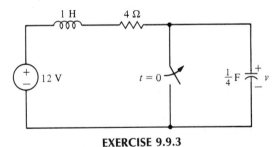

EXERCISE 9.9.3

9.9.4 Find v for $t > 0$ if (a) $C = \frac{1}{5}$ F and (b) $C = \frac{1}{10}$ F.

Answer (a) $-25e^{-t} + e^{-5t} + 24$ V; (b) $e^{-3t}(-24 \cos t - 32 \sin t) + 24$ V

EXERCISE 9.9.4

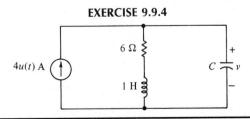

9.10

ALTERNATIVE METHODS FOR OBTAINING THE DESCRIBING EQUATIONS

We shall close this chapter by considering two methods of expediting the process of obtaining the describing equation of the circuit. In the case of the parallel and series *RLC* circuits, a single equation is required, which after differentiation, is the describing equation. However, in many second-order circuits there are two simultaneous circuit equations from which the describing equation is obtained after a tedious elimination process.

EXAMPLE 9.15   As an example, let us consider the circuit of Fig. 9.12 for $t > 0$. Taking node b as reference and writing nodal equations at nodes a and v_1 we have

$$\frac{v - v_g}{4} + \frac{v - v_1}{6} + \frac{1}{4}\frac{dv}{dt} = 0$$

$$\frac{v_1 - v}{6} + \int_0^t v_1 \, dt + i(0) = 0$$

(9.84)

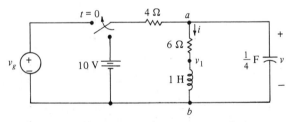

FIGURE 9.12 Circuit with two storage elements

If we are interested in finding v we must eliminate v_1 and obtain the describing equation in terms of v. The result, as the reader may verify, is

$$\frac{d^2v}{dt^2} + 7\frac{dv}{dt} + 10v = \frac{dv_g}{dt} + 6v_g$$

(9.85)

In this case the process is not overly complicated but it can be shortened by the methods we shall consider in this chapter.

The first method we shall discuss is a systematic way of obtaining a describing equation, such as (9.85), from the circuit equations, such as (9.84). To develop the method, let us first introduce the differentiation *operator D*, which is defined by

$$D = \frac{d}{dt}$$

That is, $Dx = dx/dt$, $D(Dx) = D^2x = d^2x/dt^2$, etc. Also, we have, for example,

$$a\frac{dx}{dt} + bx = aDx + bx = (aD + b)x$$

It is important here to note that x is factored out of the middle member and placed *after* the operator expression, indicating that the operation is to be performed on x. Otherwise the meaning is changed radically.

With these ideas in mind, let us rewrite (9.84) in operator form. This result, after first differentiating the second equation, is

$$\left(\frac{1}{4}D + \frac{5}{12}\right)v - \frac{1}{6}v_1 = \frac{1}{4}v_g$$

$$-\frac{1}{6}Dv + \left(\frac{1}{6}D + 1\right)v_1 = 0$$

which when cleared of fractions becomes

$$(3D + 5)v - 2v_1 = 3v_g$$
$$-Dv + (D + 6)v_1 = 0$$

(9.86)

To eliminate v_1, let us "operate" on the first equation with $D + 6$, by which we mean multiply it through by $D + 6$ on the left of each term. Then let us multiply the second equation by 2, resulting in

$$(D + 6)(3D + 5)v - 2(D + 6)v_1 = 3(D + 6)v_g$$
$$-2Dv + 2(D + 6)v_1 = 0$$

[We note that constants such as 2 commute with operators, i.e., $(D + 6)2x = 2(D + 6)x$, but variables do not.] Adding these last two equations eliminates v_1 and results in

$$[(D + 6)(3D + 5) - 2D]v = 3(D + 6)v_g \qquad (9.87)$$

Multiplying the operators as if they were polynomials, collecting terms, and dividing out the common factor 3, we have

$$(D^2 + 7D + 10)v = (D + 6)v_g$$

which is the same as (9.85):

The procedure may be carried out in a more direct manner by using determinants. For example, we may use Cramer's rule to obtain the expression for v from (9.86), given by

$$v = \frac{\Delta_1}{\Delta} \qquad (9.88)$$

where Δ is the coefficient determinant

$$\Delta = \begin{vmatrix} 3D + 5 & -2 \\ -D & D + 6 \end{vmatrix} \qquad (9.89)$$

and Δ_1 is given by

$$\Delta_1 = \begin{vmatrix} 3v_g & -2 \\ 0 & D + 6 \end{vmatrix} = 3(D + 6)v_g \qquad (9.90)$$

We note that in this last expression we must be careful to write v_g *after* the operator.

Writing (9.88) as

$$\Delta v = \Delta_1$$

we see from (9.89) and (9.90) that we have the describing equation (9.87).

The second method we shall consider is a mixture of the loop and nodal methods in which we select the inductor currents and the capacitor voltages as the unknowns, rather than the loop currents or the node voltages. We then write KVL around loops which contain only a single inductor and KCL at nodes, or generalized nodes, to which only a single capacitor is connected. In this manner each equation contains only one derivative, that of an inductor current or a capacitor voltage, and no integrals. The equations are then relatively easy to manipulate to find the describing equation.

EXAMPLE 9.16   As an example using Fig. 9.12 again, let i, the inductor current, and v, the capacitor voltage, be the unknowns. (These unknowns are sometimes called the *state variables* of the circuit.) Then the nodal equation at node a is

$$\frac{v - v_g}{4} + i + \frac{1}{4}\frac{dv}{dt} = 0 \qquad (9.91)$$

and the loop equation around the right mesh is

$$v = 6i + \frac{di}{dt} \qquad (9.92)$$

It is a relatively simple matter to solve for i in (9.91), substitute its value into (9.92), and simplify the result to obtain (9.85). The reader may note that we have applied this method, without saying so, in obtaining the describing equation of the circuit of Fig. 9.3 in Sec. 9.4.

Some advantages of this method are that no integrals appear (thus no second derivatives occur as a result of differentiation), one unknown is easily found in terms of the others, and the initial conditions on the first derivatives are easily obtained for use in determining the arbitrary constants in the general solution. For example, it may be seen from Fig. 9.12 that $i(0) = 1$ A and $v(0) = 6$ V. Thus from (9.91) and the value of $v_g(0^+)$, we have

$$\frac{dv(0^+)}{dt} = v_g(0^+) - 10$$

This last method can be greatly facilitated, particularly in the case of complex circuits, by using graph theory. Since we are looking for inductor currents and capacitor voltages (the state variables), we put the inductors in the links, whose currents constitute an independent set, and the capacitors in the tree, whose branch voltages constitute an independent set, as we recall from Chapter 6. (The tree should also contain voltage sources, and the links should contain current sources, etc., if possible.)

Each inductor L is then a link with current i, which forms a loop whose only other elements are tree branches. Therefore KVL around this loop will contain only one derivative term, $L(di/dt)$, and no integrals. This loop can easily be found since it is the only loop in the graph if the only link added to the tree is L. For example, the graph of Fig. 9.12 is shown in Fig. 9.13 with tree branches shown as solid lines and links as dashed lines. The loop containing the 1-H inductor is a, v_1, b, a through the branch labeled v, and KVL around it is (9.92).

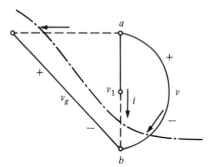

FIGURE 9.13 Graph of Fig. 9.12

Each capacitor is a tree branch whose current, together with link currents, constitutes a set of currents flowing out of a node or a generalized node, because if the capacitor is cut from the circuit, the tree is separated into two parts connected only by links. In the example of Fig. 9.13 the three currents, shown crossing the dashed line through the capacitor, labeled v, and two links, add to zero by KCL. This is stated by (9.91).

EXERCISES

9.10.1 Find the describing equation for v, $t > 0$, in Ex. 9.9.4 by (a) the first method of this section using nodal equations and (b) the second method of this section.
Answer $C(d^2v/dt^2) + 6C(dv/dt) + v = 24$

9.10.2 Solve Ex. 9.1.1 using (a) the first method of this section applied to the mesh equations and (b) the second method of this section.

9.11

SPICE FOR TRANSIENT RESPONSES OF HIGHER-ORDER CIRCUITS

SPICE, or any similar computer program, becomes very useful in solving circuits as the order of the describing differential equations increases. The application of SPICE to higher-order circuits involves essentially the same procedure as that of the first-order networks of the previous chapter.

EXAMPLE 9.17   Consider, for example, finding i and v in Fig. 9.6(b) for $R = 200$ ohms, $L = 10$ mH, $C = 1$ μF, $v(0) = 1$ V, and $i(0) = 0$ A in the interval $0 < t < 1$ ms. A circuit file for a plot of i and v is

```
PARALLEL RLC CIRCUIT OF FIG. 9.6(B) [underdamped case]
* DATA STATEMENTS
R 1 0 200OHM
L 1 0 10MH
C 1 0  1UF
* SET V(1) = 1 V and I(C) = 0 AT T = 0 USING .IC STATEMENT
.IC V(1) 1
.TRAN 0.05MS 1MS UIC
.PLOT TRAN V(1) I(L)
.END
```

In this program the .IC (initial condition) command sets node voltage $V(1) = v$, the capacitor voltage, to an initial value of 1 V. The inductor current is zero since it is not specified by an IC statement in L. The PSpice solution is shown in Fig. 9.14.

```
LEGEND:

*: V(1)
+: I(L)

   TIME        V(1)
(*)----------    -1.0000E+00   -5.0000E-01    0.0000E+00    5.0000E-01    1.0000E+00
(+)----------    -5.0000E-03    0.0000E+00    5.0000E-03    1.0000E-02    1.5000E-02
            - - - - - - - - - - - - - - - - - - - - - - - - - - - - -
  0.000E+00   1.000E+00 .                 +                 .                 .                 *
  5.000E-05   6.750E-01 .                 .                 +  .           +  *           .
  1.000E-04   2.781E-01 .                 .                 *  .       +  .  +           .
  1.500E-04  -9.158E-02 .                 .          *  .           *  .           .
  2.000E-04  -3.602E-01 .                 .    *        .    +        .           .
  2.500E-04  -4.926E-01 .                 *        +  .           .           .
  3.000E-04  -4.892E-01 .          *   +        .           .           .
  3.500E-04  -3.816E-01 .          +  .   *        .           .           .
  4.000E-04  -2.149E-01 .      +        .        *  .           .           .
  4.500E-04  -3.865E-02 .  +        .           *.           .           .
  5.000E-04   1.073E-01 .  +        .           .  *        .           .
  5.500E-04   1.982E-01 .      +  .           .      *        .           .
  6.000E-04   2.262E-01 .      +  .           .        *        .           .
  6.500E-04   1.995E-01 .      +        .           .      *        .           .
  7.000E-04   1.355E-01 .      +        .           *        .           .
  7.500E-04   5.639E-02 .      +        .    *        .           .
  8.000E-04  -1.753E-02 .           +        *        .           .
  8.500E-04  -7.149E-02 .           +    *           .           .
  9.000E-04  -9.810E-02 .       +        *        .           .
  9.500E-04  -9.783E-02 .      .+        *        .           .
  1.000E-03  -7.660E-02 .           +.        *        .           .
```

FIGURE 9.14 Transient response for circuit of Fig. 9.6(b)

EXAMPLE 9.18   As a second example, let us find v_o for $0 < t < 15$ ms in the third-order circuit of Fig. 9.15(a). Prior to time $t = 0$, $v_g = 10$ V and the switch is closed, making $i_x = 0$, which requires that the CCCS have a zero current (an open circuit). The circuit is

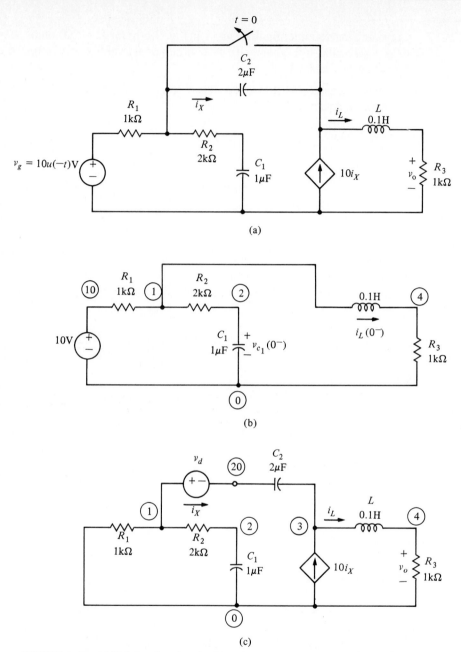

FIGURE 9.15 (a) Third-order circuit; (b) redrawn at $t = 0^-$; (c) redrawn for $t > 0$

redrawn in Fig. 9.15(b) illustrating these conditions. A circuit file for finding the initial values $v_{c1}(0^-)$ and $i_L(0^-)$ is

```
INITIAL CONDITIONS FOR CIRCUIT OF FIG. 9.15(b).
* DATA STATEMENTS
VG 10 0 DC 10
R1 10 1 1K
R2 1 2 2K
C1 2 0 1UF
L 1 4 0.1H
R3 4 0 1K
* SOLUTION CONTROL STATEMENT
.DC VG 10 10 1
* OUTPUT CONTROL STATEMENT
.PRINT DC V(C1) I(L)
.END
```

The resulting initial values are

VG	$V(C1)$	$I(L)$
$1.000E + 1$	$5.000E + 00$	$5.000E - 03$

We can now write a circuit file for $t > 0$ for Fig. 9.15(c) using these values to find $v_o = V(4)$ in the desired interval. Note that a dummy voltage source v_d has been inserted to allow i_x for the CCCS between nodes 0 and 3.

```
TRANSIENT RESPONSE FOR CIRCUIT OF FIG. 9.15 (c).
* DATA STATEMENTS
R1 0 1 1K
R2 1 2 2K
C1 2 0 1UF IC=5V
VD 1 20 DC 0
C2 20 3 2UF 1C=0
L 3 4 0.1H IC=5M
R 4 0 1K
FIX 0 3 VC 10
* SOLUTION CONTROL STATEMENT
.TRAN 1MS 15MS UIC
* OUTPUT CONTROL STATEMENT
.PLOT TRAN V(R3)
.END
```

The plot of v_o for this program is shown in Fig. 9.16.

```
       TIME        V(R3)
      (*)---------- -2.0000E+00   0.0000E+00   2.0000E+00   4.0000E+00   6.0000E+00
   0.000E+00   4.997E+00 .  - - - - - - - - - . - - - - - - - - .  *  .
   1.000E-03   1.085E+00 .              .          *       .              .
   2.000E-03   7.111E-01 .              .      *           .              .
   3.000E-03   4.493E-01 .              . *              .              .
   4.000E-03   2.655E-01 .            . *              .              .
   5.000E-03   1.376E-01 .            .*              .              .
   6.000E-03   4.932E-02 .          *                .              .
   7.000E-03  -1.149E-02 .          *                .              .
   8.000E-03  -5.260E-02 .         *                 .              .
   9.000E-03  -7.984E-02 .        *.                 .              .
   1.000E-02  -9.747E-02 .       *.                  .              .
   1.100E-02  -1.083E-01 .       *.                  .              .
   1.200E-02  -1.143E-01 .       *.                  .              .
   1.300E-02  -1.170E-01 .       *.                  .              .
   1.400E-02  -1.175E-01 .       *.                  .              .
   1.500E-02  -1.164E-01 .       *.                  .              .
                            - - - - - - - - - . - - - - - - - - .     .
```

FIGURE 9.16 Transient response for circuit of Fig. 9.15(a)

EXERCISES

9.11.1 Repeat the parallel *RLC* example for Fig. 9.6(b) with (a) $R = 50\ \Omega$; (b) $R = 25\ \Omega$.

9.11.2 Use SPICE to plot i in Fig. 9.11 in the interval $0 < t < 5$ s.

9.11.3 Use SPICE to plot v and v_g in Fig. 9.5 if $v_g = 5[u(t) - u(t - 10^{-3})]$ V in the interval $0 < t < 2$ ms.

9.11.4 Use SPICE to plot v_o and v_g in Fig. 9.14 if $v_g = -12u(-t) + 10,000t[u(t - 0.001)]$ V in the interval $0 < t < 0.01$ s.

PROBLEMS

9.1 Insert a 1-Ω resistor in series with v_g in Fig. 9.2, thereby making the source a practical rather than an ideal one. Show that in this case v_2 satisfies the second-order equation

$$5\frac{d^2v_2}{dt^2} + 11\frac{dv_2}{dt} + 4v_2 = 4\frac{dv_g}{dt} + 4v_g$$

9.2 Find i for $t > 0$ if $i(0) = 4$ A and $v(0) = 6$ V.

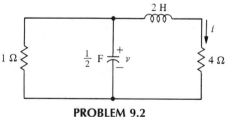

PROBLEM 9.2

9.3 Find i for $t > 0$ if the circuit is in steady state at $t = 0^-$.

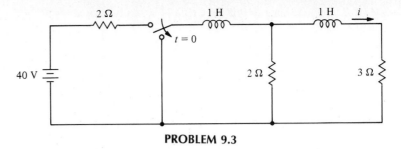

PROBLEM 9.3

9.4 Find i for $t > 0$ if $v_1(0) = 2$ V and $v_2(0) = 0$.

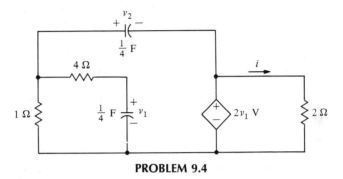

PROBLEM 9.4

9.5 Find v for $t > 0$ if the circuit is in steady state at $t = 0^-$.

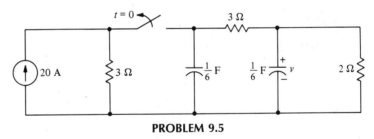

PROBLEM 9.5

9.6 Find i for $t > 0$ if the circuit is in steady state at $t = 0^-$.

PROBLEM 9.6

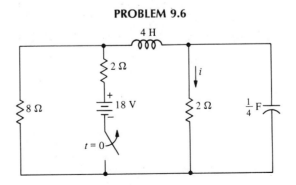

9.7 Find i for $t > 0$ if the circuit is in steady state at $t = 0^-$.

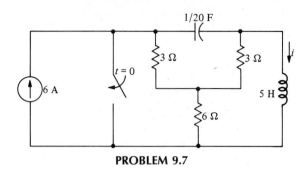

PROBLEM 9.7

9.8 The circuit is in steady state at $t = 0^-$. Find v for $t > 0$ if L is (a) 4 H, (b) 3 H, and (c) 2.4 H.

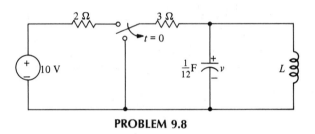

PROBLEM 9.8

9.9 Find i for $t > 0$ if the circuit is in steady state at $t = 0^-$.

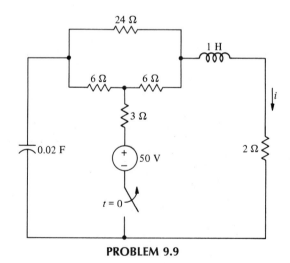

PROBLEM 9.9

9.10 Find i for $t > 0$ if $v(0) = 4$ V, $i(0) = 2$ A, and (a) $L = 2$ H, $R = 5 \ \Omega$, (b) $L = 1$ H, $R = 3 \ \Omega$, and (c) $L = 1$ H, $R = 1 \ \Omega$.

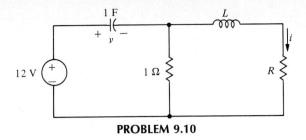

PROBLEM 9.10

9.11 Find i for $t > 0$ if the circuit is in steady state at $t = 0^-$.

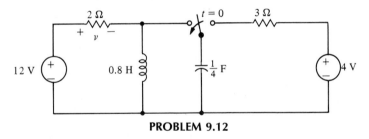

PROBLEM 9.11

9.12 Find v for $t > 0$ if the circuit is in steady state at $t = 0^-$.

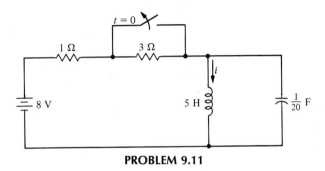

PROBLEM 9.12

9.13 Find v_1 and v_2 for $t > 0$ if the circuit is in steady state at $t = 0^-$.

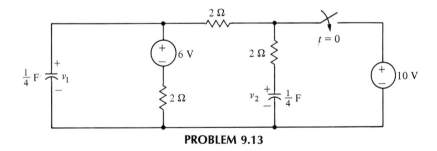

PROBLEM 9.13

9.14 Find i, $t > 0$, if there is no initial stored energy and (a) $R = \frac{1}{2} \Omega$, $\mu = 2$; (b) $R = \frac{1}{2} \Omega$, $\mu = 1$; and (c) $R = \frac{1}{4} \Omega$, $\mu = 2$.

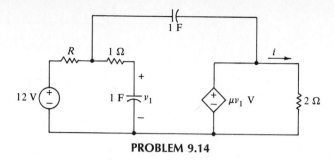

PROBLEM 9.14

9.15 Find i, $t > 0$, if there is no initial stored energy and (a) $C = \frac{1}{6}$ F, (b) $C = \frac{1}{8}$ F, and (c) $C = \frac{1}{16}$ F.

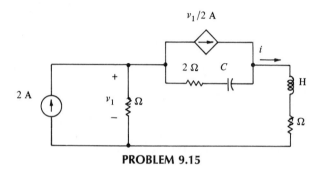

PROBLEM 9.15

9.16 Find i for $t > 0$ if $i(0) = 2$ A and $v(0) = 6$ V.

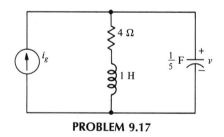

PROBLEM 9.16

9.17 Find v for $t > 0$ if (a) $i_g = 2u(t)$ A, and (b) $i_g = 2e^{-t}u(t)$ A.

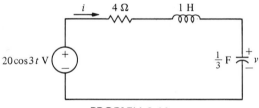

PROBLEM 9.17

9.18 Find v, $t > 0$, if $v(0) = 4$ V and $i(0) = 2$ A.

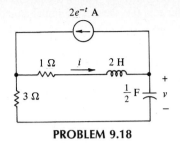

PROBLEM 9.18

9.19 (a) Find v for $t > 0$. (b) Replace the current and voltage sources by $2 \cos 2t$ A and $6 \cos 2t$ V, respectively, with the same polarities, and find v for $t > 0$ if there is no initial stored energy.

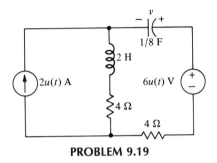

PROBLEM 9.19

9.20 Find v for $t > 0$ if $v(0) = 4$ V and $i(0) = 3$ A.

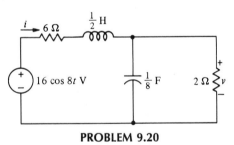

PROBLEM 9.20

9.21 (a) Find v for $t > 0$, if $v_a(0) = 0$ and $v_b(0) = 2$ V. (b) Repeat part (a) if the 4-V source is replaced by one of $26 \cos 2t$ V. (c) Repeat part (a) if the 4-V source is replaced by one of $2e^{-t}$ V.

PROBLEM 9.21

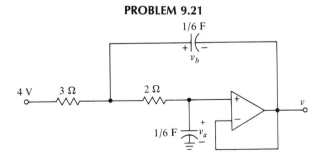

9.22 Find v for $t > 0$ if there is no initial stored energy.

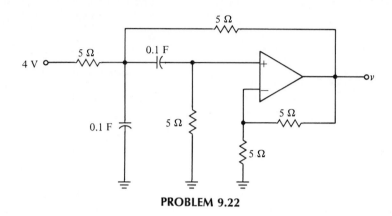

PROBLEM 9.22

9.23 Find v for $t > 0$ if there is no initial stored energy and $v_g = 5$ V.

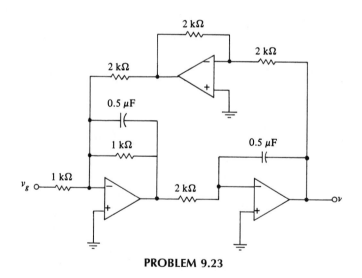

PROBLEM 9.23

9.24 Find v, $t > 0$, if (a) $v_1(0) = 4$ V, $v(0) = 0$; (b) $v_1(0) = 0$, $v(0) = 2$ V; and (c) $v_1(0) = 4$ V, $v(0) = 2$ V. (Note that the response is an unforced sinusoidal response. Such a circuit is called a *harmonic oscillator*.)

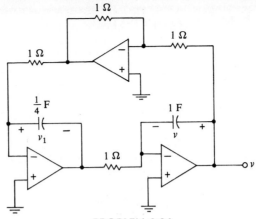

PROBLEM 9.24

9.25 Higher-order differential equations may be solved in the same manner as second-order equations. There are more natural frequencies and thus more terms in the natural response. For example, if

$$\frac{d^3x}{dt^3} + 6\frac{d^2x}{dt^2} + 11\frac{dx}{dt} + 6x = 12$$

show that the characteristic equation

$$s^3 + 6s^2 + 11s + 6 = 0$$

has the natural frequencies

$$s = -1, -2, -3$$

as its roots. Thus show that the natural response is

$$x_n = A_1e^{-t} + A_2e^{-2t} + A_3e^{-3t}$$

Show also that the forced response is

$$x_f = 2$$

and that the general solution is

$$x = x_n + x_f$$

9.26 Using the results of Prob. 9.25, find i, $t > 0$, if there is no initial stored energy.

PROBLEM 9.26

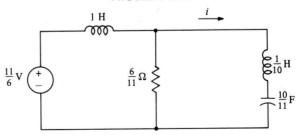

COMPUTER APPLICATION PROBLEMS

9.27 Use SPICE to plot i in Prob. 9.6 in the interval $0 < t < 1$ s.

9.28 Use SPICE to plot i in Prob. 9.16 in the interval $0 < t < 1$ s.

9.29 Replace the 2-A source with $2e^{-t}$ A in Prob. 9.15 and find v_1 in the interval $0 < t < 1$ s for (a) to (c).

9.30 Use SPICE in Prob. 9.23 to plot v and v_g in the interval $0 < t < 0.01$ s if $v_g = 12[u(t) - u(t - 0.001)]$ V.

9.31 Use SPICE to plot v in Prob. 9.24 for $0 < t < 5$ s.

9.32 Use SPICE to plot i in Prob. 9.26 if the 11/6-V source is replaced by v_g of the graph shown.

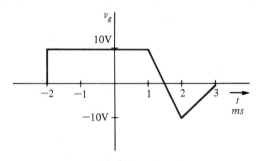

PROBLEM 9.32

10

Charles Proteus Steinmetz
1865–1923

Sinusoidal Excitation and Phasors

I have found the equation that will enable us to transmit electricity through alternating current over thousands of miles. I have reduced it to a simple problem in algebra.

Charles Proteus Steinmetz

The use of complex numbers to solve ac circuit problems—the so-called phasor method considered in this chapter—was first done by the German-Austrian mathematician and electrical engineer Charles Proteus Steinmetz in a paper presented in 1893. He is noted also for the laws of hysteresis and for his work in manufactured lightning.

Steinmetz was born in Breslau, Germany, the son of a government railway worker. He was deformed from birth and lost his mother when he was 1 year old, but this did not keep him from becoming a scientific genius. Just as his work on hysteresis later attracted the attention of the scientific community, his political activities while he was at the University at Breslau attracted the police. He was forced to flee the country just as he had finished the work for his doctorate, which he never received. He did electrical research in the United States, primarily with the General Electric Company. His paper on complex numbers revolutionized the analysis of ac circuits, though it was said at the time that no one but Steinmetz understood the method. In 1897 he also published the first book to reduce ac calculations to a science. ∎

In the two previous chapters we have analyzed circuits containing dynamic elements and have seen that the complete response is the sum of a natural and a forced response. The natural response for a given circuit is obtained from the dead circuit and therefore is independent of the sources, or excitations. The forced response, on the other hand, depends directly on the type of excitation applied to the circuit. In the case of a dc source, the forced response is a dc steady-state response, an exponential input results in an exponential forced response, and so forth.

One of the most important excitations is the sinusoidal forcing function. Sinusoids abound in nature, as, for example, in the motion of a pendulum, in the bouncing of a ball, and in the vibrations of strings and membranes. Also, as we have seen, the natural response of an underdamped second-order circuit is a damped sinusoid and in the absence of damping is a pure sinusoid.

In electrical engineering, sinusoidal functions are extremely important for a number of reasons. The carrier signals generated for communication purposes are sinusoids, and, of course, the sinusoid is the dominant signal in the electric power industry, to name two very important examples. Indeed, as we shall see later in the study of Fourier series, almost every useful signal in electrical engineering can be resolved into sinusoidal components.

Because of their importance, circuits with a sinusoidal forcing function will be considered in detail in this chapter. Since the natural response is independent of the sources and can be found by the methods of the previous chapters, we shall concentrate on finding only the forced response. This response is important in itself since it is the ac steady-state response that is left after the short time required for the transitory natural response to die.

Since we are interested only in the ac steady-state response, we shall not limit ourselves, as we did in Chapters 8 and 9, to first- and second-order circuits. As we shall see, higher-order *RLC* circuits may be handled, insofar as the ac steady-state response is concerned, in the same way as resistive circuits.

10.1

PROPERTIES OF SINUSOIDS

We shall devote this section to a review of some of the properties of sinusoidal functions. Let us begin with the sine wave,

$$v(t) = V_m \sin \omega t \qquad (10.1)$$

which is sketched in Fig. 10.1. The *amplitude* of the sinusoid is V_m, which is the maximum value that the function attains. The *radian frequency*, or *angular frequency*, is ω, measured in radians per second (rad/s).

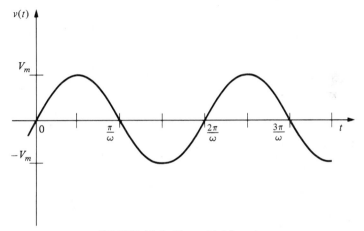

FIGURE 10.1 Sinusoidal function

The sinusoid is a *periodic* function, defined generally by the property

$$v(t + T) = v(t) \qquad (10.2)$$

where T is the *period*. That is, the function goes through a complete cycle, or period, which is then repeated, every T seconds. In the case of the sinusoid, the period is

$$T = \frac{2\pi}{\omega} \qquad (10.3)$$

as may be seen from (10.1) and (10.2). Thus in 1 s the function goes through $1/T$ cycles, or periods. Its *frequency* f is then

$$f = \frac{1}{T} = \frac{\omega}{2\pi} \qquad (10.4)$$

cycles per second, or *hertz* (abbreviated Hz). The latter term, named for the German physicist Heinrich R. Hertz (1857–1894), is now the standard unit for frequency. Some older books use the former term, but it is being discontinued. The relation between frequency and radian frequency is seen by (10.4) to be

$$\omega = 2\pi f \tag{10.5}$$

A more general sinusoidal expression is given by

$$v(t) = V_m \sin(\omega t + \phi) \tag{10.6}$$

where ϕ is the *phase angle,* or simply the *phase.* To be consistent, since ωt is in radians, ϕ should be expressed in radians. However, in electrical engineering it is often convenient to specify ϕ in degrees. For example, we may write

$$v = V_m \sin\left(2t + \frac{\pi}{4}\right)$$

or

$$v = V_m \sin(2t + 45°)$$

interchangeably, even though the latter expression contains a mathematical inconsistency.

A sketch of (10.6) is shown in Fig. 10.2 by the solid line, along with a sketch of (10.1), shown dashed. The solid curve is simply the dashed curve displaced ϕ/ω seconds, or ϕ radians to the left. Therefore, points on the solid curve, such as its peaks, occur ϕ rad, or ϕ/ω s, earlier than corresponding points on the dashed curve. Accordingly, we shall say that $V_m \sin(\omega t + \phi)$ *leads* $V_m \sin \omega t$ by ϕ rad (or degrees). In general, the sinusoid

$$v_1 = V_{m1} \sin(\omega t + \alpha)$$

leads the sinusoid

$$v_2 = V_{m2} \sin(\omega t + \beta)$$

by $\alpha - \beta$. An equivalent expression is that v_2 *lags* v_1 by $\alpha - \beta$.

FIGURE 10.2 Two sinusoids with different phases

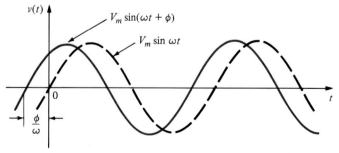

As an example, consider

$$v_1 = 4 \sin (2t + 30°)$$

and

$$v_2 = 6 \sin (2t - 12°)$$

Then v_1 leads v_2 (or v_2 lags v_1) by $30 - (-12) = 42°$.

Thus far we have considered sine functions rather than cosine functions in defining sinusoids. It does not matter which form we use since

$$\cos \left(\omega t - \frac{\pi}{2} \right) = \sin \omega t \qquad (10.7)$$

or

$$\sin \left(\omega t + \frac{\pi}{2} \right) = \cos \omega t \qquad (10.8)$$

The only difference between sines and cosines is thus the phase angle. For example, we may write (10.6) as

$$v(t) = V_m \cos \left(\omega t + \phi - \frac{\pi}{2} \right)$$

EXAMPLE 10.1   To determine how much one sinusoid leads or lags another of the same frequency, we must first express both as sine waves or as cosine waves with positive amplitudes. For example, let

$$v_1 = 4 \cos (2t + 30°)$$

and

$$v_2 = -2 \sin (2t + 18°)$$

Then, since

$$-\sin \omega t = \sin (\omega t + 180°)$$

we have

$$v_2 = 2 \sin (2t + 18° + 180°)$$

$$= 2 \cos (2t + 18° + 180° - 90°)$$

$$= 2 \cos (2t + 108°)$$

Comparing this last expression with v_1, we see that v_1 leads v_2 by $30° - 108° = -78°$, which is the same as saying that v_1 lags v_2 by $78°$.

The sum of a sine wave and a cosine wave of the same frequency is another sinusoid of that frequency. To show this, consider

$$A \cos \omega t + B \sin \omega t = \sqrt{A^2 + B^2} \left[\frac{A}{\sqrt{A^2 + B^2}} \cos \omega t + \frac{B}{\sqrt{A^2 + B^2}} \sin \omega t \right]$$

which by Fig. 10.3 may be written

$$A \cos \omega t + B \sin \omega t = \sqrt{A^2 + B^2} \, (\cos \omega t \cos \theta + \sin \omega t \sin \theta)$$

By a formula from trigonometry, this is

$$A \cos \omega t + B \sin \omega t = \sqrt{A^2 + B^2} \cos (\omega t - \theta) \tag{10.9}$$

where, by Fig. 10.3,

$$\theta = \tan^{-1} \frac{B}{A} \tag{10.10}$$

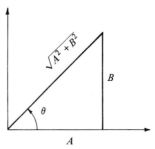

FIGURE 10.3 Triangle useful in adding two sinusoids

A similar result may be established if the sine and cosine terms have phase angles other than zero, indicating that, in general, the sum of two sinusoids of a given frequency is another sinusoid of the same frequency.

We must be clear on what is meant by (10.10), since some mathematics books take this expression as the principal value of the arctangent and place θ in a specific quadrant. We mean that the terminal side of the angle θ is in the quadrant where the point (A, B) is located.

EXAMPLE 10.2   As an example, we have

$$-5 \cos 3t + 12 \sin 3t = \sqrt{5^2 + 12^2} \cos \left[3t - \tan^{-1}\left(\frac{12}{-5}\right) \right]$$

$$= 13 \cos (3t - 112.6°)$$

since $\tan^{-1} (12/-5)$ is in the second quadrant, because $A = -5 < 0$ and $B = 12 > 0$.

EXERCISES

10.1.1   Find the period of the following sinusoids:
(a) $5 \cos (3t + 33°)$.
(b) $\cos \left(2t + \dfrac{\pi}{4} \right) + 2 \sin \left(2t - \dfrac{\pi}{6} \right)$.
(c) $4 \cos 2\pi t$.

Answer (a) $2\pi/3$; (b) π; (c) 1

10.1.2 Find the amplitude and phase of the following sinusoids:
(a) $3 \cos 2t + 4 \sin 2t$.
(b) $(4\sqrt{3} - 3) \cos (2t + 30°) + (3\sqrt{3} - 4) \cos (2t + 60°)$.
    [*Suggestion:* In (b), expand both functions and use (10.9).]
Answer (a) 5, $-53.1°$; (b) 5, $36.9°$

10.1.3 Find the frequency of the following sinusoids:
(a) $3 \cos (4\pi t - 10°)$.
(b) $4 \sin 377t$.
Answer (a) 2; (b) 60 Hz

10.2

AN RL CIRCUIT EXAMPLE

As an example of a circuit with a sinusoidal excitation, let us find the forced component i_f of the current i in Fig. 10.4. The describing equation is

$$L\frac{di}{dt} + Ri = V_m \cos \omega t \tag{10.11}$$

and, following the method of the previous chapter, let us assume the trial solution

$$i_f = A \cos \omega t + B \sin \omega t$$

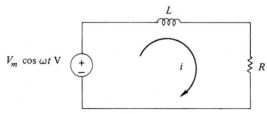

FIGURE 10.4 *RL* circuit

Substituting the trial solution into (10.11), we have

$$L(-\omega A \sin \omega t + \omega B \cos \omega t) + R(A \cos \omega t + B \sin \omega t) = V_m \cos \omega t$$

Therefore, equating coefficients of like terms, we must have

$$RA + \omega LB = V_m$$

$$-\omega LA + RB = 0$$

from which

$$A = \frac{RV_m}{R^2 + \omega^2 L^2}$$

$$B = \frac{\omega L V_m}{R^2 + \omega^2 L^2}$$

The forced response is then

$$i_f = \frac{RV_m}{R^2 + \omega^2 L^2} \cos \omega t + \frac{\omega L V_m}{R^2 + \omega^2 L^2} \sin \omega t$$

which by (10.9) and (10.10) may be written as

$$i_f = \frac{V_m}{\sqrt{R^2 + \omega^2 L^2}} \cos \left(\omega t - \tan^{-1} \frac{\omega L}{R} \right) \qquad (10.12)$$

The forced response is, therefore, a sinusoid like the excitation, as we predicted when we chose the trial solution. We may write it in the form

$$i_f = I_m \cos (\omega t + \phi) \qquad (10.13)$$

where

$$I_m = \frac{V_m}{\sqrt{R^2 + \omega^2 L^2}}$$

and

$$\phi = -\tan^{-1} \frac{\omega L}{R} \qquad (10.14)$$

Since the natural response is

$$i_n = A_1 e^{-Rt/L}$$

it is clear that after a short time $i_n \to 0$, and the current settles down to its ac steady state value, given by (10.12).

The method we have used is straightforward and conventional but, as the reader might agree, is rather laborious for such a simple problem. For a second-order circuit, the method is more tedious, as was illustrated by the example of (9.34). For very high-order circuits the procedure is, of course, even more complicated. Evidently we need a better method. One such method is developed in the remainder of this chapter, and its use allows us to treat circuits with storage elements in the same way we treated resistive circuits in Chapters 2, 4, and 5.

EXERCISES

10.2.1 Find the forced response i_f in Fig. 10.4 if $L = 4$ mH, $R = 6$ kΩ, $V_m = 5$ V, and $\omega = 2 \times 10^6$ rad/s.
Answer $0.5 \cos (2 \times 10^6 t - 53.1°)$ mA

10.2.2 Find the forced component of v.
Answer $(RI_m/\sqrt{1 + \omega^2 R^2 C^2}) \cos (\omega t - \tan^{-1} \omega RC)$ V

EXERCISE 10.2.2

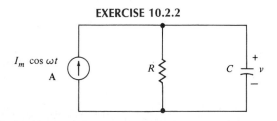

10.3

AN ALTERNATIVE METHOD USING COMPLEX NUMBERS

The alternative method of analyzing circuits with sinusoidal excitations, which we shall consider in the remainder of the chapter, relies heavily on the concept of complex numbers. The reader who is unfamiliar with complex numbers, or who needs to review the subject, should consult Appendices C and D, where complex numbers and their properties are discussed in some detail. For convenience, we shall list a few of these properties in this section before considering the alternative method of analysis.

The complex number A is written in rectangular form as

$$A = a + jb \qquad (10.15)$$

where $j = \sqrt{-1}$ and the real numbers a and b are the real and imaginary parts of A, respectively. Equivalently we may say that

$$a = \text{Re } A, \qquad b = \text{Im } A$$

where Re and Im denote *the real part of* and *the imaginary part of*.

The number A may be written also in the polar form,

$$A = |A|e^{j\alpha} = |A|\underline{/\alpha} \qquad (10.16)$$

where $|A|$ is the magnitude, given by

$$|A| = \sqrt{a^2 + b^2}$$

and α is the angle or argument, given by

$$\alpha = \tan^{-1}\frac{b}{a}$$

These relations between rectangular and polar forms are illustrated in Fig. 10.5.

EXAMPLE 10.3   As an example, suppose we have $A = 4 + j3$. Then $|A| = \sqrt{4^2 + 3^2} = 5$ and

FIGURE 10.5 Geometrical representation of a complex number A

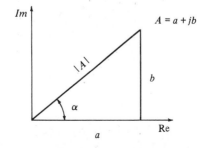

$\alpha = \tan^{-1} \frac{3}{4} = 36.9°$. Therefore the polar form is

$$A = 5\underline{/36.9°}$$

EXAMPLE 10.4   As another example, consider $A = -5 - j12$. Since both a and b are negative, the line segment representing A lies in the third quadrant, as shown in Fig. 10.6, from which we see that

$$|A| = \sqrt{5^2 + 12^2} = 13$$

and

$$\alpha = 180° + \tan^{-1} \frac{12}{5} = 247.4°$$

Thus we have $A = 13\underline{/247.4°}$.

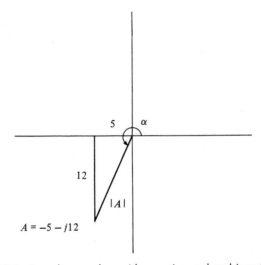

FIGURE 10.6 Complex number with negative real and imaginary parts

Other useful results are

$$j = 1\underline{/90°}$$
$$j^2 = -1 = 1\underline{/180°}$$

and so forth.

By Euler's formula, which is discussed in Appendix D and which we have used previously in Chapter 9, we know that

$$V_m \cos \omega t + jV_m \sin \omega t = V_m e^{j\omega t}$$

Therefore, we may say that

$$V_m \cos \omega t = \text{Re}(V_m e^{j\omega t}) \qquad (10.17)$$

and

$$V_m \sin \omega t = \text{Im}(V_m e^{j\omega t})$$

Returning to the *RL* circuit example of Fig. 10.4, we know that exponentials are mathematically easier to handle as excitations than sinusoids. Therefore let us see what happens if we apply the complex excitation

$$v_1 = V_m e^{j\omega t} \qquad (10.18)$$

instead of the real excitation

$$v_g = V_m \cos \omega t = \text{Re } v_1 \qquad (10.19)$$

We cannot duplicate such a complex excitation in the laboratory, but there is no reason we cannot consider it abstractly. In this case the forced component of the current, which we shall call i_1, satisfies

$$L \frac{di_1}{dt} + Ri_1 = v_1 = V_m e^{j\omega t} \qquad (10.20)$$

To solve this equation, we try

$$i_1 = A e^{j\omega t}$$

which, substituted into (10.20), yields

$$(j\omega L + R)A e^{j\omega t} = V_m e^{j\omega t}$$

from which

$$A = \frac{V_m}{R + j\omega L}$$

$$= \frac{V_m}{\sqrt{R^2 + \omega^2 L^2}} e^{-j \tan^{-1} \omega L/R}$$

using complex number division. Therefore, we have

$$i_1 = \frac{V_m}{\sqrt{R^2 + \omega^2 L^2}} e^{j(\omega t - \tan^{-1} \omega L/R)}$$

Let us now observe that

$$\text{Re } i_1 = \text{Re}\left[\frac{V_m}{\sqrt{R^2 + \omega^2 L^2}} e^{j(\omega t - \tan^{-1} \omega L/R)} \right]$$

$$= \frac{V_m}{\sqrt{R^2 + \omega^2 L^2}} \cos \left(\omega t - \tan^{-1} \frac{\omega L}{R} \right)$$

which, by (10.12), is the correct forced response of Fig. 10.4. That is,

$$i_f = \text{Re } i_1 \qquad (10.21)$$

We have established for this example the interesting result that if i_1 is the complex response to the complex forcing function v_1, then $i_f = \text{Re } i_1$ is the response to $v_g = \text{Re } v_1$. That is, v_1 yields i_1 and $\text{Re } v_1 = v_g$ yields $\text{Re } i_1 = i_f$. The reason for this is that the describing equation (10.20) contains only real coefficients. Thus, from (10.20), we have

$$\text{Re}\left(L\frac{di_1}{dt} + Ri_1\right) = \text{Re } v_1$$

or

$$L\frac{d}{dt}(\text{Re } i_1) + R(\text{Re } i_1) = V_m \cos \omega t$$

and therefore, by (10.11),

$$i = i_f = \text{Re } i_1 \tag{10.22}$$

Thus we see that it is easier to use the complex forcing function v_1 to find the complex response i_1. Then since the real forcing function is Re v_1, the real response is Re i_1. This principle holds for all our circuit analyses, since the describing equations are linear with real coefficients, as may be seen in a development analogous to that leading to (10.22).

EXERCISES

10.3.1 Replace the real forcing function $I_m \cos \omega t$ in Ex. 10.2.2 by the complex forcing function $I_m e^{j\omega t}$, find the resulting complex response v_1, and show that the real response is $v = \text{Re } v_1$.

10.3.2 Show that, for a real,

$$\text{Re}\left(a\frac{dx}{dt}\right) = a\frac{d}{dt}(\text{Re } x)$$

and use this result to establish (10.22). (*Suggestion:* Let $x = f + jg$, where f and g are real.)

10.3.3 In Ex. 10.2.2, replace the current source by $i_1 = I_m e^{j\omega t}$ A and show that the response v_1 has the property that Re $v_1 = v$, where v is the original response.

10.4

COMPLEX EXCITATIONS

Let us now generalize the results using complex excitation functions in the previous section. The excitation, as well as the forced response, may be a sinusoidal voltage or current. However, to be specific, let us consider the input to be a voltage source and the output to be a current through some element. The other cases may be considered in an analogous way.

In general, we know that if

$$v_g = V_m \cos (\omega t + \theta) \tag{10.23}$$

then the forced response is of the form

$$i = I_m \cos (\omega t + \phi) \qquad (10.24)$$

as indicated in the general circuit of Fig. 10.7. Therefore, if by some means we can find I_m and ϕ, we have our answer, since ω, θ, and V_m are known.

To solve for i in Fig. 10.7, let us apply the complex excitation

$$v_1 = V_m e^{j(\omega t + \theta)} \qquad (10.25)$$

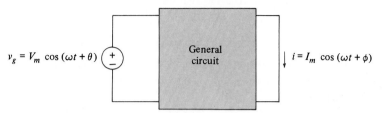

$v_g = V_m \cos (\omega t + \theta)$

General circuit

$i = I_m \cos (\omega t + \phi)$

FIGURE 10.7   General circuit with input and output

and find the complex response i_1, as shown in Fig. 10.8. Then we know from the results of the previous section that the real response of Fig. 10.7 is

$$i = \mathrm{Re}\, i_1 \qquad (10.26)$$

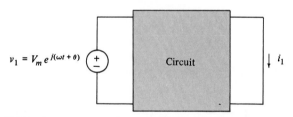

$v_1 = V_m e^{j(\omega t + \theta)}$

Circuit

i_1

FIGURE 10.8   General circuit with a complex excitation

This is a consequence of the fact that the coefficients in the describing equation are real, as pointed out previously.

The describing equation may be solved for the forced response by the method of Chapter 9. That is, since we may write the excitation as

$$v_1 = V_m e^{j\theta} e^{j\omega t} \qquad (10.27)$$

which is a constant times $e^{j\omega t}$, then the trial solution is

$$i_1 = A e^{j\omega t}$$

Comparing (10.24) and (10.26), we must have

$$I_m \cos (\omega t + \phi) = \mathrm{Re}[A e^{j\omega t}]$$

which requires that

$$A = I_m e^{j\phi}$$

and hence

$$i_1 = I_m e^{j\phi} e^{j\omega t} \qquad (10.28)$$

Taking the real part, we have the solution (10.24), of course.

As an example, let us find the forced response i_f of

$$\frac{d^2 i}{dt^2} + 2\frac{di}{dt} + 8i = 12\sqrt{2}\cos(2t + 15°)$$

First we replace the real excitation by the complex excitation,

$$v_1 = 12\sqrt{2}\,e^{j(2t+15°)}$$

where for convenience the phase is written in degrees. (This is, of course, an inconsistent mathematical expression, but as long as we interpret it correctly it should present no difficulty.) The complex response i_1 satisfies

$$\frac{d^2 i_1}{dt^2} + 2\frac{di_1}{dt} + 8i_1 = 12\sqrt{2}\,e^{j(2t+15°)}$$

and it must have the general form

$$i_1 = Ae^{j2t}$$

Therefore, we must have

$$(-4 + j4 + 8)Ae^{j2t} = 12\sqrt{2}\,e^{j2t}e^{j15°}$$

or

$$A = \frac{12\sqrt{2}\,e^{j15°}}{4 + j4} = \frac{12\sqrt{2}\,\underline{/15°}}{4\sqrt{2}\,\underline{/45°}} = 3\,\underline{/-30°}$$

which gives

$$i_1 = (3\,\underline{/-30°})e^{j2t}$$
$$= 3e^{j(2t-30°)}$$

Thus the real answer is

$$i_f = \operatorname{Re} i_1 = 3\cos(2t - 30°)$$

EXERCISES

10.4.1 (a) From the time-domain equations find the forced response v if $v_g = 10e^{j8t}$ V.
(b) Using the result in (a), find the forced response v if $v_g = 10\cos 8t$ V.
Answer (a) $2e^{j(8t - 53.1°)}$ V; (b) $2\cos(8t - 53.1°)$ V

EXERCISE 10.4.1

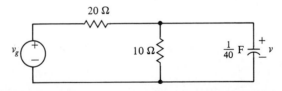

304 Chapter 10   Sinusoidal Excitation and Phasors

10.4.2 Find the forced response v in Ex. 10.4.1 if $v_g = 10 \sin 8t$ V. (*Suggestion:* $\sin 8t = \text{Im } e^{j8t}$)

Answer $2 \sin (8t - 53.1°)$ V

10.4.3 Using the method of complex excitation, find the forced response i if $v_g = 12 \cos 2t$ V.

Answer $3 \cos 2t$ A

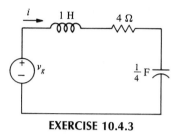

EXERCISE 10.4.3

10.4.4 Repeat Ex. 10.4.3 if $v_g = 10 \cos 4t$ V.

Answer $2 \cos (4t - 36.9°)$ A

10.5

PHASORS

The results obtained in the preceding section may be put in much more compact form by the use of quantities called *phasors*, which we shall introduce in this section. The phasor method of analyzing circuits is credited generally to Charles Proteus Steinmetz (1865–1923), a famous electrical engineer with the General Electric Company in the early part of this century.

To begin, let us recall the general sinusoidal voltage,

$$v = V_m \cos (\omega t + \theta) \qquad (10.29)$$

which, of course, is the source voltage v_g of the preceding section. If the frequency ω is known, then v is completely specified by its amplitude V_m and its phase θ. These quantities are displayed in a related complex number,

$$\mathbf{V} = V_m e^{j\theta} = V_m \underline{/\theta} \qquad (10.30)$$

which is defined as a *phasor*, or a phasor *representation*. To distinguish them from other complex numbers, phasors are printed in boldface type, as indicated.

The motivation for the phasor definition may be seen from the equivalence, by Euler's formula, of

$$V_m \cos (\omega t + \theta) = \text{Re}(V_m e^{j\theta} e^{j\omega t}) \qquad (10.31)$$

Therefore, in view of (10.29) and (10.30), we have

$$v = \text{Re}(\mathbf{V}e^{j\omega t}) \qquad (10.32)$$

EXAMPLE 10.6 As an example, suppose we have

$$v = 10 \cos (4t + 30°) \text{ V}$$

The phasor representation is then

$$\mathbf{V} = 10\underline{/30°} \text{ V}$$

since $V_m = 10$ and $\theta = 30°$. Conversely, since $\omega = 4$ rad/s is assumed to be known, v is readily obtained from $\mathbf{V}$.

In an identical fashion we define the phasor representation of the time-domain current

$$i = I_m \cos (\omega t + \theta) \tag{10.33}$$

to be

$$\mathbf{I} = I_m e^{j\phi} = I_m\underline{/\phi} \tag{10.34}$$

Thus if we know, for example, that $\omega = 6$ rad/s and that $\mathbf{I} = 2\underline{/15°}$ A, then we have

$$i = 2 \cos (6t + 15°) \text{ A}$$

We have chosen to represent sinusoids and their related phasors on the basis of cosine functions, though we could have chosen sine functions just as easily. Therefore if a function such as

$$v = 8 \sin (3t + 30°)$$

is given, we may change it to

$$v = 8 \cos (3t + 30° - 90°)$$
$$= 8 \cos (3t - 60°)$$

Then the phasor representation is

$$\mathbf{V} = 8\underline{/-60°}$$

EXAMPLE 10.7 To see how the use of phasors can greatly shorten the work, let us reconsider Fig. 10.4 and its describing equation (10.11), rewritten as

$$L\frac{di}{dt} + Ri = V_m \cos \omega t \tag{10.35}$$

Following our method, we replace the excitation $V_m \cos \omega t$ by the complex forcing function

$$v_1 = V_m e^{j\omega t}$$

which may be written

$$v_1 = \mathbf{V}e^{j\omega t}$$

since $\theta = 0$, and therefore $\mathbf{V} = V_m\underline{/0} = V_m$. Substituting this value and $i = i_1$ into (10.35), we have

$$L\frac{di_1}{dt} + Ri_1 = \mathbf{V}e^{j\omega t}$$

whose solution i_1 is related to the real solution i by

$$i = \text{Re } i_1$$

Next, trying

$$i_1 = \mathbf{I}e^{j\omega t}$$

as a solution, we have

$$j\omega L\mathbf{I}e^{j\omega t} + R\mathbf{I}e^{j\omega t} = \mathbf{V}e^{j\omega t}$$

Dividing out the factor $e^{j\omega t}$, we have the phasor equation

$$j\omega L\mathbf{I} + R\mathbf{I} = \mathbf{V} \qquad (10.36)$$

Therefore,

$$\mathbf{I} = \frac{\mathbf{V}}{R + j\omega L} = \frac{V_m}{\sqrt{R^2 + \omega^2 L^2}} \bigg/ -\tan^{-1}\frac{\omega L}{R}$$

Substituting this value into the expression for i_1, we have

$$i_1 = \frac{V_m}{\sqrt{R^2 + \omega^2 L^2}} e^{j(\omega t - \tan^{-1} \omega L/R)}$$

Taking the real part, we have $i = i_f$, obtained earlier in (10.12).

It is important to note that if we can go directly from (10.35) to (10.36), there is a vast saving of time and effort. Also, in the process we have converted the differential equation into an algebraic equation, somewhat like those encountered in resistive circuits. Indeed, the only difference is that the numbers here are complex, whereas in resistive circuits they were real. With the hand calculator as commonly available as it is today, even the complexity of the numbers presents little difficulty.

In the remainder of the chapter we shall see how to bypass all the steps between (10.35) and (10.36) by studying the phasor relationships of the circuit elements and considering Kirchhoff's laws as they pertain to phasors. Indeed, as we shall see, we may go directly from the circuit to (10.36), bypassing even the step of writing down the differential equation.

In general, the real solutions are time-domain functions, and their phasors are *frequency-domain* functions; i.e., they are functions of the frequency ω. This is illustrated by the phasor $\mathbf{I}$ of the last example. Thus to solve the time-domain problems, we may convert to phasors and solve the corresponding frequency-domain problems, which are generally much easier. Finally, we convert back to the time domain by finding the time function from its phasor representation.

EXERCISES

10.5.1 Find the phasor representation of (a) $10 \cos (2t + 45°)$, (b) $4 \cos 2t + 3 \sin 2t$, and (c) $-6 \sin (5t - 75°)$.

Answer (a) $10\underline{/45°}$; (b) $5\underline{/-36.9°}$; (c) $6\underline{/15°}$

10.5.2 Find the time-domain function represented by the phasors (a) $10\underline{/-12°}$, (b) $6 + j8$, and (c) $-j4$. In all cases $\omega = 3$.

Answer (a) $10 \cos (3t - 12°)$; (b) $10 \cos (3t + 53.1°)$;
 (c) $4 \cos (3t - 90°)$

10.6

VOLTAGE-CURRENT RELATIONSHIPS FOR PHASORS

In this section we shall show that relationships between phasor voltage and phasor current for resistors, inductors, and capacitors are very similar to Ohm's law for resistors. In fact, the phasor voltage is proportional to the phasor current, as in Ohm's law, with the proportionality factor being a constant or a function of the frequency ω.

We begin by considering the voltage-current relation for the resistor,

$$v = Ri \qquad (10.37)$$

where

$$v = V_m \cos (\omega t + \theta)$$
$$i = I_m \cos (\omega t + \phi) \qquad (10.38)$$

If we apply the complex voltage $V_m e^{j(\omega t + \theta)}$, the complex current which results is $I_m e^{j(\omega t + \phi)}$, which substituted into (10.37) yields

$$V_m e^{j(\omega t + \theta)} = RI_m e^{j(\omega t + \phi)}$$

Dividing out the factor $e^{j\omega t}$ results in

$$V_m e^{j\theta} = RI_m e^{j\phi} \qquad (10.39)$$

which, since $V_m e^{j\theta}$ and $I_m e^{j\phi}$ are the phasors **V** and **I**, respectively, reduces to

$$\boxed{\mathbf{V} = R\mathbf{I} \qquad (10.40)}$$

Thus the phasor or frequency-domain relation for the resistor is exactly like the time-domain relation. The voltage-current relations for the resistor are illustrated in Fig. 10.9.

From (10.39) we have $V_m = RI_m$ and $\theta = \phi$. Thus the sinusoidal voltage and current for a resistor have the same phase angle, in which case they are said to be *in phase*. This phase relationship is shown in Fig. 10.10, where the voltage is represented by the solid line and the current by the dashed line.

EXAMPLE 10.8 As an illustration, suppose that the voltage

$$v = 10 \cos (100t + 30°) \text{ V} \qquad (10.41)$$

is applied across a 5-Ω resistor, with the polarity indicated in Fig. 10.9(a). Then the

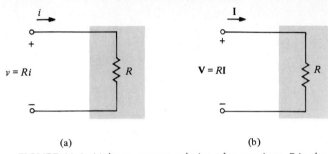

(a)   (b)

FIGURE 10.9 Voltage-current relations for a resistor R in the (a) time and (b) frequency domains

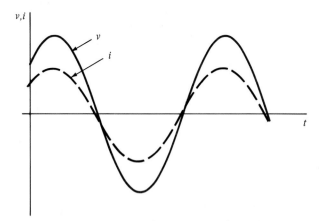

FIGURE 10.10 Voltage and current waveforms for a resistor

phasor voltage is

$$\mathbf{V} = 10\underline{/30°} \text{ V}$$

and the phasor current is

$$\mathbf{I} = \frac{\mathbf{V}}{R} = \frac{10\underline{/30°}}{5} = 2\underline{/30°} \text{ A}$$

Therefore, in the time domain we have

$$i = 2 \cos (100t + 30°) \text{ A} \qquad (10.42)$$

This is, of course, simply the result we would have obtained using Ohm's law.

In the case of the inductor, substituting the complex current and voltage into the time-domain relation,

$$v = L\frac{di}{dt}$$

gives the complex relation

$$V_m e^{j(\omega t + \theta)} = L\frac{d}{dt}[I_m e^{j(\omega t + \phi)}]$$

$$= j\omega L I_m e^{j(\omega t + \phi)}$$

Again, dividing out the factor $e^{j\omega t}$ and identifying the phasors, we obtain the phasor relation

$$\mathbf{V} = j\omega L\,\mathbf{I} \qquad\qquad (10.43)$$

Thus the phasor voltage $\mathbf{V}$, as in Ohm's law, is proportional to the phasor current $\mathbf{I}$, with the proportionality factor $j\omega L$. The voltage-current relations for the inductor are shown in Fig. 10.11.

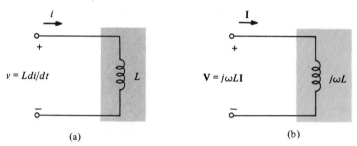

(a)

(b)

FIGURE 10.11 Voltage-current relations for an inductor L in the (a) time and (b) frequency domains

If the current in the inductor is given by the second equation of (10.38), then by (10.43), the phasor voltage is

$$\mathbf{V} = (j\omega L)(I_m\underline{/\phi})$$

$$= \omega L I_m\underline{/\phi + 90°}$$

since $j = 1\underline{/90°}$. Therefore, in the time domain we have

$$v = \omega L I_m \cos(\omega t + \phi + 90°)$$

Comparing this result with the second equation of (10.38), we see that in the case of an inductor the current *lags* the voltage by 90°. Another expression that is used is that the current and voltage are 90° out of phase. This is shown graphically in Fig. 10.12.

Finally, let us consider the capacitor. Substituting the complex current and voltage into the time-domain relation,

$$i = C\frac{dv}{dt}$$

gives the complex relation

$$I_m e^{j(\omega t + \phi)} = C\frac{d}{dt}[V_m e^{j(\omega t + \theta)}]$$

$$= j\omega C V_m e^{j(\omega t + \theta)}$$

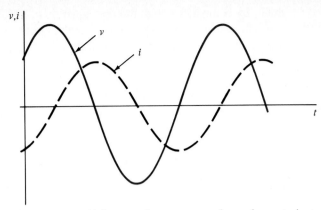

FIGURE 10.12 Voltage and current waveforms for an inductor

Again dividing by $e^{j\omega t}$ and identifying the phasors, we obtain the phasor relation

$$\mathbf{I} = j\omega C \mathbf{V} \tag{10.44}$$

or

$$\mathbf{V} = \frac{\mathbf{I}}{j\omega C} \tag{10.45}$$

Thus the phasor voltage $\mathbf{V}$ is proportional to the phasor current $\mathbf{I}$, with the proportionality factor given by $1/j\omega C$. The voltage-current relations for a capacitor in the time and frequency domains are shown in Fig. 10.13.

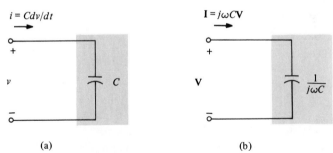

        (a)   (b)

FIGURE 10.13 Voltage-current relations for a capacitor in the (a) time and (b) frequency domains

In the general case, if the capacitor voltage is given by the first equation of (10.38), then by (10.44), the phasor current is

$$I = (j\omega C)(V_m\underline{/\theta})$$
$$= \omega C V_m\underline{/\theta + 90°}$$

Therefore, in the time domain we have

$$i = \omega C V_m \cos(\omega t + \theta + 90°)$$

which, by comparison with the first equation of (10.38), indicates that in the case of a capacitor the current and voltage are out of phase with the current *leading* the voltage by 90°. This is shown graphically in Fig. 10.14.

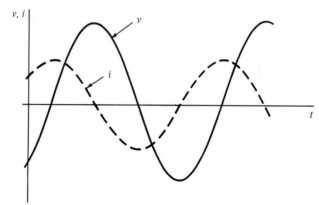

FIGURE 10.14 Voltage and current waveforms for a capacitor

EXAMPLE 10.9 As an example, if the voltage of (10.41) is applied across a 1-μF capacitor, then by (10.44) the phasor current is

$$\mathbf{I} = j(100)(10^{-6})(10\underline{/30°})\ \text{A}$$

$$= 1\underline{/120°}\ \text{mA}$$

The time-domain current is then

$$i = \cos(100t + 120°)\ \text{mA}$$

and therefore the current leads the voltage by 90°.

EXERCISES

10.6.1 Using phasors, find the ac steady-state current i if $v = 12\cos(1000t + 30°)$ V in (a) Fig. 10.9(a) for $R = 4$ kΩ, (b) Fig. 10.11(a) for $L = 15$ mH, and (c) Fig. 10.13(a) for $C = \frac{1}{2}\mu$F.
Answer (a) $3\cos(1000t + 30°)$ mA; (b) $0.8\cos(1000t - 60°)$ A;
(c) $6\cos(1000t + 120°)$ mA

10.6.2 In Ex. 10.6.1, find i in each case at $t = 1$ ms.
Answer (a) 0.142 mA; (b) 0.799 A; (c) −5.993 mA

10.7

IMPEDANCE AND ADMITTANCE

Let us now consider a general circuit of phasor quantities with two accessible terminals, as shown in Fig. 10.15. If the time-domain voltage and current at the terminals are given by (10.38), then the phasor quantities at the terminals are

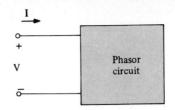

FIGURE 10.15 General phasor circuit

$$\mathbf{V} = V_m\underline{/\theta}$$

$$\mathbf{I} = I_m\underline{/\phi}$$

(10.46)

We define the ratio of the phasor voltage to the phasor current as the *impedance* of the circuit, which we denote by **Z**. That is,

$$\mathbf{Z} = \frac{\mathbf{V}}{\mathbf{I}}$$

(10.47)

which by (10.46) is

$$\mathbf{Z} = |\mathbf{Z}|\underline{/\theta_Z} = \frac{V_m}{I_m}\underline{/\theta - \phi}$$

(10.48)

where $|\mathbf{Z}|$ is the magnitude and θ_Z the angle of **Z**. Evidently,

$$|\mathbf{Z}| = \frac{V_m}{I_m}, \qquad \theta_Z = \theta - \phi$$

Impedance, as is seen from (10.47), plays the role, in a general circuit, of resistance in resistive circuits. Indeed, (10.47) looks very much like Ohm's law; also like resistance, impedance is measured in ohms, being a ratio of volts to amperes.

It is important to stress that impedance is a complex number, being the ratio of two complex numbers, but it is *not* a phasor. That is, it has no corresponding sinusoidal time-domain function of any physical meaning, as current and voltage phasors have.

The impedance **Z** is written in polar form in (10.48); in rectangular form it is generally denoted by

$$\mathbf{Z} = R + jX$$

(10.49)

where $R = \mathrm{Re}\,\mathbf{Z}$ is the *resistive component,* or simply *resistance,* and $X = \mathrm{Im}\,\mathbf{Z}$ is the *reactive component,* or *reactance.* In general, $\mathbf{Z} = \mathbf{Z}(j\omega)$ is a complex function of $j\omega$, but $R = R(\omega)$ and $X = X(\omega)$ are real functions of ω. Both R and X, like **Z**, are measured in ohms. Evidently, comparing (10.48) and (10.49) we may write

$$|\mathbf{Z}| = \sqrt{R^2 + X^2}$$

$$\theta_Z = \tan^{-1}\frac{X}{R}$$

and

$$R = |\mathbf{Z}| \cos \theta_Z$$

$$X = |\mathbf{Z}| \sin \theta_Z$$

These relations are shown graphically in Fig. 10.16.

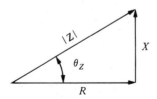

FIGURE 10.16 Graphical representation of impedance

EXAMPLE 10.10 As an example, suppose in Fig. 10.15 that $\mathbf{V} = 10\underline{/56.9°}$ V and $\mathbf{I} = 2\underline{/20°}$ A. Then we have

$$\mathbf{Z} = \frac{10\underline{/56.9°}}{2\underline{/20°}} = 5\underline{/36.9°} \ \Omega$$

In rectangular form this is

$$\mathbf{Z} = 5(\cos 36.9° + j \sin 36.9°)$$

$$= 4 + j3 \ \Omega$$

The impedances of resistors, inductors, and capacitors are readily found from their **V-I** relations of (10.40), (10.43), and (10.45). Distinguishing their impedances with subscripts R, L, and C, respectively, we have, from these equations and (10.47),

$$\mathbf{Z}_R = R$$

$$\mathbf{Z}_L = j\omega L = \omega L\underline{/90°}$$

$$\mathbf{Z}_C = \frac{1}{j\omega C} = -j\frac{1}{\omega C} = \frac{1}{\omega C}\underline{/-90°}$$

(10.50)

In the case of a resistor, the impedance is purely resistive, its reactance being zero. Impedances of inductors and capacitors are purely reactive, having zero resistive components. The *inductive reactance* is denoted by

$$X_L = \omega L$$

(10.51)

so that

$$\mathbf{Z}_L = jX_L$$

and the *capacitive reactance* is denoted by

$$X_C = -\frac{1}{\omega C} \qquad (10.52)$$

and thus

$$\mathbf{Z}_C = jX_C \qquad (10.53)$$

Since ω, L, and C are positive, we see that inductive reactance is positive and that capacitive reactance is negative. In the general case of (10.49), we may have $X = 0$, in which case the circuit appears to be resistive; $X > 0$, in which case its reactance appears to be inductive; and $X < 0$, in which case its reactance appears to be capacitive. These cases are possible when resistance, inductance, and capacitance are all present in the circuit, as we shall see. As an example, the circuit with impedance given by $\mathbf{Z} = 4 + j3$, which we have just considered, has reactance $X = 3$, which is of the inductive type. In all cases of passive circuits, as we shall see in Chapter 12, the resistance R is nonnegative.

The reciprocal of impedance, denoted by

$$\mathbf{Y} = \frac{1}{\mathbf{Z}} \qquad (10.54)$$

is called *admittance* and is analogous to conductance (the reciprocal of resistance) in resistive circuits. Evidently, since $\mathbf{Z}$ is a complex number, then so is $\mathbf{Y}$, the standard representation being

$$\mathbf{Y} = G + jB \qquad (10.55)$$

The quantities $G = \mathrm{Re}\ \mathbf{Y}$ and $B = \mathrm{Im}\ \mathbf{Y}$ are respectively called *conductance* and *susceptance* and are related to the impedance components by

$$\mathbf{Y} = G + jB = \frac{1}{\mathbf{Z}} = \frac{1}{R + jX} \qquad (10.56)$$

The units of $\mathbf{Y}$, G, and B are all mhos, since in general $\mathbf{Y}$ is the ratio of a current to a voltage phasor.

To obtain the relation between components of $\mathbf{Y}$ and $\mathbf{Z}$ we may rationalize the last member of (10.56), which results in

$$
\begin{aligned}
G + jB &= \frac{1}{R + jX} \cdot \frac{R - jX}{R - jX} \\
&= \frac{R - jX}{R^2 + X^2}
\end{aligned}
$$

Equating real and imaginary parts results in

$$G = \frac{R}{R^2 + X^2}$$

$$B = -\frac{X}{R^2 + X^2}$$

(10.57)

Therefore we note that R and G are *not* reciprocals except in the purely resistive case ($X = 0$). Similarly, X and B are never reciprocals, but in the purely reactive case ($R = 0$) they are negative reciprocals.

EXAMPLE 10.11 As an example, if we have

$$\mathbf{Z} = 4 + j3$$

then

$$\mathbf{Y} = \frac{1}{4 + j3} = \frac{4 - j3}{4^2 + 3^2} = \frac{4}{25} - j\frac{3}{25}$$

Therefore, $G = \frac{4}{25}$ and $B = -\frac{3}{25}$.

Further examples are

$$\mathbf{Y}_R = G$$

$$\mathbf{Y}_L = \frac{1}{j\omega L}$$

$$\mathbf{Y}_C = j\omega C$$

which are the admittances of a resistor, with $R = 1/G$, an inductor, and a capacitor.

EXERCISES

10.7.1 Find the impedance seen at the terminals of the source in Fig. 10.4 in both rectangular and polar form.
Answer $R + j\omega L$, $\sqrt{R^2 + \omega^2 L^2}\ \underline{/\tan^{-1} \omega L/R}$

10.7.2 Find the admittance seen at the terminals of the source in Fig. 10.4 in both rectangular and polar form.
Answer $\dfrac{R}{R^2 + \omega^2 L^2} - j\dfrac{\omega L}{R^2 + \omega^2 L^2}, \dfrac{1}{\sqrt{R^2 + \omega^2 L^2}}\ \bigg/ \underline{-\tan^{-1} \dfrac{\omega L}{R}}$

10.7.3 Find the conductance and susceptance if $\mathbf{Z}$ is (a) $6 + j8$, (b) $0.4 + j0.3$, and (c) $\dfrac{\sqrt{2}}{2}\underline{/45°}$.
Answer (a) 0.06, −0.08; (b) 1.6, −1.2; (c) 1, −1

10.8

KIRCHHOFF'S LAWS AND IMPEDANCE COMBINATIONS

Kirchhoff's laws hold for phasors as well as for their corresponding time-domain voltages or currents. We may see this by observing that if a complex excitation, say $V_m e^{j(\omega t + \theta)}$, is applied to a circuit, then complex voltages, such as $V_1 e^{j(\omega t + \theta_1)}$, $V_2 e^{j(\omega t + \theta_2)}$, etc., appear across the elements in the circuit. Since Kirchhoff's laws hold in the time domain, KVL applied around a typical loop results in an equation such as

$$V_1 e^{j(\omega t + \theta_1)} + V_2 e^{j(\omega t + \theta_2)} + \ldots + V_N e^{j(\omega t + \theta_N)} = 0$$

Dividing out the common factor $e^{j\omega t}$, we have

$$\mathbf{V}_1 + \mathbf{V}_2 + \ldots + \mathbf{V}_N = 0$$

where

$$\mathbf{V}_n = V_n \underline{/\theta_n}, \qquad n = 1, 2, \ldots, N$$

are the phasor voltages around the loop. Thus KVL holds for phasors. A similar development will also establish KCL.

In circuits having sinusoidal excitations with a common frequency ω, if we are interested only in the forced, or ac steady-state, response, we may find the phasor voltages or currents of every element and use Kirchhoff's laws to complete the analysis. The ac steady-state analysis is therefore identical to the resistive circuit analysis of Chapters 2, 4, and 5, with impedances replacing resistances and phasors replacing time-domain quantities. Once we have found the phasors, we can convert immediately to the time-domain sinusoidal answers.

EXAMPLE 10.12 As an example, consider the circuit of Fig. 10.17, which consists of N impedances connected in series. By KCL for phasors, the single phasor current $\mathbf{I}$ flows in each element. Therefore the voltages shown across each element are

FIGURE 10.17 Impedances connected in series

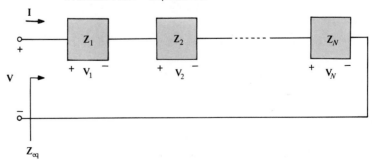

$$V_1 = Z_1 I$$

$$V_2 = Z_2 I$$

$$\vdots$$

$$V_N = Z_N I$$

and by KVL around the circuit,

$$V = V_1 + V_2 + \ldots + V_N$$

$$= (Z_1 + Z_2 + \ldots + Z_N)I$$

Since we must also have, from Fig. 10.17,

$$V = Z_{eq} I$$

where Z_{eq} is the *equivalent* impedance seen at the terminals, it follows that

$$Z_{eq} = Z_1 + Z_2 + \ldots + Z_N \tag{10.58}$$

as in the case of series resistors.

Similarly, as was the case for parallel conductances in Chapter 2, the equivalent admittance Y_{eq} of N parallel admittances is

$$Y_{eq} = Y_1 + Y_2 + \ldots + Y_N \tag{10.59}$$

In the case of two parallel elements ($N = 2$), we have

$$Z_{eq} = \frac{1}{Y_{eq}} = \frac{1}{Y_1 + Y_2} = \frac{Z_1 Z_2}{Z_1 + Z_2} \tag{10.60}$$

In like manner, voltage and current division rules hold for phasor circuits, with impedances and frequency-domain quantities, in exactly the same way that they held for resistive circuits, with resistances and time-domain quantities. The reader is asked to establish these rules in Ex. 10.8.2.

EXAMPLE 10.13 For example, let us return to the *RL* circuit considered in Sec. 10.2. The circuit and its phasor counterpart are shown in Fig. 10.18(a) and (b), respectively. By KVL in the phasor circuit we have

$$Z_L I + R I = V_m \underline{/0}$$

or

$$(j\omega L + R)I = V_m \underline{/0}$$

from which the phasor current is

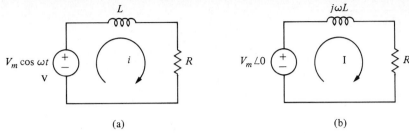

FIGURE 10.18 (a) Time-domain circuit; (b) equivalent phasor circuit

$$I = \frac{V_m \angle 0}{R + j\omega L}$$

$$= \frac{V_m}{\sqrt{R^2 + \omega^2 L^2}} \angle -\tan^{-1}\frac{\omega L}{R}$$

Therefore, in the time domain we have, as before,

$$i = \frac{V_m}{\sqrt{R^2 + \omega^2 L^2}} \cos\left(\omega t - \tan^{-1}\frac{\omega L}{R}\right)$$

An alternative method of solution is to observe that the impedance **Z** seen at the source terminals is the impedance of the inductor, $j\omega L$, and the resistor, R, connected in series. Therefore

$$\mathbf{Z} = j\omega L + R$$

and

$$\mathbf{I} = \frac{\mathbf{V}}{\mathbf{Z}} = \frac{V_m \angle 0}{R + j\omega L}$$

as obtained earlier.

EXERCISES

10.8.1 Derive (10.59).

10.8.2 Show in (a) that the voltage division rule,

$$\mathbf{V} = \frac{\mathbf{Z}_2}{\mathbf{Z}_1 + \mathbf{Z}_2}\mathbf{V}_g$$

and in (b) that the current division rule,

$$\mathbf{I} = \frac{\mathbf{Y}_2}{\mathbf{Y}_1 + \mathbf{Y}_2}\mathbf{I}_g = \frac{\mathbf{Z}_1}{\mathbf{Z}_1 + \mathbf{Z}_2}\mathbf{I}_g$$

are valid, where $\mathbf{Z}_1 = 1/\mathbf{Y}_1$ and $\mathbf{Z}_2 = 1/\mathbf{Y}_2$.

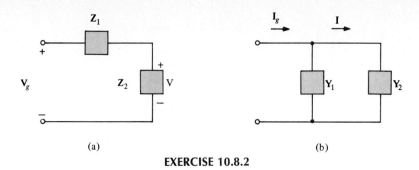

(a)   (b)

EXERCISE 10.8.2

10.8.3 Find the steady-state current i using phasors.
Answer $0.5 \cos (8t - 36.9°)$ A

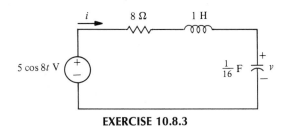

EXERCISE 10.8.3

10.8.4 Find the steady-state voltage v in Ex. 10.8.3 using phasors and voltage division.
Answer $\cos (8t - 126.9°)$ V

10.9

PHASOR CIRCUITS

As the discussion in the previous section suggests, we may omit the steps of finding the describing equation in the time domain, replacing the excitations and responses by their complex forcing functions and then dividing the equation through by $e^{j\omega t}$ to obtain the phasor equation. We may simply start with the phasor circuit, which we will now formally define as the time-domain circuit with the voltages and currents replaced by their phasors and the elements identified by their impedances, as illustrated previously in Fig. 10.18(b). The describing equation obtained from this circuit is then the phasor equation. Solving this equation yields the phasor of the answer, which then may be converted to the time-domain answer.

The procedure from starting with the phasor circuit to obtaining the phasor answer is identical to that used earlier in resistive circuits. The only difference is that the numbers are complex.

EXAMPLE 10.14 As an example, let us find the steady-state current i in Fig. 10.19(a). The phasor circuit, shown in Fig. 10.19(b), is obtained by replacing the voltage source and the currents by their phasors and labeling the elements with their impedances. In the phasor circuit the impedance seen from the source terminals is

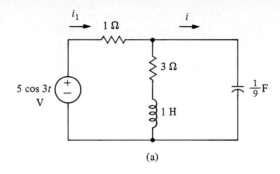

(a)

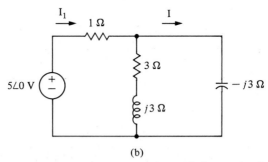

(b)

FIGURE 10.19 *RLC time-domain and phasor circuits*

$$\mathbf{Z} = 1 + \frac{(3 + j3)(-j3)}{3 + j3 - j3}$$

$$= 4 - j3 \ \Omega$$

Therefore, we have

$$\mathbf{I}_1 = \frac{5\underline{/0}}{4 - j3} = \frac{5\underline{/0}}{5\underline{/-36.9°}} = 1\underline{/36.9°}$$

and by current division,

$$\mathbf{I} = \left(\frac{3 + j3}{3 + j3 - j3}\right)\mathbf{I}_1 = \sqrt{2}\underline{/81.9°} \ \text{A}$$

In the time domain, the answer is

$$i = \sqrt{2} \cos (3t + 81.9°) \ \text{A}$$

In the case of a dependent source, such as a source kv_x volts controlled by a voltage v_x, it will appear in the phasor circuit as a source $k\mathbf{V}_x$, where $\mathbf{V}_x$ is the phasor representation of v_x, because $v_x = V_m \cos (\omega t + \phi)$ in the time domain will become $V_m e^{j(\omega t + \phi)}$ when a complex excitation is applied. Then dividing $e^{j\omega t}$ out of the equations leaves v_x represented by its phasor $V_m e^{j\phi}$. In the same way, $kv_x = kV_m \cos (\omega t + \phi)$ is represented by its phasor $kV_m e^{j\phi}$, which is k times the phasor of v_x.

EXAMPLE 10.15 As an example of a circuit containing a dependent source, let us consider Fig. 10.20(a), in which it is required to find the steady-state value of i. The corresponding phasor circuit is shown in Fig. 10.20(b). Since phasor circuits are analyzed exactly like resistive circuits, we may apply KCL at node a in Fig. 10.20(b), resulting in

$$\mathbf{I} + \frac{\mathbf{V}_1 - \frac{1}{2}\mathbf{V}_1}{-j2} = 3\underline{/0} \tag{10.61}$$

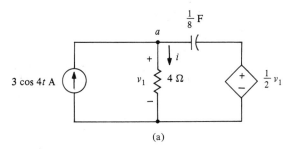

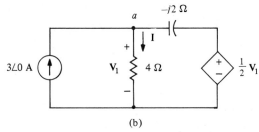

FIGURE 10.20 (a) Circuit containing a dependent source; (b) corresponding phasor circuit

By Ohm's law we have $\mathbf{V}_1 = 4\mathbf{I}$, which substituted into (10.61) yields

$$-j2\mathbf{I} + \frac{1}{2}(4\mathbf{I}) = -j6$$

or

$$\mathbf{I} = \frac{-j6}{2 - j2} = \frac{6\underline{/-90}}{2\sqrt{2}\underline{/-45}} = \frac{3}{\sqrt{2}}\underline{/-45°} \text{ A}$$

Therefore, we have

$$i = \frac{3}{\sqrt{2}} \cos (4t - 45°) \text{ A}$$

In the case of an op amp, the phasor circuit is the same as the time-domain circuit. That is, an ideal op amp in the time-domain circuit appears as an ideal op amp in the phasor circuit, because the time-domain equations

$$i = 0, \qquad v = 0$$

which characterize the current into and the voltage across the input terminals retain the identical form,

$$\mathbf{I} = 0, \qquad \mathbf{V} = 0$$

in the phasor equations.

As a final note we observe that the phasor method of finding, say i, by first finding $\mathbf{I} = \mathbf{V}/\mathbf{Z}$ and then converting $\mathbf{I}$ to i, fails to work if $\mathbf{Z}(j\omega) = 0$. This is the case when the circuit is excited at a natural frequency $j\omega$, and we must then use the method of Sec. 9.6. An example of this case is considered in Prob. 10.24.

EXERCISES

10.9.1 Solve Ex. 10.2.2 by means of the phasor circuit.

10.9.2 Find the steady-state voltage v using the phasor circuit.
Answer $2 \cos (4t - 53.1°)$ V

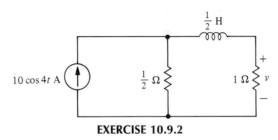

EXERCISE 10.9.2

10.9.3 Find the steady-state voltage v in Ex. 10.4.1(b) using the phasor circuit.

PROBLEMS

10.1 Determine if v_1 leads or lags v_2 and by what amount:
(a) $v_1 = 3 \cos (4t - 30°)$, $v_2 = 5 \sin 4t$.
(b) $v_1 = 20 (\cos 4t + \sqrt{3} \sin 4t)$, $v_2 = 3 \cos 4t + 4 \sin 4t$.

10.2 Find i_4, using only the properties of sinusoids, if (a) $i_1 = 5 \cos (3t + 30°)$ A, $i_2 = 5 \sin 3t$ A, and $i_3 = 5 \cos (3t + 150°)$ A and (b) $i_1 = 25 \cos (3t - 53.1°)$ A, $i_2 = 2 \sin 3t$ A, and $i_3 = 13 \cos (3t - 22.6°)$ A. (*Hint:* $\cos 22.6° = 12/13$.)

PROBLEM 10.2

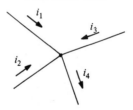

10.3 Find R and L in Fig. 10.4 if the source is 10 cos 3000t V and the response is $i =$ 4 cos (3000t − 36.9°) mA.

10.4 In the figure of Ex. 10.2.2, if the source is 3 cos 4000t mA and the output is $v =$ 9 cos (4000t − 53.1°) V, find R and C.

10.5 A voltage V_m sin ωt V, a resistor R, and an inductor L are all in series. Find the forced response i in the following two ways. (a) Noting that V_m sin $\omega t = V_m$ cos (ωt − 90°), use the complex excitation $V_m e^{j(\omega t-90°)}$ and take the real part of the current response. (b) Noting that V_m sin $\omega t = \text{Im}(V_m e^{j\omega t})$, use the complex excitation $V_m e^{j\omega t}$ and take the imaginary part of the current response.

10.6 Find v_1 from the differential equation and use the result to find the forced response v to an input voltage of 34 cos 4t V.

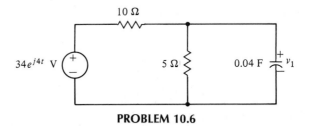

PROBLEM 10.6

10.7 Find the response v_1 to the source 2e^{j8t} A and use the result to find the response v to (a) 2 cos 8t A, and (b) 2 sin 8t A.

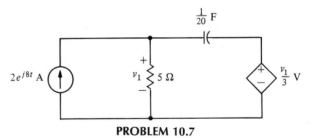

PROBLEM 10.7

10.8 A complex voltage input 10$e^{j(2t+25°)}$ V produces a current output of 5$e^{j(2t−20°)}$ A. Find the output current if the input voltage is (a) 40$e^{j(2t+60°)}$ V, (b) 20 cos 2t V, and (c) 4 sin (2t − 15°) V.

10.9 If $\omega = 10$ rad/s, find the time-domain functions represented by the phasors (a) −5 + j5, (b) −5 + j12, (c) 4 − j3, (d) −10, and (e) −j5.

10.10 Solve Prob. 10.2 using phasors.

10.11 Find the impedance of the circuit of Fig. 10.15 if the time-domain functions represented by the phasors **V** and **I** are
(a) $v = -15$ cos 2t + 8 sin 2t V, $i = 1.7$ cos (2t + 40°) A.
(b) $v = \text{Re}[je^{j2t}]$ V, $i = \text{Re}[(1 + j)e^{j(2t+30°)}]$ mA.
(c) $v = aV_m$ cos (ωt + θ) V, $i = V_m$ cos (ωt + θ − α) A.

10.12 Show that if Re **Z** = R is positive, then Re **Y** = Re $[1/\mathbf{Z}] = G$ is also positive.

10.13 A circuit has an impedance

$$\mathbf{Z} = \frac{4(1 + j\omega)(4 + j\omega)}{j\omega(2 + j\omega)}$$

Find the resistance, reactance, conductance, and susceptance at $\omega = 2$ rad/s. If the time-domain voltage applied to the circuit is 10 cos 2t V, find the steady-state current.

10.14 In Prob. 10.6, use phasors, impedance, and voltage division to find v.

10.15 Find the steady-state value of i if (a) $\omega = 4$ rad/s and (b) $\omega = 2$ rad/s. Note that in the latter case the impedance seen at the terminals of the source is purely resistive.

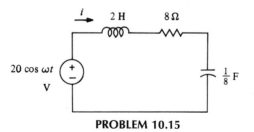

PROBLEM 10.15

10.16 Find the steady-state values of i and v.

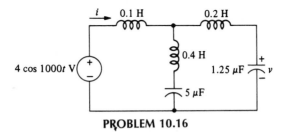

PROBLEM 10.16

10.17 Find C so that the impedance seen by the source is real. Find the power absorbed by the 6-Ω resistor in this case.

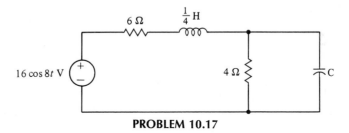

PROBLEM 10.17

10.18 Find the steady-state value of i.

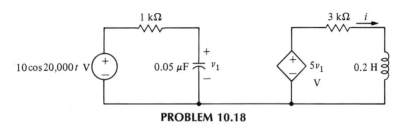

PROBLEM 10.18

10.19 Find the steady-state currents i and i_1 using phasors.

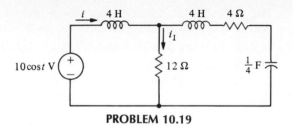

PROBLEM 10.19

10.20 Find the steady-state value of v.

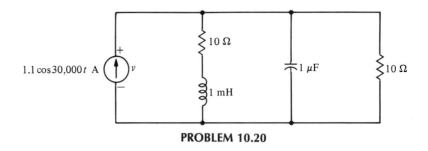

PROBLEM 10.20

10.21 Find the steady-state voltage v.

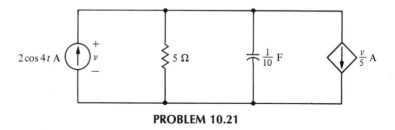

PROBLEM 10.21

10.22 Find the steady-state current i.

PROBLEM 10.22

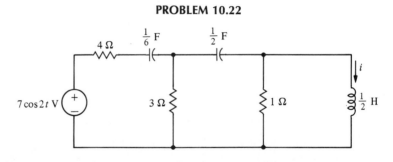

10.23 Find the steady-state voltage v if $v_g = 10 \cos 10{,}000t$ V.

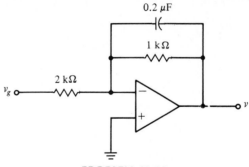

PROBLEM 10.23

10.24 Find the forced response i. [*Suggestion:* Note that $\mathbf{I} = \mathbf{V}/\mathbf{Z}$ fails to work since $\mathbf{Z}(j1) = 0$. Solve the describing equation by the method of Sec. 9.6.]

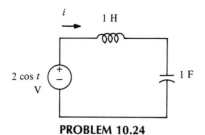

PROBLEM 10.24

10.25 Find the *complete* response i if $i(0) = 2$ A and $v(0) = 6$ V. (*Suggestion:* Use phasors to get i_f and the differential equation to get i_n.)

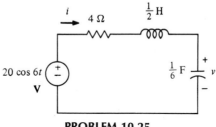

PROBLEM 10.25

10.26 Determine $i(0)$ and $v(0)$ in Prob. 10.25 so that the natural component vanishes and i is simply the forced component.

10.27 Find the steady-state voltage v if $v_g = 5 \cos 2t$ V.

PROBLEM 10.27

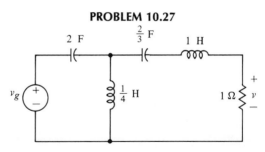

11

Samuel F. B. Morse
1791–1872

AC Steady-State Analysis

What hath God wrought! [The famous message tapped out on the first telegraph]

Samuel F. B. Morse

The first practical application of electricity is said by many to be the telegraph, developed by Samuel F. B. Morse, an American portrait painter and inventor. Morse built on the ideas of the famous American physicist Joseph Henry, using the opening and closing of relays to produce the dots and dashes (or Morse code) that represent letters and numbers.

Morse was born in Charlestown, Massachusetts, the son of a minister and author. He studied to be an artist at Yale and the Royal Academy of Arts in London, and by 1815 he was considered to be moderately successful. In 1826 he helped found and became the first president of the National Academy of Design. But the previous year his wife had died, in 1826 his father died, and in 1828 his mother died. The following year the distressed Morse went to Europe to recover and study further. In 1832, while returning home on board the passenger ship *Sully,* he met an eccentric inventor and became intrigued with developing a telegraph, the principle of which had already been considered by Henry. By 1836 Morse had a working model, and in 1837 he acquired a partner, Alfred Vail, who financed the project. Their efforts were rewarded with a patent and the financing by Congress of a telegraph in 1844, over which Morse—on May 24, 1844—sent his now-famous message, "What hath God wrought!" ■

In the previous chapter we have seen that in the case of circuits with sinusoidal inputs we may obtain the ac steady-state response by analyzing the corresponding phasor circuits. The circuits encountered were relatively simple ones which could be analyzed by the use of voltage-current relations and current and voltage division rules.

It should be clear, because of the close kinship between phasor circuits and resistive circuits, that we may extend the methods of Chapter 10 to more general circuits using nodal analysis, loop analysis, Thevenin's and Norton's theorems, super-position, etc. In this chapter we shall formally consider these more general analysis procedures, limiting ourselves to obtaining the forced, or ac steady-state, response.

11.1

NODAL ANALYSIS

As we have seen, the voltage-current relation

$$\mathbf{V} = \mathbf{ZI}$$

for passive elements is identical in form to Ohm's law, and KVL and KCL hold in phasor circuits exactly as they did in resistive circuits. Therefore the only difference in analyzing phasor circuits and resistive circuits is that the excitations and responses are complex quantities in the former case and real quantities in the latter case. Thus we may analyze phasor circuits in exactly the same manner in which we analyzed resistive circuits. Specifically, nodal and mesh, or loop, analysis methods apply. We shall illustrate nodal analysis in this section and loop analysis in the following section.

EXAMPLE 11.1   To illustrate the nodal method, let us find the ac steady-state voltages v_1 and v_2 of Fig. 11.1. First we shall obtain the phasor circuit by replacing the element values by their impedances for $\omega = 2$ rad/s and the sources and node voltages by their phasors. This results in the circuit of Fig. 11.2(a). Since we are interested in finding $\mathbf{V}_1$ and $\mathbf{V}_2$, the node voltage phasors, we may replace the two sets of parallel impedances by their equivalent impedances, resulting in the simpler, equivalent circuit of Fig. 11.2(b).

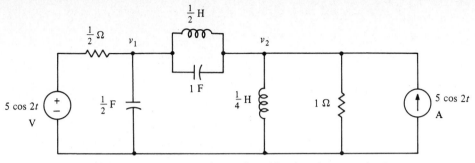

FIGURE 11.1 Circuit to be analyzed by the phasor method

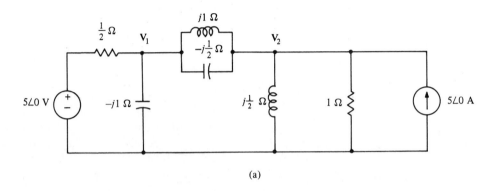

(a)

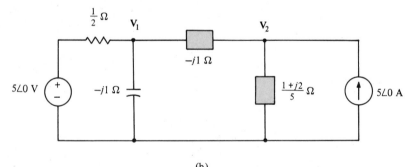

(b)

FIGURE 11.2 Two versions of the phasor circuit corresponding to Fig. 11.1

The nodal equations, from Fig. 11.2(b), are

$$2(\mathbf{V}_1 - 5\underline{/0}) + \frac{\mathbf{V}_1}{-j1} + \frac{\mathbf{V}_1 - \mathbf{V}_2}{-j1} = 0$$

$$\frac{\mathbf{V}_2 - \mathbf{V}_1}{-j1} + \frac{\mathbf{V}_2}{(1 + j2)/5} = 5\underline{/0}$$

which in simplified form are

$$(2 + j2)\mathbf{V}_1 - j1\mathbf{V}_2 = 10$$
$$-j1\mathbf{V}_1 + (1 - j1)\mathbf{V}_2 = 5$$

Solving these equations by determinants, we have

$$V_1 = \frac{\begin{vmatrix} 10 & -j1 \\ 5 & 1-j1 \end{vmatrix}}{\begin{vmatrix} 2+j2 & -j1 \\ -j1 & 1-j1 \end{vmatrix}} = \frac{10-j5}{5} = 2-j1 \text{ V}$$

$$V_2 = \frac{\begin{vmatrix} 2+j2 & 10 \\ -j1 & 5 \end{vmatrix}}{5} = \frac{10+j20}{5} = 2+j4 \text{ V}$$

In polar form these quantities are

$$V_1 = \sqrt{5}\,\underline{/-26.6°} \text{ V}$$
$$V_2 = 2\sqrt{5}\,\underline{/63.4°} \text{ V}$$

Therefore, the time-domain solutions are

$$v_1 = \sqrt{5}\cos(2t-26.6°) \text{ V}$$
$$v_2 = 2\sqrt{5}\cos(2t+63.4°) \text{ V}$$

EXAMPLE 11.2 As an example involving a dependent source, let us consider Fig. 11.3, in which it is required to find the forced response i. Taking the ground node as shown, we have the two unknown node voltages v and $v + 3000i$, as indicated. The phasor circuit in its simplest form is shown in Fig. 11.4, from which we may observe that only one nodal equation is needed. Writing KCL at the generalized node, shown dashed, we have

$$\frac{V-4}{\frac{1}{2}(10^3)} + \frac{V}{\frac{2}{3}(1-j2)(10^3)} + \frac{V+3000I}{(2-j1)(10^3)} = 0$$

Also, from the phasor circuit we have

$$I = \frac{4-V}{\frac{1}{2}(10^3)}$$

Eliminating V between these two equations and solving for I, we have

FIGURE 11.3 Circuit containing a dependent source

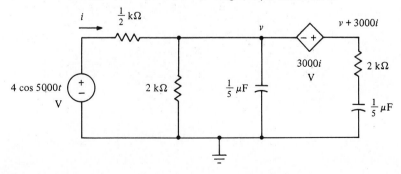

331

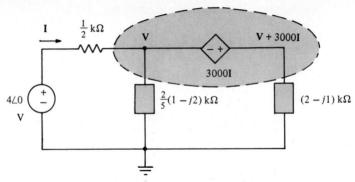

FIGURE 11.4 Phasor circuit of Fig. 11.3

$$I = 24 \times 10^{-3} \underline{/53.1°}\ A$$

$$= 24\ \underline{/53.1°}\ mA$$

Therefore, in the time domain, we have

$$i = 24\ \cos\ (5000t + 53.1°)\ mA$$

EXAMPLE 11.3   As a final example illustrating nodal analysis, let us find the forced response v in Fig. 11.5 if

$$v_g = V_m\ \cos\ \omega t\ V$$

We note first that the op amp and the two 2-kΩ resistors constitute a VCVS with gain $1 + \frac{2000}{2000} = 2$ (see Sec. 3.4). Therefore, $v = 2v_2$, or $v_2 = v/2$, as indicated by the phasor $\mathbf{V}/2$ in the phasor circuit of Fig. 11.6.

Writing nodal equations at the nodes labeled $\mathbf{V}_1$ and $\mathbf{V}/2$, we have

$$\frac{\mathbf{V}_1 - V_m\underline{/0}}{(1/\sqrt{2})(10^3)} + \frac{\mathbf{V}_1 - (\mathbf{V}/2)}{\sqrt{2}(10^3)} + \frac{\mathbf{V}_1 - \mathbf{V}}{-j10^6/\omega} = 0$$

$$\frac{(\mathbf{V}/2) - \mathbf{V}_1}{\sqrt{2}(10^3)} + \frac{\mathbf{V}/2}{-j10^6/\omega} = 0$$

Eliminating $\mathbf{V}_1$ and solving for $\mathbf{V}$ results in

$$\mathbf{V} = \frac{2V_m}{[1 - (\omega^2/10^6)] + j(\sqrt{2}\omega/10^3)}$$

In polar form this is

$$\mathbf{V} = \frac{2V_m\underline{/\theta}}{\sqrt{1 + (\omega/1000)^4}} \tag{11.1}$$

where

$$\theta = -\tan^{-1}\left[\frac{\sqrt{2}\omega/1000}{1 - (\omega/1000)^2}\right] \tag{11.2}$$

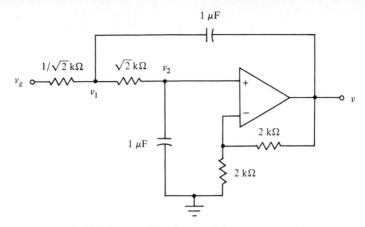

FIGURE 11.5 Circuit containing an op amp

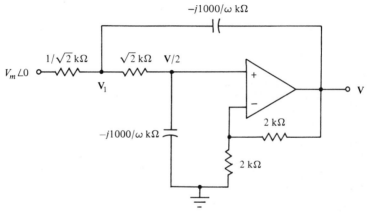

FIGURE 11.6 Phasor circuit of Fig. 11.5

In the time domain we have

$$v = \frac{2V_m}{\sqrt{1 + (\omega/1000)^4}} \cos{(\omega t + \theta)} \qquad (11.3)$$

We might note in this example that for low frequencies, say $0 < \omega < 1000$, the amplitude of the output voltage v is relatively large, and for higher frequencies, its amplitude is relatively small. Thus the circuit of Fig. 11.5 *filters* out higher frequencies and allows lower frequencies to "pass." Such a circuit is called a *filter* and will be considered in more detail in Chapter 15.

EXERCISES

11.1.1 Find the forced response v using nodal analysis.
Answer $4 \sin 6t$ V

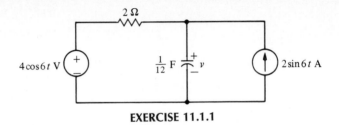

EXERCISE 11.1.1

11.1.2 Find the steady-state value of v using nodal analysis.
Answer $3\sqrt{2} \cos(4t - 135°)$ V

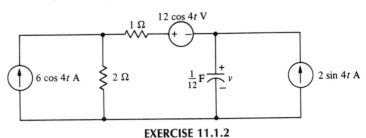

EXERCISE 11.1.2

11.1.3 Find the amplitude of v in (11.3) if $V_m = 10$ V for (a) $\omega = 0$, (b) $\omega = 1000$ rad/s, (c) $\omega = 10,000$ rad/s, and (d) $\omega = 100,000$ rad/s.
Answer (a) 20; (b) 14.14; (c) 0.2; (d) 0.002 V

11.1.4 Find the steady-state value of v using nodal analysis.
Answer $3\sqrt{2} \cos(2t - 135°)$ V

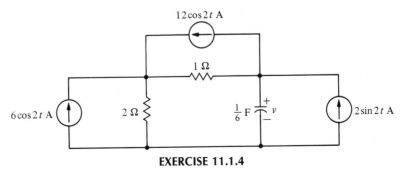

EXERCISE 11.1.4

11.2

MESH ANALYSIS

EXAMPLE 11.4 To illustrate mesh analysis of an ac steady-state circuit, let us find v_1 in Fig. 11.1, which was obtained, using nodal analysis, in the previous section. We shall use the phasor circuit of Fig. 11.2(b), which is redrawn in Fig. 11.7, with mesh currents I_1 and I_2, as indicated. Evidently, the phasor voltage V_1 may be obtained as

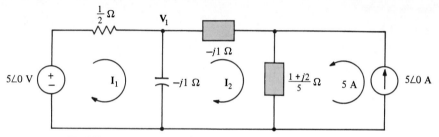

FIGURE 11.7 Circuit of Fig. 11.2 redrawn for mesh analysis

$$\mathbf{V}_1 = 5 - \frac{\mathbf{I}_1}{2} \qquad (11.4)$$

The two mesh equations are

$$\frac{1}{2}\mathbf{I}_1 - j1(\mathbf{I}_1 - \mathbf{I}_2) = 5$$

$$-j1(\mathbf{I}_2 - \mathbf{I}_1) - j1\mathbf{I}_2 + \left(\frac{1 + j2}{5}\right)(\mathbf{I}_2 + 5) = 0 \qquad (11.5)$$

Solving these equations for $\mathbf{I}_1$, we have

$$\mathbf{I}_1 = 6 + j2 \ \text{A}$$

which substituted into (11.4) yields

$$\mathbf{V}_1 = 2 - j1 \ \text{V}$$

This is the same result that was obtained in the preceding section and may be used to obtain the time-domain voltage v_1.

The same shortcut procedures for writing loop and nodal equations, discussed in Secs. 4.1 and 4.5 for resistive circuits, apply to phasor circuits. For example, in Fig. 11.7 if $\mathbf{I}_3 = -5$ is the mesh current in the right mesh in the clockwise direction, the two mesh equations are written down by inspection as

$$\left(\frac{1}{2} - j1\right)\mathbf{I}_1 - (-j1)\mathbf{I}_2 = 5$$

$$-(-j1)\mathbf{I}_1 + \left(-j1 - j1 + \frac{1 + j2}{5}\right)\mathbf{I}_2 - \left(\frac{1 + j2}{5}\right)\mathbf{I}_3 = 0$$

These are equivalent to (11.5) and are formed as in the resistive circuit case. That is, in the first equation the coefficient of the first variable is the sum of the impedances around the first mesh. The other coefficients are the negatives of the impedances common to the first mesh and the meshes whose numbers correspond to the currents. The right member is the sum of the voltage sources in the mesh with polarities consistent with the direction of the mesh current. Replacing "first" by "second" applies to the next equation, and so forth. The dual development, as described in Sec. 4.1, holds for nodal equations.

EXAMPLE 11.5 As a final example, let us consider the circuit of Fig. 11.8(a), where the response is the steady-state value of v_1. The phasor circuit is shown in Fig. 11.8(b), with the loop currents as indicated.

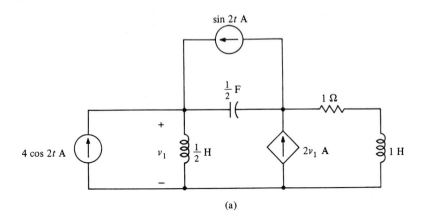

(a)

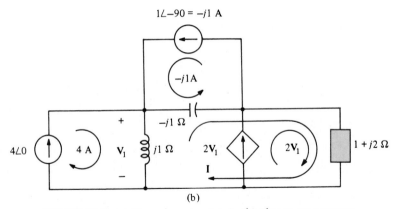

(b)

FIGURE 11.8 (a) Time-domain circuit; (b) phasor counterpart

Applying KVL around the loop labeled $\mathbf{I}$, we have

$$-\mathbf{V}_1 - j1(-j1 + \mathbf{I}) + (1 + j2)(\mathbf{I} + 2\mathbf{V}_1) = 0$$

Also, from the figure we see that

$$\mathbf{V}_1 = j1(4 - \mathbf{I})$$

Eliminating $\mathbf{I}$ from these equations and solving for $\mathbf{V}_1$, we have

$$\mathbf{V}_1 = \frac{-4 + j3}{5} = 1\underline{/143.1^\circ} \text{ V}$$

Therefore, in the time domain, the voltage is

$$v_1 = \cos(2t + 143.1^\circ) \text{ V}$$

EXERCISES

11.2.1 Find the forced response i in Fig. 11.3 using mesh analysis.

11.2.2 Solve Ex. 11.1.4 using mesh analysis.

11.2.3 Find the steady-state current i using loop analysis.
Answer $2\sqrt{2} \cos(2t - 45°)$ A

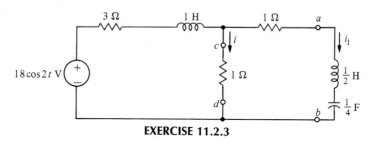

EXERCISE 11.2.3

11.3

NETWORK THEOREMS

Because the phasor circuits are exactly like the resistive circuits except for the nature of the currents, voltages, and impedances, all the network theorems discussed in Chapter 5 for resistive circuits apply to phasor circuits. In this section we shall illustrate superposition, Thevenin's theorem, Norton's theorem, and the proportionality principle, as applied to linear phasor circuits.

In the case of superposition, if a phasor circuit has two or more inputs, we may find the phasor currents or voltages due to each input acting alone (i.e., with the others dead) and add the individual corresponding time-domain responses to obtain the total. In the case of a circuit like Fig. 11.1, we may solve the corresponding phasor circuit of Fig. 11.2 by mesh or nodal analysis or by superposition because both sources are operating at the same frequency, namely $\omega = 2$ rad/s. If the sources have *different* frequencies, we *must* use superposition, because the definition of $\mathbf{Z}(j\omega)$ allows us to use only one frequency at a time, and thus we cannot even construct a phasor circuit.

EXAMPLE 11.6 To illustrate superposition, let us find the forced response i in Fig. 11.9. There are two sources, one an ac source with $\omega = 2$ rad/s and one a dc source with $\omega = 0$. Therefore,

$$i = i_1 + i_2$$

where i_1 is due to the voltage source acting alone and i_2 is due to the current source acting alone. Using phasors, we may find i_1 and i_2 by finding their respective phasor representations $\mathbf{I}_1$ and $\mathbf{I}_2$, which are the phasor currents shown in Fig. 11.10(a) and (b), respectively. Figure 11.10(a) is a phasor circuit representing the time-domain circuit with the current source killed and $\omega = 2$ rad/s. Figure 11.10(b) is also a phasor circuit, with the voltage source killed and $\omega = 0$.

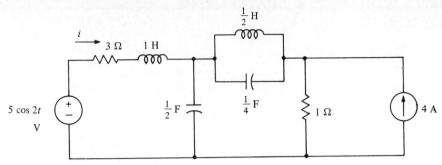

FIGURE 11.9 Circuit with an ac and a dc source

In the latter case, $\omega = 0$, the inductors are short circuits ($\mathbf{Z}_L = j0 = 0$) and the capacitor is an open circuit ($\mathbf{Z}_C = 1/j\omega C$, which becomes infinite as $\omega \to 0$). Also, since the current source is

$$i_g = 4 \cos (0t + 0) \text{ A}$$

its phasor representation is

$$\mathbf{I}_g = 4\underline{/0°} \text{ A}$$

In short, this is just the dc case considered earlier, as is evident from Fig. 11.10(b). From Fig. 11.10(a) we have,

$$\mathbf{I}_1 = \frac{5\underline{/0°}}{3 + j2 + [(1 + j2)(-j1)/(1 + j2 - j1)]}$$

$$= \sqrt{2}\underline{/-8.1°}$$

FIGURE 11.10 Phasor circuits representing Fig. 11.9

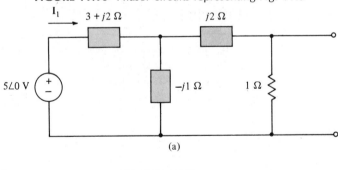

(a)

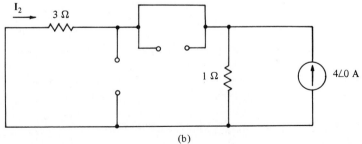

(b)

from which

$$i_1 = \sqrt{2} \cos (2t - 8.1°) \text{ A}$$

From Fig. 11.10(b) we have, by current division,

$$\mathbf{I}_2 = -\left(\frac{1}{1+3}\right)(4) = -1\underline{/0°} \text{ A}$$

from which, since $\omega = 0$.

$$i_2 = -1 \text{ A}$$

Therefore, the total forced response is

$$i = i_1 + i_2$$
$$= \sqrt{2} \cos (2t - 8.1°) - 1 \text{ A}$$

For an example of a circuit with two sinusoidal sources with nonzero frequencies, the reader is referred to Ex. 11.3.1. The procedure is, of course, exactly the same as in the preceding example.

In the case of Thevenin's and Norton's theorems the procedure is identical to that for resistive circuits. The only change is that v_{oc} and i_{sc}, the time-domain open-circuit voltage and short-circuit current, are replaced by their phasor representations, $\mathbf{V}_{oc}$ and $\mathbf{I}_{sc}$, and R_{th}, the Thevenin resistance, is replaced by $\mathbf{Z}_{th}$, the Thevenin impedance (of the dead circuit). There must be only a single frequency present, of course; otherwise we must use superposition to break the problem up into single-frequency problems, in each of which Thevenin's or Norton's theorem may be applied.

In general, the Thevenin and Norton equivalent circuits in the frequency domain are shown in Fig. 11.11. There is, of course, a close similarity with the resistive case.

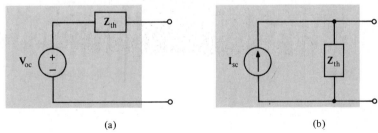

(a)   (b)

FIGURE 11.11 (a) Thevenin and (b) Norton equivalent phasor circuits

EXAMPLE 11.7   As an example, let us use Thevenin's theorem to find the forced response v of Fig. 11.12. We shall find its phasor representation $\mathbf{V}$ using the Thevenin equivalent of the phasor circuit to the left of terminals a-b. The open-circuit phasor voltage $\mathbf{V}_{oc}$ is found from Fig. 11.13(a), and the short-circuit phasor current is found from Fig. 11.13(b). Then the Thevenin impedance, as for resistive circuits, is

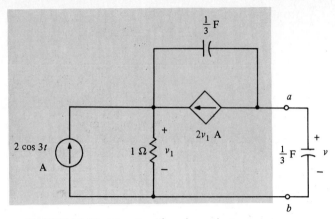

FIGURE 11.12 Circuit with a dependent current source

FIGURE 11.13 Phasor circuits for use in Thevenin's theorem

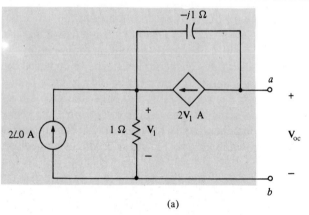

(a)

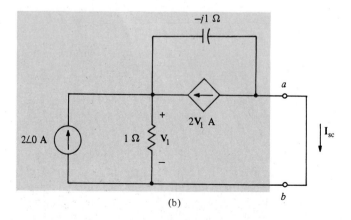

(b)

340

$$Z_{th} = \frac{V_{oc}}{I_{sc}} \tag{11.6}$$

In Fig. 11.13(a), since terminals a-b are open, the current $2\underline{/0°}$ flows in the resistor and the current $2V_1$ flows in the capacitor. Therefore, by KVL we have

$$V_{oc} = V_1 - (-j1)(2V_1)$$

where

$$V_1 = 2(1) = 2 \text{ V}$$

Thus we have

$$V_{oc} = 2 + j4 \text{ V}$$

In Fig. 11.13(b), the two nodal equations needed are

$$\frac{V_1}{1} + \frac{V_1}{-j1} - 2V_1 = 2$$

and

$$I_{sc} = -V_1 + 2$$

From these we find

$$I_{sc} = 3 + j1 \text{ A}$$

The Thevenin impedance, by (11.6), is therefore

$$Z_{th} = \frac{2 + j4}{3 + j1} = 1 + j1 \text{ } \Omega$$

The Thevenin equivalent circuit is shown in Fig. 11.14, with the $\frac{1}{3}$-F capacitor, corresponding to $-j1$ Ω, connected to terminals a-b.

It is a simple matter now to see, by voltage division, that

$$V = \left[\frac{-j1}{(1 + j1) + (-j1)} \right](2 + j4)$$

$$= 4 - j2$$

$$= 2\sqrt{5}\underline{/-26.6°} \text{ V}$$

FIGURE 11.14 Thevenin equivalent phasor circuit of Fig. 11.12

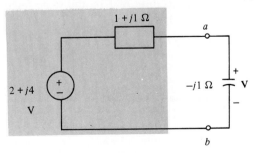

Therefore, in the time-domain, we have

$$v = 2\sqrt{5} \cos (3t - 26.6°) \text{ V}$$

EXAMPLE 11.8   As a final topic in this section, let us consider the ladder network of Fig. 11.15 and use the proportionality principle to obtain the steady-state response **V**. This requires, as in the resistive case, that we assume **V** to have some convenient value like **V** = 1 and work backward to find the corresponding V_g. Then the correct value of **V** is found by multiplying the assumed value by an appropriate constant.

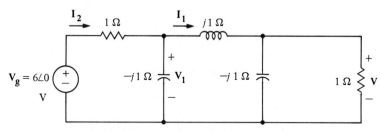

FIGURE 11.15   Phasor ladder network

Let us begin by assuming

$$\mathbf{V} = 1 \text{ V}$$

Then from the circuit we have

$$\mathbf{I}_1 = \frac{\mathbf{V}}{1} + \frac{\mathbf{V}}{-j1} = 1 + j1 \text{ A}$$

Continuing, we have

$$\mathbf{V}_1 = j1\mathbf{I}_1 + \mathbf{V} = j1(1 + j1) + 1 = j1 \text{ V}$$

$$\mathbf{I}_2 = \frac{\mathbf{V}_1}{-j1} + \mathbf{I}_1 = -1 + (1 + j1) = j1 \text{ A}$$

$$\mathbf{V}_g = 1\mathbf{I}_2 + \mathbf{V}_1 = j1 + j1 = j2 \text{ V}$$

Therefore, **V** = 1 is the response to $V_g = j2$. If we multiply this value of V_g by $6/j2$ to get the correct value of V_g, then by the proportionality principle, we must multiply the assumed response of 1 by the same factor, $6/j2$, to get the correct value of **V**. Therefore, we have

$$\mathbf{V} = \left(\frac{6}{j2}\right)(1) = -j3 \text{ V}$$

EXERCISES

11.3.1   Find the steady-state current i.
Answer $2 \cos(2t - 36.9°) + 3 \cos(t + 73.8°)$ A

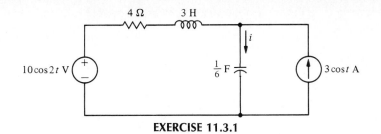

EXERCISE 11.3.1

11.3.2 For the phasor circuit corresponding to Ex. 11.2.3, replace the part to the left of terminals *a-b* by its Thevenin equivalent and find the steady-state current i_1.

Answer $\mathbf{V}_{oc} = \frac{9}{5}(2 - j1)$ V, $\mathbf{Z}_{th} = \frac{1}{10}(18 + j1)$ Ω, $i_1 = 2 \cos 2t$ A

11.3.3 Find $\mathbf{V}_1$, $\mathbf{I}_1$, and $\mathbf{I}_2$ in Fig. 11.15.

Answer 3 V, $3 - j3$ A, 3 A

11.4

PHASOR DIAGRAMS

Since phasors are complex numbers, they may be represented by vectors in a plane, where operations, such as addition of phasors, may be carried out geometrically. Such a sketch is called a *phasor diagram* and may be quite helpful in analyzing ac steady-state circuits.

EXAMPLE 11.9 To illustrate, let us consider the phasor circuit of Fig. 11.16, for which we shall draw all the voltages and currents on a phasor diagram. To begin with, let us observe that the current **I** is common to all elements and take it as our *reference* phasor, denoting it by

$$\mathbf{I} = |\mathbf{I}|\underline{/0°}$$

We have taken the angle of **I** arbitrarily to be zero, since we want **I** to be our reference. We may always adjust this assumed value to the true value by the proportionality principle discussed in the previous section.

The voltage phasors of the circuit are

FIGURE 11.16 *RLC* series phasor circuit

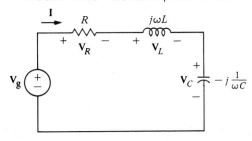

$$V_R = RI = R|\mathbf{I}|$$

$$V_L = j\omega L\mathbf{I} = \omega L|\mathbf{I}|\underline{/90°}$$

$$V_C = -j\frac{1}{\omega C}\mathbf{I} = \frac{1}{\omega C}|\mathbf{I}|\underline{/-90°}$$

and

$$\mathbf{V}_g = \mathbf{V}_R + \mathbf{V}_L + \mathbf{V}_C$$

These are shown in the phasor diagram of Fig. 11.17(a), where it is assumed that $|\mathbf{V}_L| > |\mathbf{V}_C|$. The cases, $|\mathbf{V}_L| < |\mathbf{V}_C|$ and $|\mathbf{V}_L| = |\mathbf{V}_C|$, are shown in Fig. 11.17(b) and (c), respectively. In all cases the lengths representing the units of current and voltage are not necessarily the same, so that for clarity $\mathbf{I}$ is shown longer than $\mathbf{V}_R$.

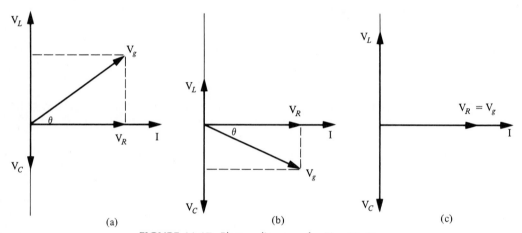

FIGURE 11.17 Phasor diagrams for Fig. 11.16

In case (a) the net reactance is inductive, and the current lags the source voltage by the angle θ that can be measured. In (b) the circuit has a net capacitive reactance, and the current leads the voltage. Finally, in (c) the current and voltage are in phase, since the inductive and capacitive reactance components exactly cancel each other. These conclusions follow also from the equation

$$\mathbf{I} = \frac{\mathbf{V}_g}{\mathbf{Z}} = \frac{\mathbf{V}_g}{R + j[\omega L - (1/\omega C)]} \qquad (11.7)$$

Case (c) is characterized by

$$\omega L - \frac{1}{\omega C} = 0$$

or

$$\omega = \frac{1}{\sqrt{LC}} \qquad (11.8)$$

If the current in Fig. 11.16 is fixed, then the real component of the voltage $\mathbf{V}_g$ is fixed, since it is $R|\mathbf{I}|$. In this case the *locus* of the phasor $\mathbf{V}_g$ (its possible location on the phasor diagram) is the dashed line of Fig. 11.18. The voltage phasor varies up and down this line as ω varies between zero and infinity. The minimum amplitude of the voltage occurs when $\omega = 1/\sqrt{LC}$, as seen from the figure. For any other frequency, a larger amplitude of voltage is required for the same current.

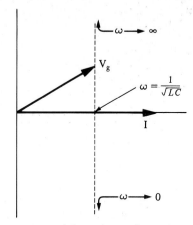

FIGURE 11.18 Locus of the voltage phasor for a fixed current

EXAMPLE 11.10 As a last example illustrating the use of phasor diagrams, let us find the locus of $\mathbf{I}$ as R varies in Fig. 11.19. The current is given by

$$\mathbf{I} = \frac{V_m}{R + j\omega L} = \frac{V_m(R - j\omega L)}{R^2 + \omega^2 L^2}$$

Therefore, if

$$\mathbf{I} = x + jy \tag{11.9}$$

we have

$$x = \operatorname{Re} \mathbf{I} = \frac{RV_m}{R^2 + \omega^2 L^2} \tag{11.10}$$

$$y = \operatorname{Im} \mathbf{I} = \frac{-\omega L V_m}{R^2 + \omega^2 L^2} \tag{11.11}$$

FIGURE 11.19 *RL* phasor circuit

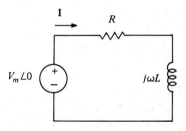

The equation of the locus is the equation satisfied by x and y as R varies; thus we need to eliminate R between these last two equations.

If we divide the first of these two equations by the second, we have

$$\frac{x}{y} = -\frac{R}{\omega L}$$

from which

$$R = -\frac{\omega L x}{y}$$

Substituting this value of R into (11.11) we have, after some simplification,

$$x^2 + y^2 = -\frac{V_m y}{\omega L}$$

This result may be rewritten as

$$x^2 + \left(y + \frac{V_m}{2\omega L}\right)^2 = \left(\frac{V_m}{2\omega L}\right)^2 \qquad (11.12)$$

which is the equation of a circle with center at $[0, -(V_m/2\omega L)]$ and radius $V_m/2\omega L$.

The circle (11.12) appears to be the locus, as R varies, of the phasor $\mathbf{I} = x + jy$. However, by (11.10), $x \geq 0$; thus the locus is actually the semicircle shown dashed on the phasor diagram of Fig. 11.20. The voltage $V_m\underline{/0}$, taken as reference, is also shown, along with the phasor $\mathbf{I}$. If $R = 0$, we have, from (11.10) and (11.11), $x = 0$ and $y = -V_m/\omega L$. If $R \to \infty$, then $x \to 0$ and $y \to 0$. Thus as R varies from 0 to ∞, the current phasor moves counterclockwise along the circle.

FIGURE 11.20 Locus of the phasor $\mathbf{I}$

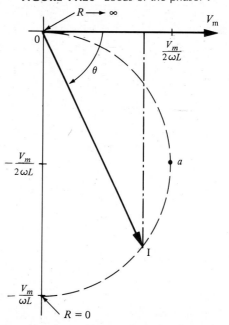

If **I** is as shown in Fig. 11.20, the current phasor may be resolved into two components, one having amplitude $I_m \cos \theta$ in phase with the voltage and one with amplitude $I_m \sin \theta$, which is 90° out of phase with the voltage. This construction is indicated by the dashed and dotted vertical line. As we shall see in the next chapter, the in-phase component of the current is important in calculating the average power delivered by the source. Thus the phasor diagram gives us a method of seeing at a glance the maximum in-phase component of current. Evidently this occurs at point a, which corresponds to $\theta = 45°$. This is the case $x = -y$, or $R = \omega L$.

EXERCISES

11.4.1 Eliminate ωL in (11.10) and (11.11) and show that, as ωL varies, the locus of the phasor $\mathbf{I} = x + jy$ is a semicircle.

$Answer \quad \left(x - \dfrac{V_m}{2R}\right)^2 + y^2 = \left(\dfrac{V_m}{2R}\right)^2, \ y \leq 0$

11.4.2 Find ωL in Ex. 11.4.1 so that Im **I** has its largest negative value. Also find **I** for this case.

$Answer \quad R, \ (V_m/\sqrt{2}\,R)\underline{/-45°}$

11.5

SPICE FOR AC STEADY-STATE CIRCUITS

SPICE is a powerful tool in the analysis of ac steady-state circuits and is particularly helpful in performing the many tedious operations of complex arithmetic associated with obtaining these solutions. The procedure is very similar to that described for the dc case using the .AC solution control statement. SPICE, in this instance, analyzes the phasor circuit for the phasor currents and voltages resulting from both independent and dependent sources, which are also expressed as phasors. Since SPICE uses the phasor circuit in the frequency domain, all ac sources must have the same frequency. The reader should review the .AC statement of Appendix E before proceeding with this section.

EXAMPLE 11.11 To illustrate the application of SPICE, consider the circuit of Fig. 11.1. Let us find the voltage of node 1 in polar form and the current of the 1-ohm resistor in rectangular form. A circuit file for calculating these values is

```
AC STEADY-STATE SOLUTION FOR CIRCUIT OF FIG. 11.1.
* DATA STATEMENTS
V1 100 0 AC 5 0
R1 100 1 .5
C1 1 0 .5
C2 1 2 1
L1 1 2 .5
L2 2 0 .25
R2 2 0 1
I1  0 2 AC 5 0
```

```
* SOLUTION CONTROL STATEMENT FOR AC ANALYSIS (f = 2/2*PI Hz)
.AC LIN 1 .3183 .3183
* OUTPUT CONTROL STATEMENT FOR V(1) & I(R2)
.PRINT AC VM(1) VP(1) IR(R2) II(R2)
.END
```

The solution printed in this case is

FREQ	VM(1)	VP(1)	IR(R2)	II(R2)
3.183E−01	2.263E+00	−2.657E+01	2.000E+00	4.000E+00

EXAMPLE 11.12 As a second example, consider finding the phasor current **I** in the circuit of Fig. 11.3. A circuit file for nodes being numbered sequentially clockwise, beginning with 1 at the positive terminal of the sinusoidal source, is

```
AC STEADY-STATE SOLUTION FOR FIG. 11.3.
* DATA STATEMENTS
V 1 0 AC 4 0
R1 1 2 0.5K
R2 2 0 2K
C1 2 0 0.2UF
H 3 2 V −3000
R3 3 4 2K
C2 4 0 0.2UF
* SOLUTION CONTROL STATEMENT FOR f = 5000/(2*3.1416)
.AC LIN 1 795.77 795.77
.PRINT AC IM(R1) IP(R1)
.END
```

The solution is

FREQ	IM(R1)	IP(R1)
7.958E+02	2.400E−02	5.313E+01

EXAMPLE 11.13 For a final example of the utility of SPICE, let us find the phasor output voltage of the op amp circuit of Fig. 11.5 if the input voltage is $v_g = 10 \cos(1000t + 30°)$ V. A circuit file for the nodes of the op amp inverting input, op amp output, and input source, assigned as 3, 4, and 10, respectively, with nodes 1 and 2 as shown, is

```
AC STEADY-STATE SOLUTION OF FIG. 11.5.
* DATA STATEMENTS USING OPAMP.CKT OF CHAPTER 4
.LIB OPAMP.CKT
VG 10 0 AC 10 30
R1 10 1 0.707K
R2 1 2 1.414K
C1 1 4 1UF
C2 2 0 1UF
R3 3 0 2K
R4 3 4 2K
XOPAMP 3 2 4 OPAMP
* SOLUTION CONTROL STATEMENT [f = 1000/(2*3.1416) Hz]
```

```
.AC LIN 1 159.15 159.15
.PRINT AC VM(4) VP(4)
.END
```

This circuit file gives a solution

FREQ	VM(4)	VP(4)
1.592E+02	1.414E+01	−5.999E+01

Our discussion has been limited to solutions for circuits at a single frequency. Plots of responses for varying frequency inputs will be discussed in Chapter 15.

EXERCISES

11.5.1 Use SPICE to find the phasor representation of v in Ex. 11.1.4 for $\omega = 5$ rad/s.
Answer $2.224 \underline{/-158.2°}$ V

11.5.2 Use SPICE to find the phasor current of the 1-H inductor in Fig. 11.8(a) for $\omega = 6.283$ rad/s.
Answer $0.331 \underline{/47.8°}$ A

PROBLEMS

11.1 Find the steady-state voltage v using nodal analysis.

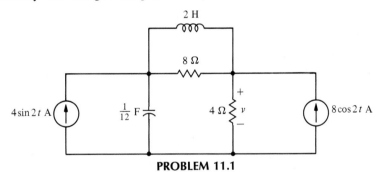

PROBLEM 11.1

11.2 Use nodal analysis to find the steady-state value of v if $v_g = V_m \cos \omega t$ and find the amplitude of v for the cases $\omega =$ (a) 0, (b) 1000, and (c) 10^6 rad/s. Compare the results for this circuit with those for Fig. 11.6.

PROBLEM 11.2

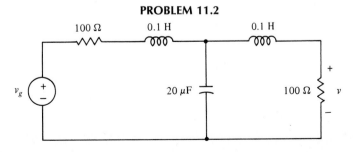

11.3 Find the steady-state voltage v using nodal analysis.

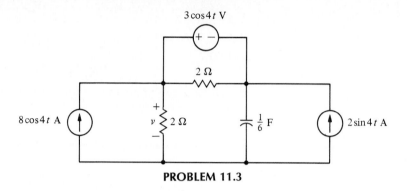

PROBLEM 11.3

11.4 Find the steady-state current i_1 using nodal analysis.

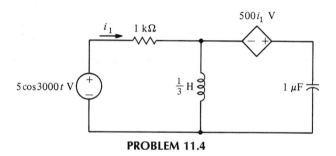

PROBLEM 11.4

11.5 Solve Ex. 11.1.2 using mesh analysis.

11.6 Find the steady-state voltage v using mesh analysis if $i_{g1} = 6 \cos 4t$ A and $i_{g2} = 2 \cos 4t$ A.

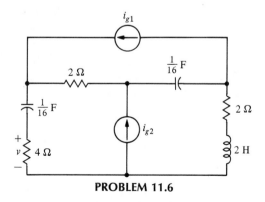

PROBLEM 11.6

11.7 Find the steady-state voltage v.

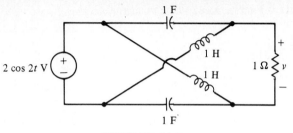

PROBLEM 11.7

11.8 Find the steady-state current i.

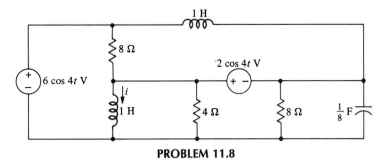

PROBLEM 11.8

11.9 Find the steady-state current i.

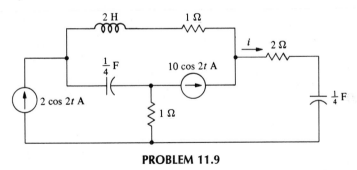

PROBLEM 11.9

11.10 Find the steady-state voltage v.

PROBLEM 11.10

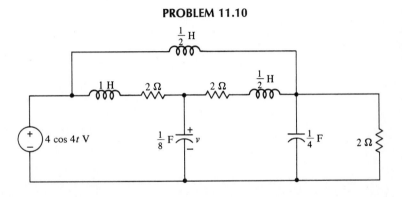

11.11 Find the steady-state voltage v.

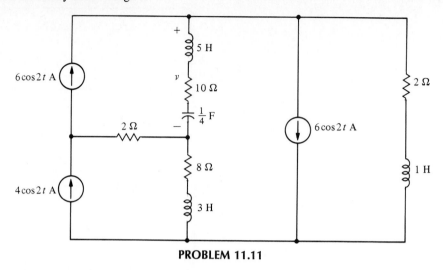

PROBLEM 11.11

11.12 Find the forced response i if $v_g = 4 \cos 1000t$ V.

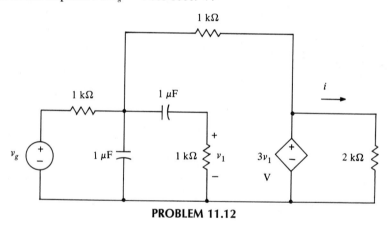

PROBLEM 11.12

11.13 Find the steady-state voltage v.

PROBLEM 11.13

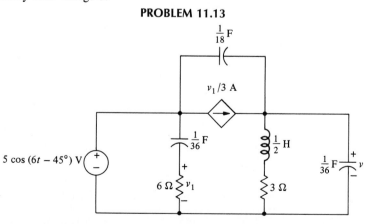

11.14 Show that if $Z_1 Z_4 = Z_2 Z_3$ in the "bridge" circuit shown, then $I = V = 0$ and therefore all the other currents and voltages remain unchanged for any value of Z_5. Thus it may be replaced by an open circuit, a short circuit, etc. In this case, the circuit is said to be a *balanced bridge*.

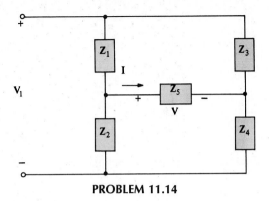

PROBLEM 11.14

11.15 Show that the circuit of Prob. 11.10 is a balanced bridge, with the series combination of 2 Ω and $\frac{1}{2}$ H constituting Z_5 in Prob. 11.14. Replace Z_5 by a short, and show that the same v is obtained as before.

11.16 Find the steady-state current i if $v_g = 4 \cos 3000t$ V.

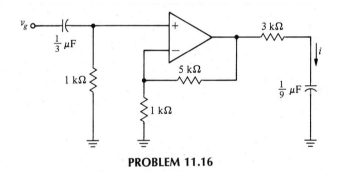

PROBLEM 11.16

11.17 Find the steady-state voltage v if $v_g = 2 \cos 2t$ V.

PROBLEM 11.17

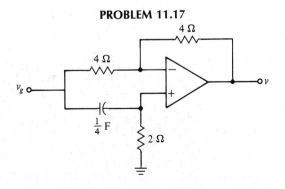

11.18   Find the steady-state voltage v if $v_g = 5 \cos 3t$ V.

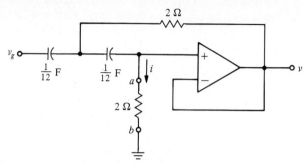

PROBLEM 11.18

11.19   Find the steady-state voltage v if $v_g = 3 \cos 2t$ V.

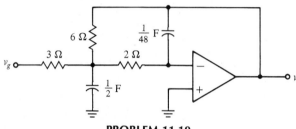

PROBLEM 11.19

11.20   Find the steady-state voltage v.

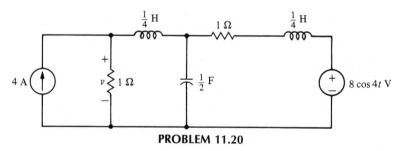

PROBLEM 11.20

11.21   Find the steady-state current i if $i_g = 9 - 20 \cos t - 39 \cos 2t + 18 \cos 3t$ A.

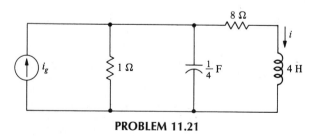

PROBLEM 11.21

11.22   Find the steady-state voltage v.

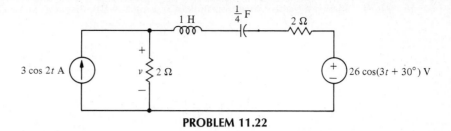

PROBLEM 11.22

11.23 For the phasor circuit corresponding to Ex. 11.2.3, replace everything except the 1-Ω resistor between terminals *c-d* by its Thevenin equivalent and find the steady-state current *i*.

11.24 Replace the circuit to the left of terminals *a-b* by its Norton equivalent, and find **V**.

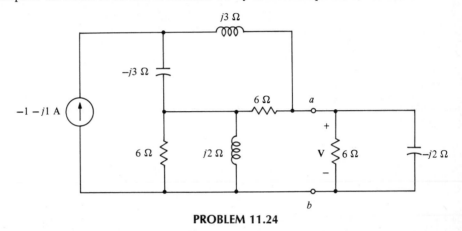

PROBLEM 11.24

11.25 In the phasor circuit corresponding to Prob. 11.7, replace everything except the 1-Ω resistor by its Thevenin equivalent and use the result to find the steady-state voltage *v*.

11.26 In the phasor circuit corresponding to Prob. 11.18, replace everything except the 2-Ω resistor between terminals *a-b* by its Thevenin equivalent and use the result to find the steady-state current *i*.

11.27 Solve Prob. 11.2 for the steady-state voltage *v* by applying the proportionality principle to the corresponding phasor circuit. Assume that $v_g = 2 \cos 1000t$ V. (*Suggestion:* Assume the phasor of *v* is 100 V.)

11.28 Find the steady-state current *i* using the principle of proportionality.

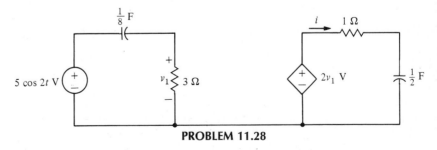

PROBLEM 11.28

11.29 Find the steady-state current *i* in Prob. 11.18 using the proportionality principle.

11.30 Find **I**, the phasor representation of i, using a phasor diagram. Show the phasors of i_R, i_C, and i_L, with the phasor of the source voltage as reference.

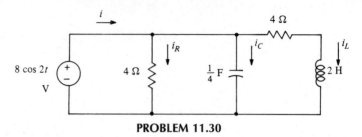

PROBLEM 11.30

11.31 Show that the locus of V_C in Fig. 11.16 is a circle if $R = 2\ \Omega$, $C = \frac{1}{4}\ \text{F}$, $V_g = 1\underline{/0°}\ \text{V}$, $\omega = 2\ \text{rad/s}$, and L varies from 0 to ∞. Select the value of L that gives the maximum amplitude of the time-domain voltage v_C, and find v_C in this case.

11.32 Find the locus of V_C in Prob. 11.31 if $L = 1\ \text{H}$, $C = \frac{1}{2}\ \text{F}$, $V_g = 2\underline{/0°}\ \text{V}$, $\omega = 2\ \text{rad/s}$, and R varies from 0 to ∞. Select the value of R for which Im V_C has its largest negative value, and find v_C for this case.

COMPUTER APPLICATION PROBLEMS

11.33 Using SPICE, find v in Prob. 11.10.

11.34 Using SPICE, find the Thevenin equivalent circuit in Prob. 11.13 for the network to the left of the 1/36-F capacitor. (*Hint:* Find V_{oc} and I_{sc}).

11.35 Using SPICE, find v in Prob. 11.19 if output v of the circuit of Prob. 11.17 supplies v_g to the 3-ohm resistor of this network.

11.36 In the circuit shown, find v if $v_g = 12\cos(10^4 t + 45°)\ \text{V}$.

PROBLEM 11.36

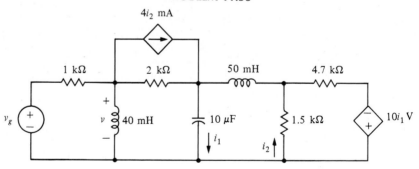

12

James Prescott Joule
1818–1889

AC Steady-State Power

The heating of a conductor depends upon its resistance and the square of the current passing through it.

James P. Joule

The man to whom we are indebted for the familiar expression i^2R for the power dissipated in a resistor is the English physicist James Prescott Joule, who published the result as Joule's law in 1841. He also shared in the famous discovery of the conservation of energy.

Joule was born in Salford, England, the second of five children of a wealthy brewer. He taught himself electricity and magnetism at home as a young boy and obtained his formal education at nearby Manchester University. His experiments on heat were conducted in his home laboratory, and to maintain the accuracy of his measurements he was forced to develop his own system of units. His chief claim to fame is that he did more than any other person to establish the idea that heat is a form of energy. Throughout most of his life Joule was an isolated amateur scientist, but toward the end of his years his work was recognized by honorary doctorates from Dublin and Oxford. In his honor the unit of energy was named the *joule*. ∎

In this chapter we shall consider power relationships for networks that are excited by periodic currents and voltages. We shall concern ourselves primarily with sinusoidal currents and voltages since nearly all electrical power is generated in this form. Instantaneous power, as we now well know, is the rate at which energy is absorbed by an element, and it varies as a function of time. The instantaneous power is an important quantity in engineering applications because its maximum value must be limited for all physical devices. For this reason, the maximum instantaneous power, or *peak power,* is a commonly used specification for characterizing electrical devices. In an electronic amplifier, for instance, if the specified peak power at the input is exceeded, the output signal will be distorted. Greatly exceeding this input rating may even permanently damage the amplifier.

A more important measure of power, particularly for periodic currents and voltages, is that of *average power*. The average power is equal to the average rate at which energy is absorbed by an element, and it is independent of time. This power, for example, is what is monitored by the electric company in determining monthly electricity bills. Average powers are encountered which range from a few picowatts, in applications such as satellite communications, to millions of watts, in applications such as supplying the electrical needs of a large city.

Our discussion will begin with a study of average power. We shall then consider superposition once again and introduce a mathematical measure for characterizing periodic currents or voltages, known as effective or rms values. We shall then consider the power factor associated with a load and present a complex power. Finally, we shall describe the measurement of power.

12.1

AVERAGE POWER

In linear networks which have inputs that are periodic functions of time, the steady-state currents and voltages produced are periodic, each having identical periods. Consider an instantaneous power

$$p = vi \qquad (12.1)$$

where v and i are periodic of period T. That is, $v(t + T) = v(t)$, and $i(t + T) = i(t)$. In this case

$$p(t + T) = v(t + T)i(t + T)$$
$$= v(t)i(t)$$
$$= p(t) \qquad (12.2)$$

Therefore, the instantaneous power is also periodic of period T. That is, p repeats itself every T seconds.

The *fundamental* period T_1 of p (the *minimum* time in which p repeats itself) is not necessarily equal to T, however, but T must contain an integral number of periods T_1. In other words,

$$T = nT_1 \qquad (12.3)$$

where n is a positive integer.

EXAMPLE 12.1   As an example, suppose that a resistor R carries a current $i = I_m \cos \omega t$ with period $T = 2\pi/\omega$. Then

$$p = Ri^2$$
$$= RI_m^2 \cos^2 \omega t$$
$$= \frac{RI_m^2}{2}(1 + \cos 2\omega t)$$

Evidently, $T_1 = \pi/\omega$, and, therefore, $T = 2T_1$. Thus, for this case, $n = 2$ in (12.3). This is illustrated by the graph of p and i shown in Fig. 12.1(a).

If we now take $i = I_m(1 + \cos \omega t)$, then

$$p = RI_m^2(1 + \cos \omega t)^2$$

In this case, $T_1 = T = 2\pi/\omega$, and $n = 1$ in (12.3). This may be seen also from the graph of the function in Fig. 12.1(b).

Mathematically, the average value of a periodic function is defined as the time integral of the function over a complete period, divided by the period. Therefore the average power P for a periodic instantaneous power p is given by

$$P = \frac{1}{T_1} \int_{t_1}^{t_1+T_1} p \, dt \qquad (12.4)$$

where t_1 is arbitrary.

A periodic instantaneous power p is shown in Fig. 12.2. It is clear that if we integrate over an integral number of periods, say mT_1 (where m is a positive integer), then the total area is simply m times that of the integral in (12.4). Thus we may write

$$P = \frac{1}{mT_1} \int_{t_1}^{t_1+mT_1} p \, dt \qquad (12.5)$$

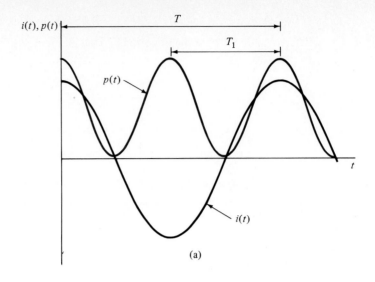

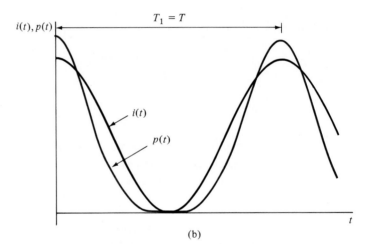

(b)

FIGURE 12.1 Graphs of p and i

FIGURE 12.2 Periodic instantaneous power

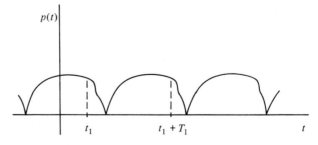

If we select m such that $T = mT_1$ (the period of v or i), then

$$P = \frac{1}{T} \int_{t_1}^{t_1+T} p \, dt \qquad (12.6)$$

Therefore we may obtain the average power by integrating over the period of v or i.

Let us now consider several examples of the average power associated with sinusoidal currents and voltages. A number of very important integrals which often occur are tabulated in Table 12.1. Verification of these integrals is left as an exercise (see Ex. 12.1.1).

TABLE 12.1 Integrals for Sinusoidal Functions and Their Products

$f(t)$	$\int_0^{2\pi/\omega} f(t) \, dt, \; \omega \neq 0$
1. $\sin(\omega t + \alpha)$, $\cos(\omega t + \alpha)$	0
2. $\sin(n\omega t + \alpha)$, $\cos(n\omega t + \alpha)$*	0
3. $\sin^2(\omega t + \alpha)$, $\cos^2(\omega t + \alpha)$	π/ω
4. $\sin(m\omega t + \alpha)\cos(n\omega t + \alpha)$*	0
5. $\cos(m\omega t + \alpha)\cos(n\omega t + \beta)$*	$0, \; m \neq n$
	$\pi \cos(\alpha - \beta)/\omega, \; m = n$

*m and n are integers.

First, let us consider the general two-terminal device of Fig. 12.3, which is assumed to be in ac steady state. If, in the frequency domain,

$$\mathbf{Z} = |\mathbf{Z}|\underline{/\theta}$$

is the input impedance of the device, then for

$$v = V_m \cos(\omega t + \phi) \qquad (12.7)$$

we have

$$i = I_m \cos(\omega t + \phi - \theta) \qquad (12.8)$$

where

$$I_m = \frac{V_m}{|\mathbf{Z}|}$$

FIGURE 12.3 General two-terminal device

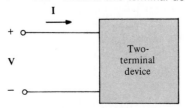

The average power delivered to the device, taking $t_1 = 0$ in (12.6), is

$$P = \frac{\omega V_m I_m}{2\pi} \int_0^{2\pi/\omega} \cos(\omega t + \phi) \cos(\omega t + \phi - \theta)\, dt$$

Referring to Table 12.1, entry 5, we find, for $m = n = 1$, $\alpha = \phi$, and $\beta = \phi - \theta$,

$$P = \frac{V_m I_m}{2} \cos \theta \qquad (12.9)$$

Thus the average power absorbed by a two-terminal device is determined by the amplitudes V_m and I_m and the angle θ by which the voltage v leads the current i.

In terms of the phasors of v and i,

$$\mathbf{V} = V_m \underline{/\phi} = |\mathbf{V}| \underline{/\phi}$$

$$\mathbf{I} = I_m \underline{/\phi - \theta} = |\mathbf{I}| \underline{/\phi - \theta}$$

we have, from (12.9),

$$P = \frac{1}{2} |\mathbf{V}| |\mathbf{I}| \cos(\text{ang } \mathbf{V} - \text{ang } \mathbf{I}) \qquad (12.10)$$

where

$$\text{ang } \mathbf{V} = \phi, \qquad \text{ang } \mathbf{I} = \phi - \theta$$

are the angles of the phasors $\mathbf{V}$ and $\mathbf{I}$.

If the two-terminal device is a resistor R, then $\theta = 0$, and $V_m = RI_m$, so that (12.9) becomes

$$P_R = \frac{1}{2} R I_m^2$$

It is worth noting at this point that if $i = I_{dc}$, a constant (dc) current, then $\omega = \phi = \theta = 0$, and $I_m = I_{dc}$ in (12.8). In this special case Table 12.1 does not apply, but by (12.6) we have

$$P_R = R I_{dc}^2$$

In the case of an inductor, $\theta = 90°$, and in the case of a capacitor, $\theta = -90°$, and thus for either one, by (12.9), $P = 0$. Therefore an inductor or a capacitor, or for that matter any network composed entirely of ideal inductors and capacitors, in any combination, dissipates zero average power. For this reason, ideal inductors and capacitors are sometimes called *lossless* elements. Physically, lossless elements store energy during part of the period and release it during the other part, so that the average delivered power is zero.

A very useful alternative form of (12.9) may be obtained by recalling that

$$\mathbf{Z} = \text{Re } \mathbf{Z} + j \text{ Im } \mathbf{Z} = |\mathbf{Z}| \underline{/\theta}$$

and therefore

$$\cos \theta = \frac{\text{Re } \mathbf{Z}}{|\mathbf{Z}|}$$

Substituting this value into (12.9) and noting that $V_m = |\mathbf{Z}| I_m$, we have

$$P = \frac{1}{2} I_m^2 \text{ Re } \mathbf{Z} \qquad (12.11)$$

Let us now consider this result if the device is a passive load. We know from the definition of passivity in (1.7) that the net energy delivered to a passive load is nonnegative. Since the average power is the average rate at which energy is delivered to a load, it follows that the average power is nonnegative. That is, $P \geq 0$. This requires, by (12.11), that

$$\text{Re } \mathbf{Z}(j\omega) \geq 0$$

or, equivalently,

$$-\frac{\pi}{2} \leq \theta \leq \frac{\pi}{2}$$

If $\theta = 0$, the device is equivalent to a resistance, and if $\theta = \pi/2$ (or $-\pi/2$), the device is equivalent to an inductance (or capacitance). For $-\pi/2 < \theta < 0$, the device is equivalent to an RC combination, whereas for $0 < \theta < \pi/2$, it is equivalent to an RL combination.

Finally, if $|\theta| > \pi/2$, then $P < 0$, and the device is active rather than passive. In this case the device is delivering power from its terminals and, of course, acts like a source.

EXAMPLE 12.2   As an example, let us find the power delivered by the source of Fig. 12.4. The impedance across the source is

$$\mathbf{Z} = 100 + j100$$
$$= 100\sqrt{2}\underline{/45°}\ \Omega$$

The maximum current is

$$I_m = \frac{V_m}{|\mathbf{Z}|} = \frac{1}{\sqrt{2}}\ \text{A}$$

FIGURE 12.4   *RL* circuit in the ac steady state

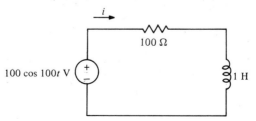

Therefore, from (12.9), the power delivered to $\mathbf{Z}$ is

$$P = \frac{100}{2\sqrt{2}} \cos 45° = 25 \text{ W}$$

Alternatively, from (12.11),

$$P = \frac{1}{2}\left(\frac{1}{\sqrt{2}}\right)^2 (100) = 25 \text{ W}$$

We may also note that the power absorbed by the 100-Ω resistor is

$$P_R = \frac{RI_m^2}{2} = \frac{(100)(1/\sqrt{2})^2}{2} = 25 \text{ W}$$

This power, of course, is equal to that delivered to $\mathbf{Z}$ since the inductor absorbs no power.

The power absorbed by the source is

$$P_g = -\frac{V_m I_m}{2} \cos \theta = -25 \text{ W}$$

where the minus sign is used because the current is flowing out of the positive terminal of the source. The source therefore is delivering 25 W to $\mathbf{Z}$. We note that the power flowing from the source is equal to that absorbed by the load, which illustrates the principle of conservation of power.

EXERCISES

12.1.1 Verify the integrals of Table 12.1.

12.1.2 For a capacitor of C farads carrying a current $i = I_m \cos \omega t$, verify that the average power is zero from (12.6). Repeat this for an inductor of L henrys.

12.1.3 Find the average power delivered to a 100-Ω resistor carrying a current given by (a) $10|\cos 10t|$ mA, (b) a square wave for which one cycle consists of 10 mA for 1 s followed by −10 mA for 1 s, and (c) a triangular wave for which one cycle consists of a current that increases linearly from 0 to 10 mA in 1 s.
Answer (a) 5 mW; (b) 10 mW; (c) 3.33 mW

12.1.4 Find the average power absorbed by the inductor, the 1-Ω resistor in parallel with the inductor, the other resistor, and the source.
Answer 0, 10, 20, −30 W

EXERCISE 12.1.4

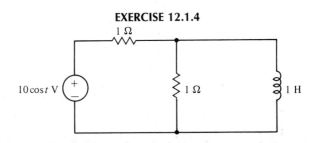

12.1.5 If $f_1(t)$ is periodic of period T_1 and $f_2(t)$ is periodic of period T_2, show that $f_1(t) + f_2(t)$ is periodic of period T if positive integers m and n exist such that

$$T = mT_1 = nT_2$$

Extend this result to the function $(1 + \cos \omega t)^2$, considered in this section, to find its period, $T = 2\pi/\omega$.

12.2

SUPERPOSITION AND POWER

In this section we shall consider the power in networks containing two or more sources, such as the simple circuit of Fig. 12.5. By superposition, we know that

$$i = i_1 + i_2$$

where i_1 and i_2 are the currents in R due to v_{g1} and v_{g2}, respectively. The instantaneous power is

$$p = R(i_1 + i_2)^2$$
$$= Ri_1^2 + Ri_2^2 + 2Ri_1i_2$$
$$= p_1 + p_2 + 2Ri_1i_2$$

where p_1 and p_2 are the instantaneous powers, respectively, due to v_{g1} acting alone and to v_{g2} acting alone. In general, $2Ri_1i_2 \neq 0$; thus $p \neq p_1 + p_2$, and superposition does *not* apply for instantaneous power.

In the case of p periodic with period T, the average power is

$$P = \frac{1}{T} \int_0^T p \, dt$$

$$= \frac{1}{T} \int_0^T (p_1 + p_2 + 2Ri_1i_2) \, dt$$

$$= P_1 + P_2 + \frac{2R}{T} \int_0^T i_1i_2 \, dt$$

where P_1 and P_2 are the average powers from v_{g1} and v_{g2}, respectively, acting alone.

FIGURE 12.5 Simple circuit with two sources

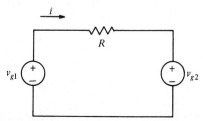

(We are assuming, of course, that each component of p is periodic of period T.) Superposition for average power applies when

$$P = P_1 + P_2 \qquad (12.12)$$

Clearly, this condition holds if

$$\int_0^T i_1 i_2 \, dt = 0 \qquad (12.13)$$

The most important case in which this equation is satisfied is when $i(t)$ is composed of sinusoidal components of different frequencies. Suppose, for instance, that

$$i_1 = I_{m1} \cos (\omega_1 t + \phi_1)$$

and

$$i_2 = I_{m2} \cos (\omega_2 t + \phi_2)$$

Since we are assuming that $i = i_1 + i_2$ is periodic of period T, we must have

$$I_{m1} \cos [\omega_1(t + T) + \phi_1] + I_{m2} \cos [\omega_2(t + T) + \phi_2]$$
$$= I_{m1} \cos (\omega_1 t + \phi_1) + I_{m2} \cos (\omega_2 t + \phi_2)$$

which requires that

$$\omega_1 T = 2\pi m, \qquad \omega_2 T = 2\pi n$$

where m and n are positive integers. Therefore, if ω is a number such that $T = 2\pi/\omega$, then $\omega_1 = m\omega$ and $\omega_2 = n\omega$. In this case the integral in (12.13) becomes, using Table 12.1,

$$\int_0^T i_1 i_2 \, dt = I_{m1} I_{m2} \int_0^{2\pi/\omega} \cos (m\omega t + \phi_1) \cos (n\omega t + \phi_2) \, dt$$

$$= \frac{I_{m1} I_{m2} \pi \cos (\phi_1 - \phi_2)}{\omega}, \qquad m = n$$

$$= 0, \qquad m \neq n$$

Thus if $m = n$ ($\omega_1 = \omega_2$), superposition *does not* apply. However, if $m \neq n$, superposition *does* apply. We may generalize this result to the case of a periodic sinusoid with any number of sinusoidal components of different frequencies. *The average power due to the sum of the components is the sum of the average powers due to each component acting alone.*

It can be shown that superposition of average power holds for sinusoids whose frequencies are not integral multiples of some frequency ω, provided we generalize the definition of average power to

$$P = \lim_{\tau \to \infty} \frac{1}{\tau} \int_0^\tau p \, dt$$

This generalization applies to the periodic case just considered as well as to the case $i = i_1 + i_2$, where

$$i_1 = \cos t$$

$$i_2 = \cos \pi t$$

In this case i is not even periodic (the ratio $\omega_1/\omega_2 = 1/\pi$ is not a rational number m/n), but

$$\lim_{\tau \to \infty} \frac{1}{\tau} \int_0^\tau i_1 i_2 \, dt = 0$$

EXAMPLE 12.3 As an example, suppose in Fig. 12.5 that $v_{g1} = 100 \cos (377t + 60°)$ V, $v_{g2} = 50 \cos 377t$ V, and $R = 100$ Ω. Since $\omega_1 = \omega_2$, we cannot apply superposition to power. However, we can use superposition to find the current and subsequently find the power. The phasor currents due to the respective sources are

$$\mathbf{I}_1 = 1\underline{/60°} \text{ A}$$

$$\mathbf{I}_2 = -0.5 \text{ A}$$

Therefore,

$$\mathbf{I} = \mathbf{I}_1 + \mathbf{I}_2 = j0.866 \text{ A}$$

so that $I_m = 0.866$ A. From (12.11), we find

$$P = \frac{1}{2}(100)(0.866)^2 = 37.5 \text{ W}$$

EXAMPLE 12.4 Let us now repeat the above example with $v_{g2} = 50$ V dc. Since v_{g1} and v_{g2} are sinusoids with $\omega_1 = 377$ and $\omega_2 = 0$ rad/s, respectively, superposition for the average power is applicable. Proceeding as before, we find

$$\mathbf{I}_1 = 1\underline{/60°} \qquad \text{for } \omega = 377$$

$$\mathbf{I}_2 = -0.5 \qquad \text{for } \omega = 0$$

where $\mathbf{I}_2$ is now a dc current. Therefore,

$$P_1 = \frac{RI_{m1}^2}{2} = 50 \text{ W}$$

$$P_2 = RI_{m2}^2 = 25 \text{ W}$$

and the average power is

$$P = P_1 + P_2 = 75 \text{ W}$$

This example illustrates a very important case for electronic amplifiers with sinusoidal inputs. These amplifiers contain dc power supplies that produce dc currents which provide the energy for the amplified ac signals. Thus superposition is very useful in finding the average power associated with each frequency, including $\omega = 0$.

Extending the above procedure to a periodic current which is the sum of $N + 1$ sinusoids of *different* frequencies,

$$i = I_{dc} + I_{m1} \cos(\omega_1 t + \phi_1) + I_{m2} \cos(\omega_2 t + \phi_2)$$
$$+ \ldots + I_{mN} \cos(\omega_N t + \phi_N) \tag{12.14}$$

we find the average power delivered to a resistance R is

$$P = RI_{dc}^2 + \frac{R}{2}(I_{m1}^2 + I_{m2}^2 + \ldots + I_{mN}^2) \tag{12.15}$$

The first term, RI_{dc}^2, in which the factor $\frac{1}{2}$ is missing, is the special case of zero frequency and must be considered separately, as in Sec. 12.1. That is, by (12.6),

$$\frac{1}{T} \int_0^T RI_{dc}I_{mi} \cos(\omega_i t + \phi_i)\, dt = 0, \qquad i = 1, 2, 3, \ldots, N$$

and

$$\frac{1}{T} \int_0^T RI_{dc}^2\, dt = RI_{dc}^2$$

This last expression is P_{dc}, the average power delivered by I_{dc}. Thus, if we denote the other terms in (12.15) by P_1 (due to i_1), P_2 (due to i_2), and so on, we have superposition of power in the form

$$P = P_{dc} + P_1 + P_2 + \ldots + P_N$$

EXERCISES

12.2.1 Find the average power delivered to the resistor in Fig. 12.5 if $R = 10\ \Omega$ and
(a) $v_{g1} = 10 \cos 100t$ and $v_{g2} = 20 \cos(100t + 60°)$ V.
(b) $v_{g1} = 100 \cos(t + 60°)$ and $v_{g2} = 50 \sin(2t - 30°)$ V.
(c) $v_{g1} = 50 \cos(t + 30°)$ and $v_{g2} = 100 \sin(t + 30°)$ V.
(d) $v_{g1} = 20 \cos(t + 25°)$ and $v_{g2} = 30 \sin(5t - 35°)$ V.
Answer (a) 15 W; (b) 625 W; (c) 625 W; (d) 65 W

12.2.2 Find the average power absorbed by each resistor and each source.
Answer 3.5 W, 3.5 W, −5 W, −2 W

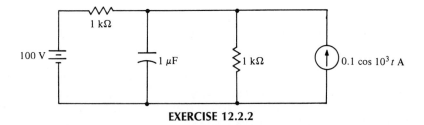

EXERCISE 12.2.2

12.2.3 Find the average power absorbed by the 1-Ω resistor.
Answer 16 W

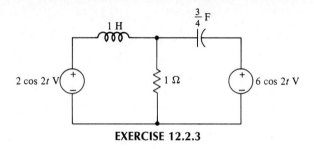

EXERCISE 12.2.3

12.3

RMS VALUES

We have seen in the previous sections that periodic currents and voltages deliver an average power to resistive loads. The amount of power that is delivered depends on the characteristics of the particular waveform. A method of comparing the power delivered by different waveforms is therefore very useful. One such method is the use of *rms* or *effective* values for periodic currents or voltages.

The *rms* value of a periodic current (voltage) is a constant that is equal to the dc current (voltage) that would deliver the same average power to a resistance R. Thus, if I_{rms} is the *rms* value of i, we may write

$$P = RI_{rms}^2 = \frac{1}{T} \int_0^T Ri^2 \, dt$$

from which the rms current is

$$I_{rms} = \sqrt{\frac{1}{T} \int_0^T i^2 \, dt} \qquad (12.16)$$

In a similar manner, it is easily shown that the rms voltage is

$$V_{rms} = \sqrt{\frac{1}{T} \int_0^T v^2 \, dt}$$

The term rms is an abbreviation for *root-mean-square*. Inspecting (12.16), we see that we are indeed taking the square *root* of the average, or *mean*, value of the *square* of the current.

From our definition, the rms value of a constant (dc) is simply the constant itself. The dc case is a special case ($\omega = 0$) of the most important type of waveform, the sinusoidal current or voltage.

Suppose we now consider a sinusoidal current $i = I_m \cos(\omega t + \phi)$. Then, from (12.16) and Table 12.1, we find

$$I_{rms} = \sqrt{\frac{\omega I_m^2}{2\pi} \int_0^{2\pi/\omega} \cos^2(\omega t + \phi) \, dt}$$

$$= \frac{I_m}{\sqrt{2}}$$

Thus a sinusoidal current having an amplitude I_m delivers the same average power to a resistance R as does a dc current which is equal to $I_m/\sqrt{2}$. We also see that the rms current is independent of the frequency ω or the phase ϕ of the current i. Similarly, in the case of a sinusoidal voltage, we find that

$$V_{rms} = \frac{V_m}{\sqrt{2}}$$

Substituting these values into the important power relations of (12.9) and (12.11) for the two-terminal network, we have

$$P = V_{rms} I_{rms} \cos \theta \qquad (12.17)$$

and

$$P = I_{rms}^2 \, \text{Re } \mathbf{Z} \qquad (12.18)$$

In practice, rms values are usually used in the fields of power generation and distribution. For instance, the nominal 115-V ac power which is commonly used for household appliances is an rms value. Thus the power supplied to our homes is provided by a 60-Hz voltage having a maximum value of $115\sqrt{2} \approx 163$ V. On the other hand, maximum values are more commonly used in electronics and communications.

Finally, let us consider the rms value of the current in (12.14), which is made up of sinusoids of *different* frequencies. In terms of rms currents, (12.15) becomes

$$P = R(I_{dc}^2 + I_{1rms}^2 + I_{2rms}^2 + \ldots + I_{Nrms}^2) \qquad (12.19)$$

Since $P = RI_{rms}^2$, we see that the rms value of a sinusoidal current consisting of *different* frequencies is

$$I_{rms} = \sqrt{I_{dc}^2 + I_{1rms}^2 + I_{2rms}^2 + \ldots + I_{Nrms}^2} \qquad (12.20)$$

Similarly,

$$V_{rms} = \sqrt{V_{dc}^2 + V_{1rms}^2 + V_{2rms}^2 + \ldots + V_{Nrms}^2}$$

These results are particularly important in the study of *noise* in electrical networks, a subject of later courses.

EXERCISES

12.3.1 Find the rms value of a periodic current for which one period is defined by

(a) $i = I$, $\qquad 0 \le t \le 1$
$\quad = 0$, $\qquad 1 < t < 4 \; (T = 4)$.

(b) $i = 3t$, $\qquad 0 < t \le T$.

(c) $i = I_m \sin \omega t$, $\qquad 0 \le t \le \pi/\omega$
$\quad = 0$, $\qquad \pi/\omega \le t \le 2\pi/\omega (T = 2\pi/\omega)$.

Answer (a) $I/2$, (b) $\sqrt{3} \, T$; (c) $I_m/2$

12.3.2 Find the rms values of (a) $i = 10 \sin \omega t + 20 \cos (\omega t + 30°)$, (b) $i = 8 \sin \omega t + 6 \cos (2\omega t + 10°)$, and (c) $i = I(1 + \cos 377t)$.

Answer (a) 12.25; (b) 7.07; (c) $I\sqrt{\frac{3}{2}}$

12.3.3 Find $\mathbf{I}_{\text{rms}}$.

Answer $\sqrt{2}$ A

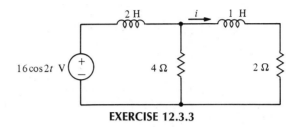

EXERCISE 12.3.3

12.4

POWER FACTOR

The average power delivered to a load in the ac steady state, repeating (12.17), is

$$P = V_{\text{rms}} I_{\text{rms}} \cos \theta$$

The power is thus equal to the product of the rms voltage, the rms current, and the cosine of the angle between the voltage and current phasors. In practice, the rms current and voltage are easily measured and their product, $V_{\text{rms}} I_{\text{rms}}$, is called the *apparent power*. The apparent power is usually referred to in terms of its units, voltamperes (VA) or kilovoltamperes (kVA), in order to avoid confusing it with the unit of average power, the watt. It is clear that the average power can never be greater than the apparent power.

The ratio of the average power to the apparent power is defined as the *power factor*. Thus if we denote the power factor by *pf*, then in the sinusoidal case

$$pf = \frac{P}{V_{\text{rms}} I_{\text{rms}}} = \cos \theta \qquad (12.21)$$

which, of course, is dimensionless. The angle θ, in this case, is often referred to as the *pf angle*. We also recognize it as the angle of the impedance $\mathbf{Z}$ of the load.

In the case of purely resistive loads, the voltage and current are in phase. Therefore, $\theta = 0$, $pf = 1$, and the average and apparent powers are equal. A unity power factor ($pf = 1$) can also exist for loads which contain inductors and capacitors if the reactances of these elements are such that they cancel one another. Adjusting the reactance of loads so as to approximate this condition is very important in electrical power systems, as we shall see shortly.

In a purely reactive load, $\theta = \pm 90°$, $pf = 0$, and the average power is zero. In this case, the equivalent load is an inductance ($\theta = 90°$) or a capacitance ($\theta = -90°$), and the current and voltage differ in phase by 90°.

A load for which $-90° < \theta < 0$ is equivalent to an RC combination, whereas one having $0 < \theta < 90°$ is an equivalent RL combination. Since $\cos \theta = \cos(-\theta)$, it is evident that the pf for an RC load having $\theta = -\theta_1$, where $0 < \theta_1 < 90°$, is equal to that of an RL load with $\theta = \theta_1$. To avoid this difficulty in identifying such loads, the pf is characterized as *leading* or *lagging* by the *phase of the current with respect to that of the voltage*. Therefore an RC load has a leading pf and an RL load has a lagging pf. For example, the series connection of a 100-Ω resistor and a 0.1-H inductor at 60 Hz has $\mathbf{Z} = 100 + j37.7 = 106.9\underline{/20.66°}\ \Omega$ and has a pf of $\cos 20.66° = 0.936$ lagging.

EXAMPLE 12.5
In practice, the power factor of a load is very important. In industrial applications, for instance, loads may require thousands of watts to operate, and the power factor greatly affects the electric bill. Suppose, for example, a mill consumes 100 kW from a 220-V rms line. At a pf of 0.85 lagging, we see that the rms current into the mill is

$$I_{\text{rms}} = \frac{P}{V_{\text{rms}} pf} = \frac{10^5}{(220)(0.85)} = 534.8 \text{ A}$$

which means that the apparent power supplied is

$$V_{\text{rms}} I_{\text{rms}} = (220)(534.8) \text{ VA} = 117.66 \text{ kVA}$$

Now suppose that the pf by some means is increased to 0.95 lagging. Then

$$I_{\text{rms}} = \frac{10^5}{(220)(0.95)} = 478.5 \text{ A}$$

and the apparent power is reduced to

$$V_{\text{rms}} I_{\text{rms}} = 105.3 \text{ kVA}$$

Comparing the latter case with the former, we see that I_{rms} was reduced by 56.3 A (10.5%). Therefore the generating station must generate a larger current in the case of the lower pf. Since the transmission lines supplying the power have resistance, the generator must produce a larger average power to supply the 100 kW to the load. If the resistance is 0.1 Ω, for instance, then the power generated by the source must be

$$P_g = 10^5 + 0.1 I_{\text{rms}}^2$$

Therefore, we find

$$P_g = 128.6 \text{ kW}, \qquad pf = 0.85$$
$$= 122.9 \text{ kW}, \qquad pf = 0.95$$

which requires that the power station produce 5.7 kW (4.64%) more power to supply the lower *pf* load. It is for this reason that power companies encourage a *pf* exceeding say 0.9 and impose a penalty on large industrial users who do not comply.

Let us now consider a method of correcting the power factor of a load having an impedance

$$\mathbf{Z} = R + jX$$

We may alter the power factor by connecting an impedance $\mathbf{Z}_1$ in parallel with $\mathbf{Z}$, as shown in Fig. 12.6. For this connection, it is clear that the load voltage does not change. Since $\mathbf{Z}$ is fixed, $\mathbf{I}$ does not change, and the power delivered to the load is not affected. The current $\mathbf{I}_1$ supplied by the generator, however, does change.

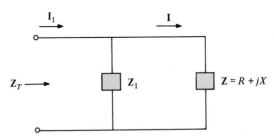

FIGURE 12.6 Circuit for correcting the power factor

Let us denote the impedance of the parallel combination by

$$\mathbf{Z}_T = \frac{\mathbf{Z}\mathbf{Z}_1}{\mathbf{Z} + \mathbf{Z}_1}$$

In general, we select $\mathbf{Z}_1$ so that (1) $\mathbf{Z}_1$ absorbs zero average power, and (2) $\mathbf{Z}_T$ satisfies the desired power factor $pf = PF$. The first condition requires that $\mathbf{Z}_1$ be purely reactive. That is,

$$\mathbf{Z}_1 = jX_1$$

The second condition requires that

$$\cos\left[\tan^{-1}\left(\frac{\operatorname{Im}\mathbf{Z}_T}{\operatorname{Re}\mathbf{Z}_T}\right)\right] = PF$$

Substituting $\mathbf{Z}_T$ in terms of R, X, and X_1 into this equation, we find that (see Prob. 12.16)

$$X_1 = \frac{R^2 + X^2}{R\tan(\cos^{-1}PF) - X} \tag{12.22}$$

where we note that $\tan(\cos^{-1}PF)$ is positive if PF is lagging and negative if PF is leading.

EXAMPLE 12.6 As an example of the application of (12.22), let us change the power factor for the circuit of Fig. 12.4 to 0.95 lagging. We have already found that

$$\mathbf{Z} = 100 + j100 = 141.4\underline{/45°}$$

Therefore, before a parallel reactance is added across $\mathbf{Z}$, the power factor is

$$pf = \cos\theta = \cos 45° = 0.707 \text{ lagging}$$

Since we desire a power factor of 0.95 lagging, $\tan(\cos^{-1} PF)$ is positive, so that by (12.22) we have

$$X_1 = \frac{100^2 + 100^2}{100 \tan(\cos^{-1} 0.95) - 100} = -297.92 \ \Omega$$

Since $X_1 < 0$, the reactance is a capacitance $C = -1/\omega X_1 = 33.6 \ \mu\text{F}$. The load impedance now becomes

$$\mathbf{Z}_T = \frac{(100 + j100)(-j297.92)}{100 + j100 - j297.92} = 190.0\underline{/18.2°}$$

Therefore, the power to the corrected load is

$$P = \frac{100^2}{2(190.0)} \cos(18.2°) = 25 \text{ W}$$

which is the same as that delivered to $\mathbf{Z}$ in Fig. 12.4. The current, however, is

$$I_{\text{rms}} = \frac{100}{190\sqrt{2}} = 0.372 \text{ A}$$

as compared to that of Fig. 12.4, given by

$$I_{\text{rms}} = \frac{I_m}{\sqrt{2}} = 0.5 \text{ A}$$

We see, therefore, that the current has been reduced by 0.128 A, or 25.6%.

EXERCISES

12.4.1 Find the apparent power for (a) a load that requires 30 A rms from a 230-V rms line and (b) a load consisting of a 100-Ω resistor in parallel with a 25-μF capacitor connected to a 120-V rms 60-Hz source.
Answer (a) 6.9 kVA; (b) 197.9 VA

12.4.2 Find the power factor for (a) a load consisting of a series connection of a 100-Ω resistor and a 20-μF capacitor operating at 60 Hz, (b) one that is capacitive and requires 50 A rms and 5 kW at 110 V rms, and (c) one consisting of a 5-kW load at a power factor of 0.85 leading, connected in parallel with a 10-kW load at a power factor of 0.9 lagging.
Answer (a) 0.602 leading; (b) 0.909 leading; (c) 0.993 lagging

12.4.3 Use (12.22) to correct the power factor seen by the source of Fig. 12.4 to 0.95 leading. Compare the result with that of the example given following (12.22).
Answer $C = 66.4 \ \mu\text{F}$

12.5

COMPLEX POWER

We shall now introduce a *complex power* in the ac steady state, which is very useful for determining and correcting power factors associated with interconnected loads. Let us begin by defining rms phasors for general sinusoidal voltages and currents. The phasor representations for (12.7) and (12.8) are

$$\mathbf{V} = V_m e^{j\phi}$$

$$\mathbf{I} = I_m e^{j(\phi - \theta)}$$

The rms phasors for these quantities are defined as

$$\mathbf{V}_{\text{rms}} = \frac{\mathbf{V}}{\sqrt{2}} = V_{\text{rms}} e^{j\phi}$$

$$\mathbf{I}_{\text{rms}} = \frac{\mathbf{I}}{\sqrt{2}} = I_{\text{rms}} e^{j(\phi - \theta)} \tag{12.23}$$

Let us now consider the average power given in (12.17). Using Euler's formula, we may write

$$P = V_{\text{rms}} I_{\text{rms}} \cos \theta = \text{Re}(V_{\text{rms}} I_{\text{rms}} e^{j\theta})$$

Next, inspecting (12.23), we see that

$$\mathbf{V}_{\text{rms}} \mathbf{I}_{\text{rms}}^* = V_{\text{rms}} I_{\text{rms}} e^{j\theta}$$

where $\mathbf{I}_{\text{rms}}^*$ is the complex conjugate of $\mathbf{I}_{\text{rms}}$. Thus

$$P = \text{Re}(\mathbf{V}_{\text{rms}} \mathbf{I}_{\text{rms}}^*) \tag{12.24}$$

and the product $\mathbf{V}_{\text{rms}} \mathbf{I}_{\text{rms}}^*$ is a complex power whose real part is the average power. Denoting this complex power by $\mathbf{S}$, we have

$$\mathbf{S} = \mathbf{V}_{\text{rms}} \mathbf{I}_{\text{rms}}^* = P + jQ \tag{12.25}$$

where Q is the *reactive power*. Dimensionally, P and Q have the same units; however, the unit of Q is defined as the *var* (voltampere reactive) to distinguish it from the watt. The magnitude of the complex power is

$$|\mathbf{S}| = |\mathbf{V}_{\text{rms}} \mathbf{I}_{\text{rms}}^*| = |\mathbf{V}_{\text{rms}}| |\mathbf{I}_{\text{rms}}^*| = V_{\text{rms}} I_{\text{rms}}$$

which, of course, is equal to the apparent power.

From (12.25), we see that

$$Q = \operatorname{Im} \mathbf{S} = V_{rms} I_{rms} \sin \theta \qquad (12.26)$$

For an impedance $\mathbf{Z}$, we know that $\sin \theta = (\operatorname{Im} \mathbf{Z})/|\mathbf{Z}|$, so that

$$Q = V_{rms} I_{rms} \frac{\operatorname{Im} \mathbf{Z}}{|\mathbf{Z}|}$$

Therefore, since $V_{rms}/|\mathbf{Z}| = I_{rms}$, we see that

$$Q = I_{rms}^2 \operatorname{Im} \mathbf{Z}$$
$$= V_{rms}^2 \frac{\operatorname{Im} \mathbf{Z}}{|\mathbf{Z}|^2} \qquad (12.27)$$

A phasor diagram of $\mathbf{V}_{rms}$ and $\mathbf{I}_{rms}$ is shown in Fig. 12.7. We see that the phasor current can be resolved into the two components $I_{rms} \cos \theta$ and $I_{rms} \sin \theta$. The component $I_{rms} \cos \theta$ is in phase with $\mathbf{V}_{rms}$, and it produces the real power P. In contrast, $I_{rms} \sin \theta$ is 90° out of phase with $\mathbf{V}_{rms}$, and it causes the reactive power Q. Since $I_{rms} \sin \theta$ is 90° out of phase with $\mathbf{V}_{rms}$, it is called the *quadrature component* of $\mathbf{I}_{rms}$. As a consequence, the reactive power is sometimes referred to as the *quadrature power*.

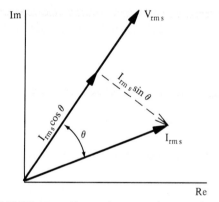

FIGURE 12.7 Phasor diagram of $\mathbf{V}_{rms}$ and $\mathbf{I}_{rms}$

It is often convenient to view the complex power in terms of a diagram such as that of Fig. 12.8. It is evident that for an inductive load (lagging *pf*), $0 < \theta \leq 90°$, Q is positive, and $\mathbf{S}$ lies in the first quadrant. For a capacitive load (leading *pf*), $-90° \leq \theta < 0$, Q is negative, and $\mathbf{S}$ lies in the fourth quadrant. A load having a unity power factor requires that $Q = 0$ since $\theta = 0$. In general, we see that

$$\theta = \tan^{-1}\left(\frac{Q}{P}\right) \qquad (12.28)$$

Let us now consider the complex power associated with a load consisting of two impedances $\mathbf{Z}_1$ and $\mathbf{Z}_2$, as shown in Fig. 12.9. The complex power delivered to the combined impedances is

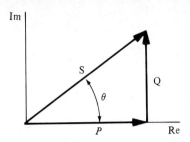

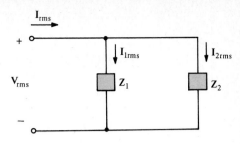

FIGURE 12.8 Diagram for the complex power

FIGURE 12.9 Load consisting of $\mathbf{Z}_1$ and $\mathbf{Z}_2$ in parallel

$$\mathbf{S} = \mathbf{V}_{rms}\mathbf{I}_{rms}^* = \mathbf{V}_{rms}(\mathbf{I}_{1rms} + \mathbf{I}_{2rms})^*$$

$$= \mathbf{V}_{rms}\mathbf{I}_{1rms}^* + \mathbf{V}_{rms}\mathbf{I}_{2rms}^*$$

Therefore, the complex power delivered by the source to the interconnected loads is the sum of that delivered to each individual load, and the complex power is thus conserved. This statement is true no matter how many individual loads there are or how they are interconnected, because it depends only on Kirchhoff's laws and the definition of complex power. This principle is known as *conservation of complex power*.

Conservation of complex power may be used in a straightforward manner to correct the power factor. As an example, let us consider the circuit of Fig. 12.6 once again. The complex power to the uncorrected load $\mathbf{Z}$ is

$$\mathbf{S} = P + jQ$$

Connecting a pure reactance $\mathbf{Z}_1$ in parallel with $\mathbf{Z}$ results in a complex power to $\mathbf{Z}_1$ of

$$\mathbf{S}_1 = jQ_1$$

Therefore, from conservation of complex power, for the composite load we have

$$\mathbf{S}_T = \mathbf{S} + \mathbf{S}_1$$

$$= P + j(Q + Q_1)$$

It is evident that the addition of $\mathbf{Z}_1$ does not affect the average power P delivered to the load. It does, however, affect the net reactive power. We can therefore select Q_1 to obtain our desired power factor, which, of course, is usually raised. This causes a reduction in the current required to produce P, as discussed previously.

EXAMPLE 12.7   Let us again consider the circuit of Fig. 12.4 and change the power factor to $PF = 0.95$ lagging. The complex power of the uncorrected load is

$$\mathbf{S} = \mathbf{V}_{rms}\mathbf{I}_{rms}^* = P + jQ = 25 + j25$$

since

$$\mathbf{V}_{rms} = 70.7 \text{ V}$$

$$\mathbf{I}_{rms} = \frac{\mathbf{V}_{rms}}{\mathbf{Z}} = 0.3535(1 - j1) \text{ A}$$

From (12.28), we see that $Q_T = Q + Q_1$ must satisfy

$$\theta = \tan^{-1}\left(\frac{Q_T}{P}\right)$$

Therefore,

$$\cos\theta = PF = \cos\left(\tan^{-1}\frac{Q_T}{P}\right)$$

and

$$Q_T = P \tan(\cos^{-1} PF)$$
$$= 25 \tan 18.2° = 8.22 \text{ vars}$$

The required Q_1 is

$$Q_1 = Q_T - Q = 8.22 - 25 = -16.78 \text{ vars}$$

Since $Q_1 = V_{rms}^2(\text{Im } \mathbf{Z}_1)/|\mathbf{Z}_1|^2$ and $\mathbf{Z}_1 = jX_1$, we may write

$$Q_1 = \frac{V_{rms}^2}{X_1}$$

Solving for X_1 in the case under consideration, we have

$$X_1 = \frac{(70.7)^2}{-16.78} = -297.9 \ \Omega$$

which represents a capacitance $C = -1/\omega X_1 = 33.6 \ \mu\text{F}$. This is identical to our previous result in Sec. 12.4.

EXAMPLE 12.8   As a final example, let us find the power factor of two loads connected in parallel, as shown in Fig. 12.9. Suppose $\mathbf{Z}_1$ represents a 10-kW load with a power factor $pf_1 = 0.9$ lagging and $\mathbf{Z}_2$ a 5-kW load with $pf_2 = 0.95$ leading. For $\mathbf{Z}_1$ we have

$$\mathbf{S}_1 = P_1 + jQ_1$$

where $P_1 = 10^4$ W, $\theta_1 = \cos^{-1} pf_1 = 25.84°$, and from (12.28),

$$Q_1 = P_1 \tan\theta_1 = 4843 \text{ vars}$$

Similarly, for $\mathbf{Z}_2$ we have

$$\mathbf{S}_2 = P_2 + jQ_2$$

where $P_2 = 5 \times 10^3$ W, $\theta_2 = -\cos^{-1} pf_2 = -18.2°$, and

$$Q_2 = 5 \times 10^3 \tan\theta_2 = -1643 \text{ vars}$$

The total complex power is

$$\mathbf{S}_T = P_1 + P_2 + j(Q_1 + Q_2)$$
$$= 1.5 \times 10^4 + j3200$$

Therefore, for the combined loads.

$$\theta = \tan^{-1}\left(\frac{3200}{1.5 \times 10^4}\right) = 12.04°$$

and

$$pf = 0.978 \text{ lagging}$$

EXERCISES

12.5.1 Find the complex power delivered to a load which has a 0.91 lagging power factor and (a) absorbs 1 kW, (b) 1 kvar, and (c) 1 kVA.
Answer (a) 1099$\underline{/24.5°}$; (b) 2412$\underline{/24.5°}$; (c) 1000$\underline{/24.5°}$ VA

12.5.2 Find the impedance of the loads in Ex. 12.5.1 if $V_{rms} = 120$ V.
Answer (a) 13.10$\underline{/24.5°}$; (b) 5.97$\underline{/24.5°}$ Ω; (c) 14.40$\underline{/24.5°}$ Ω

12.5.3 Repeat Ex. 12.4.2 (c) using the concept of complex power.

12.6

POWER MEASUREMENT

A *wattmeter* is a device which measures the average power that is delivered to a load. It contains a rotating high-resistance *voltage, or potential, coil,* connected in parallel with the load and a fixed low-resistance *current coil* which is connected in series with the load. The device has four terminals, a pair to accommodate each coil. A typical connection is shown in Fig. 12.10. We see that the current coil responds to the load current, whereas the voltage coil responds to the load voltage. For frequencies above a few hertz, the meter movement responds to the average power. Ideally, the voltage

FIGURE 12.10 Typical connection of a wattmeter

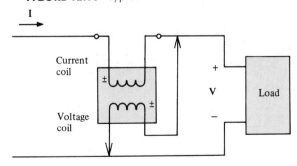

across the current coil and the current in the voltage coil are both zero, so that the presence of the meter does not influence the power it is measuring.

One terminal on each of the coils is marked $\pm$ so that if the current enters the $\pm$ terminal of the current coil and the $\pm$ terminal of the voltage coil is positive with respect to its other terminal, then the meter gives a positive, or upscale, reading. In Fig. 12.10 this corresponds to the load absorbing power. If the terminal connections of either the current coil or voltage coil (but not both) are reversed, a negative, or downscale, reading is indicated. Most meters cannot read downscale—the pointer simply rests on the downscale stop. Thus such a reading requires reversing the connections of one of the coils, usually the voltage coil. Reversing the connections of both coils does not affect the reading.

The wattmeter of Fig. 12.10, represented by the rectangle and the two coils, is connected to read

$$P = |\mathbf{V}| \cdot |\mathbf{I}| \cos \theta$$

where $\mathbf{V}$ and $\mathbf{I}$ are as indicated and θ is the angle between $\mathbf{V}$ and $\mathbf{I}$ (or, equivalently, the angle of the load impedance). This, of course, is the power delivered to the load.

Other types of meters are available for measuring the apparent power and the reactive power. An *apparent power* or VA meter simply measures the product of the rms current and rms voltage. The *varmeter,* on the other hand, measures the reactive power.

EXERCISE

12.6.1 Determine the power reading of each wattmeter after assigning the terminal markings required for a positive reading.
Answer 87.5 W, 9.375 W, 0 W

EXERCISE 12.6.1

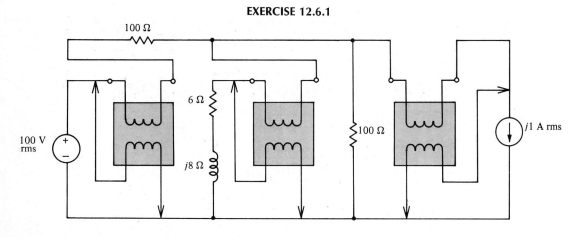

PROBLEMS

12.1   One cycle of a periodic current is given by

$$i = 10t \text{ A}, \qquad 0 \le t < 1 \text{ ms}$$

If the current flows in a 300-Ω resistor, find the average power.

12.2   Determine the average power for $p(t) = RI_m^2(1 + \cos \omega t)^2$.

12.3   Find the average power absorbed by the 6-Ω resistor.

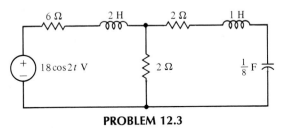

PROBLEM 12.3

12.4   Find the average power absorbed by the 10-Ω resistor and by the dependent source.

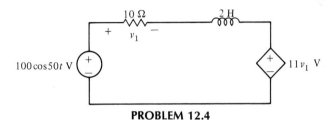

PROBLEM 12.4

?.5   Find the average power delivered to R if $R = 0.4 \ \Omega$.

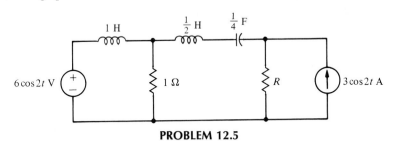

PROBLEM 12.5

12.6   For a Thevenin equivalent circuit consisting of a voltage source $\mathbf{V}_g$ and an impedance $\mathbf{Z}_g = R_g + jX_g$, (a) show that the circuit delivers maximum average power to a load $\mathbf{Z}_L = R_L + jX_L$ when $R_L = R_g$ and $X_L = -X_g$, and (b) show that the maximum average power is delivered to a load R_L when $R_L = |\mathbf{Z}_g|$. This is the *maximum power transfer theorem* for ac circuits. (In both cases, $\mathbf{V}_g$ and $\mathbf{Z}_g$ are fixed, and the load is variable.)

12.7   Using the result of Prob. 12.6, replace R in Prob. 12.5 by (a) a resistive load that will draw the maximum power and (b) a general load (resistive and reactance elements) that will draw the maximum power. In both cases, find the resulting power.

12.8 Find the average power delivered to the $\frac{1}{2}$-Ω resistor.

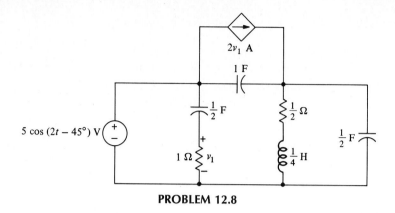

PROBLEM 12.8

12.9 Find the average power delivered to R.

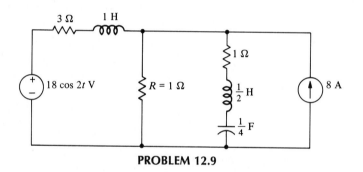

PROBLEM 12.9

12.10 Find the average power delivered to the 2-Ω resistor.

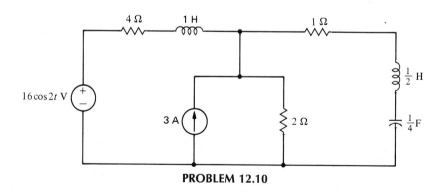

PROBLEM 12.10

12.11 Find the average power delivered to the 1-Ω resistor.

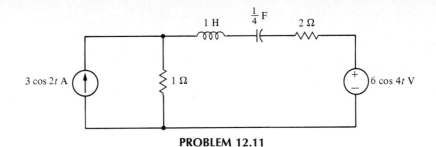

PROBLEM 12.11

12.12 We have defined the average power for an instantaneous power $p(t)$, which is not necessarily periodic, by

$$P = \lim_{\tau \to \infty} \frac{1}{\tau} \int_0^\tau p(t)\, dt$$

Show for v and i of (12.7)–(12.8) that this definition yields the same result as (12.6).

12.13 Given: $i = i_1 + i_2$, where

$$i_1 = I_{1m} \cos \omega_1 t$$

$$i_2 = I_{2m} \cos \omega_2 t$$

is the current flowing in a resistor R. Using the definition of Prob. 12.12, show for $\omega_1 \neq \omega_2$ that

$$P = P_1 + P_2$$

where P_1 and P_2 are the average powers associated with i_1 and i_2, respectively, acting alone. Note that this includes the case of $p(t)$ being nonperiodic.

12.14 Find the rms value of the voltage (a) $v = 8 + 6\sqrt{2} \cos t$ V, (b) $v = 4 \cos 3t + \sqrt{2} \cos 6t + 12 \cos (8t - 60°)$ V, and (c) $v = 4 \cos 2t - 8 \cos 5t + 6\sqrt{2} \cos (3t - 45°)$ V.

12.15 Find the rms value of a periodic current for which one cycle is given by

(a) $i = 3t$ A, $\qquad$ $0 \leq t \leq 2$ s

$\quad\ \ = 0,$ $\qquad\qquad$ $2 < t \leq 4$ s.

(b) $i = 6$ A, $\qquad$ $0 < t < 2$ s

$\quad\ \ = 0,$ $\qquad\qquad$ $2 < t < 8$ s.

(c) $i = I_m \sin \dfrac{2\pi t}{T}$ A, $\qquad$ $0 < t < \dfrac{T}{2}$ s

$\quad\ \ = 0,$ $\qquad\qquad\quad$ $\dfrac{T}{2} < t < T$ s.

(d) $i = I_m \sin \dfrac{\pi t}{T}$ A, $\qquad$ $0 < t < T$ s.

12.16 Derive (12.22).

12.17 Find the rms value of the steady-state current i and the power factor seen from the source terminals. What element connected in parallel with the source will correct the power factor to 0.8 lagging?

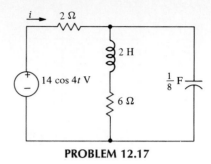

PROBLEM 12.17

12.18 Find the power factor seen from the terminals of the source and the reactance necessary to connect in parallel with the source to change the power factor to unity.

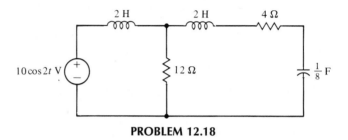

PROBLEM 12.18

12.19 Find the reactive element to be placed in parallel with the source to correct the power factor seen by the source to 0.8 leading.

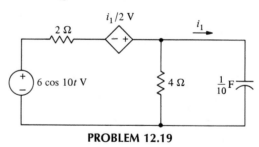

PROBLEM 12.19

12.20 Find the real power, the reactive power, and the complex power delivered by the source.

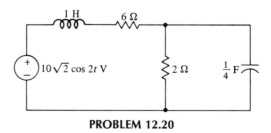

PROBLEM 12.20

12.21 Find the complex power delivered by the source and the power factor seen by the source.

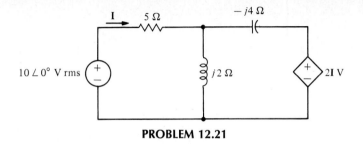

PROBLEM 12.21

12.22 Two loads in parallel draw respectively 210 W at a power factor of 0.6 lagging and 40 W at a power factor of 0.8 leading. If the voltage source across the parallel combination is $V_g = 25\underline{/15°}$ V rms, find the current **I** delivered by the source.

12.23 Three parallel passive loads, $\mathbf{Z}_1$, $\mathbf{Z}_2$, and $\mathbf{Z}_3$, are receiving complex power values of $3 - j4$, $2 + j5$, and $3 + j5$ VA, respectively. If a voltage source of $20\underline{/0°}$ V rms is connected across these loads, find the rms value of the current that flows from the source and the power factor seen by the source.

12.24 A load is supplied the complex power $\mathbf{S} = 6 + j8$ VA by a voltage source of $10 \cos 100t$ V. Find the capacitance that should be connected in parallel with the load so that the power factor seen by the source is (a) unity and (b) 0.8 lagging.

12.25 Find the wattmeter reading.

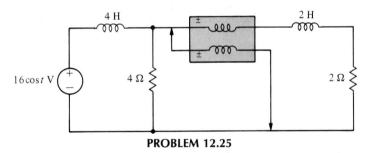

PROBLEM 12.25

12.26 Find the wattmeter reading if $v_g = 4 \cos 1000t$ V.

PROBLEM 12.26

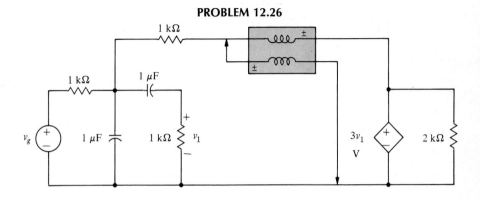

12.27 Find the wattmeter reading.

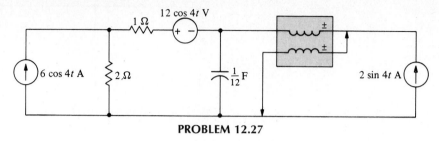

PROBLEM 12.27

13

Thomas Alva Edison
1847–1931

Three-Phase Circuits

Genius is one per cent inspiration and ninety-nine per cent perspiration.
                    Thomas A. Edison

The greatest American inventor and perhaps the greatest inventor in history was Thomas Alva Edison, who changed the lives of people everywhere with such inventions as the electric light and the phonograph. He patented over 1100 inventions of his own and improved many other persons' inventions, such as the telephone, the typewriter, the electric generator, and the motion picture. Perhaps most importantly of all, he was one of the first to organize research, at one time employing some 3000 helpers.

Edison was born in Milan, Ohio, the youngest of seven children. He had only 3 months of formal education because his mother took him out of school and taught him herself. He asked too many questions to get along with the schoolmaster. He was exempt from military service because of deafness, and during the Civil War he roamed from city to city as a telegraph operator. During this time he patented improvements on the stock ticker and sold the patents for the then astounding price of $40,000. In 1876 he moved to Menlo Park, New Jersey, and from there his steady stream of inventions made him world famous. The electric light was his greatest invention, but to supply it to the world he also designed the first electric power station. His discovery of the Edison effect, the movement of electrons in the vacuum of his light bulb, also marked the beginning of the age of electronics. ∎

As we have already noted, one very important use of ac steady-state analysis is its application to power systems, most of which are alternating current systems. One principal reason for this is that it is economically feasible to transmit power over long distances only if the voltages involved are very high, and it is easier to raise and lower voltages in ac systems than in dc systems. Alternating voltage can be stepped up for transmission and stepped down for distribution with transformers, as we shall see in Chapter 16. Transformers have no moving parts and are relatively simple to construct, whereas with the present technology, rotating machines are generally needed to raise and lower dc voltages.

Also, for reasons of economics and performance, almost all electric power is produced by *polyphase* sources (those generating voltages with more than one phase). In a single-phase circuit, the instantaneous power delivered to a load is pulsating, even if the current and voltage are in phase. A polyphase system, on the other hand, is somewhat like a multicylinder automobile engine in that the power delivered is steadier. Consequently, there is less vibration in the rotating machinery, which, in turn, performs more efficiently. An economic advantage is that the weight of the conductors and associated components required in a polyphase system is appreciably less than that required in a single-phase system that delivers the same power. Virtually all the power produced in the world is polyphase power at 50 or 60 Hz. In the United States 60 Hz is the standard frequency.

In this chapter we shall begin with single-phase, three-wire systems, but we shall concentrate on three-phase circuits, which are by far the most common of the polyphase systems. In the latter case the sources are three-phase generators that produce a *balanced* set of voltages, by which we mean three sinusoidal voltages having the same amplitude and frequency but displaced in phase by 120°. Thus the three-phase source is equivalent to three interconnected single-phase sources, each generating a voltage with a different phase. If the three currents drawn from the sources also constitute a balanced set, then the system is said to be a *balanced* three-phase system. This is the case to which we shall restrict ourselves, for the most part.

SINGLE-PHASE, THREE-WIRE SYSTEMS

Before proceeding to the three-phase case, let us digress in this section to establish our notation and consider an example of a single-phase system that is in common household use. This example will also serve to give us some practice with a single-phase system, with which we are already familiar and which will serve as an introduction to the three-phase systems to be considered next.

In this chapter we shall find extremely useful the double-subscript notation introduced in Chapter 1 for voltages. In the case of phasors, the notation is $\mathbf{V}_{ab}$ for the voltage of point a with respect to point b. We shall also use a double-subscript notation for current, taking, for example, $\mathbf{I}_{ab}$ as the current flowing in the *direct* path from point a to point b. These quantities are illustrated in Fig. 13.1, where the direct path from a to b is distinguished from the alternative path from a to b through c.

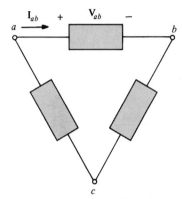

FIGURE 13.1 Illustration of double-subscript notation

Because of the simpler expressions for average power that result, we shall use rms values of voltage and current throughout this chapter. (These are also the values read by most meters.) That is, if

$$\mathbf{V} = |\mathbf{V}|\underline{/0°} \text{ V rms}$$
$$\mathbf{I} = |\mathbf{I}|\underline{/-\theta} \text{ A rms}$$

(13.1)

are the phasors associated with an element having impedance,

$$\mathbf{Z} = |\mathbf{Z}|\underline{/\theta} \ \Omega$$

(13.2)

the average power delivered to the element is

$$P = |\mathbf{V}| \cdot |\mathbf{I}| \cos \theta$$
$$= |\mathbf{I}|^2 \text{ Re } \mathbf{Z} \text{ W}$$

(13.3)

In the time domain the voltage and current are

$$v = \sqrt{2}\,|\mathbf{V}|\,\cos \omega t \text{ V}$$
$$i = \sqrt{2}\,|\mathbf{I}|\,\cos (\omega t - \theta) \text{ A}$$

EXAMPLE 13.1   The use of double subscripts makes it easier to handle phasors both analytically and geometrically. For example, in Fig. 13.2(a), the voltage $\mathbf{V}_{ab}$ is

$$\mathbf{V}_{ab} = \mathbf{V}_{an} + \mathbf{V}_{nb}$$

This is evident without referring to a circuit since by KVL the voltage between two points a and b is the same regardless of the path, which in this case is the path a, n, b. Also, since $\mathbf{V}_{nb} = -\mathbf{V}_{bn}$, we have

$$\mathbf{V}_{ab} = \mathbf{V}_{an} - \mathbf{V}_{bn}$$
$$= 100 - 100\underline{/-120°}$$

which, after simplification, is

$$\mathbf{V}_{ab} = 100\sqrt{3}\underline{/30°} \text{ V rms}$$

These steps are shown graphically in Fig. 13.2(b).

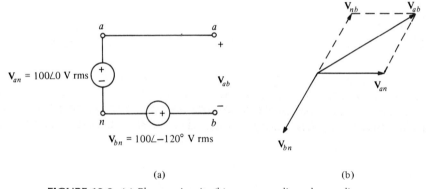

(a)   (b)

FIGURE 13.2 (a) Phasor circuit; (b) corresponding phasor diagram

A single-phase, three-wire source, as shown in Fig. 13.3, is one having three output terminals a, b, and a *neutral* terminal n, for which the terminal voltages are equal. That is,

$$\mathbf{V}_{an} = \mathbf{V}_{nb} = \mathbf{V}_1 \qquad (13.4)$$

This is a common arrangement in a normal house supplied with both 115 V and 230 V rms, since if $|\mathbf{V}_{an}| = |\mathbf{V}_1| = 115$ V, then $|\mathbf{V}_{ab}| = |2\mathbf{V}_1| = 230$ V.

Let us now consider the source of Fig. 13.3 loaded with two identical loads, both having an impedance $\mathbf{Z}_1$, as shown in Fig. 13.4. The currents in the lines aA and bB are

$$\mathbf{I}_{aA} = \frac{\mathbf{V}_{an}}{\mathbf{Z}_1} = \frac{\mathbf{V}_1}{\mathbf{Z}_1}$$

Solving for the currents, we have

$$\mathbf{I}_1 = 16.32\underline{/\text{...}}$$

$$\mathbf{I}_2 = 15.73\underline{/\text{...}}$$

$$\mathbf{I}_3 = 14.46\underline{/\text{...}}$$

Therefore, the neutral current is

$$\mathbf{I}_{nN} = \mathbf{I}_2 - \mathbf{I}_1 =$$

and, of course, is not zero.

EXERCISES

13.1.1 Derive (13.5) by superposition applied to

13.1.2 Find the power P_{40}, P_{60}, and P_{10+j10} d
$10 + j10$ Ω, respectively, of Fig. 13.6.
Answer 249, 181, 2091 W

13.1.3 Find the power P_{aA}, P_{bB}, and P_{nN} lost in t
Answer 266.3, 247.4, 1.2 W

13.1.4 Find the power P_{an} and P_{nb} delivered by the
in Exs. 13.1.2 and 13.1.3 for conservati
Answer 1561.4, 1474.5 W

13.2

THREE-PHASE Y-Y SYSTEMS

Let us consider the three-phase source of

FIGURE 13.7 Two represen

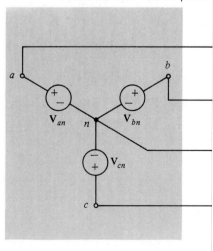

(a)

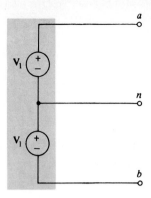

FIGURE 13.3 Single-phase, three-wire source

and

$$\mathbf{I}_{bB} = \frac{\mathbf{V}_{bn}}{\mathbf{Z}_1} = -\frac{\mathbf{V}_1}{\mathbf{Z}_1} = -\mathbf{I}_{aA}$$

Therefore, the current in the neutral wire, nN, by KCL is

$$\mathbf{I}_{nN} = -(\mathbf{I}_{aA} + \mathbf{I}_{bB}) = 0$$

Thus the neutral could be removed without changing any current or voltage in the system.

If the lines aA and bB are not perfect conductors but have equal impedances $\mathbf{Z}_2$, then $\mathbf{I}_{nN}$ is still zero because we may simply add the series impedances $\mathbf{Z}_1$ and $\mathbf{Z}_2$ and have essentially the same situation as in Fig. 13.4. Indeed, in the more general case shown in Fig. 13.5, the neutral current $\mathbf{I}_{nN}$ is still zero. This may be seen by writing the two mesh equations

$$(\mathbf{Z}_1 + \mathbf{Z}_2 + \mathbf{Z}_3)\mathbf{I}_{aA} + \mathbf{Z}_3\mathbf{I}_{bB} - \mathbf{Z}_1\mathbf{I}_3 = \mathbf{V}_1$$

$$\mathbf{Z}_3\mathbf{I}_{aA} + (\mathbf{Z}_1 + \mathbf{Z}_2 + \mathbf{Z}_3)\mathbf{I}_{bB} + \mathbf{Z}_1\mathbf{I}_3 = -\mathbf{V}_1$$

FIGURE 13.4 Single-phase, three-wire system with two identical loads

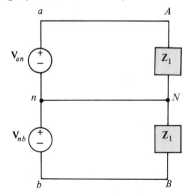

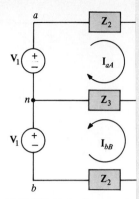

FIGURE 13.5 Symmetric[...]

and adding the result, which yields

$$(\mathbf{Z}_1 + \mathbf{Z}_2 + \mathbf{Z}_3)(\mathbf{I}_{aA}$$

or

$$\mathbf{I}_{aA}$$

Since by KCL the left member of the la[...]
This is, of course, a consequence of th[...]

If the symmetry of Fig. 13.5 is d[...]
A-N and *N-B* or unequal line impedan[...]
neutral current.

EXAMPLE 13.2   For example, let us consider the situatio[...]
at approximately 115 V and one at app[...]

$$43\mathbf{I}$$

$$-2\mathbf{I}_1$$

$$-40\mathbf{I}_1 - 60\mathbf{I}_2 +$$

FIGURE 13.6 Unsymmetric[...]

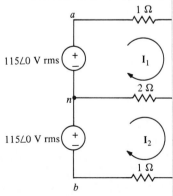

---

and *c* and a *neutral* terminal *n*. In this case, the source is said to be **Y**-*connected* (connected in a **Y,** as shown). An equivalent representation is that of Fig. 13.7(b), which is somewhat easier to draw.

The voltages $\mathbf{V}_{an}$, $\mathbf{V}_{bn}$, and $\mathbf{V}_{cn}$ between the line terminals and the neutral terminal are called *phase voltages* and in most cases we shall consider are given by

$$\mathbf{V}_{an} = V_p\underline{/0°}$$
$$\mathbf{V}_{bn} = V_p\underline{/-120°} \tag{13.6}$$
$$\mathbf{V}_{cn} = V_p\underline{/120°}$$

or

$$\mathbf{V}_{an} = V_p\underline{/0°}$$
$$\mathbf{V}_{bn} = V_p\underline{/120°} \tag{13.7}$$
$$\mathbf{V}_{cn} = V_p\underline{/-120°}$$

In both cases, each phase voltage has the same rms magnitude V_p, and the phases are displaced 120°, with $\mathbf{V}_{an}$ arbitrarily selected as the reference phasor. Such a set of voltages is called a *balanced* set and is characterized by

$$\mathbf{V}_{an} + \mathbf{V}_{bn} + \mathbf{V}_{cn} = 0 \tag{13.8}$$

as may be seen from (13.6) or (13.7).

The sequence of voltages in (13.6) is called the *positive sequence,* or *abc sequence,* while that of (13.7) is called the *negative,* or *acb,* sequence. Phasor diagrams of the two sequences are shown in Fig. 13.8, where we may see by inspection that (13.8) holds. Evidently, the only difference between the positive and

FIGURE 13.8 (a) Positive and (b) negative phase sequence

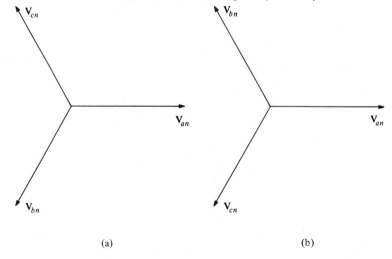

(a)   (b)

negative sequence is the arbitrary choice of the terminal labels, *a, b,* and *c.* Thus without loss in generality we shall consider only the positive sequence.

By (13.6), the voltages in the *abc* sequence may each be related to $\mathbf{V}_{an}$. The relationships, which will be useful later, are

$$\mathbf{V}_{bn} = \mathbf{V}_{an}\underline{/-120°}$$
$$\mathbf{V}_{cn} = \mathbf{V}_{an}\underline{/120°} \tag{13.9}$$

The *line-to-line* voltages, or simply *line* voltages, in Fig. 13.7 are $\mathbf{V}_{ab}$, $\mathbf{V}_{bc}$, and $\mathbf{V}_{ca}$, which may be found from the phase voltages. For example,

$$\mathbf{V}_{ab} = \mathbf{V}_{an} + \mathbf{V}_{nb}$$
$$= V_p\underline{/0°} - V_p\underline{/-120°}$$
$$= \sqrt{3}\ V_p\underline{/30°}$$

In like manner,

$$\mathbf{V}_{bc} = \sqrt{3}\ V_p\underline{/-90°}$$
$$\mathbf{V}_{ca} = \sqrt{3}\ V_p\underline{/-210°}$$

If we denote the magnitude of the line voltages by V_L, then we have

$$V_L = \sqrt{3}\ V_p \tag{13.10}$$

and thus

$$\mathbf{V}_{ab} = V_L\underline{/30°}, \qquad \mathbf{V}_{bc} = V_L\underline{/-90°}, \qquad \mathbf{V}_{ca} = V_L\underline{/-210°} \tag{13.11}$$

These results also may be obtained graphically from the phasor diagram shown in Fig. 13.9.

FIGURE 13.9 Phasor diagram showing phase and line voltages

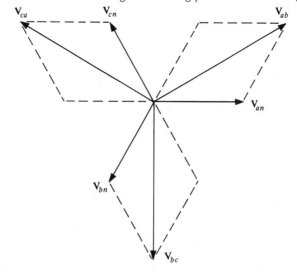

Let us now consider the system of Fig. 13.10, which is a *balanced* **Y-Y**, three-phase, four-wire system, if the source voltages are given by (13.6). The term **Y-Y** applies since both the source and the load are **Y**-connected. The system is said to be balanced since the source voltages constitute a balanced set and the load is balanced (each *phase impedance* is equal—in this case—to $\mathbf{Z}_p$). The fourth wire is the neutral line *n-N*, which may be omitted to form a three-phase, three-wire system.

The line currents of Fig. 13.10 are evidently

$$\mathbf{I}_{aA} = \frac{\mathbf{V}_{an}}{\mathbf{Z}_p}$$

$$\mathbf{I}_{bB} = \frac{\mathbf{V}_{bn}}{\mathbf{Z}_p} = \frac{\mathbf{V}_{an}\underline{/-120°}}{\mathbf{Z}_p} = \mathbf{I}_{aA}\underline{/-120°} \tag{13.12}$$

$$\mathbf{I}_{cC} = \frac{\mathbf{V}_{cn}}{\mathbf{Z}_p} = \frac{\mathbf{V}_{an}\underline{/120°}}{\mathbf{Z}_p} = \mathbf{I}_{aA}\underline{/120°}$$

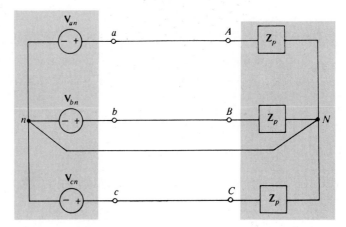

FIGURE 13.10 Balanced Y-Y system

The last two results are a consequence of (13.9) and show that the line currents also form a balanced set. Therefore their sum is

$$-\mathbf{I}_{nN} = \mathbf{I}_{aA} + \mathbf{I}_{bB} + \mathbf{I}_{cC} = 0$$

Thus the neutral carries no current in a balanced **Y-Y** four-wire system.

In the case of **Y**-connected loads, the currents in the lines *aA*, *bB*, and *cC* are also the *phase currents* (the currents carried by the phase impedances). If the magnitudes of the phase and line currents are I_p and I_L, respectively, then $I_L = I_p$, and (13.12) becomes

$$\mathbf{I}_{aA} = I_L\underline{/-\theta} = I_p\underline{/-\theta}$$

$$\mathbf{I}_{bB} = I_L\underline{/-\theta - 120°} = I_p\underline{/-\theta - 120°} \tag{13.13}$$

$$\mathbf{I}_{cC} = I_L\underline{/-\theta + 120°} = I_p\underline{/-\theta + 120°}$$

where θ is the angle of $\mathbf{Z}_p$.

The average power P_p delivered to each phase of Fig. 13.10 is

$$P_p = V_p I_p \cos \theta$$

$$= I_p^2 \operatorname{Re} \mathbf{Z}_p$$

and the total power delivered to the load is

$$P = 3P_p$$

The angle θ of the phase impedance is thus the power factor angle of the three-phase load as well as that of a single phase.

Suppose now that an impedance $\mathbf{Z}_L$ is inserted in each of the lines aA, bB, and cC and that an impedance $\mathbf{Z}_N$ not necessarily equal to $\mathbf{Z}_L$, is inserted in line nN. In other words, the lines are not to be perfect conductors but are to contain impedances. Evidently, the line impedances, except for $\mathbf{Z}_N$, are in series with the phase impedances, and the two sets of impedances may be combined to form perfect conducting lines aA, bB, and cC with a load impedance $\mathbf{Z}_p + \mathbf{Z}_L$ in each phase. Therefore, except for $\mathbf{Z}_N$ in the neutral, the equivalent system has perfect conducting lines and a balanced load. If the impedance $\mathbf{Z}_N$ were not present in the neutral, it would be a perfect conductor, and the system would be balanced as in Fig. 13.10. In this case, as we have seen, points n and N are at the same potential, and there is no neutral current. Thus it does not matter what is in the neutral line. It may be a short circuit or an open-circuit or contain an impedance such as $\mathbf{Z}_N$, and still no neutral current would flow and no voltage would appear across nN. Obviously, then, the presence of equal line impedances in aA, bB, and cC and an impedance in the neutral does not change the fact that the line currents form a balanced set.

EXAMPLE 13.3 As an example, let us find the line currents in Fig. 13.11. We may combine the 1-Ω line impedance and $(3 + j3)$-Ω phase impedance to obtain

$$\mathbf{Z}_p = 4 + j3 = 5\underline{/36.9°}\ \Omega$$

as the effective phase load. Since by the foregoing discussion there is no neutral

FIGURE 13.11 Balanced system with line impedances

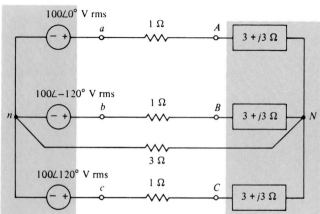

current, we have

$$\mathbf{I}_{aA} = \frac{100\underline{/0^\circ}}{5\underline{/36.9^\circ}} = 20\underline{/-36.9^\circ} \text{ A rms}$$

The currents form a balanced, positive sequence set, so that we also have

$$\mathbf{I}_{bB} = 20\underline{/-156.9^\circ} \text{ A rms}, \qquad \mathbf{I}_{cC} = 20\underline{/-276.9^\circ} \text{ A rms}$$

This example was solved on a "per-phase" basis. Since the impedance in the neutral is immaterial in a balanced **Y-Y** system, we may imagine the neutral line to be a short circuit. We may do this if it contains an impedance or even if the neutral wire is not present (a three-wire system). We may then look at only one phase, say "phase A," consisting of the source $\mathbf{V}_{an}$ in series with $\mathbf{Z}_L$ and $\mathbf{Z}_p$, as shown in Fig. 13.12. (The line nN is replaced by a short circuit.) The line current $\mathbf{I}_{aA}$, the phase voltage $\mathbf{I}_{aA}\,\mathbf{Z}_p$, and the voltage drop in the line $\mathbf{I}_{aA}\,\mathbf{Z}_L$ may all be found from this single-phase analysis. The other voltages and currents in the system may be found similarly, or from the previous results, since the system is balanced.

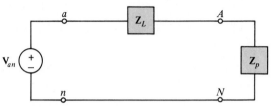

FIGURE 13.12 Single phase for a per-phase analysis

EXAMPLE 13.4   As another example, suppose we have a balanced **Y**-connected source, having line voltage $V_L = 200$ V rms, which is supplying a balanced **Y**-connected load with $P = 900$ W at a power factor of 0.9 lagging. Let us find the line current I_L and the phase impedance $\mathbf{Z}_p$. Since the power supplied to the load is 900 W, the power supplied to each phase is $P_p = \frac{900}{3} = 300$ W, and from

$$P_p = V_p I_p \cos\theta$$

we have

$$300 = \left(\frac{200}{\sqrt{3}}\right) I_p (0.9)$$

Therefore since for a **Y**-connected load the phase current is also the line current, we have

$$I_L = I_p = \frac{3\sqrt{3}}{2(0.9)} = 2.89 \text{ A rms}$$

The magnitude of $\mathbf{Z}_p$ is given by

$$|\mathbf{Z}_p| = \frac{V_p}{I_p} = \frac{200/\sqrt{3}}{3\sqrt{3}/(2)(0.9)} = 40 \text{ }\Omega$$

and since $\theta = \cos^{-1}0.9 = 25.84^\circ$ is the angle of $\mathbf{Z}_p$, we have

$$\mathbf{Z}_p = 40\underline{/25.84°} \; \Omega$$

If the load is unbalanced but there is a neutral wire which is a perfect conductor, then we may still use the per-phase method of solution for each phase. However, if this is not the case, this shortcut method does not apply. There is a very useful method employing so-called *symmetrical components* which is applicable to unbalanced systems and which the reader may encounter in a course on power systems. It is important to note, in any case, that a three-phase circuit is still a circuit and, balanced or unbalanced, may always be solved by general analysis procedures.

EXERCISES

13.2.1 Given: $\mathbf{V}_{ab} = 200\underline{/0°}$ V rms is a line voltage of a balanced **Y**-connected three-phase source. If the phase sequence is *abc*, find the phase voltages.
Answer $115.5\underline{/-30°}$, $115.5\underline{/-150°}$, $115.5\underline{/-270°}$ V rms

13.2.2 In Fig. 13.10 the source voltages are determined by Ex. 13.2.1, and the load in each phase is a series combination of a 40-Ω resistor, a 100-μF capacitor, and a 0.1-H inductor. The frequency is $\omega = 200$ rad/s. Find the line currents and the power delivered to the load.
Answer $2.31\underline{/6.9°}$, $2.31\underline{/-113.1°}$, $2.31\underline{/-233.1°}$ A rms, 640 W

13.2.3 Show that if a balanced three-phase, three-wire system has two balanced three-phase loads connected in parallel, as shown, then the load is equivalent to that of Fig. 13.10 with

$$\mathbf{Z}_p = \frac{\mathbf{Z}_1\mathbf{Z}_2}{\mathbf{Z}_1 + \mathbf{Z}_2}$$

EXERCISE 13.2.3

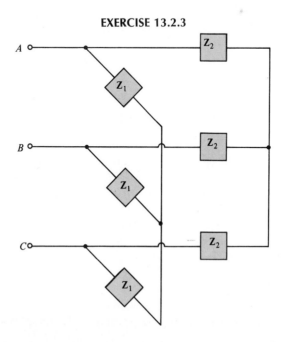

13.2.4 If in Ex. 13.2.3 $\mathbf{Z}_1 = 3 - j4\ \Omega$, $\mathbf{Z}_2 = 3 + j4\ \Omega$, and the line voltage is $V_L = 100\sqrt{3}$ V rms, find the current I_L in each line.

Answer 24 A rms

13.3

THE DELTA CONNECTION

Another method of connecting a three-phase load to a line is the *delta* connection, or Δ connection. A balanced Δ-connected load (with equal phase impedances) is shown in Fig. 13.13(a), in a way that resembles a Δ, and in an equivalent way in Fig. 13.13(b). If the source is **Y**-or Δ-connected, then the system is a **Y**-Δ or a Δ-Δ system.

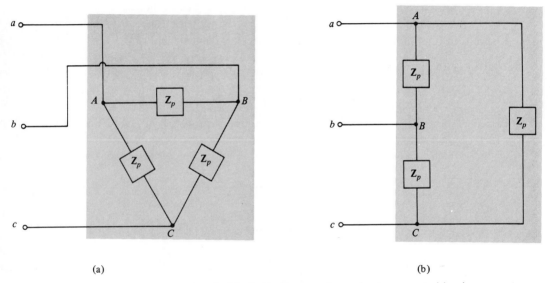

(a)   (b)

FIGURE 13.13 Two versions of a Δ-connected load

An advantage of a Δ-connected load over a **Y**-connected load is that loads may be added or removed more readily on a single phase of a Δ, since the loads are connected directly across the lines. This may not be possible in the **Y** connection, since the neutral may not be accessible. Also, for a given power delivered to the load the phase currents in a Δ are smaller than those in a **Y**. On the other hand, the Δ phase voltages are higher than those of the **Y** connection. Sources are rarely Δ-connected, because if the voltages are not perfectly balanced, there will be a net voltage, and consequently a circulating current, around the delta. This, of course, causes undesirable heating effects in the generating machinery. Also, the phase voltages are lower in the **Y**-connected generator, and thus less insulation is required. Obviously, systems with Δ-connected loads are three-wire systems, since there is no neutral connection.

From Fig. 13.13 we see that in the case of a Δ-connected load the line voltages are the same as the phase voltages. Therefore if the line voltages are given by (13.11), as before, then the phase voltages are

$$\mathbf{V}_{AB} = V_L\,\underline{/30°}, \qquad \mathbf{V}_{BC} = V_L\,\underline{/-90°}, \qquad \mathbf{V}_{CA} = V_L\,\underline{/-210°} \qquad (13.15)$$

where

$$V_L = V_p \qquad (13.16)$$

If $\mathbf{Z}_p = |\mathbf{Z}_p|\,\underline{/\theta}$, then the phase currents are

$$\mathbf{I}_{AB} = \frac{\mathbf{V}_{AB}}{\mathbf{Z}_p} = I_p\,\underline{/30° - \theta}$$

$$\mathbf{I}_{BC} = \frac{\mathbf{V}_{BC}}{\mathbf{Z}_p} = I_p\,\underline{/-90° - \theta} \qquad (13.17)$$

$$\mathbf{I}_{CA} = \frac{\mathbf{V}_{CA}}{\mathbf{Z}_p} = I_p\,\underline{/-210° - \theta}$$

where

$$I_p = \frac{V_L}{|\mathbf{Z}_p|} \qquad (13.18)$$

The current in line aA is

$$\mathbf{I}_{aA} = \mathbf{I}_{AB} - \mathbf{I}_{CA}$$

which after some simplification is

$$\mathbf{I}_{aA} = \sqrt{3}\,I_p\,\underline{/-\theta}$$

The other line currents, obtained similarly, are

$$\mathbf{I}_{bB} = \sqrt{3}\,I_p\,\underline{/-120° - \theta}$$
$$\mathbf{I}_{cC} = \sqrt{3}\,I_p\,\underline{/-240° - \theta}$$

Evidently the relation between the line and phase current magnitudes in the Δ case is

$$I_L = \sqrt{3}\,I_p \qquad (13.19)$$

and the line currents are thus

$$\mathbf{I}_{aA} = I_L\,\underline{/-\theta}, \qquad \mathbf{I}_{bB} = I_L\,\underline{/-120° - \theta}, \qquad \mathbf{I}_{cC} = I_L\,\underline{/-240° - \theta} \qquad (13.20)$$

Thus the currents and voltages are balanced sets, as expected. The relations between line and phase currents for the Δ-connected load are summed up in the phasor diagram of Fig. 13.14.

EXAMPLE 13.5   As an example of a three-phase circuit with a Δ-connected load, let us find the line current I_L in Fig. 13.13 if the line voltage is 250 V rms and the load draws 1.5 kW at a lagging power factor of 0.8. For one phase, $P_p = \frac{1500}{3} = 500$ W, and thus

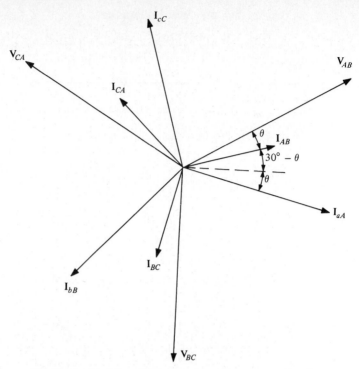

FIGURE 13.14 Phasor diagram for a Δ-connected load

$$500 = 250I_p(0.8)$$

or

$$I_p = 2.5 \text{ A rms}$$

Therefore, we have

$$I_L = \sqrt{3}\,I_p = 4.33 \text{ A rms}$$

Finally, in this section, let us derive a formula for the power delivered to a balanced three-phase load with a power factor angle θ. Whether the load is **Y**-connected or Δ-connected, we have

$$P = 3P_p = 3V_pI_p \cos \theta$$

In the **Y**-connected case, $V_p = V_L/\sqrt{3}$ and $I_p = I_L$, and in the Δ-connected case, $V_p = V_L$ and $I_p = I_L/\sqrt{3}$. In either case, then,

$$P = 3\frac{V_LI_L}{\sqrt{3}} \cos \theta$$

or

$$P = \sqrt{3}\,V_LI_L \cos \theta \qquad\qquad (13.21)$$

As a check on the previous example, (13.21) yields

$$1500 = \sqrt{3}(250)I_L(0.8)$$

or, as before,

$$I_L = 4.33 \text{ A rms}$$

EXERCISES

13.3.1 Solve Ex. 13.2.2 if the source and load are unchanged except that the load is Δ-connected. [*Suggestion:* Note that in (13.15), (13.17), and (13.20) 30° must be subtracted from every angle.]
Answer $2\sqrt{3}/-66.9°$, $2\sqrt{3}/-186.9°$, $2\sqrt{3}/-306.9°$ A rms, 480 W

13.3.2 A balanced Δ-connected load has $\mathbf{Z}_p = 3 + j4$ Ω, and the line voltage is $V_L = 100$ V rms at the load terminals. Find the total power delivered to the load.
Answer 3.6 kW

13.3.3 A balanced Δ-connected load has a line voltage of $V_L = 200$ V rms at the load terminals and absorbs a total power of 4.8 kW. If the power factor of the load is 0.8 lagging, find the phase impedance.
Answer $16 + j12$ Ω

13.4

Y-Δ TRANSFORMATIONS

In many power systems applications it is important to be able to convert from a Y-connected load to an equivalent Δ-connected load and vice versa. For example, suppose we have a Y-connected load in parallel with a Δ-connected load, as shown in Fig. 13.15, and wish to replace the combination by an equivalent three-phase load. If both loads were Δ-connected, this would be relatively easy since corresponding phase impedances would be in parallel. Also, as we saw in Ex. 13.2.3, if both loads

FIGURE 13.15 Y-connected and Δ-connected loads in parallel

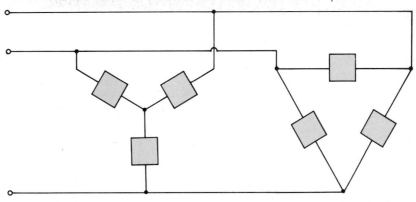

are **Y**-connected and balanced, the phase impedances may also be combined as parallel impedances.

To obtain **Y**-to-Δ or Δ-to-**Y** conversion formulas, let us consider the **Y** and Δ connections of Fig. 13.16. To effect a **Y**-Δ transformation we need expressions for $\mathbf{Y}_{ab}$, $\mathbf{Y}_{bc}$, and $\mathbf{Y}_{ca}$ of the Δ in terms of $\mathbf{Y}_a$, $\mathbf{Y}_b$, and $\mathbf{Y}_c$ of the **Y** so that the Δ connection is equivalent to the **Y** connection at the terminals A, B, and C. That is, if the **Y** is replaced by the Δ, the same node voltages $\mathbf{V}_A$, $\mathbf{V}_B$, and $\mathbf{V}_C$ will appear, and the same currents $\mathbf{I}_1$ and $\mathbf{I}_2$ will flow. Conversely, a Δ-**Y** transformation is an expression of the **Y** parameters in terms of the Δ parameters.

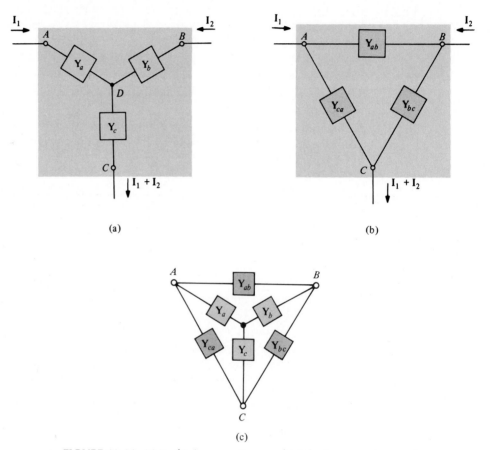

(a) (b)

(c)

FIGURE 13.16 (a) **Y**, (b) Δ connection, and (c) the two superimposed

Let us begin by writing nodal equations for both circuits. If node C is taken as reference, in the case of the **Y** network we have

$$\mathbf{Y}_a \mathbf{V}_A - \mathbf{Y}_a \mathbf{V}_D = \mathbf{I}_1$$

$$\mathbf{Y}_b \mathbf{V}_B - \mathbf{Y}_b \mathbf{V}_D = \mathbf{I}_2$$

$$-\mathbf{Y}_a \mathbf{V}_A - \mathbf{Y}_b \mathbf{V}_B + (\mathbf{Y}_a + \mathbf{Y}_b + \mathbf{Y}_c)\mathbf{V}_D = 0$$

Solving for $\mathbf{V}_D$ in the third equation and substituting its value into the first two equations, we have, after simplification,

$$\left(\frac{Y_a Y_b + Y_a Y_c}{Y_a + Y_b + Y_c}\right)V_A - \left(\frac{Y_a Y_b}{Y_a + Y_b + Y_c}\right)V_B = I_1$$

$$-\left(\frac{Y_a Y_b}{Y_a + Y_b + Y_c}\right)V_A + \left(\frac{Y_a Y_b + Y_b Y_c}{Y_a + Y_b + Y_c}\right)V_B = I_2$$

(13.22)

The nodal equations for the Δ circuit are

$$(Y_{ab} + Y_{ca})V_A - Y_{ab}V_B = I_1$$

$$-Y_{ab}V_A + (Y_{ab} + Y_{bc})V_B = I_2$$

Equating coefficients of like terms in these equations and (13.22) and solving for the admittances of the Δ circuit, we have the Y-Δ transformation

$$Y_{ab} = \frac{Y_a Y_b}{Y_a + Y_b + Y_c}$$

$$Y_{bc} = \frac{Y_b Y_c}{Y_a + Y_b + Y_c}$$

(13.23)

$$Y_{ca} = \frac{Y_c Y_a}{Y_a + Y_b + Y_c}$$

If we imagine the **Y** and Δ circuits superimposed on a single diagram as in Fig. 13.16(c), then Y_a and Y_b are *adjacent* to Y_{ab}, Y_b and Y_c are *adjacent* to Y_{bc}, etc. Thus we may state (13.23) in words, as follows:

The admittance of an arm of the Δ is equal to the product of the admittances of the adjacent arms of the **Y** divided by the sum of the **Y** admittances.

To obtain the Δ-**Y** transformation we may solve (13.23) for the **Y** admittances, a difficult task, or we may write two sets of loop equations for the **Y** and Δ circuits. In the latter case we shall have the dual of the procedure which led to (13.23). In either case, as the reader is asked to show in Prob. 13.19, the Δ-**Y** transformation is

$$Z_a = \frac{Z_{ab}Z_{ca}}{Z_{ab} + Z_{bc} + Z_{ca}}$$

$$Z_b = \frac{Z_{bc}Z_{ab}}{Z_{ab} + Z_{bc} + Z_{ca}}$$

(13.24)

$$Z_c = \frac{Z_{ca}Z_{bc}}{Z_{ab} + Z_{bc} + Z_{ca}}$$

where the **Z**'s are the reciprocals of the **Y**'s of Fig. 13.16. The rule is as follows:

The impedance of an arm of the **Y** is equal to the product of the impedances of the adjacent arms of the Δ divided by the sum of the Δ impedances.

(By *adjacent* here we mean "on each side of and terminating on the same node as." For example, in the superimposed drawing of the **Y** and **Δ**, $\mathbf{Z}_a$ lies between $\mathbf{Z}_{ab}$ and $\mathbf{Z}_{ca}$ and all three have a common terminal A. Thus $\mathbf{Z}_{ab}$ and $\mathbf{Z}_{ca}$ are *adjacent* arms of $\mathbf{Z}_a$.)

EXAMPLE 13.6 As an example, let us find the input impedance **Z** of Fig. 13.17(a). This is a problem which would have required us to write loop or nodal equations in the past, because we cannot simplify the circuit by combining series and/or parallel impedances. Replacing the 6-, 3-, and 2-Ω resistors, which constitute a **Y**, by their equivalent **Δ**, as shown in Fig. 13.17(b), however, enables us to solve the problem readily.

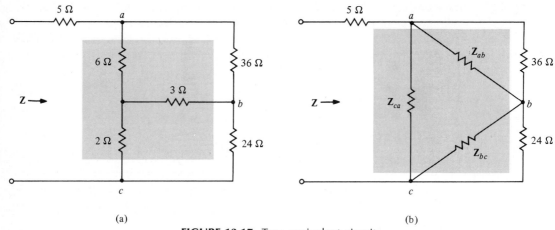

FIGURE 13.17 Two equivalent circuits

Comparing Figs. 13.17(a) and 13.16(a), we see that $\mathbf{Y}_a = \frac{1}{6}$, $\mathbf{Y}_b = \frac{1}{3}$, and $\mathbf{Y}_c = \frac{1}{2}$ ℧. Therefore, from (13.23), we have

$$\mathbf{Y}_{ab} = \frac{\frac{1}{6}\left(\frac{1}{3}\right)}{\frac{1}{6} + \frac{1}{3} + \frac{1}{2}} = \frac{1}{18}$$

$$\mathbf{Y}_{bc} = \frac{\frac{1}{3}\left(\frac{1}{2}\right)}{\frac{1}{6} + \frac{1}{3} + \frac{1}{2}} = \frac{1}{6}$$

$$\mathbf{Y}_{ca} = \frac{\frac{1}{2}\left(\frac{1}{6}\right)}{\frac{1}{6} + \frac{1}{3} + \frac{1}{2}} = \frac{1}{12}$$

Therefore, in Fig. 13.17(b) we have

$$\mathbf{Z}_{ab} = 18 \; \Omega, \qquad \mathbf{Z}_{bc} = 6 \; \Omega, \qquad \mathbf{Z}_{ca} = 12 \; \Omega$$

Thus Fig. 13.17(b) may be simplified by combining parallel and series resistors to obtain

$$\mathbf{Z} = 12 \; \Omega$$

EXAMPLE 13.7 As another example, suppose we have a balanced **Y**-connected load with phase impedance $\mathbf{Z}_y$ and wish to convert it to an equivalent **Δ**-connected load. By (13.23), since Y_a, Y_b, and Y_c are all equal to Y_y, the reciprocal of Z_y, then the equivalent Δ-connected load is also balanced because

$$Y_{ab} = Y_{bc} = Y_{ca} = \frac{Y_y^2}{3Y_y} = \frac{Y_y}{3}$$

Thus if Z_d is the phase impedance of the equivalent balanced Δ-connected load, then

$$Z_d = 3Z_y \qquad (13.25)$$

which may be used to convert from Y to Δ and vice versa.

EXERCISES

13.4.1 Find the input impedance seen by the source using a Y-Δ or Δ-Y transformation to simplify the circuit. From this result find the average power delivered by the source.
Answer $(1 + j2)/5\ \Omega$, 8 W

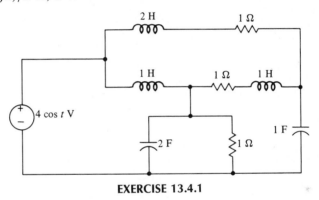

EXERCISE 13.4.1

13.4.2 A balanced three-phase source with $V_L = 200$ V rms is delivering power to a balanced Y-connected load with phase impedance $Z_1 = 6 + j8\ \Omega$ in parallel with a balanced Δ-connected load with phase impedance $Z_2 = 9 + j12\ \Omega$. Find the power delivered by the source.
Answer 7.2 kW

13.4.3 Show that the Y-Δ transformation of (13.23) is equivalent to

$$Z_{ab} = \frac{Z_a Z_b + Z_b Z_c + Z_c Z_a}{Z_c}$$

$$Z_{bc} = \frac{Z_a Z_b + Z_b Z_c + Z_c Z_a}{Z_a}$$

$$Z_{ca} = \frac{Z_a Z_b + Z_b Z_c + Z_c Z_a}{Z_b}$$

or in words:

The impedance of an arm of the Δ is equal to the sum of the products of the impedances of the Y, taken two at a time, divided by the impedance of the *opposite* arm of the Y.

13.4.4 If the lines in Ex. 13.3.2 each have a resistance of 0.1 Ω, find the power lost in the lines.

Answer 360 W

13.5
POWER MEASUREMENT

It appears to be a simple matter to measure the power delivered to a three-phase load by using one wattmeter for each of the three phases. This is illustrated for the three-wire **Y**-connected load in Fig. 13.18. Each wattmeter has its current coil in series with one phase of the load and its potential coil across one phase of the load. The connections are theoretically correct but may be useless in practice because the neutral point N may not be accessible (as, for example, in the case of a Δ-connected load). It would be better, in general, to be able to make the measurements using only the lines a, b, and c. In this section we shall show that this is, in fact, possible and that, moreover, only two wattmeters are required, instead of three. The method is general and is applicable to unbalanced as well as balanced systems.

Let us consider the three-wire **Y**-connected load of Fig. 13.19, which has three wattmeters connected so that each has its current coil in one line and its potential coil between that line and a common point x. If T is the period of the source voltages and i_a, i_b, and i_c are the time-domain line currents, directed into the loads, then the total power P_x indicated by the three meters is

FIGURE 13.18 Power measurement using three wattmeters

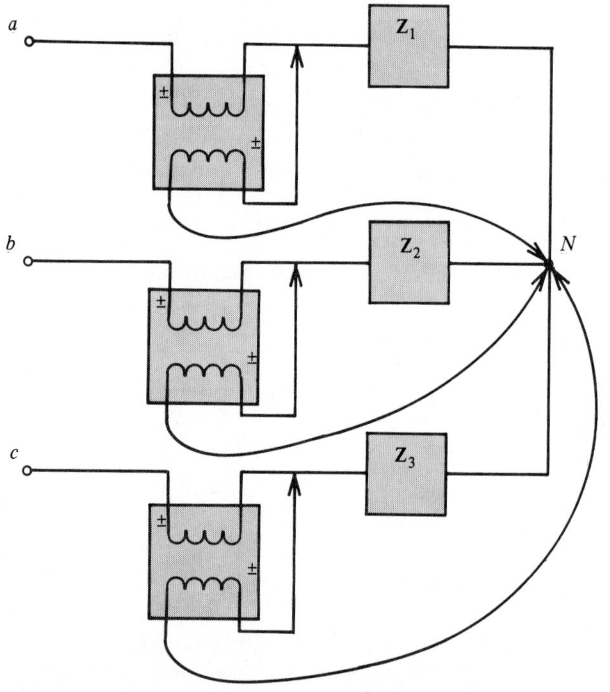

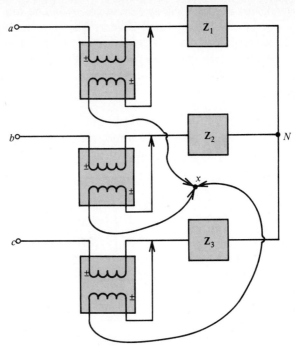

FIGURE 13.19 Three wattmeters connected to a common point

$$P_x = \frac{1}{T} \int_0^T (v_{ax}i_a + v_{bx}i_b + v_{cx}i_c)\, dt \qquad (13.26)$$

Regardless of the point x, which is completely arbitrary, we have

$$v_{ax} = v_{aN} + v_{Nx}$$

$$v_{bx} = v_{bN} + v_{Nx}$$

$$v_{cx} = v_{cN} + v_{Nx}$$

Substituting these results into (13.26) and rearranging, we obtain

$$P_x = \frac{1}{T} \int_0^T (v_{aN}i_a + v_{bN}i_b + v_{cN}i_c)\, dt + \frac{1}{T} \int_0^T v_{Nx}(i_a + i_b + i_c)\, dt$$

By KCL we have

$$i_a + i_b + i_c = 0$$

so that

$$P_x = \frac{1}{T} \int_0^T (v_{aN}i_a + v_{bN}i_b + v_{cN}i_c)\, dt \qquad (13.27)$$

Thus the sum of the three wattmeter readings is precisely the total average power delivered to the three-phase load, since the three terms in the integrand of (13.27) are the instantaneous phase powers.

Since the point x in Fig. 13.19 is arbitrary, we may place it on one of the lines.

Then the meter whose current coil is in that line will read zero because the voltage across its potential coil is zero. Therefore the total power delivered to the load is measured by the other two meters, and the meter reading zero is unnecessary. For example, the point x is placed on line b in Fig. 13.20, and the total power delivered to the load is

$$P = P_A + P_C$$

where P_A and P_C are the readings of meters A and C. It is important to note that one or the other of the two wattmeters may indicate a negative reading, and thus the sum of the two readings is the *algebraic* sum.

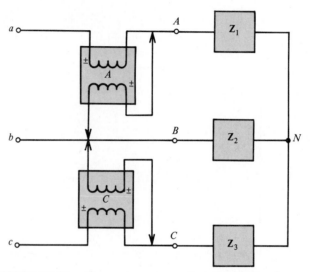

FIGURE 13.20 Two wattmeters reading the total load power

The proof of the two-wattmeter method of measuring the total three-phase power has been carried out for a **Y**-connected load. However, it holds also for a Δ-connected load, as the reader is asked to show in Prob. 13.21.

EXAMPLE 13.8 As an example, in Fig. 13.20 let the line voltages be a balanced *abc* sequence with

$$\mathbf{V}_{ab} = 100\sqrt{3}\underline{/0°} \text{ V rms}$$

and phase impedances given by

$$\mathbf{Z}_1 = \mathbf{Z}_2 = \mathbf{Z}_3 = 10 + j10 \text{ } \Omega$$

Then we have

$$\mathbf{V}_{cb} = -\mathbf{V}_{bc} = -100\sqrt{3}\underline{/-120°} = 100\sqrt{3}\underline{/60°} \text{ V rms}$$

$$\mathbf{I}_{aA} = \frac{\mathbf{V}_{AN}}{\mathbf{Z}_1} = \frac{(100\sqrt{3}/\sqrt{3})\underline{/-30°}}{10\sqrt{2}\underline{/45°}} = 5\sqrt{2}\underline{/-75°} \text{ A rms}$$

$$\mathbf{I}_{cC} = 5\sqrt{2}\underline{/-315°} \text{ A rms}$$

The meter readings are thus

$$\begin{aligned}
P_A &= |\mathbf{V}_{ab}||\mathbf{I}_{aA}| \cos (\text{ang } \mathbf{V}_{ab} - \text{ang } \mathbf{I}_{aA}) \\
&= (100\sqrt{3})(5\sqrt{2}) \cos (0° + 75°) \\
&= 317 \text{ W}
\end{aligned}$$

and

$$\begin{aligned}
P_C &= |\mathbf{V}_{cb}||\mathbf{I}_{cC}| \cos (\text{ang } \mathbf{V}_{cb} - \text{ang } \mathbf{I}_{cC}) \\
&= (100\sqrt{3})(5\sqrt{2}) \cos (60° + 315°) \\
&= 1183 \text{ W}
\end{aligned}$$

with a total of

$$P = 1500 \text{ W}$$

As a check, the power delivered to phase A is

$$\begin{aligned}
P_p &= |\mathbf{V}_{AN}||\mathbf{I}_{aA}| \cos (\text{ang } \mathbf{V}_{AN} - \text{ang } \mathbf{I}_{aA}) \\
&= \left(\frac{100\sqrt{3}}{\sqrt{3}}\right)(5\sqrt{2}) \cos (-30° + 75°) \\
&= 500 \text{ W}
\end{aligned}$$

Since the system is balanced, the total power is

$$P = 3P_p = 1500 \text{ W}$$

which agrees with the previous result.

EXERCISES

13.5.1 In Fig. 13.18, let $\mathbf{Z}_1 = \mathbf{Z}_2 = \mathbf{Z}_3 = 10\underline{/60°}$ Ω, and let the line voltages be a balanced *abc* sequence set, with $\mathbf{V}_{ab} = 200\underline{/0°}$ V rms. Find the reading of each meter.
Answer $\frac{2}{3}$ kW

13.5.2 If the power delivered to the load of Ex. 13.5.1 is measured by the two wattmeters *A* and *C* connected as shown in Fig. 13.20, find the wattmeter readings. Check for consistency with the answer of Ex. 13.5.1.
Answer 0, 2 kW

13.5.3 Find the wattmeter readings P_A and P_C and the total power P in Fig. 13.20 if the line voltages are as given in Ex. 13.5.1 and $\mathbf{Z}_1 = \mathbf{Z}_2 = \mathbf{Z}_3 = 10\underline{/75°}$ Ω. Check the answer by using $P = 3P_p$.
Answer −598, 1633, 1035 W

13.6

SPICE FOR THREE-PHASE CIRCUIT ANALYSIS

The analysis of three-phase networks presented previously has been restricted to balanced systems whose solutions can be expressed in terms of a single phase. The application of SPICE is easily used for both balanced or unbalanced systems when applied to the entire network. All principles necessary for using SPICE have been presented in previous chapters.

EXAMPLE 13.9   As a first example, consider finding the line voltage and phase current at the load for phase A of the balanced Y-Y system of Fig. 13.21. The transmission line for interconnecting the generator and load has losses that are represented by 2-ohm resistors.

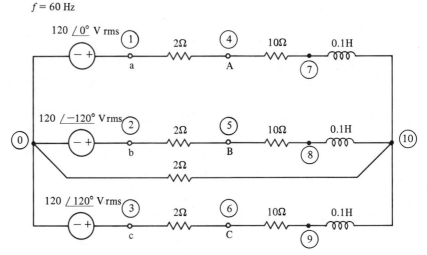

FIGURE 13.21   Balanced Y-Y system for SPICE analysis

A circuit file for this network is

```
3-PHASE Y-Y SYSTEM WITH TRANSMISSION LINE LOSSES
* DATA STATEMENTS VOLTAGES EXPRESSED IN RMS
VAN 1 0 AC 120 0
VBN 2 0 AC 120 -120
VCN 3 0 AC 120 120
RLOSSA 1 4 2
RLOSSB 2 5 2
RLOSSC 3 6 2
RLOSSN 10 0 2
RA 4 7 10
LA  7 10 0.1
RB  5 8 10
LB  8 10 0.1
RC  6 9 10
LC  9 10 0.1
```

* SOLUTION CONTROL STATEMENT FOR f = 60 Hz
.AC LIN 1 60 60
* OUTPUT CONTROL STATEMENT
.PRINT AC VM(4,5) VP(4,5) IM(VAN) IP(VAN)
.END

The computer output yields

FREQ	VM(4,5)	VP(4,5)	IM(VAN)	IP(VAN)
6.000E+01	2.049E+02	3.280E+01	3.033E+00	1.077E+02

EXAMPLE 13.10 As a second example, let us find the line voltage and line current for the unbalanced Y-Δ system of Fig. 13.22. The transmission line losses are denoted by the 1-, 2-, and 3-ohm resistors, respectively. A circuit file is

```
3-PHASE Y-D UNBALANCED SYSTEM WITH T-LINE LOSSES
* DATA STATEMENTS WITH VOLTAGES IN RMS VALUES
VAN 1 0 AC 120 0
VBN 2 0 AC 120 −120
VCN 3 0 AC 120 120
RLOSSA 1 4 1
RLOSSB 2 5 2
RLOSSC 3 6 3
RAB 4 5 12
RBC 5 7 8
CBC 6 7 1000UF
RAC 4 8 10
LAC 6 8 0.05
* OUTPUT CONTROL STATEMENT FOR f = 400 Hz
.AC LIN 1 400 400
* OUTPUT CONTROL STATEMENT
.PRINT AC VM(4.5) VP(4.5) IM(VAN) IP(VAN)
.END
```

FIGURE 13.22   Unbalanced Y-Δ system

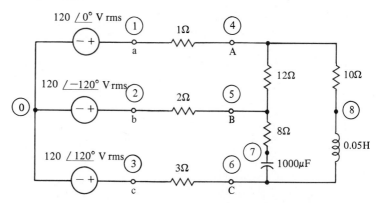

Section 13.6   SPICE for Three-Phase Circuit Analysis

413

The computer output contains

FREQ	VM(4,5)	VP(4,5)	IM(VAN)	IP(VAN)
4.000E+02	1.601E+02	2.219E+01	1.255E+01	−1.627E+02

EXERCISES

13.6.1 Find the line voltage and phase current for the load of phase A of the system of Fig. 13.21 if the load of phase C is short-circuited (called a phase fault).
Answer 204.9 $\underline{/32.8°}$ V, 4.07 $\underline{/90.54°}$ A

13.6.2 Repeat Ex. 13.6.1 if the 2-ohm neutral line between nodes 0 and 10 is removed.
Answer 204.9 $\underline{/32.8°}$ V, 5.33 $\underline{/82.41°}$ A

PROBLEMS

13.1 If in Fig. 13.4 $\mathbf{V}_{an} = \mathbf{V}_{nb} = 100\underline{/0°}$ V rms, the impedance between terminals A-N is $10\underline{/-30°}$ Ω, and that between terminals N-B is $10\underline{/30°}$ Ω, find the neutral current $\mathbf{I}_{nN}$.

13.2 In Fig. 13.5, let $\mathbf{V}_1 = 100\underline{/0°}$ V rms, $\mathbf{Z}_1 = 5 + j5$ Ω, $\mathbf{Z}_2 = 0.5$ Ω, $\mathbf{Z}_3 = 1$ Ω, and $\mathbf{Z}_4 = 10 - j5$ Ω. Find the average power absorbed by the loads, lost in the lines, and delivered by the sources.

13.3 In Fig. 13.10 the line currents form a balanced, positive sequence set with $\mathbf{I}_{aA} = 20\underline{/-30°}$ A rms and $\mathbf{V}_{ab} = 60\underline{/30°}$ V rms. Find $\mathbf{Z}_p$ and the power delivered to the three-phase load.

13.4 A balanced $\mathbf{Y}$-connected load is present on a 240-V rms (line-to-line) three-phase system. If the phase impedance is $3\sqrt{3}\underline{/30°}$ Ω, find the total average power delivered to the load.

13.5 A balanced three-phase $\mathbf{Y}$-connected load draws 6 kW at a power factor of 0.8 lagging. If the line voltages are a balanced 200-V rms set, find the line current I_L.

13.6 In Fig. 13.10, the source is balanced, with positive phase sequence, and $\mathbf{V}_{an} = 100\underline{/0°}$ V rms. Find $\mathbf{Z}_p$ if the source delivers 2.4 kW at a power factor of 0.8 lagging.

13.7 A balanced $\mathbf{Y}$-$\mathbf{Y}$ three-wire, positive sequence system has $\mathbf{V}_{an} = 200\underline{/0°}$ V rms and $\mathbf{Z}_p = 3 - j4$ Ω. The lines each have a resistance of 1 Ω. Find the line current I_L and the power delivered to the load.

13.8 A balanced $\mathbf{Y}$-connected source, $\mathbf{V}_{an} = 240\underline{/0°}$ V rms, positive sequence, is connected by four perfect conductors (having zero impedance) to an unbalanced $\mathbf{Y}$-connected load, $\mathbf{Z}_{AN} = 10$ Ω, $\mathbf{Z}_{BN} - 10 - j5$ Ω, and $\mathbf{Z}_{CN} = j20$ Ω. Find the four line currents.

13.9 A balanced three-phase $\mathbf{Y}$-connected load draws 3 kW at a power factor of 0.8 lagging. A balanced $\mathbf{Y}$ of capacitors is to be placed in parallel with the load so that the power factor of the combination is 0.85 lagging. If the frequency is 60 Hz and the line voltages are a balanced 200-V rms set, find the capacitances required.

13.10 Repeat Prob. 13.6 if the load is a balanced Δ.

13.11 In the $\mathbf{Y}$-Δ system shown, the source is positive sequence with $\mathbf{V}_{an} = 200\underline{/0°}$ V rms and the phase impedance is $\mathbf{Z}_p = 4 + j3$ Ω. Find the line voltage V_L, the line current I_L, and the power delivered to the load.

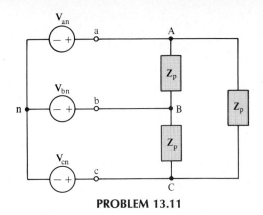

PROBLEM 13.11

13.12 In the Y-Δ system of Prob. 13.11 the source voltage $\mathbf{V}_{an} = 100\underline{/0°}$ V rms and $\mathbf{Z}_p = 10\underline{/60°}$ Ω. Find the line voltage, the line current, and the load current magnitudes, and the power delivered to the load.

13.13 Find the line current I_L.

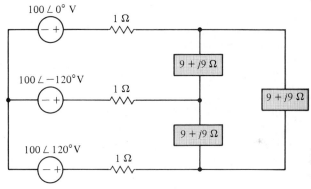

PROBLEM 13.13

13.14 Find the power delivered to the load in Prob. 13.13 if the magnitude of the source voltages is 120 V rms and $\mathbf{Z}_p = 6 - j9$ Ω.

13.15 A balanced three-phase, positive-sequence source with $\mathbf{V}_{ab} = 200\underline{/0°}$ V rms is supplying a Δ-connected load, $\mathbf{Z}_{AB} = 3 - j4$ Ω, $\mathbf{Z}_{BC} = 20\underline{/60°}$ Ω, and $\mathbf{Z}_{CA} = 50\underline{/30°}$ Ω. Find the phasor line currents. (Assume perfectly conducting lines.)

13.16 A balanced three-phase, positive-sequence source with $\mathbf{V}_{ab} = 200\underline{/0°}$ V rms is supplying a Δ-connected load, $\mathbf{Z}_{AB} = 50$ Ω, $\mathbf{Z}_{BC} = 20 + j20$ Ω, and $\mathbf{Z}_{CA} = 30 - j40$ Ω. Find the line currents.

13.17 A balanced three-phase positive-sequence source with $\mathbf{V}_{ab} = 240\underline{/0°}$ V rms is supplying a parallel combination of a Y-connected load and a Δ-connected load. If the Y and Δ loads are balanced with phase impedances of $8 + j8$ Ω and $24 - j24$ Ω, respectively, find the line current I_L and the power supplied by the source, assuming perfectly conducting lines.

13.18 Find the line currents I_{aA}, I_{bB}, and I_{cC}.

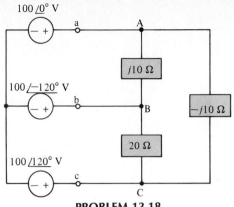

100 $\underline{/0°}$ V

a

A

$j10\ \Omega$

100 $\underline{/-120°}$ V

b

B

$-j10\ \Omega$

$20\ \Omega$

100 $\underline{/120°}$ V

c

C

PROBLEM 13.18

13.19 Derive (13.24).

13.20 For a balanced three-phase system, if the phase voltages are

$$v_a(t) = V_m \cos \omega t$$

$$v_b(t) = V_m \cos (\omega t - 120°)$$

$$v_c(t) = V_m \cos (\omega t - 240°)$$

then the phase currents are

$$i_a(t) = \overline{I_m} \cos (\omega t - \theta)$$

$$i_b(t) = I_m \cos (\omega t - \theta - 120°)$$

$$i_c(t) = I_m \cos (\omega t - \theta - 240°)$$

Show that the total instantaneous power,

$$p(t) = v_a i_a + v_b i_b + v_c i_c$$

is a constant given by

$$p(t) = \frac{3}{2} V_m I_m \cos \theta$$

which is also P, the total average power. [*Suggestion:* Recall that $\cos \alpha + \cos (\alpha - 120°) + \cos (\alpha - 240°) = 0$.]

13.21 Show that P_x given by (13.26) is equal to the total average power delivered to a Δ-connected load.

13.22 In the system of Fig. 13.20, the line voltages are a balanced, positive-sequence set, with $\mathbf{V}_{ab} = 100\underline{/0°}$ V rms, and $\mathbf{Z}_1 = \mathbf{Z}_2 = \mathbf{Z}_3 = 10\underline{/30°}\ \Omega$. Find the power delivered to the load (a) by finding the readings of the two wattmeters and (b) by $P = 3P_p$.

13.23 Repeat Prob. 13.22 if $\mathbf{Z}_1 = \mathbf{Z}_2 = \mathbf{Z}_3 = 10\underline{/60°}\ \Omega$. Note that in this case wattmeter C reads the total power since wattmeter A reads zero.

13.24 Show that in Fig. 13.20 if $\mathbf{V}_{ab} = V_L\underline{/\alpha}$, $\mathbf{V}_{bc} = V_L\underline{/\alpha - 120°}$, $\mathbf{V}_{ca} = V_L\underline{/\alpha - 240°}$ V rms, and $\mathbf{Z}_1 = \mathbf{Z}_2 = \mathbf{Z}_3 = |\mathbf{Z}|\underline{/60°}$, then wattmeter A reads zero and wattmeter C reads the total average power delivered to the load,

$$P = \frac{V_L^2}{2|\mathbf{Z}|}$$

This is a generalization of the result of Prob. 13.23.

13.25 Find the readings P_A and P_B of the wattmeters and the total power delivered to the load if the source is a balanced **Y**-connected abc sequence source with $\mathbf{V}_{an} = 100\underline{/0°}$ V rms, and the phase impedances are each $\mathbf{Z}_p = 30\underline{/30°}$ Ω.

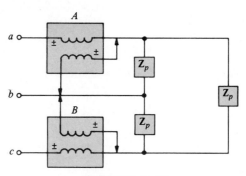

PROBLEM 13.25

13.26 For the system shown, the line voltages are a balanced positive sequence set with $\mathbf{V}_{ab} = 300\underline{/0°}$ V rms. Find the meter readings P_A and P_B and the total power delivered to the load if $\mathbf{Z}_p = 10\underline{/30°}$ Ω.

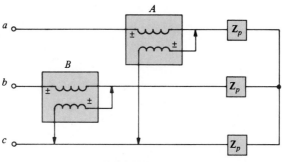

PROBLEM 13.26

13.27 Show that if a balanced, positive-sequence source is connected by three perfectly conducting wires to a balanced load having phase impedance $\mathbf{Z}_p = |\mathbf{Z}_p|\underline{/\theta}$, then the two wattmeters, connected as in Fig. 13.20, read, respectively, P_A and P_C, where

$$\frac{P_A}{P_C} = \frac{\cos(30° + \theta)}{\cos(30° - \theta)}$$

From this result show that

$$\tan \theta = \sqrt{3}\,\frac{P_C - P_A}{P_C + P_A}$$

Thus we may find the power factor, $\cos \theta$, of the load from the two wattmeter readings.

13.28 In Prob. 13.27, find the *pf* of the load if (a) $P_A = P_C$, (b) $P_A = -P_C$, (c) $P_A = 0$, (d) $P_C = 0$, and (e) $P_A = 2P_C$.

COMPUTER APPLICATION PROBLEMS

13.29 Use SPICE to solve the system of Prob. 13.16 for $f = 60$ Hz.

13.30 Use SPICE to find the current of the phase A source if phase voltage C is zero (phase C generator fault) in the Y-Y system of Fig. 13.21.

13.31 Use SPICE to find the current in the 12-ohm load of Fig. 13.22 if the load between phases B and C (8 ohm and 1000 μF) is short-circuited.

13.32 Repeat Prob. 13.30 for the system of Fig. 13.22.

14

Complex Frequency and Network Functions

Nikola Tesla
1856–1943

If Thomas A. Edison has a rival for the title of the world's greatest inventor, it is certainly the Croatian-American engineer, Nikola Tesla. When the tall, lanky Tesla arrived in the United States in 1884, the country was in the middle of the "battle of the currents" between Thomas A. Edison, promoting dc, and George Westinghouse, leading the ac forces. Tesla quickly settled the argument in favor of ac with his marvelous inventions, such as the polyphase ac power system, the induction motor, the Tesla coil, and fluorescent lights.

Tesla was born in Smiljan, Austria-Hungary (now Yugoslavia), the son of a clergyman of the Greek Orthodox Church. As a boy Tesla had a talent for mathematics and an incredible memory, with the ability to recite by heart entire books and poems. He spent 2 years at the Polytechnic Institute of Graz, Austria, where he conceived the idea of the rotating magnetic field that was the later basis for his induction motor. At this point in Tesla's life his father died, and he decided to leave school, taking a job in Paris with the Continental Edison Company. Two years later he came to America, where he remained until his death. During his remarkable lifetime he held over 700 patents, settled the ac versus dc dispute, and was primarily responsible for the selection of 60 Hz as the standard ac frequency in the United States and throughout much of the world. After his death he was honored by the choice of *tesla* as the unit of magnetic flux density. ∎

In the previous chapters we have considered resistive circuit analysis, natural and forced responses of circuits containing storage elements, and, in particular, ac steady-state analysis. The excitations we have considered were, for the most part, constants, exponentials and sinusoids. In this chapter we shall consider an excitation, the damped sinusoid, which includes all these excitations as special cases. From this function we shall develop generalized phasors and general network functions which include, as special cases, the phasors and impedances of Chapters 10-13.

The network functions will be expressed in terms of a complex frequency that includes the frequency $j\omega$ as a special case. The concepts of complex frequency and general network functions will enable us to combine all of our earlier results into one common procedure. Both the natural and forced responses of a circuit may be found from its excitation and its network function, as we shall see. In addition, the network function may be used to determine the frequency-domain properties of the circuit, which will be the subject of Chapter 15.

14.1

THE DAMPED SINUSOID

In this section we shall consider the *damped sinusoid,*

$$v = V_m e^{\sigma t} \cos (\omega t + \phi) \tag{14.1}$$

which is a sinusoid, like those of the previous chapters, multiplied by a damping factor $e^{\sigma t}$. The constant σ (the Greek letter sigma) is real and is usually negative or zero, which accounts for the term *damping factor*.

The damped sinusoid contains, as special cases, most of the functions we have considered thus far. For example, if $\sigma = 0$, we have the pure sinusoid

$$v = V_m \cos (\omega t + \phi) \tag{14.2}$$

of the previous chapters. If $\omega = 0$, we have the exponential function

$$v = V_0 e^{\sigma t} \qquad (14.3)$$

and if $\sigma = \omega = 0$, we have the constant (dc) case

$$v = V_0 \qquad (14.4)$$

where, in both (14.3) and (14.4), $V_0 = V_m \cos \phi$. These functions are sketched in Fig. 14.1 for the various cases of σ and ω.

As we see from Fig. 14.1, $\sigma > 0$ represents growing oscillations (b) or a growing exponential (e), and $\sigma < 0$ represents decaying oscillations (a) or a decaying

FIGURE 14.1 Various cases of (14.1)

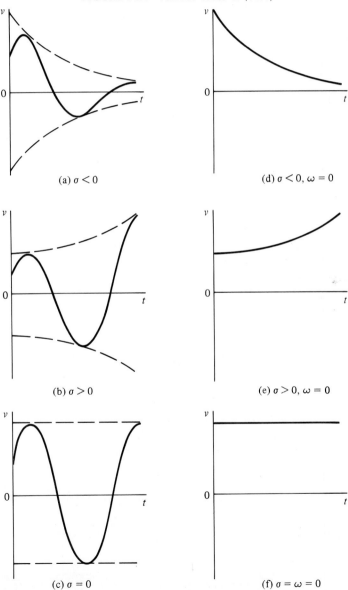

(a) $\sigma < 0$

(d) $\sigma < 0, \omega = 0$

(b) $\sigma > 0$

(e) $\sigma > 0, \omega = 0$

(c) $\sigma = 0$

(f) $\sigma = \omega = 0$

exponential (d). Finally, $\sigma = 0$ represents ac steady state (c) or dc steady state (f).

The units of ω are radians per second and of ϕ are radians or degrees, as before. Since σt is dimensionless, σ is in units of 1 per sec (1/s). This unit was encountered earlier, in connection with natural frequencies, in Chapter 9, where the dimensionless unit, *neper* (Np), was used for σt. Thus σ is the *neper frequency* in nepers per second (Np/s). The neper was chosen in honor of the Scottish mathematician John Napier (1550–1617), who invented logarithms.

EXAMPLE 14.1 As an example of a circuit having a damped sinusoidal excitation, let us find the forced response i in Fig. 14.2. The loop equation is

$$2\frac{di}{dt} + 5i = 25e^{-t} \cos 2t \qquad (14.5)$$

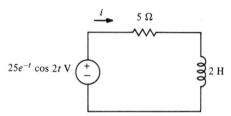

FIGURE 14.2 Circuit excited by a damped sinusoid

Trying as a forced response

$$i = e^{-t}(A \cos 2t + B \sin 2t)$$

which is the excitation and all its possible derivatives, we have

$$2e^{-t}(-2A \sin 2t - A \cos 2t + 2B \cos 2t - B \sin 2t)$$
$$+ 5e^{-t}(A \cos 2t + B \sin 2t) = 25e^{-t} \cos 2t$$

Since this must be an identity, we have

$$3A + 4B = 25$$
$$-4A + 3B = 0$$

or $A = 3$ and $B = 4$. Therefore, the forced response is

$$i = e^{-t}(3 \cos 2t + 4 \sin 2t) \text{ A}$$

or equivalently,

$$i = 5e^{-t} \cos (2t - 53.1°) \text{ A} \qquad (14.6)$$

As expected, the forced response is, like the excitation, a damped sinusoid.

EXERCISES

14.1.1 Excite the circuit of Fig. 14.2 by the complex function

$$v_1 = 25e^{-t}e^{j2t}$$
$$= 25e^{(-1+j2)t}$$

and show that the forced response is

$$i_1 = (5\underline{/-53.1°})e^{(-1+j2)t}$$

14.1.2 Show that if i_1 is the response to v_1 in

$$2\frac{di_1}{dt} + 5i_1 = v_1$$

then Re i_1 is the response to Re v_1. Apply this result to the functions of Ex. 14.1.1 to obtain i given by (14.6).

14.1.3 Find the forced response v using the methods of Exs. 14.1.1 and 14.1.2 if

$$\frac{d^3v}{dt^3} + 6\frac{d^2v}{dt^2} + 11\frac{dv}{dt} + 6v = i$$

where

$$i = 4e^{-2t}\cos(t - 60°)$$

Answer $2e^{-2t}\cos(t + 30°)$

14.2

COMPLEX FREQUENCY AND GENERALIZED PHASORS

In Chapter 10 we considered circuits with sinusoidal excitations such as

$$v_1 = V_m \cos(\omega t + \phi) \tag{14.7}$$

which we also wrote in the equivalent form

$$v_1 = \text{Re}(V_m e^{j\phi}e^{j\omega t})$$

Using the phasor representation of v_1,

$$\mathbf{V} = V_m e^{j\phi} = V_m\underline{/\phi} \tag{14.8}$$

we may also write

$$v_1 = \text{Re}(\mathbf{V}e^{j\omega t}) \tag{14.9}$$

The phasor representation was extremely useful in solving ac steady-state circuits in which the voltages and currents were sinusoids of the form of (14.7).

In this chapter we wish to consider circuits in which the excitations and forced responses are damped sinusoids, such as

$$v = V_m e^{\sigma t}\cos(\omega t + \phi) \tag{14.10}$$

A good question at this point is, Can we define phasor representations of damped sinusoids that will work as well for us as the phasors, such as (14.8), of undamped sinusoids? It seems plausible that we can, because the properties of sinusoids that made the phasors possible are shared by damped sinusoids. That is, the sum or difference of two damped sinusoids is a damped sinusoid, and the derivative, indefinite integral, or constant multiple of a damped sinusoid is a damped sinusoid. In all these operations only V_m and ϕ may change. This is exactly the case with undamped sinusoids.

Let us see what follows if we write (14.10) in the form analogous to (14.9). That is,

$$
\begin{aligned}
v &= V_m e^{\sigma t} \cos(\omega t + \phi) \\
&= \mathrm{Re}[V_m e^{\sigma t} e^{j(\omega t + \phi)}] \\
&= \mathrm{Re}[V_m e^{j\phi} e^{(\sigma + j\omega)t}]
\end{aligned}
$$

If we define the quantity

$$
s = \sigma + j\omega \tag{14.11}
$$

then we have

$$
v = \mathrm{Re}(\mathbf{V}e^{st}) \tag{14.12}
$$

where $\mathbf{V}$ is the phasor of (14.8).

Evidently, since $\mathbf{V}$ is the same for both, the undamped sinusoid given by (14.9) is identical in form to the damped sinusoid of (14.12). The only difference is that the number $j\omega$ is used in one case and the number s is used in the other. Obviously, then, we may do anything with the damped sinusoids that we did with undamped sinusoids as long as we use s instead of $j\omega$. We may define the phasor $\mathbf{V}$ of (14.8) to be the phasor representation of v in (14.10) and use it for damped sinusoidal circuit problems in exactly the same way that we used it for the sinusoidal problems. In the damped sinusoidal case we may wish to write the phasor as $\mathbf{V}(s)$ to distinguish it from the undamped sinusoidal case, $\mathbf{V}(j\omega)$.

EXAMPLE 14.2 As an example, the damped sinusoid

$$
v = 25e^{-t} \cos 2t \ \mathrm{V} \tag{14.13}
$$

of Fig. 14.2 has the phasor representation

$$
\mathbf{V} = \mathbf{V}(s) = 25\underline{/0^\circ} \tag{14.14}
$$

where $s = -1 + j2$. Conversely, if $\mathbf{V}$ is given by (14.14) and s is as specified, then v is given by (14.13).

Some authors prefer to call $\mathbf{V}(s)$, corresponding to $v(t)$ in (14.10), a *generalized phasor*, even though it is identical to the phasors of sinusoidal functions. It is, however, a function of a *generalized frequency*, namely s, given by (14.11). Since s is a complex number, it is more often called a *complex frequency*. Its components are

$$
\sigma = \mathrm{Re}\ s\ \mathrm{Np/s}
$$

$$
\omega = \mathrm{Im}\ s\ \mathrm{rad/s}
$$

which have units of frequency, and, indeed, s is the coefficient of t in an exponential function, as was the case in Chapter 9 for the natural frequencies of a circuit. The units of s are 1/second and are sometimes called complex nepers per second or complex radians per second.

It might be worth noting at this point that a function which can be written in the form

$$f(t) = K_1 e^{s_1 t} + K_2 e^{s_2 t} + \ldots + K_n e^{s_n t}$$

where the K_i and s_i are independent of t, may be said to be characterized by the complex frequencies $s_1, s_2, \ldots, s_n$. For example, writing (14.10) in the form

$$v = V_m e^{\sigma t} \left(\frac{e^{j(\omega t + \phi)} + e^{-j(\omega t + \phi)}}{2} \right)$$

results in

$$v = K_1 e^{(\sigma + j\omega)t} + K_2 e^{(\sigma - j\omega)t}$$

where $K_1 = V_m e^{j\phi}/2$ and $K_2 = V_m e^{-j\phi}/2 = K_1^*$ (the complex conjugate). Thus v possesses not one but two complex frequencies, namely $s_1 = \sigma + j\omega$ and $s_2 = s_1^* = \sigma - j\omega$. This concept of complex frequency is consistent with that of the natural frequencies considered in Chapter 9.

EXERCISES

14.2.1 Find the complex frequencies associated with (a) $6 + 4e^{-2t}$, (b) $\cos \omega t$, (c) $\sin (\omega t + \theta)$, (d) $6e^{-3t} \sin (4t + 10°)$, and (e) $e^{-t}(1 + \cos 2t)$.
Answer (a) $0, -2$; (b) $\pm j\omega$; (c) $\pm j\omega$; (d) $-3 \pm j4$; (e) $-1, -1 \pm j2$

14.2.2 Show that if i is a damped sinusoid,

$$i = I_m e^{\sigma t} \cos (\omega t + \phi)$$

then v, defined by

$$v = L \frac{di}{dt} + Ri$$

is also a damped sinusoid of the same complex frequency.
Answer

$$v = I_m \sqrt{(R + \sigma L)^2 + \omega^2 L^2} \; e^{\sigma t} \cos \left(\omega t + \phi + \tan^{-1} \frac{\omega L}{R + \sigma L} \right)$$

14.2.3 Find s and $V(s)$ if $v(t)$ is given by (a) 6, (b) $6e^{-2t}$, (c) $6e^{-3t} \cos(4t + 10°)$, and (d) $6 \cos(2t + 10°)$.
Answer (a) $0, 6\underline{/0°}$; (b) $-2, 6\underline{/0°}$; (c) $-3 + j4, 6\underline{/10°}$; (d) $j2, 6\underline{/10°}$

14.2.4 Find $v(t)$ if (a) $\mathbf{V} = 4\underline{/0°}$, $s = -3$; (b) $\mathbf{V} = 5\underline{/15°}$, $s = j2$; and (c) $\mathbf{V} = 5\underline{/30°}$, $s = -3 + j2$.
Answer (a) $4e^{-3t}$; (b) $5 \cos(2t + 15°)$; (c) $5e^{-3t} \cos(2t + 30°)$

IMPEDANCE AND ADMITTANCE

Because of the identical form of the phasor representations for sinusoids and damped sinusoids, we may find forced responses to damped sinusoidal excitations using phasors precisely as we did in the previous chapters. All the concepts and rules, such as impedance, admittance, KCL, KVL, Thevenin's theorem, Norton's theorem, superposition, etc., carry over to the damped sinusoidal case exactly. We need only to use $s = \sigma + j\omega$ rather than $j\omega$.

It follows that, in the s domain, the phasor current $\mathbf{I}(s)$ and voltage $\mathbf{V}(s)$, associated with a two-terminal device, are related by

$$\mathbf{V}(s) = \mathbf{Z}(s)\mathbf{I}(s)$$

where $\mathbf{Z}(s)$ is the *generalized impedance,* or simply impedance, of the device. We may obtain $\mathbf{Z}(s)$ from $\mathbf{Z}(j\omega)$, the impedance in the ac steady-state case, by simply replacing $j\omega$ by s.

For a resistance R, the impedance is, therefore,

$$\mathbf{Z}_R(s) = R$$

In the case of an inductance L, the impedance is

$$\mathbf{Z}_L(s) = sL$$

and for a capacitance C, it is

$$\mathbf{Z}_C(s) = \frac{1}{sC}$$

In a similar manner the admittances are, respectively,

$$\mathbf{Y}_R(s) = \frac{1}{R} = G, \qquad \mathbf{Y}_L(s) = \frac{1}{sL}, \qquad \mathbf{Y}_C(s) = sC$$

EXAMPLE 14.3 Impedances in series or parallel are combined in exactly the same way as in the ac steady-state case, since we merely replace $j\omega$ by s. For example, let us reconsider the circuit of Fig. 14.2, whose phasor circuit is shown in Fig. 14.3, where $s = -1 + j2$. The impedance seen from the source terminals consists of the impedances $2s$ and 5 in series and thus is given by

$$\mathbf{Z}(s) = 2s + 5 \ \Omega$$

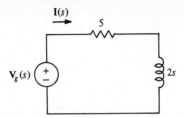

FIGURE 14.3 Phasor circuit of Fig. 14.2

Therefore, since the input voltage phasor is

$$\mathbf{V}_g(s) = 25\underline{/0°} \text{ V}$$

we have

$$\mathbf{I}(s) = \frac{\mathbf{V}_g(s)}{2s + 5} = \frac{25\underline{/0°}}{2(-1 + j2) + 5} = 5\underline{/-53.1°} \text{ A}$$

Thus in the time domain the forced response, as before, is

$$i(t) = 5e^{-t} \cos(2t - 53.1°) \text{ A}$$

EXAMPLE 14.4   As another example, let us consider the time-domain circuit of Fig. 14.4(a), where it is required to find the forced response $v_o(t)$ for a given damped sinusoidal input $v_g(t)$. The phasor circuit is shown in Fig. 14.4(b), from which we have the nodal equations,

$$\left(\frac{1}{2} + 1 + \frac{1}{4}s\right)\mathbf{V}_1 - \frac{1}{2}\mathbf{V}_g - \mathbf{V}_2 - \frac{1}{4}s\mathbf{V}_o = 0$$

$$\left(1 + \frac{1}{4}s\right)\mathbf{V}_2 - \mathbf{V}_1 = 0$$

We note that $\mathbf{V}_2 = \mathbf{V}_o/2$ since the op amp and its two connected resistors constitute a VCVS with a gain of 2. Therefore, eliminating $\mathbf{V}_1$ and $\mathbf{V}_2$, we have

$$\mathbf{V}_o(s) = \frac{16}{s^2 + 2s + 8}\mathbf{V}_g(s) \tag{14.15}$$

If we have

$$v_g(t) = e^{-2t} \cos 4t \text{ V}$$

then $s = -2 + j4$ and $\mathbf{V}_g(s) = 1\underline{/0°}$. Thus from (14.15) we have

$$\mathbf{V}_o(s) = \sqrt{2}\underline{/135°} \text{ V}$$

and therefore

$$v_o(t) = \sqrt{2} e^{-2t} \cos(4t + 135°) \text{ V}$$

EXAMPLE 14.5   As a final example in this section let us find the forced response i in Fig. 14.5 if

$$v_{g1} = 8e^{-t} \cos t \text{ V}$$

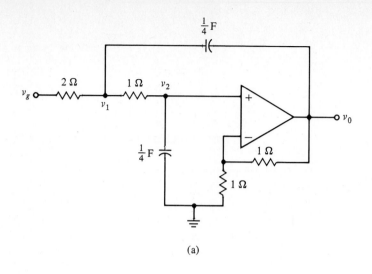

(a)

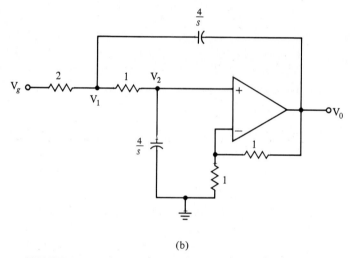

(b)

FIGURE 14.4 (a) Time-domain circuit; (b) its phasor circuit

FIGURE 14.5 Circuit with two sources

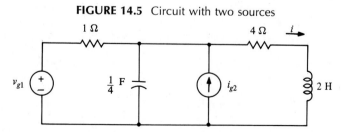

428

and
$$i_{g2} = 2e^{-5t} \text{ A}$$

Since the complex frequencies of v_{g1} and i_{g2} are different, we must use superposition. That is,

$$i = i_1 + i_2$$

where i_1 is due to v_{g1} acting alone and i_2 is due to i_{g2} acting alone.

The phasor circuits for $i_{g2} = 0$ and $v_{g1} = 0$ are shown in Figs. 14.6(a) and (b), respectively. The phasor currents are $\mathbf{I}_1$ and $\mathbf{I}_2$, as shown. In Fig. 14.6(a), using current division, we have

$$\mathbf{I}_1 = \frac{8/0°}{1 + \{[(2s + 4)(4/s)]/(2s + 4 + 4/s)\}} \times \frac{4/s}{2s + 4 + 4/s}$$

which, since $s = -1 + j1$ for this case, becomes

$$\mathbf{I}_1 = 2\sqrt{2}/{-45°} \text{ A}$$

Therefore, the forced component due to v_{g1} alone is

$$i_1 = 2\sqrt{2}\, e^{-t} \cos (t - 45°) \text{ A}$$

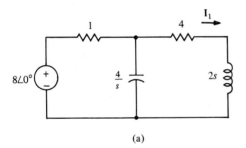 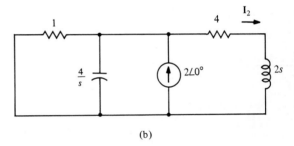

(a) (b)

FIGURE 14.6 Phasor circuits associated with Fig. 14.5

In Fig. 14.6(b), again by current division, we obtain

$$\mathbf{I}_2 = \left(\frac{[1(4/s)]/(1 + 4/s)}{\{[1(4/s)]/(1 + 4/s)\} + 2s + 4} \right)(2/0°)$$

which, since $s = -5$, is

$$\mathbf{I}_2 = 0.8/0° \text{ A}$$

Therefore, the forced component due to i_{g2} alone is

$$i_2 = 0.8e^{-5t} \text{ A}$$

The complete forced response of Fig. 14.5 is, therefore,

$$i = 2\sqrt{2}\, e^{-t} \cos (t - 45°) + 0.8e^{-5t} \text{ A}$$

EXERCISES

14.3.1 Find the forced response v if $v_{g1} = 4e^{-2t}\cos(t - 45°)$ V and $i_{g2} = 2e^{-t}$ A.
Answer $2\sqrt{2}\, e^{-2t}\cos(t + 90°) + 4e^{-t}$ V

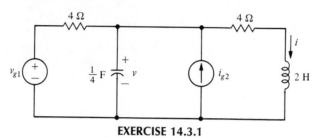

EXERCISE 14.3.1

14.3.2 For the circuit of Ex. 14.3.1, take $i_{g2} = e^{-2t}\cos(t + 45°)$ A, leaving v_{g1} as it is, and find v by (a) using superposition, (b) writing a single node equation, and (c) using source transformations.
Answer $4e^{-2t}\cos(t + 135°)$ V

14.3.3 In the phasor circuit for Ex. 14.3.1 with $i_{g2} = 0$ and v_{g1} as it is, replace everything except the capacitor by its Thevenin equivalent and use the result to find $\mathbf{V}(s)$, the phasor of v. Note that the result leads to the first component of v given in Ex. 14.3.1.
Answer

$$\mathbf{V}_{oc}(s) = \frac{s + 2}{s + 4}\mathbf{V}_g(s), \quad \mathbf{Z}_{th}(s) = \frac{4(s + 2)}{s + 4}\Omega, \quad \mathbf{V}(s) = \frac{s + 2}{s^2 + 3s + 4}\mathbf{V}_g(s)$$

14.4

NETWORK FUNCTIONS

A generalization of impedance and admittance is the so-called *network function*, which, in the case of a single excitation and response, is defined as the ratio of the response phasor to the excitation phasor. For example, if $\mathbf{V}(s)$ and $\mathbf{I}(s)$ are the voltage and current phasors associated with a two-terminal network, then the input impedance

$$\mathbf{Z}(s) = \frac{\mathbf{V}(s)}{\mathbf{I}(s)}$$

is the network function if $\mathbf{I}(s)$ is the excitation and $\mathbf{V}(s)$ is the response. On the other hand, if $\mathbf{V}(s)$ is the input and $\mathbf{I}(s)$ is the output, then the input admittance

$$\mathbf{Y}(s) = \frac{\mathbf{I}(s)}{\mathbf{V}(s)}$$

is the network function.

The foregoing examples are special cases in that both the input and output are measured at the same pair of terminals. In general, for a given input current or voltage at a specified pair of terminals, the output may be a current or voltage at any place in the circuit. Thus the network function, which we designate in general as $\mathbf{H}(s)$, may be a ratio of a voltage to a current (in which case its units are ohms), a current to a voltage (with units of mhos), a voltage to a voltage, or a current to a current. In the last two cases, $\mathbf{H}(s)$ is a dimensionless quantity.

EXAMPLE 14.6 As an example, consider the circuit of Fig. 14.4(b), which was analyzed in the previous section. If $\mathbf{V}_o(s)$ is the output phasor and $\mathbf{V}_g(s)$ is the input phasor, then, by (14.15), the network function is

$$\mathbf{H}(s) = \frac{\mathbf{V}_o(s)}{\mathbf{V}_g(s)} = \frac{16}{s^2 + 2s + 8}$$

EXAMPLE 14.7 As another example, if the source $\mathbf{V}_g = 8\underline{/0°}$ V is the input and $\mathbf{I}_1$ is the output of Fig. 14.6(a), then we may show that

$$\mathbf{H}(s) = \frac{\mathbf{I}_1(s)}{\mathbf{V}_g(s)} = \frac{2}{s^2 + 6s + 10} \qquad (14.16)$$

In the former example $\mathbf{H}(s)$ is dimensionless and in the latter case its units are mhos.

The network function $\mathbf{H}(s)$ is independent of the input, being a function only of the network elements and their interconnections. Of course, when the input is specified this determines the value of s to be used in a given application. From a knowledge of the network function and the input function, we may then find the output phasor and subsequently the time-domain output. To be specific, suppose $\mathbf{V}_i(s)$ is the input and $\mathbf{V}_o(s)$ is the output. Then

$$\mathbf{H}(s) = \frac{\mathbf{V}_o(s)}{\mathbf{V}_i(s)} \qquad (14.17)$$

from which

$$\mathbf{V}_o(s) = \mathbf{H}(s)\mathbf{V}_i(s) \qquad (14.18)$$

In general, since s is complex, $\mathbf{H}(s)$ is complex. Therefore, we may write, in polar form,

$$\mathbf{H}(s) = |\mathbf{H}(s)|\underline{/\theta} \qquad (14.19)$$

where $|\mathbf{H}(s)|$ is the amplitude and θ the phase of $\mathbf{H}(s)$. Therefore, if

$$\mathbf{V}_i(s) = V_m\underline{/\phi} \qquad (14.20)$$

then

$$\mathbf{V}_o(s) = V_m|\mathbf{H}(s)|\underline{/\phi + \theta} \qquad (14.21)$$

Thus the amplitude of $\mathbf{V}_o$ is $V_m|\mathbf{H}(s)|$, and its phase is $\phi + \theta$. That is, the amplitude of the output is that of the input times that of the network function, and the phase of the output is that of the input plus that of the network function.

EXAMPLE 14.8 As an example, $\mathbf{I}_1(s)$ in (14.16) is given by

$$\mathbf{I}_1(s) = \mathbf{H}(s)\mathbf{V}_g(s)$$

where $\mathbf{V}_g(s) = 8\underline{/0°}$ V and $s = -1 + j1$, as given in the previous section. Therefore,

$$\mathbf{H}(-1 + j1) = \frac{2}{(-1 + j1)^2 + 6(-1 + j1) + 10}$$

$$= \frac{\sqrt{2}}{4}\underline{/-45°}$$

and

$$\mathbf{I}_1 = \left(\frac{\sqrt{2}}{4}\right)(8)\underline{/0 - 45°}$$

$$= 2\sqrt{2}\underline{/-45°} \text{ A}$$

as was obtained earlier.

EXAMPLE 14.9 As a final note in this section, if there are two or more inputs, we may use super-position to define a network function relating the output to each individual input, with the other inputs made zero. For example, in the circuit of Fig. 14.5, we may find, from Fig. 14.6(a) (with $\mathbf{I}_{g2} = 0$), the network function

$$\mathbf{H}_1(s) = \frac{\mathbf{I}_1}{\mathbf{V}_{g1}}$$

where $s = -1 + j1$ and $\mathbf{V}_{g1} = 8\underline{/0°}$ V. Then from Fig. 14.6(b) (with $\mathbf{V}_{g1} = 0$), we have the network function

$$\mathbf{H}_2(s) = \frac{\mathbf{I}_2}{\mathbf{I}_{g2}}$$

where $s = -5$ and $\mathbf{I}_{g2} = 2\underline{/0°}$ A. We may then find $\mathbf{I}_1$ and $\mathbf{I}_2$ and subsequently the corresponding time-domain functions, i_1 and i_2. These are, by superposition, the components of i, as before.

EXERCISES

14.4.1 Given the network function

$$\mathbf{H}(s) = \frac{4(s + 5)}{s^2 + 4s + 5}$$

and the input

$$\mathbf{V}_i(s) = 2\underline{/0°}$$

find the forced response $v_o(t)$ if (a) $s = -2$, (b) $s = -4 + j1$, and (c) $s = -2 + j3$.

Answer (a) $24e^{-2t}$; (b) $-2e^{-4t}\sin t$; (c) $3\sqrt{2}e^{-2t}\cos(3t - 135°)$

14.4.2 Find $\mathbf{H}(s)$ if the response is (a) $\mathbf{I}_1(s)$, (b) $\mathbf{I}_0(s)$, and (c) $\mathbf{V}_0(s)$.

Answer

(a) $\dfrac{s^2 + 3s + 1}{(s + 1)(s + 2)(s + 3)}$; (b) $\dfrac{1}{(s + 1)(s + 2)(s + 3)}$;

(c) $\dfrac{1}{(s + 1)(s + 2)}$

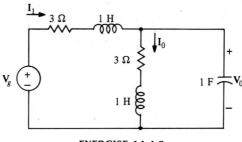

EXERCISE 14.4.2

14.4.3 Find $\mathbf{H}(s)$ if the response is v. Use the result to find the forced response if $v_g = 5\cos t$ V.

Answer $\dfrac{-8s}{s^2 + 4s + 4}$, $8\cos(t - 143.1°)$ V

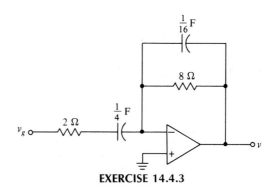

EXERCISE 14.4.3

14.5

POLES AND ZEROS

In general, the network function is a ratio of polynomials in s with real coefficients that are independent of the excitation. To illustrate this, let us consider the example of Fig. 14.2, described by

$$2\frac{di}{dt} + 5i = v$$

where i is the output and v is the input. Using the same technique with complex forcing functions that we used in Chapter 10, we note that if $v = \mathbf{V}e^{st}$, then the output must have the same form, namely $i = \mathbf{I}e^{st}$, where, of course, $\mathbf{V} = \mathbf{V}(s)$ and $\mathbf{I} = \mathbf{I}(s)$ are the phasor representations of v and i. Substituting these values into the differential equation, we have

$$(2s + 5)\mathbf{I}e^{st} = \mathbf{V}e^{st}$$

from which the network function is

$$\frac{\mathbf{I}}{\mathbf{V}} = \frac{1}{2s + 5}$$

In the general case, if the input and output of the circuit are $v_i(t)$ and $v_o(t)$, respectively, then the describing equation is

$$
\begin{aligned}
a_n\frac{d^n v_o}{dt^n} &+ a_{n-1}\frac{d^{n-1}v_o}{dt^{n-1}} + \ldots + a_1\frac{dv_o}{dt} + a_0 v_o \\
&= b_m\frac{d^m v_i}{dt^m} + b_{m-1}\frac{d^{m-1}v_i}{dt^{m-1}} + \ldots + b_1\frac{dv_i}{dt} + b_0 v_i
\end{aligned}
\tag{14.22}
$$

The a's and b's, of course, are real constants and are independent of v_i.

As before, if $v_i = \mathbf{V}_i e^{st}$, then the output must have the form $v_o = \mathbf{V}_o e^{st}$, where $\mathbf{V}_i(s)$ and $\mathbf{V}_o(s)$ are the phasor representations of v_i and v_o. Substituting these values into (14.22), we have

$$
\begin{aligned}
(a_n s^n + a_{n-1}s^{n-1} &+ \ldots + a_1 s + a_0)\mathbf{V}_o e^{st} \\
&= (b_m s^m + b_{m-1}s^{m-1} + \ldots + b_1 s + b_0)\mathbf{V}_i e^{st}
\end{aligned}
\tag{14.23}
$$

From this we obtain the network function

$$\frac{\mathbf{V}_o(s)}{\mathbf{V}_i(s)} = \mathbf{H}(s) = \frac{b_m s^m + b_{m-1}s^{m-1} + \ldots + b_1 s + b_0}{a_n s^n + a_{n-1}s^{n-1} + \ldots + a_1 s + a_0} \tag{14.24}$$

which is a ratio of polynomials in s.

We may also write the network function (14.24) in the factored form

$$\mathbf{H}(s) = \frac{b_m(s - z_1)(s - z_2)\ldots(s - z_m)}{a_n(s - p_1)(s - p_2)\ldots(s - p_n)} \tag{14.25}$$

In this case the numbers $z_1, z_2, \ldots, z_m$ are called *zeros* of the network function because they are values of s for which the function becomes zero. The numbers p_1, $p_2, \ldots, p_n$ are values of s for which the function becomes infinite and are called *poles* of the network function. The values of the poles and zeros, along with the values of the factors a_n and b_m, uniquely determine the network function.

EXAMPLE 14.10 As an example, the network function

$$\mathbf{H}(s) = \frac{6(s + 1)(s^2 + 2s + 2)}{s(s + 2)(s^2 + 4s + 13)}$$

$$= \frac{6(s + 1)(s + 1 + j1)(s + 1 - j1)}{s(s + 2)(s + 2 + j3)(s + 2 - j3)} \qquad (14.26)$$

has zeros at -1, $-1 + j1$, and $-1 - j1$ and poles at 0, -2, $-2 + j3$, and $-2 - j3$. As is generally the case, because the a's and b's of (14.24) are real, complex poles or zeros always occur in conjugate pairs. Since the network function is a ratio of a third-degree to a fourth-degree polynomial, it approaches zero as s becomes infinite. Thus we also have a zero at $s = \infty$. If the numerator were higher in degree than the denominator, $s = \infty$ would be a pole.

The poles and zeros of a network function may be sketched as a *pole-zero plot*, which is simply the *s-plane*, consisting of σ- and $j\omega$-axes, with the poles and zeros located on it. Zeros are represented by a small circle and poles by a small cross. As an example, the pole-zero plot of (14.26) is shown in Fig. 14.7. The values of σ are on the horizontal (σ) axis and those of ω are on the vertical (so-called $j\omega$) axis.

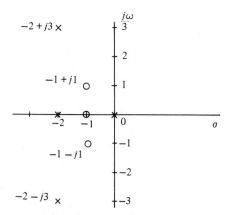

FIGURE 14.7 Pole-zero plot

As we shall see in Chapter 15, pole-zero plots are useful in considering frequency-domain properties of circuits.

EXERCISES

14.5.1 If the zeros of $\mathbf{H}(s)$ are $s = -1$, $-1 \pm j1$, the poles are $s = -2$, $-1 \pm j2$, and $\mathbf{H}(0) = 4$, find $\mathbf{H}(s)$.

Answer $\dfrac{20(s + 1)(s^2 + 2s + 2)}{(s + 2)(s^2 + 2s + 5)}$

14.6

THE NATURAL RESPONSE FROM THE NETWORK FUNCTION

As we know from our previous experience, an output of a circuit consists of the sum of a natural and a forced response. In the last few chapters we have been concerned entirely with finding the forced response, and we have seen that the phasor technique enables us to do this in a very easy, straightforward way in the cases of sinusoidal, damped sinusoidal, and exponential excitations. In power systems studies the forced response is, of course, an ac steady-state response and is always present. Therefore, the forced response is usually of more interest than the natural response, which is transient and gone after a very short time.

With damped sinusoidal excitations, on the other hand, both the forced and natural responses are transients. (In an actual circuit the natural response must be a transient, for otherwise we would have either a sustained or a growing response without an external excitation.) Therefore, the natural response assumes more importance, relative to the forced response, than it does in ac steady-state cases. In finding natural responses in Chapters 8 and 9 we considered only first- and second-order circuits, for the simple reason that our methods were applied to the describing differential equations and became more difficult to use as the order of the circuit increased. However, as we shall see in this section, the natural response may be obtained relatively easily from the phasor representation.

The results of the previous section illustrate that we may obtain the network function quite readily from the describing equation. For example, if (14.22) is the describing equation, then the network function (14.24) is

$$\mathbf{H}(s) = \frac{\mathbf{V}_o(s)}{\mathbf{V}_i(s)} = \frac{N(s)}{D(s)} \tag{14.27}$$

with the numerator

$$N(s) = b_m s^m + b_{m-1} s^{m-1} + \ldots + b_1 s + b_0 \tag{14.28}$$

and the denominator

$$D(s) = a_n s^n + a_{n-1} s^{n-1} + \ldots + a_1 s + a_0 \tag{14.29}$$

The last two expressions are merely the right member and left member, respectively, of (14.22) with the derivatives replaced by powers of s.

EXAMPLE 14.11 For example, if the describing equation is

$$\frac{d^2 v_o}{dt^2} + 4\frac{dv_o}{dt} + 3v_o = 2\frac{dv_i}{dt} + v_i \tag{14.30}$$

then the network function is

$$\mathbf{H}(s) = \frac{\mathbf{V}_o(s)}{\mathbf{V}_i(s)} = \frac{2s + 1}{s^2 + 4s + 3} \tag{14.31}$$

The process is reversible if there has been no cancellation of common terms in the network function. For example, it is a simple matter to write down (14.30), given (14.31).

In general, from (14.28) and (14.29) we may reconstruct the describing equation (14.22). In the special case of (14.29) equated to zero

$$D(s) = 0 \qquad (14.32)$$

replacing powers of s by corresponding derivatives of v_o results in the homogeneous equation of the system

$$a_n \frac{d^n v_o}{dt^n} + a_{n-1} \frac{d^{n-1} v_o}{dt^{n-1}} + \ldots + a_1 \frac{dv_o}{dt} + a_0 v_o = 0$$

as discussed in Chapter 9. Therefore (14.32) is the characteristic equation, and its roots are the natural frequencies of the circuit. Since these roots, by (14.29), are also the poles of the network function, we see that the natural response of the circuit is

$$v_n = A_1 e^{p_1 t} + A_2 e^{p_2 t} + \ldots + A_n e^{p_n t} \qquad (14.33)$$

where the natural frequencies $p_1, p_2, \ldots, p_n$ are the poles of the network function and $A_1, A_2, \ldots, A_n$ are arbitrary constants. Modifications, of course, must be made, as described in Chapter 9, if the natural frequencies are not distinct.

We now have a very simple method, based on phasors, for finding the complete response of a circuit. All we need to find is the network function from which by (14.27) we may obtain the output phasor. The forced response is found from the phasor response in the usual way, and the natural response is given by (14.33), where the natural frequencies are the poles of the network function.

If the input phasor is of the form

$$\mathbf{V}_i(s) = V_m \underline{/\phi}$$

where V_m and ϕ are constants, then, by (14.27),

$$\mathbf{V}_o(s) = (V_m \underline{/\phi}) \mathbf{H}(s)$$

Therefore, since $\mathbf{V}_i(s)$ has no poles, the poles of $\mathbf{H}(s)$ are the poles of $\mathbf{V}_o(s)$. Thus the complete response $v_o(t)$ may be obtained from its phasor representation $\mathbf{V}_o(s)$. The forced response is obtained as before, and the natural frequencies are the poles of $\mathbf{V}_o(s)$, from which the natural response may be constructed.

EXAMPLE 14.12 As an example, let us find the complete response i of the circuit of Fig. 14.8. Using current division, the phasor $\mathbf{I}(s)$ of i is given by

$$\mathbf{I}(s) = \frac{\mathbf{V}_g(s)}{12 + \{[3s(2s + 6)]/(3s + 2s + 6)\}} \times \frac{3s}{3s + 2s + 6}$$

$$= \frac{s}{(s + 1)(s + 12)}$$

since $\mathbf{V}_g = 2 \underline{/0°}$. Since the poles of $\mathbf{I}$ are $s = -1, -12$, the natural response is

$$i_n(t) = A_1 e^{-t} + A_2 e^{-12t} \text{ A}$$

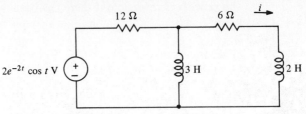

FIGURE 14.8 Second-order circuit

For $s = -2 + j1$, the phasor representation is

$$\mathbf{I} = 0.16\underline{/12.7°}$$

so that the forced response is

$$i_f(t) = 0.16e^{-2t} \cos (t + 12.7°) \text{ A}$$

The complete response is therefore

$$i(t) = A_1e^{-t} + A_2e^{-12t} + 0.16e^{-2t} \cos (t + 12.7°) \text{ A}$$

The arbitrary constants may be evaluated, as in Chapter 9, if we know the initial energy conditions.

EXERCISES

14.6.1 Find the natural response in Ex. 14.4.1, assuming that there is no cancellation in the network function.
Answer $e^{-2t}(A_1 \cos t + A_2 \sin t)$

14.6.2 Find the complete response in Ex. 14.4.1(c) if the natural response is as given in Ex. 14.6.1 and

$$v_o(0^+) = \frac{dv_o(0^+)}{dt} = 0$$

Answer $e^{-2t} [3 \cos t - 9 \sin t + 3\sqrt{2} \cos (3t - 135°)]$

14.6.3 Find the complete response i if $i_g = 10 \cos 2t$ A, $i(0) = 0$, and $di(0^+)/dt = 8$ A/s.
Answer $2e^{-t} - 2e^{-4t} + \sin 2t$ A

EXERCISE 14.6.3

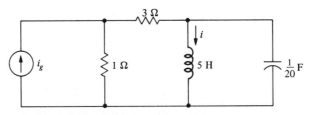

14.7

NATURAL FREQUENCIES

As we have seen, the natural frequencies are the poles of the network function if there is no cancellation of a common pole and zero. Also, as discussed in Chapter 9, the natural frequencies are the same for any response in a given circuit unless one portion of the circuit is physically separated from the rest. Thus if we are looking only for the natural frequencies, we may consider any response, and it is obviously better to choose one which is easier to find.

EXAMPLE 14.13 For example, let us consider the circuit of Fig. 14.8 with the source killed (since the natural response corresponds to a zero source). The natural frequencies are the poles of any network function. Therefore, we may excite the circuit in some proper manner and, for some chosen output, determine the network function. Figure 14.9 illustrates the two proper means of applying an excitation to the dead circuit. We may insert a voltage source in series with an element as in x-x' of Fig. 14.9(a), or we may place a current source across an element, as across y-y' of Fig. 14.9(b). The first entry into the circuit is sometimes called the pliers entry since we cut a wire and insert a voltage source. The second entry is a soldering entry since we solder the source across two nodes. Any other method of entry would be improper because killing the inserted sources would result in a different dead circuit.

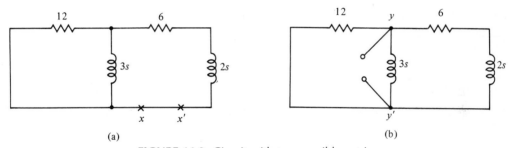

(a)   (b)

FIGURE 14.9 Circuit with two possible entries

In the pliers entry, if $V_x(s)$ is the voltage source inserted and $I_x(s)$, the current into the circuit at the source, is the response, then the network function is

$$Y_x(s) = \frac{I_x(s)}{V_x(s)}$$

Thus the natural frequencies are the poles of $Y_x(s)$ or the zeros of its reciprocal, $Z_x(s)$, the input impedance at x-x'. In this case we have

$$Z_x(s) = 2s + 6 + \frac{12(3s)}{12 + 3s}$$

$$= \frac{2(s + 1)(s + 12)}{s + 4}$$

Therefore, the natural frequencies are -1, -12, as before.

In the case of Fig. 14.9(b), if a current source $\mathbf{I}_y(s)$ is placed across y-y' and the voltage $\mathbf{V}_y(s)$ across the source is the output, then the network function is

$$\mathbf{Z}_y(s) = \frac{\mathbf{V}_y(s)}{\mathbf{I}_y(s)}$$

Therefore, the natural frequencies are the poles of $\mathbf{Z}_y(s)$ or the zeros of its reciprocal, $\mathbf{Y}_y(s)$, given by

$$\mathbf{Y}_y(s) = \frac{1}{12} + \frac{1}{3s} + \frac{1}{2s + 6}$$

$$= \frac{(s + 1)(s + 12)}{12s(s + 3)}$$

Again, the natural frequencies are -1, -12.

EXERCISES

14.7.1 Find the natural frequencies of the circuit of Fig. 14.5 by killing the sources and using (a) a pliers entry in series with the capacitor and (b) a soldering entry across the capacitor.
Answer $-3 \pm j1$

14.7.2 Find the natural frequencies of the circuit of Ex. 14.4.2 by killing the source and using (a) a pliers entry in series with the capacitor and (b) a soldering entry across the capacitor. (*Note:* Do not cancel the common pole and zero, $s = -3$.)
Answer -1, -2, -3

14.8
TWO-PORT NETWORKS

One of the most important applications of the network function concept is to networks for which the input and output signals are measured at different pairs of terminals. The simplest and most-often-encountered circuit for which this is possible is the *two-port* network, a port being defined as a pair of terminals at which a signal may enter or leave. A general two-port network is symbolized by Fig. 14.10(b), as contrasted with the *one-port* network of Fig. 14.10(a).

In general, as seen in Fig. 14.10(b), a two-port network has four terminals. It

FIGURE 14.10 (a) One-port and (b) two-port networks

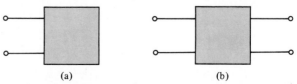

(a)   (b)

is possible, of course, for two of the terminals to be the same, in which case we have a three-terminal, or *grounded*, network. A general example of this case is shown in Fig. 14.11.

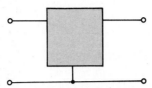

FIGURE 14.11 Three-terminal two-port network

We may associate two pairs of currents and voltages with a general two-port network, as shown in the frequency-domain case of Fig. 14.12, with variables $V_1(s)$, $I_1(s)$, $V_2(s)$, and $I_2(s)$ as indicated. In case the network is linear, these variables may be related in a number of ways. For example, if I_1 and I_2 are inputs and V_1 and V_2 are outputs, by superposition V_1 and V_2 each have components proportional to I_1 and I_2. That is,

$$
V_1 = z_{11}I_1 + z_{12}I_2
$$
$$
V_2 = z_{21}I_1 + z_{22}I_2
$$

(14.34)

where the z's are proportionality factors, which in general are functions of s.

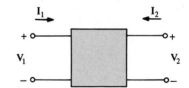

FIGURE 14.12 General two-port network

Since the z's multiply currents to yield voltages, they must be measured in ohms. Therefore, they are impedance functions. We may find the z's from the network by open-circuiting either port 1 (with variables V_1 and I_1) or port 2 (with V_2 and I_2). For example, if port 2 is open-circuited ($I_2 = 0$), then from (14.34) we have

$$
z_{11} = \left. \frac{V_1}{I_1} \right|_{I_2=0}
$$
$$
z_{21} = \left. \frac{V_2}{I_1} \right|_{I_2=0}
$$

(14.35)

Similarly, if port 1 is open-circuited ($I_1 = 0$), we have

$$z_{12} = \frac{V_1}{I_2} \bigg|_{I_1=0}$$

$$z_{22} = \frac{V_2}{I_2} \bigg|_{I_1=0}$$

$$(14.36)$$

Accordingly, the z's are called *open-circuit impedances,* or *open-circuit parameters,* or simply *z-parameters.* In any case, they are examples of network functions.

By (14.35) and (14.36), we see that z_{11} is the impedance seen looking in the *primary* port (port 1) when the *secondary* port (port 2) is open, and z_{22} is that seen at the secondary when the primary is open. The parameters z_{12} and z_{21} are *transfer* impedances, which are ratios of a voltage at one port to a current at another.

EXAMPLE 14.14 As an example, let us consider the three-terminal network of Fig. 14.13. Because of its shape, it is sometimes called a T network. Evidently, it is simply a **Y** network, as discussed in Chapter 13. To find z_{11} and z_{21}, we open port 2 and excite port 1 with a current source I_1. Since $I_2 = 0$, we have

$$V_1 = (Z_1 + Z_3)I_1$$

and

$$V_2 = Z_3 I_1$$

Therefore

$$z_{11} = \frac{V_1}{I_1} = Z_1 + Z_3$$

$$z_{21} = \frac{V_2}{I_1} = Z_3$$

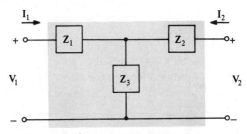

FIGURE 14.13 T network

The other two parameters are found in a similar manner with the primary open ($I_1 = 0$) and the secondary excited with a source I_2. The result is

$$z_{11} = Z_1 + Z_3$$

$$z_{12} = z_{21} = Z_3 \qquad (14.37)$$

$$z_{22} = Z_2 + Z_3$$

EXAMPLE 14.15 As another example, let us find the z-parameters for the phasor circuit of Fig. 14.14. We will illustrate the procedure in this case by simply writing the two loop equations,

$$V_1 = 2I_2 + \left(4 + \frac{4}{s}\right)I_1 + \frac{2}{s}(I_1 + I_2)$$

$$V_2 = \left(6 + \frac{2}{s}\right)I_2 + \frac{2}{s}(I_1 + I_2)$$

or

$$V_1 = \left(4 + \frac{6}{s}\right)I_1 + \left(2 + \frac{2}{s}\right)I_2$$

$$V_2 = \frac{2}{s}I_1 + \left(6 + \frac{4}{s}\right)I_2$$

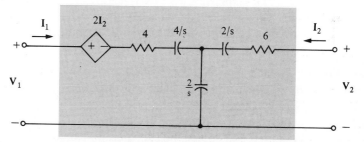

FIGURE 14.14 Two-port network

Comparing these results with the definition (14.34), the z-parameters, the coefficients of the currents in the last two equations, are given by

$$z_{11} = 4 + \frac{6}{s}$$

$$z_{12} = 2 + \frac{2}{s}$$

$$z_{21} = \frac{2}{s}$$

$$z_{22} = 6 + \frac{4}{s}$$

We may find another set of parameters considering V_1 and V_2 as inputs and I_1 and I_2 as outputs in Fig. 14.12. In this case we have, by superposition,

$$I_1 = y_{11}V_1 + y_{12}V_2$$
$$I_2 = y_{21}V_1 + y_{22}V_2$$

$$(14.38)$$

where, evidently, the proportionality factors are admittances given by

$$y_{11} = \frac{I_1}{V_1}\bigg|_{V_2=0}$$

$$y_{12} = \frac{I_1}{V_2}\bigg|_{V_1=0}$$

(14.39)

$$y_{21} = \frac{I_2}{V_1}\bigg|_{V_2=0}$$

$$y_{22} = \frac{I_2}{V_2}\bigg|_{V_1=0}$$

Accordingly, the $\mathbf{y}$'s are *short-circuit admittances* ($\mathbf{V}_1 = 0$ or $\mathbf{V}_2 = 0$), or *short-circuit parameters,* or simply *y-parameters.*

The parameters $\mathbf{y}_{11}$ and $\mathbf{y}_{22}$ are admittances seen looking in one port with the other port short circuited, and $\mathbf{y}_{12}$ and $\mathbf{y}_{21}$ are transfer admittances, or ratios of the current at one port to the voltage at another, under the appropriate short-circuit conditions.

EXAMPLE 14.16 As an example, let us find the *y*-parameters of the three-terminal circuit of Fig. 14.15. Because of its shape, it is sometimes called a π network. Evidently, it is simply a Δ network, as considered in Chapter 13. If the secondary is short circuited ($\mathbf{V}_2 = 0$), $\mathbf{Y}_c$ is shorted out, so that $\mathbf{Y}_a$ and $\mathbf{Y}_b$ are in parallel. Therefore

$$\mathbf{y}_{11} = \mathbf{Y}_a + \mathbf{Y}_b$$

Also in this case, $-\mathbf{I}_2$ flows to the right through $\mathbf{Y}_b$ so that

$$-\mathbf{I}_2 = \mathbf{Y}_b\mathbf{V}_1$$

or

$$\mathbf{y}_{21} = \frac{\mathbf{I}_2}{\mathbf{V}_1} = -\mathbf{Y}_b$$

The other two $\mathbf{y}$'s may be found in a similar way with the primary short circuited. The result is

FIGURE 14.15 π network

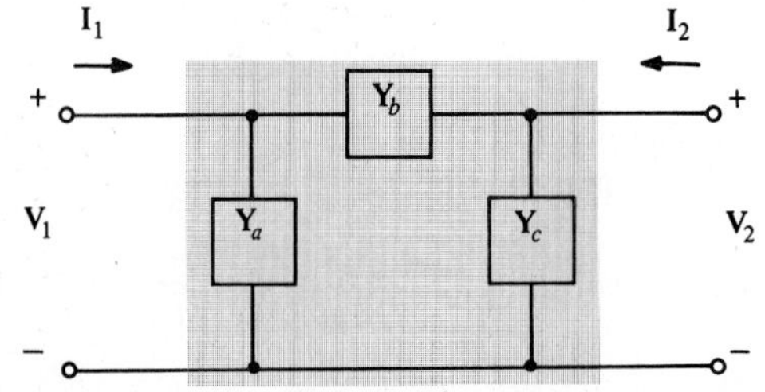

$$\mathbf{y}_{11} = \mathbf{Y}_a + \mathbf{Y}_b$$
$$\mathbf{y}_{12} = \mathbf{y}_{21} = -\mathbf{Y}_b \qquad (14.40)$$
$$\mathbf{y}_{22} = \mathbf{Y}_b + \mathbf{Y}_c$$

If $\mathbf{z}_{12} = \mathbf{z}_{21}$ (or, equivalently, $\mathbf{y}_{12} = \mathbf{y}_{21}$), the network is a *reciprocal* network. This is always the case when the elements inside the network box of Fig. 14.12 are resistors, inductors, and capacitors. Examples, as we note from (14.37) and (14.40), are the circuits of Figs. 14.13 and 14.15. The circuit of Fig. 14.14 contains a dependent source, and is *nonreciprocal,* since $\mathbf{z}_{12} \neq \mathbf{z}_{21}$.

The z- and y-parameters are but two sets of parameters associated with a two-port network. Two other sets are the *hybrid parameters,* $\mathbf{h}_{11}$, $\mathbf{h}_{12}$, $\mathbf{h}_{21}$, $\mathbf{h}_{22}$ and $\mathbf{g}_{11}$, $\mathbf{g}_{12}$, $\mathbf{g}_{21}$, $\mathbf{g}_{22}$, which are also very important, especially in electronic circuits. They are defined, respectively, by

$$\mathbf{V}_1 = \mathbf{h}_{11}\mathbf{I}_1 + \mathbf{h}_{12}\mathbf{V}_2$$
$$\mathbf{I}_2 = \mathbf{h}_{21}\mathbf{I}_1 + \mathbf{h}_{22}\mathbf{V}_2 \qquad (14.41)$$

and

$$\mathbf{I}_1 = \mathbf{g}_{11}\mathbf{V}_1 + \mathbf{g}_{12}\mathbf{I}_2$$
$$\mathbf{V}_2 = \mathbf{g}_{21}\mathbf{V}_1 + \mathbf{g}_{22}\mathbf{I}_2 \qquad (14.42)$$

They are called hybrid parameters because they relate a mixture of a current and voltage to another current and voltage, rather than two currents to two voltages, or vice versa. From (14.41) and (14.42), we see that

$$\mathbf{h}_{11} = \left.\frac{\mathbf{V}_1}{\mathbf{I}_1}\right|_{\mathbf{V}_2=0}$$
$$\mathbf{h}_{12} = \left.\frac{\mathbf{V}_1}{\mathbf{V}_2}\right|_{\mathbf{I}_1=0}$$
$$\mathbf{h}_{21} = \left.\frac{\mathbf{I}_2}{\mathbf{I}_1}\right|_{\mathbf{V}_2=0} \qquad (14.43)$$
$$\mathbf{h}_{22} = \left.\frac{\mathbf{I}_2}{\mathbf{V}_2}\right|_{\mathbf{I}_1=0}$$

and

$$g_{11} = \left.\frac{I_1}{V_1}\right|_{I_2=0}$$

$$g_{12} = \left.\frac{I_1}{I_2}\right|_{V_1=0}$$

$$g_{21} = \left.\frac{V_2}{V_1}\right|_{I_2=0} \tag{14.44}$$

$$g_{22} = \left.\frac{V_2}{I_2}\right|_{V_1=0}$$

Therefore h_{11} is the impedance at the primary when the secondary is short circuited, h_{12} is the voltage ratio transfer function when the primary is open circuited, h_{21} is the current ratio transfer function when the secondary is short circuited, and h_{22} is the admittance at the secondary when the primary is open circuited. We can analyze the **g**'s in the same manner, of course.

EXAMPLE 14.17 As an example, let us find the h-parameters for the circuit of Fig. 14.13, if $\mathbf{Z}_1 = 6\ \Omega$, $\mathbf{Z}_2 = 8\ \Omega$, and $\mathbf{Z}_3 = 10\ \Omega$. From the **z**'s found in (14.37), we know that

$$\begin{aligned} V_1 &= 16I_1 + 10I_2 \\ V_2 &= 10I_1 + 18I_2 \end{aligned} \tag{14.45}$$

If we substitute the value of I_2 from the second of these into the first, we have

$$V_1 = 16I_1 + 10\left[\frac{V_2 - 10I_1}{18}\right]$$

which simplifies to

$$V_1 = \frac{94}{9}I_1 + \frac{5}{9}V_2$$

If we solve the second of (14.45) for I_2, we have

$$I_2 = -\frac{5}{9}I_1 + \frac{1}{18}V_2$$

Comparing these last two results with the definition (14.41) of the **h**'s, we have

$$h_{11} = \frac{94}{9}$$

$$h_{12} = \frac{5}{9}$$

$$h_{21} = -\frac{5}{9} \tag{14.46}$$

$$\mathbf{h}_{22} = \frac{1}{18} \qquad\qquad (14.46 \; continued)$$

Still another set of two-port parameters is the *transmission* parameters, **A, B, C,** and **D**, defined for the general two-port network by

$$\mathbf{V}_1 = \mathbf{A}\mathbf{V}_2 - \mathbf{B}\mathbf{I}_2$$
$$\mathbf{I}_1 = \mathbf{C}\mathbf{V}_2 - \mathbf{D}\mathbf{I}_2$$

$$(14.47)$$

These are important to transmission engineers because they express the primary (sending end) variables $\mathbf{V}_1$ and $\mathbf{I}_1$ in terms of the secondary (receiving end) variables $\mathbf{V}_2$ and $-\mathbf{I}_2$. (The negative of $\mathbf{I}_2$ is used because that is the current entering the receiving end load.) Evidently, from (14.47), we may write

$$\mathbf{A} = \frac{\mathbf{V}_1}{\mathbf{V}_2}\bigg|_{\mathbf{I}_2=0}$$

$$\mathbf{B} = \frac{\mathbf{V}_1}{-\mathbf{I}_2}\bigg|_{\mathbf{V}_2=0}$$

$$\mathbf{C} = \frac{\mathbf{I}_1}{\mathbf{V}_2}\bigg|_{\mathbf{I}_2=0}$$

$$\mathbf{D} = \frac{\mathbf{I}_1}{-\mathbf{I}_2}\bigg|_{\mathbf{V}_2=0}$$

$$(14.48)$$

Therefore **A** is a voltage ratio and **C** is a transfer admittance, with the secondary open, and **B** is a transfer impedance and **D** is a current ratio, with the secondary short circuited.

A second set of transmission parameters may be defined giving $\mathbf{V}_2$ and $\mathbf{I}_2$ in terms of $\mathbf{V}_1$ and $-\mathbf{I}_1$. They are illustrated in Ex. 14.8.5.

The derivation of the h-parameters of (14.46) suggests a general method of obtaining one set of parameters from another. For example, if we solve (14.38) for the **V**'s, we have

$$\mathbf{V}_1 = \frac{\mathbf{y}_{22}}{\Delta_Y}\mathbf{I}_1 + \frac{-\mathbf{y}_{12}}{\Delta_Y}\mathbf{I}_2$$

$$\mathbf{V}_2 = \frac{-\mathbf{y}_{21}}{\Delta_Y}\mathbf{I}_1 + \frac{\mathbf{y}_{11}}{\Delta_Y}\mathbf{I}_2$$

$$(14.49)$$

where Δ_Y is the determinant,

$$\Delta_Y = \mathbf{y}_{11}\mathbf{y}_{22} - \mathbf{y}_{12}\mathbf{y}_{21} \qquad\qquad (14.50)$$

Comparing this result with (14.34), we have the z-parameters in terms of the y-parameters:

$$\mathbf{z}_{11} = \frac{\mathbf{y}_{22}}{\Delta_Y} \qquad \mathbf{z}_{12} = \frac{-\mathbf{y}_{12}}{\Delta_Y}$$

$$\mathbf{z}_{21} = \frac{-\mathbf{y}_{21}}{\Delta_Y} \qquad \mathbf{z}_{22} = \frac{\mathbf{y}_{11}}{\Delta_Y}$$

(14.51)

Table 14.1 is a compilation of conversion formulas relating one set of two-port parameters to the other sets for the four most important cases. The parameters are collected in matrices, and each row in the table gives one set of parameters in terms of the other three. For example, (14.51) is represented by the second matrix in the first row. Comparing corresponding entries in the first matrix with the second matrix of that row yields (14.51). The determinants Δ_Z, Δ_Y, Δ_H, and Δ_T are those of the z, y, h, and transmission matrices, respectively.

TABLE 14.1 Two-port Parameter Conversion Formulas

$$
\begin{bmatrix} \mathbf{z}_{11} & \mathbf{z}_{12} \\ \mathbf{z}_{21} & \mathbf{z}_{22} \end{bmatrix}
\quad
\begin{bmatrix} \dfrac{\mathbf{y}_{22}}{\Delta_Y} & \dfrac{-\mathbf{y}_{12}}{\Delta_Y} \\ \dfrac{-\mathbf{y}_{21}}{\Delta_Y} & \dfrac{\mathbf{y}_{11}}{\Delta_Y} \end{bmatrix}
\quad
\begin{bmatrix} \dfrac{A}{C} & \dfrac{\Delta_T}{C} \\ \dfrac{1}{C} & \dfrac{D}{C} \end{bmatrix}
\quad
\begin{bmatrix} \dfrac{\Delta_H}{\mathbf{h}_{22}} & \dfrac{\mathbf{h}_{12}}{\mathbf{h}_{22}} \\ \dfrac{-\mathbf{h}_{21}}{\mathbf{h}_{22}} & \dfrac{1}{\mathbf{h}_{22}} \end{bmatrix}
$$

$$
\begin{bmatrix} \dfrac{\mathbf{z}_{22}}{\Delta_Z} & \dfrac{-\mathbf{z}_{12}}{\Delta_Z} \\ \dfrac{-\mathbf{z}_{21}}{\Delta_Z} & \dfrac{\mathbf{z}_{11}}{\Delta_Z} \end{bmatrix}
\quad
\begin{bmatrix} \mathbf{y}_{11} & \mathbf{y}_{12} \\ \mathbf{y}_{21} & \mathbf{y}_{22} \end{bmatrix}
\quad
\begin{bmatrix} \dfrac{D}{B} & \dfrac{-\Delta_T}{B} \\ -\dfrac{1}{B} & \dfrac{A}{B} \end{bmatrix}
\quad
\begin{bmatrix} \dfrac{1}{\mathbf{h}_{11}} & \dfrac{-\mathbf{h}_{12}}{\mathbf{h}_{11}} \\ \dfrac{\mathbf{h}_{21}}{\mathbf{h}_{11}} & \dfrac{\Delta_H}{\mathbf{h}_{11}} \end{bmatrix}
$$

$$
\begin{bmatrix} \dfrac{\mathbf{z}_{11}}{\mathbf{z}_{21}} & \dfrac{\Delta_Z}{\mathbf{z}_{21}} \\ \dfrac{1}{\mathbf{z}_{21}} & \dfrac{\mathbf{z}_{22}}{\mathbf{z}_{21}} \end{bmatrix}
\quad
\begin{bmatrix} \dfrac{-\mathbf{y}_{22}}{\mathbf{y}_{21}} & \dfrac{-1}{\mathbf{y}_{21}} \\ \dfrac{-\Delta_Y}{\mathbf{y}_{21}} & \dfrac{-\mathbf{y}_{11}}{\mathbf{y}_{21}} \end{bmatrix}
\quad
\begin{bmatrix} A & B \\ C & D \end{bmatrix}
\quad
\begin{bmatrix} \dfrac{-\Delta_H}{\mathbf{h}_{21}} & \dfrac{-\mathbf{h}_{11}}{\mathbf{h}_{21}} \\ \dfrac{-\mathbf{h}_{22}}{\mathbf{h}_{21}} & \dfrac{-1}{\mathbf{h}_{21}} \end{bmatrix}
$$

$$
\begin{bmatrix} \dfrac{\Delta_Z}{\mathbf{z}_{22}} & \dfrac{\mathbf{z}_{12}}{\mathbf{z}_{22}} \\ \dfrac{-\mathbf{z}_{21}}{\mathbf{z}_{22}} & \dfrac{1}{\mathbf{z}_{22}} \end{bmatrix}
\quad
\begin{bmatrix} \dfrac{1}{\mathbf{y}_{11}} & \dfrac{-\mathbf{y}_{12}}{\mathbf{y}_{11}} \\ \dfrac{\mathbf{y}_{21}}{\mathbf{y}_{11}} & \dfrac{\Delta_Y}{\mathbf{y}_{11}} \end{bmatrix}
\quad
\begin{bmatrix} \dfrac{B}{D} & \dfrac{\Delta_T}{D} \\ -\dfrac{1}{D} & \dfrac{C}{D} \end{bmatrix}
\quad
\begin{bmatrix} \mathbf{h}_{11} & \mathbf{h}_{12} \\ \mathbf{h}_{21} & \mathbf{h}_{22} \end{bmatrix}
$$

EXERCISES

14.8.1 Find the z-parameters and the *ABCD* parameters of the circuit shown.
Answer $\mathbf{z}_{11} = 6$, $\mathbf{z}_{12} = \mathbf{z}_{21} = 4$, $\mathbf{z}_{22} = 10\ \Omega$; $A = 3/2$, $B = 11\ \Omega$, $C = 1/4\ \mho$, $D = 5/2$

EXERCISE 14.8.1

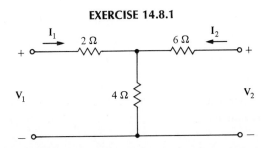

14.8.2 Show that the y-parameters may be obtained from the z-parameters by

$$\mathbf{y}_{11} = \frac{\mathbf{z}_{22}}{\Delta_z}, \qquad \mathbf{y}_{12} = -\frac{\mathbf{z}_{12}}{\Delta_z}$$

$$\mathbf{y}_{21} = -\frac{\mathbf{z}_{21}}{\Delta_z}, \qquad \mathbf{y}_{22} = \frac{\mathbf{z}_{11}}{\Delta_z}$$

where $\Delta_z = \mathbf{z}_{11}\mathbf{z}_{22} - \mathbf{z}_{12}\mathbf{z}_{21}$. (This is given in the first entry of the second row of Table 14.1.) Use this result to find the y-parameters of the circuit of Ex. 14.8.1.
Answer $\mathbf{y}_{11} = \frac{5}{22}$, $\mathbf{y}_{12} = \mathbf{y}_{21} = -\frac{1}{11}$, $\mathbf{y}_{22} = \frac{3}{22}$ ℧

14.8.3 Find the h-parameters and the g-parameters of the network of Ex. 14.8.1.
Answer $\mathbf{h}$'s: $\frac{44}{10}$, $\frac{4}{10}$, $-\frac{4}{10}$, $\frac{1}{10}$ $\mathbf{g}$'s: $\frac{1}{6}$, $-\frac{4}{6}$, $\frac{4}{6}$, $\frac{44}{6}$

14.8.4 Using Table 14.1, find the conditions on (a) the h-parameters and (b) the transmission parameters that a circuit be reciprocal ($\mathbf{z}_{12} = \mathbf{z}_{21}$).
Answer (a) $\mathbf{h}_{12} = -\mathbf{h}_{21}$, (b) $\Delta_T = 1$

14.8.5 A second set of transmission parameters may be defined, expressing the output variables in terms of the input variables, by

$$\mathbf{V}_2 = \mathbf{a}\mathbf{V}_1 - \mathbf{b}\mathbf{I}_1$$

$$\mathbf{I}_2 = \mathbf{c}\mathbf{V}_1 - \mathbf{d}\mathbf{I}_1$$

Find the parameters $\mathbf{a}$, $\mathbf{b}$, $\mathbf{c}$, and $\mathbf{d}$ in terms of the other transmission parameters $\mathbf{A}$, $\mathbf{B}$, $\mathbf{C}$, and $\mathbf{D}$.
Answer $\mathbf{a} = \dfrac{\mathbf{D}}{\Delta_T}$, $\mathbf{b} = \dfrac{\mathbf{B}}{\Delta_T}$, $\mathbf{c} = \dfrac{\mathbf{C}}{\Delta_T}$, $\mathbf{d} = \dfrac{\mathbf{A}}{\Delta_T}$

14.9

APPLICATIONS OF TWO-PORT PARAMETERS

The two-port parameters of the previous section are useful in many ways, as we will see in this section. Our first use of them will be to obtain various network functions.

EXAMPLE 14.18 For example, if port 2 is open ($\mathbf{I}_2 = 0$), then from (14.34) we may find the voltage ratio function

$$\frac{\mathbf{V}_2}{\mathbf{V}_1} = \frac{\mathbf{z}_{21}}{\mathbf{z}_{11}} \tag{14.52}$$

Also, for a short-circuited secondary ($\mathbf{V}_2 = 0$), we may write, using (14.38),

$$\frac{\mathbf{I}_2}{\mathbf{I}_1} = \frac{\mathbf{y}_{21}}{\mathbf{y}_{11}} \tag{14.53}$$

EXAMPLE 14.19 As another example, let us find the current ratio function for the two-port loaded with 1 Ω, as shown in Fig. 14.16. Evidently, we have $\mathbf{V}_2 = -\mathbf{I}_2$, which substituted into the second equation of (14.34) yields, after some rearrangement,

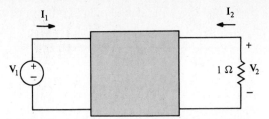

FIGURE 14.16 Two-port loaded with a 1-Ω resistor

$$\frac{I_2}{I_1} = \frac{-z_{21}}{1 + z_{22}} \tag{14.54}$$

If we make the same substitution ($-I_2$ for V_2) in (14.38), we have the voltage ratio

$$\frac{V_2}{V_1} = \frac{-y_{21}}{1 + y_{22}} \tag{14.55}$$

EXAMPLE 14.20 As another example, let us find the transfer function I_2/I_1 for the circuit of Fig. 14.14 if the secondary is terminated in a 1-Ω resistor. We have previously found

$$z_{21} = \frac{2}{s}$$

$$z_{22} = 6 + \frac{4}{s}$$

so that by (14.54) the transfer function is

$$\frac{I_2}{I_1} = \frac{-2/s}{1 + \left(6 + \dfrac{4}{s}\right)} = \frac{-2}{7s + 4}$$

EXAMPLE 14.21 As another example, for the π network of Fig. 14.17 we may write down by inspection the parameters,

$$y_{21} = -\frac{1}{2s}$$

$$y_{22} = s + \frac{1}{2s}$$

so that by (14.55) the voltage ratio is

$$\frac{V_2}{V_1} = \frac{1/2s}{1 + \left(s + \dfrac{1}{2s}\right)} = \frac{1}{2s^2 + 2s + 1}$$

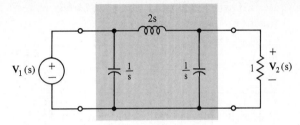

FIGURE 14.17 LC two-port terminated in 1 Ω

Next we will consider a *doubly terminated* two-port network as shown in Fig. 14.18. The load impedance at the output port is $\mathbf{Z}_L$, and $\mathbf{Z}_g$ at the input port is the internal impedance of the source $\mathbf{V}_g$. If $\mathbf{Z}_g = 0$ and $\mathbf{Z}_L = 1$, we have the case of Fig. 14.16.

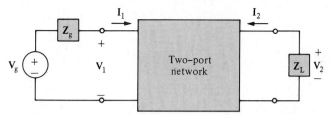

FIGURE 14.18 Doubly terminated two-port network

From the circuit we may write

$$\mathbf{I}_2 = \frac{-\mathbf{V}_2}{\mathbf{Z}_L}$$

and

$$\mathbf{V}_1 = \mathbf{V}_g - \mathbf{Z}_g\mathbf{I}_1$$

which substituted into (14.34) yields

$$\mathbf{V}_g - \mathbf{Z}_g\mathbf{I}_1 = \mathbf{z}_{11}\mathbf{I}_1 - \frac{\mathbf{z}_{12}}{\mathbf{Z}_L}\mathbf{V}_2$$

$$\mathbf{V}_2 = \mathbf{z}_{21}\mathbf{I}_1 - \frac{\mathbf{z}_{22}}{\mathbf{Z}_L}\mathbf{V}_2$$

(14.56)

If we solve each of these for $\mathbf{I}_1$, we have

$$\mathbf{I}_1 = \frac{\mathbf{V}_g + (\mathbf{z}_{12}/\mathbf{Z}_L)\mathbf{V}_2}{\mathbf{z}_{11} + \mathbf{Z}_g} = \frac{\mathbf{V}_2 + (\mathbf{z}_{22}/\mathbf{Z}_L)\mathbf{V}_2}{\mathbf{z}_{21}}$$

From the last two members, we may solve for the voltage ratio function,

$$\frac{\mathbf{V}_2}{\mathbf{V}_g} = \frac{\mathbf{z}_{21}\mathbf{Z}_L}{(\mathbf{z}_{11} + \mathbf{Z}_g)(\mathbf{z}_{22} + \mathbf{Z}_L) - \mathbf{z}_{12}\mathbf{z}_{21}}$$

(14.57)

If we divide both sides of (14.57) by $-\mathbf{Z}_L$, we have the transfer admittance function,

$$\frac{\mathbf{I}_2}{\mathbf{V}_g} = \frac{-\mathbf{z}_{21}}{(\mathbf{z}_{11} + \mathbf{Z}_g)(\mathbf{z}_{22} + \mathbf{Z}_L) - \mathbf{z}_{12}\mathbf{z}_{21}} \tag{14.58}$$

Others that may be readily derived in terms of the z-parameters are the current ratio function,

$$\frac{\mathbf{I}_2}{\mathbf{I}_1} = \frac{-\mathbf{z}_{21}}{\mathbf{z}_{22} + \mathbf{Z}_L} \tag{14.59}$$

the transfer impedance function,

$$\frac{\mathbf{V}_2}{\mathbf{I}_1} = \frac{\mathbf{z}_{21}\mathbf{Z}_L}{\mathbf{z}_{22} + \mathbf{Z}_L} \tag{14.60}$$

and the input impedance,

$$\mathbf{Z}_{\text{in}} = \frac{\mathbf{V}_g}{\mathbf{I}_1} = \mathbf{z}_{11} + \mathbf{Z}_g - \frac{\mathbf{z}_{12}\mathbf{z}_{21}}{\mathbf{z}_{22} + \mathbf{Z}_L} \tag{14.61}$$

These same functions, along with a number of others, may also be derived in terms of the other two-port parameters. We will consider some of these in the exercises and problems.

EXAMPLE 14.22 To illustrate the use of (14.57), let us find the voltage ratio function for the circuit of Fig. 14.19. By inspection we have

$$\mathbf{z}_{11} = s + \frac{2}{3s}$$

$$\mathbf{z}_{12} = \mathbf{z}_{21} = \frac{2}{3s}$$

$$\mathbf{z}_{22} = 2s + \frac{2}{3s}$$

$$\mathbf{Z}_g = 1$$

$$\mathbf{Z}_L = 2$$

which substituted into (14.57) gives

$$\frac{\mathbf{V}_2}{\mathbf{V}_g} = \frac{2/3}{s^3 + 2s^2 + 2s + 1}$$

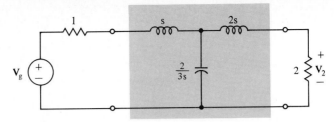

FIGURE 14.19 Two-port network terminated in two resistors

Another use for the two-port parameters is in the construction of equivalent circuits. For example, in the first equation of (14.34), V_1 is a sum of two terms, $z_{11}I_1$ and $z_{12}I_2$. The first may be obtained by an impedance z_{11} carrying a current I_1, and the second may be obtained by a dependent voltage source controlled by I_2. In a similar way we may interpret the second equation of (14.34) and draw the circuit representing these equations. The result, as may be verified by inspection, is shown in Fig. 14.20. We say that this circuit is equivalent (that is, at the terminals) to that of Fig. 14.12, because they both have the same two-port parameters. Thus if the port currents are the same, then the port voltages are the same, and vice versa, for both networks.

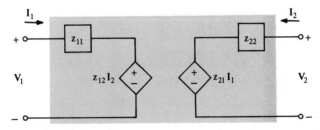

FIGURE 14.20 Equivalent circuit for Fig. 14.12

If the circuit of Fig. 14.12 is a reciprocal network ($z_{12} = z_{21}$), then a simpler, passive equivalent circuit can be found by solving (14.37) for $\mathbf{Z}_1$, $\mathbf{Z}_2$, $\mathbf{Z}_3$. The result,

$$\mathbf{Z}_3 = z_{12}$$

$$\mathbf{Z}_1 = z_{11} - \mathbf{Z}_3 = z_{11} - z_{12}$$

$$\mathbf{Z}_2 = z_{22} - \mathbf{Z}_3 = z_{22} - z_{12}$$

substituted into Fig. 14.13 yields the equivalent passive circuit shown in Fig. 14.21. It is an equivalent circuit for a general reciprocal three-terminal network.

Many other circuits equivalent to the general two-port network may be obtained from the defining equations of the two-port parameters. For example, the circuit of Fig. 14.22 is another such equivalent circuit using the h-parameters, as the reader is asked to verify in Ex. 14.9.4. This is a popular equivalent circuit, often used to represent a transistor.

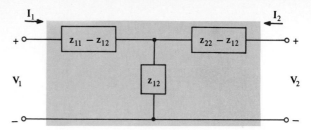

FIGURE 14.21 Equivalent circuit of a general reciprocal three-terminal network

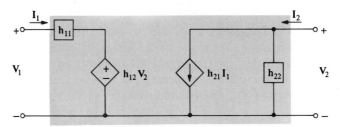

FIGURE 14.22 Equivalent circuit of the general three-terminal two-port, using h-parameters

EXERCISES

14.9.1 Find the voltage ratio transfer function for the two-port terminated in 1 Ω, shown in Fig. 14.16, with z-parameters $\mathbf{z}_{11} = 6\ \Omega$, $\mathbf{z}_{12} = \mathbf{z}_{21} = 4\ \Omega$, and $\mathbf{z}_{22} = 10\ \Omega$.
Answer 2/25

14.9.2 Show that the given circuit is equivalent to the general reciprocal three-terminal network by showing that (14.38) holds.

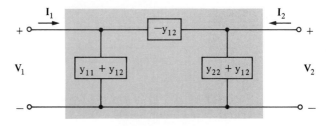

EXERCISE 14.9.2

14.9.3 Show that the given circuit is equivalent to the general two-port network.

EXERCISE 14.9.3

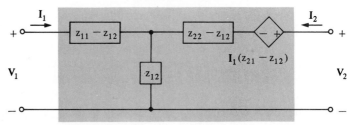

14.9.4 Verify that the circuit of Fig. 14.22 is equivalent to the general three-terminal network by showing that (14.41) holds.

14.10

INTERCONNECTIONS OF TWO-PORT NETWORKS

The two-port networks of the previous sections may be used as building blocks to design more complicated circuits. That is, subsections may be designed as two-port networks and then interconnected to form the overall circuit. In this section we will consider some of these interconnections and see how the overall circuit may be analyzed by analyzing its component two-port parts.

The first interconnection we consider is that of Fig. 14.23(a), which is called a *parallel* connection for reasons that will be clear later. The parallel connection of two grounded two-port networks is shown in Fig. 14.23(b). In the work to follow, we want each subnetwork to retain its integrity as a two-port network, which will be true in Fig. 14.23(a) if the currents into the top leads of each network come out the bottom leads. To ensure this in some cases, it may be necessary to place an *ideal transformer* (to be discussed in Chapter 16) at one of the four ports. In any case, the integrity of the two-port networks in the grounded case of Fig. 14.23(b) is always maintained, because the lower (or grounded) terminals are common to the subnetworks and the overall network.

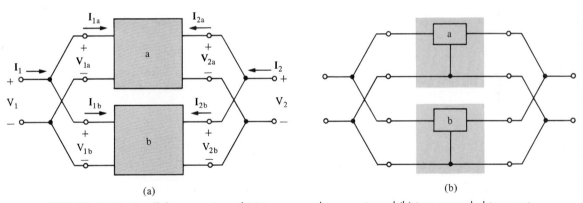

(a)   (b)

FIGURE 14.23 Parallel connection of (a) two general two-ports and (b) two grounded two-ports

By Fig. 14.23(a), we may write

$$I_{1a} = y_{11a}V_{1a} + y_{12a}V_{2a}$$

$$I_{2a} = y_{21a}V_{1a} + y_{22a}V_{2a}$$

and

$$I_{1b} = y_{11b}V_{1b} + y_{12b}V_{2b}$$

$$I_{2b} = y_{21b}V_{1b} + y_{22b}V_{2b}$$

where

$$V_1 = V_{1a} = V_{1b}$$
$$V_2 = V_{2a} = V_{2b}$$
$$I_1 = I_{1a} + I_{1b}$$
$$I_2 = I_{2a} + I_{2b}$$

Combining these results we have

$$I_1 = (y_{11a} + y_{11b})V_1 + (y_{12a} + y_{12b})V_2$$
$$I_2 = (y_{21a} + y_{21b})V_1 + (y_{22a} + y_{22b})V_2$$

Therefore we see that the y-parameters of the interconnection are the sums of the y-parameters of each subnetwork. That is,

$$\begin{array}{ll} y_{11} = y_{11a} + y_{11b}, & y_{12} = y_{12a} + y_{12b} \\ y_{21} = y_{21a} + y_{21b}, & y_{22} = y_{22a} + y_{22b} \end{array} \tag{14.62}$$

The fact that the admittances add, as they do for parallel circuit elements, is motivation for the name *parallel connection*.

EXAMPLE 14.23 As an example, let us find the transfer function V_2/V_g in the circuit of Fig. 14.24, which is a parallel connection of two two-ports terminated in a 1-Ω resistor. By inspection we have

$$y_{21} = y_{21a} + y_{21b} = -\frac{1}{s} - s$$

$$y_{22} = y_{22a} + y_{22b} = s + \frac{1}{s} + 1 + s = 2s + 1 + \frac{1}{s}$$

Therefore by (14.55) we have

FIGURE 14.24 Parallel connection terminated in 1 Ω

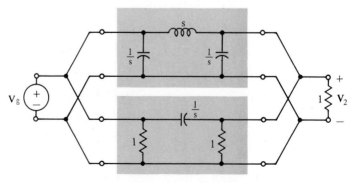

$$\frac{V_2}{V_g} = \frac{(1/s) + s}{1 + \left(2s + 1 + \dfrac{1}{s}\right)} = \frac{s^2 + 1}{2s^2 + 2s + 1}$$

The connection of Fig. 14.25(a) is a *series connection* of the two networks a and b, so called because, as we will see, the z-parameters add, as the y-parameters do in the parallel connection. For the two networks a and b we have

$$V_{1a} = z_{11a}I_{1a} + z_{12a}I_{2a}$$
$$V_{2a} = z_{21a}I_{1a} + z_{22a}I_{2a}$$

and

$$V_{1b} = z_{11b}I_{1b} + z_{12b}I_{2b}$$
$$V_{2b} = z_{21b}I_{1b} + z_{22b}I_{2b}$$

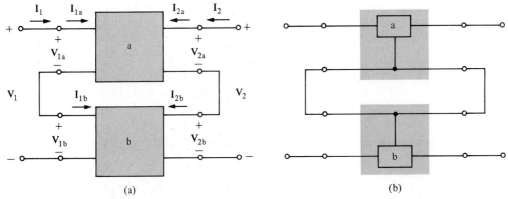

FIGURE 14.25 Series connection of (a) two general two-ports and (b) two grounded two-ports

Also, from the figure we see that

$$I_1 = I_{1a} = I_{1b}$$
$$I_2 = I_{2a} = I_{2b}$$

and

$$V_1 = V_{1a} + V_{1b} = (z_{11a} + z_{11b})I_1 + (z_{12a} + z_{12b})I_2$$
$$V_2 = V_{2a} + V_{2b} = (z_{21a} + z_{21b})I_1 + (z_{22a} + z_{22b})I_2$$

Therefore the z-parameters of the overall network are given by

$$\boxed{\begin{aligned}
z_{11} &= z_{11a} + z_{11b}, & z_{12} &= z_{12a} + z_{12b} \\
z_{21} &= z_{21a} + z_{21b}, & z_{22} &= z_{22a} + z_{22b}
\end{aligned}} \qquad (14.63)$$

Again, we are assuming that the two-ports of Fig. 14.25(a) maintain their integrity as two-port networks. This is always the case in Fig. 14.25(b), where the subnetworks are grounded two-ports.

The last interconnection we will consider is the *cascade* connection of Fig. 14.26, in which the output port of network a is the input port of network b. From the figure and the definition of the transmission parameters, we see that

$$\mathbf{V}_1 = \mathbf{V}_{1a} = \mathbf{A}_a\mathbf{V}_{2a} - \mathbf{B}_a\mathbf{I}_{2a}$$

$$= \mathbf{A}_a\mathbf{V}_{1b} + \mathbf{B}_a\mathbf{I}_{1b}$$

$$= \mathbf{A}_a(\mathbf{A}_b\mathbf{V}_{2b} - \mathbf{B}_b\mathbf{I}_{2b}) + \mathbf{B}_a(\mathbf{C}_b\mathbf{V}_{2b} - \mathbf{D}_b\mathbf{I}_{2b})$$

$$= (\mathbf{A}_a\mathbf{A}_b + \mathbf{B}_a\mathbf{C}_b)\mathbf{V}_{2b} - (\mathbf{A}_a\mathbf{B}_b + \mathbf{B}_a\mathbf{D}_b)\mathbf{I}_{2b}$$

or

$$\mathbf{V}_1 = (\mathbf{A}_a\mathbf{A}_b + \mathbf{B}_a\mathbf{C}_b)\mathbf{V}_2 - (\mathbf{A}_a\mathbf{B}_b + \mathbf{B}_a\mathbf{D}_b)\mathbf{I}_2$$

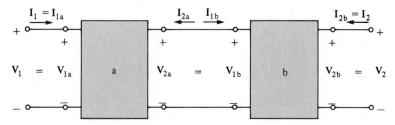

FIGURE 14.26 Cascade connection of two two-port networks

In a similar manner, using the transmission parameter equation for $\mathbf{I}_1$, we may obtain

$$\mathbf{I}_1 = (\mathbf{C}_a\mathbf{A}_b + \mathbf{D}_a\mathbf{C}_b)\mathbf{V}_2 - (\mathbf{C}_a\mathbf{B}_b + \mathbf{D}_a\mathbf{D}_b)\mathbf{I}_2$$

Comparing these last two equations with (14.47) for the overall network, we have for the cascade connection,

$$
\begin{aligned}
\mathbf{A} &= \mathbf{A}_a\mathbf{A}_b + \mathbf{B}_a\mathbf{C}_b \\
\mathbf{B} &= \mathbf{A}_a\mathbf{B}_b + \mathbf{B}_a\mathbf{D}_b \\
\mathbf{C} &= \mathbf{C}_a\mathbf{A}_b + \mathbf{D}_a\mathbf{C}_b \\
\mathbf{D} &= \mathbf{C}_a\mathbf{B}_b + \mathbf{D}_a\mathbf{D}_b
\end{aligned}
\tag{14.64}
$$

A reader familiar with matrix multiplication will recognize this result to be a statement that the transmission matrix of the overall network is the product of the transmission matrices of networks a and b. That is,

$$\begin{bmatrix} \mathbf{A} & \mathbf{B} \\ \mathbf{C} & \mathbf{D} \end{bmatrix} = \begin{bmatrix} \mathbf{A}_a & \mathbf{B}_a \\ \mathbf{C}_a & \mathbf{D}_a \end{bmatrix}\begin{bmatrix} \mathbf{A}_b & \mathbf{B}_b \\ \mathbf{C}_b & \mathbf{D}_b \end{bmatrix} \tag{14.65}$$

EXAMPLE 14.24 As an example, we may show that the transmission parameters of the simple two-ports of Fig. 14.27(a) and (b) are, respectively,

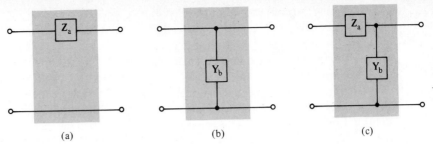

FIGURE 14.27 Two simple two-port networks, and their cascade connection

$$\mathbf{A}_a = 1 \qquad \mathbf{B}_a = \mathbf{Z}_a$$
$$\mathbf{C}_a = 0 \qquad \mathbf{D}_a = 1 \tag{14.66}$$

and

$$\mathbf{A}_b = 1 \qquad \mathbf{B}_b = 0$$
$$\mathbf{C}_b = \mathbf{Y}_b \qquad \mathbf{D}_b = 1 \tag{14.67}$$

and therefore those of their cascade connection (c) are, by (14.64),

$$\mathbf{A} = 1 + \mathbf{Z}_a \mathbf{Y}_b, \qquad \mathbf{B} = \mathbf{Z}_a$$
$$\mathbf{C} = \mathbf{Y}_b, \qquad \mathbf{D} = 1 \tag{14.68}$$

EXAMPLE 14.25 We may use the result in (14.68) to find the voltage ratio transfer function $\mathbf{V}_2/\mathbf{V}_1$ for the circuit of Fig. 14.28. As the reader is asked to show in Prob. 14.32 for the general two-port network terminated in 1 Ω, the voltage transfer function is

$$\frac{\mathbf{V}_2}{\mathbf{V}_1} = \frac{1}{A + B} \tag{14.69}$$

The two port in this case is like that of Fig. 14.27(c) with $\mathbf{Z}_a = 2s$ and $\mathbf{Y}_b = s$. Therefore we have

$$A = 1 + 2s^2$$
$$B = 2s$$

so that

FIGURE 14.28 Two-port network with a 1-Ω termination

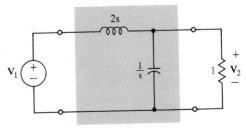

$$\frac{V_2}{V_1} = \frac{1}{2s^2 + 2s + 1}$$

There are two other interconnections of two-port networks that we will mention but will not consider further. The first is the *series-parallel* connection in which the primary is connected like the primary of the series connection of Fig. 14.25 and the secondary is connected like the secondary of the parallel connection of Fig. 14.23. The second is the parallel-series connection, in which the primary is parallel connected and the secondary is series connected. If the integrity of each two-port network is maintained with the interconnections, the *h*-parameters add to give the overall *h*-parameters in the series-parallel connection, and the *g*-parameters add to give the overall *g*-parameters in the parallel-series connection.

EXERCISES

14.10.1   Find V_2/V_g for the circuit shown.

Answer $\dfrac{4s^2 + 6s}{2s^4 + 15s^3 + 35s^2 + 28s + 3}$

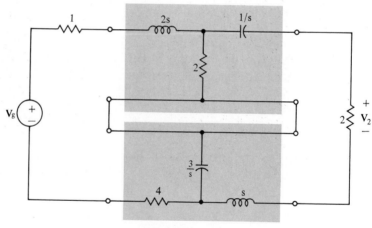

EXERCISE 14.10.1

14.10.2   Use the *ABCD* parameters to find V_2/V_1 for the two-port network terminated in a 1-Ω resistor if the two-port is the T network of Fig. 14.13 with $Z_1 = Z_2 = 2s$ and $Z_3 = 1/s$.

(*Suggestion:* Note that the two port is that of Fig. 14.28 with the 2*s* impedance added at the output.)

Answer $\dfrac{1}{4s^3 + 2s^2 + 4s + 1}$

PROBLEMS

14.1 Find the phasor representation $\mathbf{V}(s)$ of (a) $v(t) = 5e^{-2t}\cos(10t - 30°)$, (b) $v(t) = e^{-t}\sin(5t + 45°)$, (c) $v(t) = 2e^{-3t}(3\cos 3t + 4\sin 3t)$, and (d) $v(t) = 5e^{-8t}$.

14.2 Find $v(t)$ if (a) $\mathbf{V}(s) = 6\underline{/10°}$ with $s = -2 + j8$, (b) $\mathbf{V}(s) = 5\underline{/0°}$ with $s = -10$, (c) $\mathbf{V}(s) = 4 + j3$ with $s = -1 + j2$, and (d) $\mathbf{V}(s) = -j6$ with $s = j4$.

14.3 Find the complex frequencies which characterize (a) $v = e^{-t}(\cos 2t + \sin 2t)$, (b) $v = e^{-3t} + e^{-4t} + 2\cos t$, and (c) $v = 10$.

14.4 Find the impedance $\mathbf{Z}(s)$ seen by the source and locate its poles and zeros. If the source is $v_g = 10e^{-t}\cos 2t$ V, find the forced component of current it delivers.

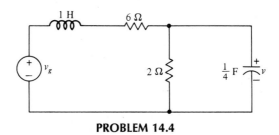

PROBLEM 14.4

14.5 Find the forced component of v in Prob. 14.4 if $v_g = 16e^{-4t}\cos 2t$ V.

14.6 Show that

$$\mathbf{Y} = \mathbf{Y}_1 + \cfrac{1}{\mathbf{Z}_2 + \cfrac{1}{\mathbf{Y}_3 \pm \cfrac{1}{\mathbf{Z}_4 + \cfrac{1}{\mathbf{Y}_5}}}}$$

(*Suggestion:* Note that $\mathbf{Y} = \mathbf{Y}_1 + \mathbf{Y}_a = \mathbf{Y}_1 + 1/\mathbf{Z}_a = \mathbf{Y}_1 + 1/(\mathbf{Z}_2 + \mathbf{Z}_b) = \ldots$)

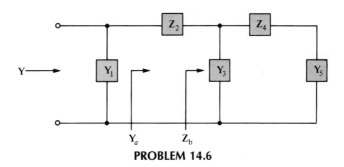

PROBLEM 14.6

14.7 Use the method of Prob. 14.6 to find the impedance seen by the source, and use the result to find the forced response v.

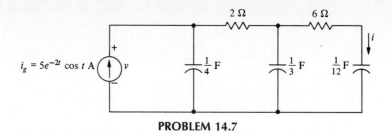

PROBLEM 14.7

14.8   In Prob. 14.7 find the network function $\mathbf{H}(s) = \mathbf{I}(s)/\mathbf{I}_g(s)$ by using the proportionality principle. Use the result to find the forced response i.

14.9   Use the principle of proportionality to find the network function $\mathbf{V}_2/\mathbf{V}_1$, and use the result to find the forced response $v_2(t)$ when $v_1(t) = 4 \cos t$ V.

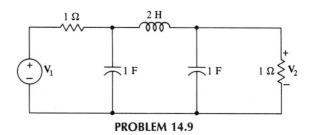

PROBLEM 14.9

14.10   If the voltage source in Prob. 9.14 is $v_g(t) = 6e^{-t} \cos 2t$ V, find the network function $\mathbf{I}(s)/\mathbf{V}_g(s)$ and the forced response i in each case.

14.11   Find $\mathbf{H}(s) = \mathbf{V}(s)/\mathbf{V}_g(s)$ and use the result to find the forced response v if $v_g = e^{-t} \cos t$ V.

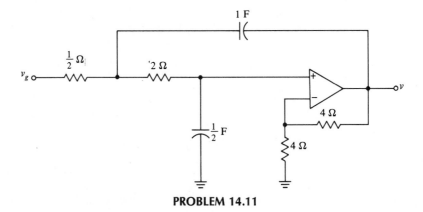

PROBLEM 14.11

14.12   Replace in the corresponding phasor circuit everything except the inductor by its Thevenin equivalent circuit and use the result to find the forced response i if $v_g = 4e^{-2t} \cos 2t$ V.

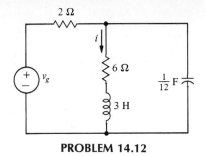

PROBLEM 14.12

14.13 For the corresponding phasor circuit, replace everything to the left of terminals *a-b* by its Thevenin equivalent and use the result to find the forced response *i*.

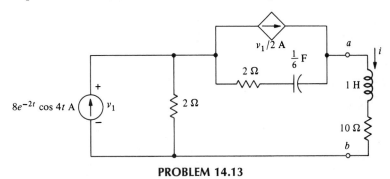

PROBLEM 14.13

14.14 Find the network function $\mathbf{V}(s)/\mathbf{V}_g(s)$ and use the result to find the forced response v if $v_g = 6e^{-2t} \cos 4t$ V.

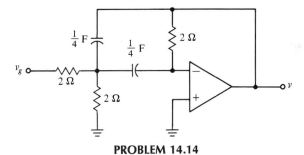

PROBLEM 14.14

14.15 Find the network function and the forced response $v(t)$ if $v_g(t) = 6e^{-2t}$ V.

PROBLEM 14.15

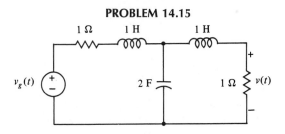

14.16 Given the network function

$$\mathbf{H}(s) = \frac{\mathbf{V}_o(s)}{\mathbf{V}_i(s)} = \frac{4s(s + 2)}{(s + 1)(s + 3)}$$

If no cancellation has occurred, find the complete response $v_o(t)$, for $t > 0$, if $v_i(t) = 6e^{-t} \cos 2t$ V and $v_o(0^+) = dv_o(0^+)/dt = 0$.

14.17 Repeat Prob. 14.16 if $v_i(t) = 6e^{-t}$ V. (*Suggestion:* Observe the describing differential equation.)

14.18 Find the complete response v for $t > 0$ using the network function, if $v(0^+)$ and $dv(0^+)/dt$ are both zero.

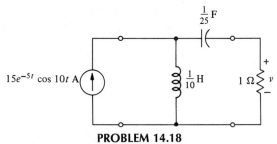

PROBLEM 14.18

14.19 Find the network function $\mathbf{V}(s)/\mathbf{V}_g(s)$ and the forced response v if $v_g = 2e^{-2t} \cos t$ V.

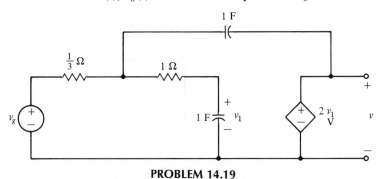

PROBLEM 14.19

14.20 Find the network function $\mathbf{V}(s)/\mathbf{V}_g(s)$ and the forced response v if (a) $v_g = 6e^{-4t} \cos 2t$ V and (b) $v_g = 6 \cos 2t$ V.

PROBLEM 14.20

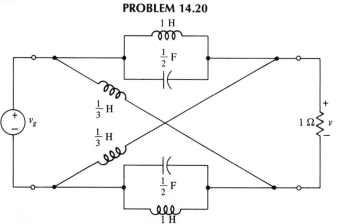

14.21   Find the complete response v if $v(0) = 0$. Use network functions and superposition.

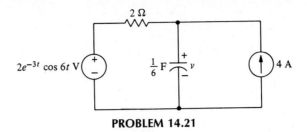

PROBLEM 14.21

14.22   Find the natural frequencies in v in Prob. 14.18 by killing the source and using (a) the pliers entry with a voltage source inserted between the capacitor and the resistor, and (b) the soldering entry with a current source across the resistor.

14.23   Find the network function $\mathbf{V}(s)/\mathbf{V}_g(s)$ and the forced response v if $v_g = 6e^{-2t} \cos t$ V.

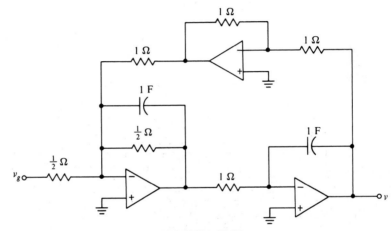

PROBLEM 14.23

14.24   Considering the figure for Prob. 14.18 as a two-port network with terminals as shown, find the z- and y-parameters as functions of s.

14.25   Find the y-parameters in Prob. 14.20.

14.26   Show that the hybrid parameters defined in (14.42) may be obtained from the z-parameters by

$$g_{11} = \frac{1}{\mathbf{z}_{11}}, \qquad g_{12} = -\frac{\mathbf{z}_{12}}{\mathbf{z}_{11}}$$

$$g_{21} = \frac{\mathbf{z}_{21}}{\mathbf{z}_{11}}, \qquad g_{22} = \frac{\Delta_z}{\mathbf{z}_{11}}$$

where $\Delta_z = \mathbf{z}_{11}\mathbf{z}_{22} - \mathbf{z}_{12}\mathbf{z}_{21}$.

14.27   Find the h- and g-parameters of the two-port network of Prob. 14.24.

14.28   Show that the transmission parameters, **A, B, C,** and **D,** defined for the two-port network of Fig. 14.12 by

$$V_1 = AV_2 - BI_2$$

$$I_1 = CV_2 - DI_2$$

are given by

$$A = \frac{\mathbf{z}_{11}}{\mathbf{z}_{21}}, \qquad B = \frac{\Delta_z}{\mathbf{z}_{21}}$$

$$C = \frac{1}{\mathbf{z}_{21}}, \qquad D = \frac{\mathbf{z}_{22}}{\mathbf{z}_{21}}$$

(This is the first entry in the third row of Table 14.1.)

14.29 Find the transmission parameters of the two-port network of Prob. 14.24.

14.30 Derive the $\mathbf{Y}$-Δ transformation of Chapter 13 by making Figs. 14.13 and 14.14 equivalent at the terminals (that is, equal two-port parameters).

14.31 Find the y-parameters of the network shown. Terminate the output port with a 1-Ω resistor, and find the resulting network function $\mathbf{V}_2/\mathbf{V}_1$.

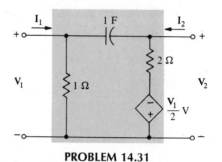

PROBLEM 14.31

14.32 Show that for Fig. 14.16, in terms of the transmission parameters we have

$$\frac{\mathbf{V}_2}{\mathbf{V}_1} = \frac{1}{\mathbf{A} + \mathbf{B}}$$

Check the result, using this formula, for the terminated network of Prob. 14.31.

14.33 Let $\mathbf{h}_{11} = 1 \text{ k}\Omega$, $\mathbf{h}_{12} = 10^{-4}$, $\mathbf{h}_{21} = 100$, and $\mathbf{h}_{22} = 10^{-4} \, \mho$ in Fig. 14.22 and find the network function $\mathbf{V}_2/\mathbf{V}_1$ if port 2 is open circuited.

14.34 Show for the doubly terminated network of Fig. 14.18 that the voltage ratio transfer function is given by the three equations

$$\frac{\mathbf{V}_2}{\mathbf{V}_g} = \frac{\mathbf{y}_{21} \mathbf{Z}_L}{\mathbf{y}_{12} \mathbf{y}_{21} \mathbf{Z}_g \mathbf{Z}_L - (1 + \mathbf{y}_{11} \mathbf{Z}_g)(1 + \mathbf{y}_{22} \mathbf{Z}_L)}$$

$$\frac{\mathbf{V}_2}{\mathbf{V}_g} = \frac{-\mathbf{h}_{21} \mathbf{Z}_L}{(\mathbf{h}_{11} + \mathbf{Z}_g)(1 + \mathbf{h}_{22} \mathbf{Z}_L) - \mathbf{h}_{12} \mathbf{h}_{21} \mathbf{Z}_L}$$

and

$$\frac{\mathbf{V}_2}{\mathbf{V}_g} = \frac{\mathbf{Z}_L}{(\mathbf{A} + \mathbf{C} \mathbf{Z}_g)\mathbf{Z}_L + \mathbf{B} + \mathbf{D} \mathbf{Z}_g}$$

14.35 Use the second of the formulas of Prob. 14.34 to find the voltage ratio transfer function for a doubly terminated network with $\mathbf{h}_{11} = 1 \text{ k}\Omega$, $\mathbf{h}_{12} = 10^{-4}$, $\mathbf{h}_{21} = 100$, $\mathbf{h}_{22} = 10^{-4} \, \mho$, $\mathbf{Z}_g = 360 \, \Omega$, and $\mathbf{Z}_L = 1 \text{ k}\Omega$.

14.36 Show that the given circuit is equivalent to the general two-port network. Note how it differs from the equivalent circuit of Ex. 14.9.3.

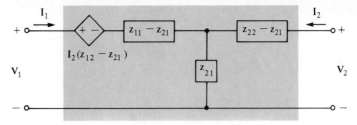

PROBLEM 14.36

14.37 Show that the given circuits are equivalent to the general two-port network. Note how the two circuits differ.

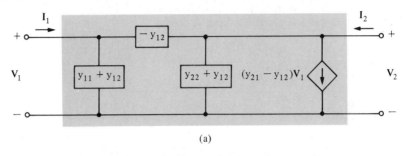

(a)

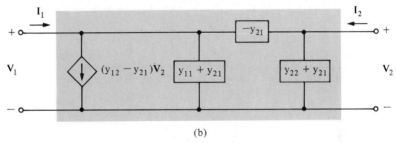

(b)

PROBLEM 14.37

14.38 The circuit shown is a *lattice* with *series arms* both equal to $\mathbf{Z}_a$ and *cross arms* both equal to $\mathbf{Z}_b$. It is called a *symmetrical* lattice because the series arms are equal and the cross arms are equal. Show that the z- and y-parameters are given by

$$\mathbf{z}_{11} = \mathbf{z}_{22} = \frac{1}{2}(\mathbf{Z}_b + \mathbf{Z}_a)$$

$$\mathbf{z}_{12} = \mathbf{z}_{21} = \frac{1}{2}(\mathbf{Z}_b - \mathbf{Z}_a)$$

and

$$\mathbf{y}_{11} = \mathbf{y}_{22} = \frac{1}{2}(\mathbf{Y}_b + \mathbf{Y}_a)$$

$$\mathbf{y}_{12} = \mathbf{y}_{21} = \frac{1}{2}(\mathbf{Y}_b - \mathbf{Y}_a)$$

where $\mathbf{Y}_a = 1/\mathbf{Z}_a$ and $\mathbf{Y}_b = 1/\mathbf{Z}_b$.

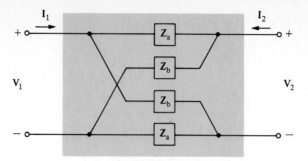

PROBLEM 14.38

14.39 The symmetrical lattice of Prob. 14.38 is terminated in 1 Ω and Z_a and Z_b are as shown. Find the voltage ratio transfer function.

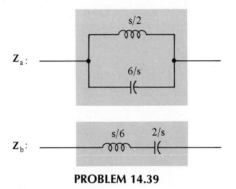

PROBLEM 14.39

14.40 Repeat Prob. 14.39 if Z_a and Z_b are as shown.

PROBLEM 14.40

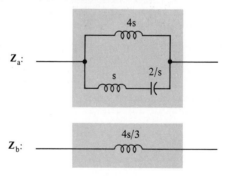

Alexander Graham Bell
1847–1922

15

Frequency Response

Mr. Watson, come here. I want you.
Alexander Graham Bell

Undoubtedly the most common and most widely used electrical instrument is the telephone, invented by the Scottish-American scientist Alexander Graham Bell. The date was June 2, 1875, when Bell and his assistant Thomas Watson transmitted a musical note. The first intelligible telephone sentences, "Mr. Watson, come here. I want you," were spoken inadvertently by Bell himself on March 10, 1876, when he called Watson to come to an adjoining room to help with some spilled acid.

Bell was born in Edinburgh, Scotland. His father, Alexander Melville Bell, was a well-known speech teacher and his grandfather, Alexander Bell, was also a speech teacher. Young Bell, after attending the University of Edinburgh and the University of London, also became a speech teacher. In 1866 Bell became interested in trying to transmit speech electrically after reading a book describing how vowel sounds could be made with tuning forks. Shortly afterwards, Bell's two brothers died of tuberculosis and Melville Bell moved his family to Canada for health reasons. In 1873 young Graham became a professor at Boston University and began his electrical experiments in his spare time. It was there that he formed his partnership with Watson and went on to his great invention. Bell's telephone patent was the most valuable one ever issued, and the telephone opened a new age in the development of civilization. ■

Frequency-domain functions are very useful, as we have seen, for finding corresponding time-domain functions. The frequency response of a circuit, however, is extremely useful in its own right, as we shall see in this chapter. For example, if we are interested in which frequencies are dominant in an output signal, say $\mathbf{V}(j\omega)$, then we need only consider the amplitude $|\mathbf{V}(j\omega)|$. The dominant frequencies correspond to relatively large amplitudes, and frequencies that are virtually suppressed correspond to relatively small amplitudes.

There are many applications in which frequency responses are important. One very common application is in the design of electric filters, which are networks that pass signals of certain frequencies and block signals of other frequencies. That is, if the output signal of the filter has amplitude $|\mathbf{V}(j\omega)|$, then ω_1 passes if $|\mathbf{V}(j\omega_1)|$ is relatively large and is blocked if $|\mathbf{V}(j\omega_1)|$ is relatively small. There are many examples of electric filters in our modern society, some of the more common being those in our television sets, which allow us to tune in a certain channel by passing its band of frequencies while filtering out those of the other channels.

In this chapter we shall consider frequency responses, both amplitude and phase. We shall also define resonance and quality factor and show how they are related to the frequency responses. Finally, we shall consider methods of scaling networks to yield a given frequency response with practical circuit element values.

15.1

AMPLITUDE AND PHASE RESPONSES

A network function $\mathbf{H}(j\omega)$, as well as any phasor quantity, is in general a complex function, having a real and an imaginary part. That is,

$$\mathbf{H}(j\omega) = \text{Re } \mathbf{H}(j\omega) + j \text{ Im } \mathbf{H}(j\omega) \tag{15.1}$$

As we know, we may also write the network function in the polar form

$$\mathbf{H}(j\omega) = |\mathbf{H}(j\omega)|e^{j\phi(\omega)} \qquad (15.2)$$

where $|\mathbf{H}(j\omega)|$ is the *amplitude,* or *magnitude, response* and $\phi(\omega)$ is the *phase response,* given, respectively, by

$$|\mathbf{H}(j\omega)| = \sqrt{\text{Re}^2\,\mathbf{H}(j\omega) + \text{Im}^2\,\mathbf{H}(j\omega)} \qquad (15.3)$$

and

$$\phi(\omega) = \tan^{-1}\frac{\text{Im }\mathbf{H}(j\omega)}{\text{Re }\mathbf{H}(j\omega)} \qquad (15.4)$$

The amplitude and phase responses are, of course, special cases of *frequency responses.*

EXAMPLE 15.1   As an example, suppose the network function of the *RLC* parallel circuit of Fig. 15.1 is the input impedance

$$\mathbf{H}(s) = \frac{\mathbf{V}_2(s)}{\mathbf{I}_1(s)} = \mathbf{Z}(s) = \frac{1}{(1/R) + sC + (1/sL)} \qquad (15.5)$$

or

$$\mathbf{H}(s) = \frac{(1/C)s}{s^2 + (1/RC)s + (1/LC)} \qquad (15.6)$$

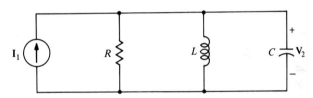

FIGURE 15.1   *RLC* parallel circuit

For $s = j\omega$ we have, from (15.5),

$$\mathbf{H}(j\omega) = \frac{1}{(1/R) + j[\omega C - (1/\omega L)]} \qquad (15.7)$$

so that the amplitude and phase responses are

$$|\mathbf{H}(j\omega)| = \frac{1}{\sqrt{(1/R^2) + [\omega C - (1/\omega L)]^2}} \qquad (15.8)$$

and

$$\phi(\omega) = -\tan^{-1} R\left(\omega C - \frac{1}{\omega L}\right) \qquad (15.9)$$

Since *R, L,* and *C* are constants, the maximum amplitude occurs at the frequency

471

$\omega = \omega_0$ for which the denominator in (15.8) is a minimum. Evidently this occurs when

$$\omega C - \frac{1}{\omega L} = 0$$

or

$$\omega_0 = \frac{1}{\sqrt{LC}} \qquad\qquad (15.10)$$

Thus

$$|\mathbf{H}(j\omega)|_{\max} = |\mathbf{H}(j\omega_0)| = R$$

Also, it is clear that $|\mathbf{H}(j\omega)| \to 0$ as $\omega \to 0$ and $\omega \to \infty$. Therefore, the amplitude response has the form shown in Fig. 15.2(a). In a like manner we may sketch the phase response, shown in Fig. 15.2(b), since $\phi(\omega_0) = 0$, $\phi(\omega) \to \pi/2$ as $\omega \to 0$, and $\phi(\omega) \to -\pi/2$ as $\omega \to \infty$.

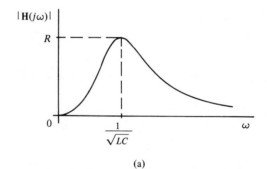

(a)

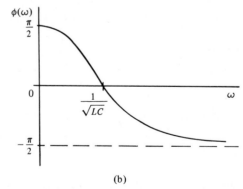

(b)

FIGURE 15.2 (a) Amplitude and (b) phase response of (15.7)

If the input of Fig. 15.1 is the time-domain function

$$i_1(t) = I_m \cos \omega t$$

then the input phasor is $\mathbf{I}_1 = I_m\underline{/0°}$, and the output phasor is $\mathbf{V}_2 = I_m\mathbf{Z} = I_m\mathbf{H}$. Thus the amplitude of the output is simply that of the network function multiplied by a constant. Therefore, we may obtain as much information from the network function response as from the output response. For this reason and the reason that the network function depends only on the network and not on how it is excited, we shall usually consider the frequency response of the network function.

EXERCISES

15.1.1 Let $R = 2\ \Omega$, $L = \frac{1}{9}$ H, and $C = \frac{1}{4}$ F in Fig. 15.1, and find the maximum amplitude and the point where it occurs. Also, sketch the amplitude and phase responses.
Answer $|\mathbf{H}|_{max} = 2$, $\omega_0 = 6$

15.1.2 For the *RLC* series circuit with a voltage source v_g, let the network function be $\mathbf{H} = \mathbf{I}/\mathbf{V}_g$, where $\mathbf{I}$ is the phasor current. Show that the amplitude and phase responses are similar to those of Fig. 15.2 with $|\mathbf{H}|_{max} = 1/R$ and $\omega_0 = 1/\sqrt{LC}$.

15.1.3 Let the network function of Fig. 15.1 be $\mathbf{H} = \mathbf{I}_L/\mathbf{I}_1$, where $\mathbf{I}_L$ is the inductor phasor current directed downward. Show that

$$\mathbf{H}(s) = \frac{1/LC}{s^2 + (1/RC)s + (1/LC)}$$

and that $|\mathbf{H}|_{max} = 1$, occurring at $\omega_0 = 0$, provided

$$2R^2C \le L$$

15.2

FILTERS

With reference to Fig. 15.2(a) we see that frequencies clustered around $\omega_0 = 1/\sqrt{LC}$ rad/s, or $f_0 = 1/(2\pi\sqrt{LC})$ Hz, correspond to relatively large amplitudes, while those near zero and larger than ω_0 correspond to relatively small amplitudes. Thus Fig. 15.1 is an example of a *bandpass filter*, which passes the *band* of frequencies centered around ω_0. In the general amplitude case, shown in Fig. 15.3, we say that ω_0, the frequency at which the maximum amplitude occurs, is the *center frequency*. The band of frequencies that passes, or the *passband*, is *defined* to be

$$\omega_{c_1} \le \omega \le \omega_{c_2}$$

where ω_{c_1} and ω_{c_2} are called the *cutoff* points and are defined as the frequencies at which the amplitude is $1/\sqrt{2} = 0.707$ times the maximum amplitude. The width of the passband, given by

$$B = \omega_{c_2} - \omega_{c_1} \qquad (15.11)$$

is called the *bandwidth*.

As we shall see in Sec. 15.4, the bandwidth in the case of Fig. 15.1 is $B = 1/RC$.

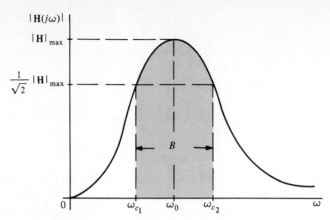

FIGURE 15.3 General bandpass amplitude response

EXAMPLE 15.2 As another example, let us consider the circuit of Fig. 15.4. Analyzing the circuit, we may readily obtain the voltage-ratio function.

$$H(s) = \frac{V_2(s)}{V_1(s)} = \frac{2}{s^2 + 2s + 2} \tag{15.12}$$

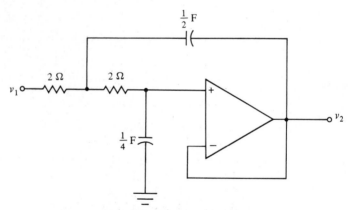

FIGURE 15.4 Low-pass filter

Letting $s = j\omega$ and calculating the amplitude response, we have

$$|H(j\omega)| = \frac{2}{\sqrt{(2 - \omega^2)^2 + 4\omega^2}}$$

or, after simplification,

$$|H(j\omega)| = \frac{1}{\sqrt{1 + (\omega^4/4)}}$$

The amplitude function continuously decreases as ω increases, because its numerator is constant and its denominator continuously increases with frequency. Therefore, the amplitude response attains its maximum of $|H|_{max} = 1$ at $\omega_0 = 0$ and thus

474 Chapter 15 Frequency Response

has the shape of Fig. 15.5(a). From this we see that the circuit of Fig. 15.4 is a *low-pass* filter. That is, it passes low frequencies (relatively large amplitudes) and rejects high frequencies (relatively small amplitudes).

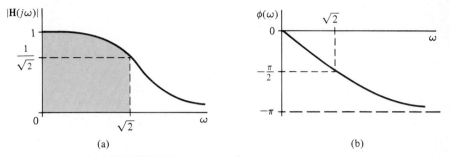

FIGURE 15.5 Low-pass frequency responses

There is only one cutoff point, as indicated in the figure. This follows from the definition

$$|\mathbf{H}(j\omega_c)| = \frac{1}{\sqrt{2}}|\mathbf{H}(j\omega)|_{\text{max}}$$

$$= \frac{1}{\sqrt{2}} \quad (1)$$

$$= \frac{1}{\sqrt{1 + (\omega_c^4/4)}}$$

whose only real positive answer is $\omega_c = \sqrt{2}$. Thus the band of frequencies which passes is the low-frequency band

$$0 \le \omega \le \sqrt{2}$$

The phase response for Fig. 15.4 may be easily shown from $\mathbf{H}(j\omega)$ to be

$$\phi(\omega) = -\tan^{-1}\frac{2\omega}{2 - \omega^2}$$

which is sketched in Fig. 15.5(b).

An example of a passive, low-pass filter is that of Fig. 15.1, where the network function is as defined in Ex. 15.1.3. In fact, if $R = 1\ \Omega$, $L = 1$ H, and $C = \frac{1}{2}$ F, the network function is the same as that of Fig. 15.4.

There are many types of filters other than low-pass and bandpass. Two of the more common are *high-pass* filters, which pass high frequencies and reject low frequencies, and *band-reject* filters, which pass all frequencies except a single band. Typical amplitude responses are shown in Fig. 15.6, and examples are considered in Exs. 15.2.2 and 15.2.3. In Fig. 15.6(a), ω_c is the cutoff point, and the passband is $\omega > \omega_c$. In Fig. 15.6(b), ω_0 is the center frequency of the rejected band of bandwidth $B = \omega_{c_2} - \omega_{c_1}$.

In general, the *order* of a filter is the degree of the denominator polynomial of

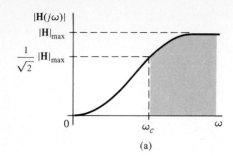 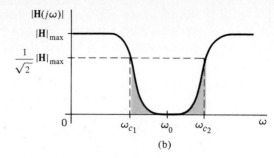

(a)   (b)

FIGURE 15.6 (a) High-pass and (b) band-reject amplitude response

its network function. Thus the filters of Figs. 15.1 and 15.4 are second-order filters, as seen from (15.6) and (15.12). Higher-order filters are more expensive to construct but they have better frequency responses than lower-order filters. This will be seen for a special type of filter in Prob. 15.12.

EXERCISES

15.2.1   Show that

$$\mathbf{H}(s) = \frac{2s}{s^2 + 0.2s + 1}$$

is the network function of a bandpass filter and find ω_0, ω_{c_1}, ω_{c_2}, and B.
Answer 1; $(\mp 0.2 + \sqrt{4.04})/2 = 0.905$, 1.105; 0.2

15.2.2   Show that

$$\mathbf{H}(s) = \frac{2s^2}{s^2 + s + 0.5}$$

is the network function of a high-pass filter, and find $|\mathbf{H}(j\omega)|_{max}$ and ω_c.
Answer $|\mathbf{H}|_{max} = 2$, $\omega_c = 1/\sqrt{2}$

15.2.3   Show that

$$\mathbf{H}(s) = \frac{3(s^2 + 25)}{s^2 + s + 25}$$

is the network function of a band-reject filter, and find $|\mathbf{H}(j\omega)|_{max}$, ω_0, ω_{c_1}, and ω_{c_2}.
Answer $|\mathbf{H}|_{max} = 3$, $\omega_0 = 5$, $\omega_{c_1,c_2} = (\mp 1 + \sqrt{101})/2 = 4.525$, 5.525

15.2.4   An *all-pass* filter is one whose amplitude response is constant. (Thus it passes *all* frequencies equally well.) It can be cascaded with another filter to keep a desired amplitude response but *shift* the phase. Show by finding the network function $\mathbf{V}_2/\mathbf{V}_1$, the amplitude response, and the phase response that the given circuit is a first-order all-pass filter.
Answer $\dfrac{s - 2}{s + 2}$, 1, $180° - 2 \arctan \omega/2$

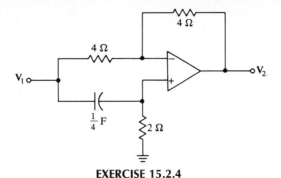

EXERCISE 15.2.4

RESONANCE

A physical system which has a sinusoidal type of natural response reacts vigorously, and sometimes violently, when it is excited at, or near, one of its natural frequencies. This effect may have been noticed by the reader in Sec. 9.6, particularly in the case of Ex. 9.6.3. The system in this respect is somewhat like all of us. When urged to do what it naturally wants to do, it responds with enthusiasm.

This phenomenon is known as *resonance,* and its side effects may be good or they may be bad. As an example, a singer may break a crystal goblet with his voice alone by properly producing a note at precisely the right frequency. Also, a bridge may be destroyed if it is subjected to a periodic force with the same frequency as one of its natural frequencies. This is why no thoughtful troop commander will march his men in step across a bridge. On the other hand, without resonance there could be no electric filters.

We shall define a sinusoidally excited network to be in resonance when the amplitude of the network function attains a pronounced maximum or minimum value. The frequency at which this occurs is called the *resonant frequency.* As an example, the *RLC* parallel circuit of Fig. 15.1 is in resonance when the frequency of the driving function is $\omega_0 = 1/\sqrt{LC}$. This was shown in Sec. 15.1 where it was demonstrated that the maximum network function amplitude occurred at ω_0. The amplitude response of Fig. 15.2(a) is typical, with its relatively high peak at the resonant frequency. The parallel *RLC* circuit is so important that the term *parallel resonance* is reserved for its resonant condition.

The reader may recall encountering the term *resonant frequency* earlier in Sec. 9.8 in connection with the underdamped case of the parallel *RLC* circuit. The two frequencies, there and here, are exactly the same.

The natural frequencies of the parallel *RLC* circuit are the poles of the network function, given from (15.6) by

$$s_{1,2} = -\alpha \pm j\omega_d \tag{15.13}$$

where

$$\alpha = \frac{1}{2RC} \tag{15.14}$$

and

$$\omega_d = \sqrt{\omega_0^2 - \alpha^2} \tag{15.15}$$

From the pole-zero plot, shown in Fig. 15.7, we see that the resonant frequency ω_0, or $s = j\omega_0$, is very near the natural frequency $s = -\alpha + j\omega_d$, since, by (15.15), ω_0 is the radius of the dashed semicircle. If R is made larger, then α is made smaller and the resonant frequency is even closer to the natural frequency. In this case, as is clear in Fig. 15.2(a), the peak is more pronounced. Of course, if R were infinite (open-circuited), then the resonant frequency would coincide with the natural frequency and the amplitude would be infinite.

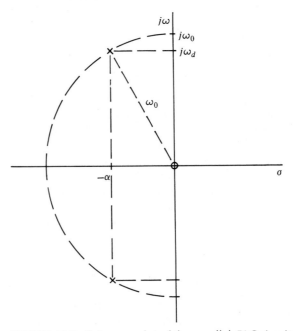

FIGURE 15.7 Pole-zero plot of the parallel RLC circuit

Before leaving the parallel RLC circuit, let us note that the network function is actually the input impedance, as stated in (15.5). The resonant frequency $\omega_0 = 1/\sqrt{LC}$ is the frequency for which the input impedance is purely real, as seen in (15.7). Indeed, many authors define the resonant frequency precisely this way in the case of a two-terminal network. In the general case, maximum amplitude does not always occur exactly at the frequency of real impedance, but usually there is very little difference.

In the case of the RLC series circuit excited by a voltage source $\mathbf{V}_g$, if the phasor current $\mathbf{I}$ is the output, then

$$\mathbf{H}(j\omega) = \frac{\mathbf{I}(j\omega)}{\mathbf{V}_g(j\omega)} = \mathbf{Y}(j\omega) = \frac{1}{R + j[\omega L - (1/\omega C)]}$$

where $\mathbf{Y}$ is the input admittance seen at the source terminals. Evidently series resonance occurs at $\omega_0 = 1/\sqrt{LC}$, yielding a maximum amplitude of $1/R$. As in the parallel resonance case, the resonant frequency is also the frequency of real input impedance or admittance. At resonance the effect of the storage elements exactly cancels, and the source sees only the resistance.

EXERCISES

15.3.1 Show that the resonant frequency in the case of the function of Ex. 15.2.1 coincides with the frequency for which the function is real.

15.3.2 Find the resonant frequency for the parallel RLC circuit described by (a) $R = 2$ kΩ, $L = 4$ mH, and $C = 0.1$ μF; (b) $\omega_d = 5$ rad/s and $\alpha = 12$ Np/s; and (c) $\alpha = 1$ Np/s, $R = 4$ Ω, and $L = 2$ H.
Answer (a) 50,000; (b) 13; (c) 2 rad/s

15.3.3 In Ex. 15.3.2(a), find the amplitude of the voltage across the combination if the current source is $i_g = \cos \omega t$ mA and ω is (a) 10^4, (b) 5×10^4, and (c) 25×10^4 rad/s.
Answer (a) 0.0417; (b) 2; (c) 0.0417 V

15.4

BANDPASS FUNCTIONS AND QUALITY FACTOR

The network function (15.6) is a special case of the general second-order bandpass function

$$\mathbf{H}(s) = \frac{Ks}{s^2 + as + b} \tag{15.16}$$

where K, $a > 0$, and $b > 0$ are real constants. To see that the function is of the bandpass type, let us consider its amplitude

$$|\mathbf{H}(j\omega)| = \frac{|K\omega|}{\sqrt{(b - \omega^2)^2 + a^2\omega^2}}$$

$$= \frac{|K|}{\sqrt{a^2 + [(b - \omega^2)/\omega]^2}}$$

Evidently the maximum value is

$$|\mathbf{H}(j\omega)|_{\max} = \frac{|K|}{a}$$

occurring at the center, or resonant, frequency ω_0, satisfying

$$b = \omega_0^2 \tag{15.17}$$

At a cutoff frequency ω_c, we must have

$$|\mathbf{H}(j\omega_c)| = \frac{1}{\sqrt{2}}|\mathbf{H}(j\omega)|_{\text{max}}$$

or

$$\frac{|K|}{\sqrt{a^2 + [(b - \omega_c^2)/\omega_c]^2}} = \frac{|K|}{\sqrt{2}a}$$

which evidently holds if

$$\frac{b - \omega_c^2}{\omega_c} = \pm a$$

Thus we have

$$\omega_c^2 \pm a\omega_c - b = 0 \tag{15.18}$$

which, because of the double-sign possibility, has four solutions. Using the positive sign, we have, by the quadratic formula,

$$\omega_{c_1} = \frac{-a + \sqrt{a^2 + 4b}}{2} \tag{15.19}$$

We have discarded the negative sign on the radical because this gives a negative cutoff frequency. Using the negative sign in (15.18) and again suppressing the negative root, we have the other cutoff frequency,

$$\omega_{c_2} = \frac{a + \sqrt{a^2 + 4b}}{2} \tag{15.20}$$

Evidently, from (15.19) and (15.20) we have the bandwidth,

$$B = \omega_{c_2} - \omega_{c_1} = a \tag{15.21}$$

Thus in view of (15.17) and (15.21) we may write the network function as

$$\mathbf{H}(s) = \frac{Ks}{s^2 + Bs + \omega_0^2} \tag{15.22}$$

which is the general network function of a second-order bandpass filter having center frequency ω_0 and bandwidth B. The amplitude response is shown, of course, in Fig. 15.3.

Another result worth noting from (15.17), (15.19), and (15.20) is

$$\omega_0^2 = \omega_{c_1}\omega_{c_2} \tag{15.23}$$

which demonstrates that the center frequency ω_0 is the geometric mean $\sqrt{\omega_{c_1}\omega_{c_2}}$ of the cutoff points.

A good measure of *selectivity* or *sharpness of peak* in a resonant circuit is the so-called *quality factor Q*, which is defined as the ratio of the resonant frequency to the bandwidth. That is,

$$Q = \frac{\omega_0}{B} \qquad (15.24)$$

(The letter Q is also our symbol for reactive power, as the reader will recall. However, the two quantities will never be used in the same context so there should be no confusion.) Evidently, since $B = \omega_0/Q$, a low Q corresponds to a relatively large bandwidth, and a high Q (sometimes arbitrarily taken as 5 or more) indicates a small bandwidth, or a more selective circuit.

With this definition of Q we may write (15.22) in the form

$$\mathbf{H}(s) = \frac{Ks}{s^2 + (\omega_0/Q)s + \omega_0^2} \qquad (15.25)$$

so that we see at a glance the center frequency, the quality factor, and the bandwidth.

EXAMPLE 15.3   As an example, the filter described in Ex. 15.2.1 is a bandpass filter with $\omega_0 = 1$ rad/s, $B = 0.2$ rad/s, and $Q = 5$. Another example is the parallel RLC circuit with the network function given in (15.6). In that case, $\omega_0 = 1/\sqrt{LC}$, $B = 1/RC$, and $Q = \omega_0/B$, which has the equivalent values

$$Q = \omega_0 RC$$

$$= R\sqrt{\frac{C}{L}} \qquad (15.26)$$

$$= \frac{R}{\omega_0 L}$$

We should mention that the quantity we have called Q is defined as *selectivity* by some authors, who reserve for Q the definition

$$Q = 2\pi \frac{\text{Total energy stored at resonance}}{\text{Energy dissipated per cycle at resonance}} \qquad (15.27)$$

In the examples we have considered, the two definitions are the same, as the reader is asked to show in Exs. 15.4.1 and 15.4.2. However, in general, there is a slight difference.

Finally, let us consider the effect of Q on resonance. Incorporating (15.19) and (15.20) into one equation and replacing a and b by their values, we have

$$\omega_{c_1,c_2} = \mp \frac{\omega_0}{2Q} + \sqrt{\omega_0^2 + \left(\frac{\omega_0}{2Q}\right)^2}$$

or

$$\omega_{c_1,c_2} = \left(\mp \frac{1}{2Q} + \sqrt{\left(\frac{1}{2Q}\right)^2 + 1} \right)\omega_0 \qquad (15.28)$$

Evidently if Q is high, then we may neglect $(1/2Q)^2$ in comparison with 1, and write

$$\omega_{c_1,c_2} \approx \mp \frac{\omega_0}{2Q} + \omega_0$$

or, approximately,

$$\omega_{c_1} = \omega_0 - \frac{B}{2}$$

$$\omega_{c_2} = \omega_0 + \frac{B}{2}$$

(15.29)

Thus as Q increases, the amplitude response approaches arithmetic symmetry. That is, the cutoff points are half the bandwidth above and below the center frequency.

In the example of Ex. 15.2.1, we have $\omega_0 = 1$ and $Q = 5$, which we consider as high. Thus by (15.29), the approximate cutoff frequencies are $\omega_{c_1} = 0.9$ and $\omega_{c_2} = 1.1$ rad/s. The exact values are

$$\omega_{c_1,c_2} = 0.905, \qquad 1.105$$

EXERCISES

15.4.1 For the *RLC* parallel circuit in resonance, show that the total energy stored is

$$w(t) = \frac{1}{2}R^2 C I_m^2 \cos^2 \omega_0 t + \frac{R^2 I_m^2}{2\omega_0^2 L} \sin^2 \omega_0 t$$

$$= \frac{1}{2}R^2 C I_m^2$$

where the excitation is $i = I_m \cos \omega_0 t$.

15.4.2 Show that for the circuit of Ex. 15.4.1 the energy dissipated per cycle is

$$\int_0^{2\pi/\omega_0} R I_m^2 \cos^2 \omega_0 t \, dt = \frac{\pi R I_m^2}{\omega_0}$$

and thus by definition (15.27) and the result of Ex. 15.4.1 that

$$Q = \omega_0 R C$$

15.4.3 For the *RLC* series circuit with excitation $v_g = V_m \cos \omega t$ V, show that $\omega_0 = 1/\sqrt{LC}$ and $Q = \omega_0 L/R$. The response is the loop current.

15.4.4 In Ex. 14.4.3, replace the 8-Ω resistor by 2 Ω, show that the circuit is a bandpass filter, and find ω_0 and Q.
Answer 4 rad/s, 0.4

15.5

USE OF POLE-ZERO PLOTS

The pole-zero plot of a network function may often be used to readily sketch the frequency responses. To see how this may be done, let us write the network function in the form

$$\mathbf{H}(s) = \frac{K(s - z_1)(s - z_2)\ldots(s - z_m)}{(s - p_1)(s - p_2)\ldots(s - p_n)}$$ (15.30)

where, as before, the z's and p's are the zeros and poles. Each of the factors is a complex number of the form $s - s_1$ and may be represented in the s-plane by a vector, drawn from s_1 to s, as shown in Fig. 15.8. This is true since by vector addition the vector s is clearly the sum of the vectors s_1 and $s - s_1$.

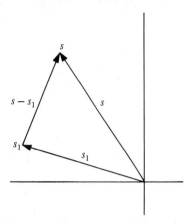

FIGURE 15.8 Vector representation of $s - s_1$

The typical vector $s - s_1$, drawn from s_1 to s, may be written in polar form where its magnitude is its length and its phase is the angle it makes with the positive real axis. If $s = j\omega$, the point s in Fig. 15.8 is on the $j\omega$-axis, and the factors of (15.30) may be written

$$j\omega - z_i = N_i e^{j\alpha_i}, \qquad i = 1, 2, \ldots, m$$

$$j\omega - p_k = M_k e^{j\beta_k}, \qquad k = 1, 2, \ldots, n$$

Therefore, the network function is

$$\mathbf{H}(j\omega) = |\mathbf{H}(j\omega)| e^{j\phi(\omega)}$$

where, for $K > 0$, the amplitude is

$$|\mathbf{H}(j\omega)| = \frac{KN_1N_2 \ldots N_m}{M_1M_2 \ldots M_n}$$ (15.31)

and the phase is

$$\phi(\omega) = (\alpha_1 + \alpha_2 + \ldots + \alpha_m) - (\beta_1 + \beta_2 + \ldots + \beta_n)$$ (15.32)

both of which may be measured directly from the pole-zero plot. If $K < 0$, then $K = |K|\underline{/180°}$, which must be accounted for in the amplitude and phase.

Thus for any point $j\omega$ we simply draw vectors from all the poles and zeros to $j\omega$, measure their lengths and angles, and calculate the amplitude and phase from (15.31) and (15.32). Quite often we need only a few points to sketch the responses, as we shall see.

EXAMPLE 15.4 As an example, suppose we have

$$\mathbf{H}(s) = \frac{4s}{s^2 + 2s + 401}$$

which has a zero at 0 and poles at $-1 \pm j20$. These are shown on the pole-zero plots of Fig. 15.9. By (15.31) and (15.32) we have

FIGURE 15.9 Steps in the construction of the frequency responses. (Figures not drawn to scale.)

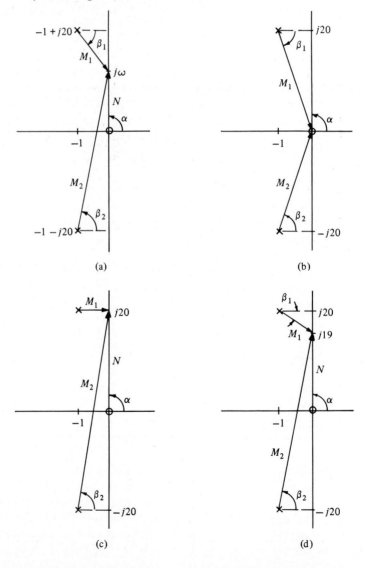

$$|\mathbf{H}(j\omega)| = \frac{4N}{M_1 M_2}$$

$$\phi(\omega) = \alpha - (\beta_1 + \beta_2)$$

whose components are identified in Fig. 15.9(a). First we note that if ω varies from 0^+ to $+\infty$, then $\alpha = 90°$. For $\omega = 0$ we have, from Fig. 15.9(b),

$$|\mathbf{H}| = \frac{4(0)}{\sqrt{401}\sqrt{401}} = 0$$

$$\phi = 90° - (-\tan^{-1} 20 + \tan^{-1} 20) = 90°$$

For $\omega = 20$ [Fig. 15.9(c)] we have

$$|\mathbf{H}| = \frac{4(20)}{(1)\sqrt{1601}} \approx 2$$

$$\phi = 90° - (0 + \tan^{-1} 40) \approx 0$$

This point is in the region where the amplitude changes the fastest, since $M_1 = 1$ is a minimum and is changing percentage-wise much faster than N or M_2. The amplitude therefore will reach its peak near this point. Actually we know from our previous work that $\omega_0 = \sqrt{401}$ yields the peak amplitude of 2.

If $\omega = \omega_h$, a very high value, say, such as 10^6, then all three vectors will be essentially vertical, so that

$$N \approx M_1 \approx M_2 \approx \omega_h$$

$$\alpha \approx \beta_1 \approx \beta_2 \approx 90°$$

Therefore we have $|\mathbf{H}| \approx 4/\omega_h$, a very small value, and $\phi \approx -90°$.

Sketching the functions, we have the forms of Fig. 15.2, as expected, with $\omega_0 \approx 20$ and $|\mathbf{H}|_{max} \approx 2$.

We may get a rough idea of the cutoff points from Fig. 15.9(d). In this figure $M_1 = \sqrt{2}$, which is $\sqrt{2}$ times its value at the approximate peak, represented by Fig. 15.9(c). Since M_2 and N have changed by much smaller percentages, the amplitude is then approximately $1/\sqrt{2}$ times its peak, so that $\omega_{c_1} \approx 19$. By a similar argument at $\omega = 21$, we have $\omega_{c_2} \approx 21$. The exact values, by (15.19) and (15.20), are $\omega_{c_1,c_2} = 19.05, 21.05$. In this example, by comparing the network function with (15.25), we see that $Q = 10.01$. Thus we have a high Q so that the poles are very near the $j\omega$-axis. Thus ω_d is very near ω_0, and the approximations we have made are very near the exact values.

EXERCISES

15.5.1 Use the method of this section to sketch the amplitude and phase responses for

$$\mathbf{H}(s) = \frac{16s}{s^2 + 4s + 2504}$$

Answer $\omega_0 \approx 50$, $Q \approx 12.5$, $B = 4$, $\omega_{c_1,c_2} \approx 48, 52$

15.5.2 Find the exact values for the answers given in Ex. 15.5.1.
Answer 50.04, 12.51, 4, 48.08, 52.08

15.6

SCALING THE NETWORK FUNCTION

The reader may have noticed that in many of our examples we have used network elements such as 1 Ω, 1 F, 2 H, etc., which are extremely nice numbers to have when we are analyzing a network. For example, in Fig. 15.4 we had elements of 2 Ω, $\frac{1}{4}$ F, and $\frac{1}{2}$ F, and the network was a low-pass filter with a cutoff frequency of $\sqrt{2}$ rad/s, or 0.23 Hz. However, such element values as these are not very practical, to say the least, and there is very little demand for filters that pass frequencies between 0 and 0.23 Hz. (As an illustration, a 1-F capacitor, constructed of two parallel plates 1 cm apart with air as the dielectric, would require a face area of 1.13×10^9 m^2.)

It would be an ideal situation if the circuits we analyze or design contain simple element values such as 1 F, etc., while the circuits we build have practical element values such as 0.047 μF, and useful characteristics such as cutoff points of 100 Hz. As we shall see in this section, *network scaling* allows us to have it both ways.

We shall consider two types of network scaling, namely *impedance scaling* and *frequency scaling*. To illustrate the former, let us suppose that the network function is an impedance given by

$$\mathbf{Z}'(s) = sL' + R' + \frac{1}{sC'}$$

The network has been impedance-scaled by an *impedance scale factor* k_i, if the impedance $\mathbf{Z}(s)$ of the scaled network is $\mathbf{Z}(s) = k_i \mathbf{Z}'(s)$. In other words, we must have

$$\mathbf{Z}(s) = k_i\left(sL' + R' + \frac{1}{sC'}\right)$$

$$= s(k_iL') + k_iR' + \frac{1}{s(C'/k_i)} \tag{15.33}$$

If the impedance of the scaled network is

$$\mathbf{Z}(s) = sL + R + \frac{1}{sC}$$

then we see by comparison with (15.33) that

$$L = k_iL'$$
$$R = k_iR' \tag{15.34}$$
$$C = \frac{C'}{k_i}$$

In summary, to impedance-scale a network by the factor k_i, we multiply the L's and R's by k_i and divide the C's by k_i. We have illustrated this for a special case, but it may be shown to hold in general. Also, if there are dependent sources, the scaling is accomplished by multiplying gain constants having units of ohms by k_i dividing those with units of mhos by k_i, and leaving unchanged those that are dimensionless. The scaling multiplies by k_i, divides by k_i, or leaves unchanged network functions that have units of ohms or mhos or are dimensionless.

EXAMPLE 15.5 To illustrate, let us impedance-scale the network of Fig. 15.4 by $k_i = 5$. The 2-Ω resistors become $2 \times 5 = 10 \ \Omega$, the $\frac{1}{2}$-F capacitor becomes $\frac{1}{2} \div 5 = 0.1$ F, and the $\frac{1}{4}$-F capacitor becomes $\frac{1}{4} \div 5 = 0.05$ F. The op amp, an infinite gain device, remains an infinite gain device, and thus is unchanged. The network function $\mathbf{V}_2(s)/\mathbf{V}_1(s)$, being dimensionless, is also unchanged.

To frequency-scale a network function by a *frequency scale factor* k_f, we simply replace s by s/k_f. That is, if the unscaled network has the network function $\mathbf{H}'(S)$, then the scaled network function $\mathbf{H}(s)$ is obtained by letting $S = s/k_f$, resulting in

$$\mathbf{H}(s) = \mathbf{H}'\left(\frac{s}{k_f}\right) \tag{15.35}$$

Thus if the unscaled network had a property, such as center frequency, when $S = j1$, then the scaled network has this property at $s = jk_f$. This is clear from (15.35), which gives

$$\mathbf{H}(jk_f) = \mathbf{H}'(j1)$$

Another way to consider frequency scaling is to note that $s = k_f S$, so that if $s = j\omega$ corresponds to $S = j\Omega$, then $\omega = k_f \Omega$. Thus the values on the frequency axis have been multiplied by the scale factor k_f, without affecting values on the vertical axis of a frequency response.

The scaling of the network to effect the transformation of (15.35) is quite simple. If $\mathbf{Z}'(S)$, given by

$$\mathbf{Z}'(S) = SL' + R' + \frac{1}{SC'}$$

is any impedance in the unscaled network, then the corresponding impedance in the scaled network is

$$\mathbf{Z}(s) = sL + R + \frac{1}{sC}$$

$$= \mathbf{Z}'\left(\frac{s}{k_f}\right)$$

$$= s\left(\frac{L'}{k_f}\right) + R' + \frac{1}{s(C'/k_f)}$$

Comparing these results, we have

$$L = \frac{L'}{k_f}$$

$$R = R' \qquad\qquad (15.36)$$

$$C = \frac{C'}{k_f}$$

Therefore, to frequency-scale a network by a factor k_f, we divide the L's and C's by k_f and leave the R's unchanged. If there are dependent sources with constant gains, these also are left unchanged.

EXAMPLE 15.6   To illustrate, the circuit of Fig. 15.4 is a low-pass filter with a cutoff frequency of $\omega_c = \sqrt{2}$ rad/s, as we have previously seen. Suppose we want to frequency-scale the network so that the cutoff frequency is 2 rad/s. Then the scale factor k_f is given by

$$\sqrt{2}\, k_f = 2$$

or $k_f = \sqrt{2}$. Dividing the C's in Fig. 15.4 by k_f we have, in the scaled network, capacitances of $1/2\sqrt{2}$ F and $1/4\sqrt{2}$ F. The rest of the circuit remains unchanged, there being no inductors to scale.

Of course, we may perform both impedance and frequency scaling on a network. To obtain a network with a practical property such as center or cutoff frequency, we may first apply frequency scaling with the proper factor k_f. Then to make the resulting element values more practical we may impedance-scale by a factor k_i.

EXAMPLE 15.7   For example, if the parallel RLC network of Fig. 15.1 contains a 1-Ω resistor, a 2-H inductor, and a $\frac{1}{2}$-F capacitor, then for the network function of (15.5) we have an amplitude response as in Fig. 15.2(a), with a resonant frequency of 1 rad/s and a peak amplitude of 1 Ω. Suppose we amplitude- and frequency-scale the network to obtain a resonant frequency of 10^6 rad/s using a capacitor of 1 nF. Then we have $k_f = 10^6$, and the new capacitance is

$$C = 10^{-9} = \frac{1/2}{k_i k_f} = \frac{1}{2k_i \times 10^6}$$

Therefore, $k_i = 500$, and the new inductance and resistance are

$$L = \frac{2k_i}{k_f} = 10^{-3}\ \text{H} = 1\ \text{mH}$$

$$R = 1k_i = 500\ \Omega$$

The scaled network is shown in Fig. 15.10(a), and its amplitude response, a scaled version of Fig. 15.2(a), is shown in Fig. 15.10(b).

488   Chapter 15   Frequency Response

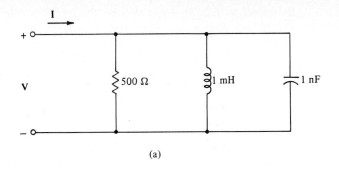

(a)

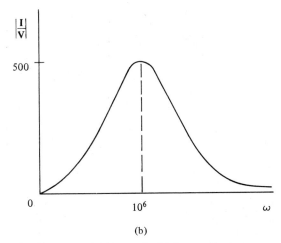

(b)

FIGURE 15.10 (a) Network; (b) its amplitude response

EXERCISE

15.6.1 Frequency- and impedance-scale the circuit of Fig. 15.4 to obtain $\omega_c = 2000\pi$ rad/s ($f_c = 1000$ Hz), using capacitors of 0.01 and 0.005 μF.
Answer $k_f = (1000\sqrt{2})\pi$, $k_i = 10^5/2\sqrt{2}\pi$, R's $= 22.5$ kΩ each

---

15.7

THE DECIBEL

If the output of a network is used to provide some quantity that is to be sensed by a human being, the function is sensed in a noncontinuous manner. For example, if the network provides sound, as in the case of the telephone, the listener cannot detect continuous changes in intensity. Moreover, if the sound intensity is 1 (on some arbitrary scale) and it must be increased to 1.1 before the listener detects any change, then the same listener can detect no change, if the original level is 2, until it is increased to 2.2. In other words, the ear is not a linear device but more like a *logarithmic* device, since the differences,

$$\log 1.1 - \log 1 = \log 1.1$$

and

$$\log 2.2 - \log 2 = \log\left(\frac{2.2}{2}\right)$$

in the two cases are the same.

For this reason, among others, the amplitude response,

$$|\mathbf{H}(j\omega)| = \left|\frac{\mathbf{V}_2(j\omega)}{\mathbf{V}_1(j\omega)}\right|$$

is more commonly expressed in *decibels,* abbreviated dB and defined by

$$\text{number of dB} = 20\log_{10}|\mathbf{H}(j\omega)| \qquad (15.37)$$

Historically, the logarithmic unit, now known as the *bel,* was defined originally by Alexander Graham Bell (1847–1922), who, of course, invented the telephone, as the power unit,

$$\text{number of bels} = \log_{10}\frac{P_2}{P_1}$$

However, this proved to be a large unit, so that the unit decibel ($\frac{1}{10}$ bel) became common. That is,

$$\text{number of dB} = 10\log_{10}\frac{P_2}{P_1}$$

If the two average powers P_1 and P_2 are referred to equal impedances, the last expression may be given in terms of the corresponding voltages by

$$\text{number of dB} = 10\log_{10}\left|\frac{\mathbf{V}_2(j\omega)}{\mathbf{V}_1(j\omega)}\right|^2$$

which, of course, is equivalent to (15.37). In any case, (15.37) is taken as the standard definition.

In practice, frequencies are not ideally blocked or *attenuated* in the filtering process, as may be seen in the low-pass response of Fig. 15.5(a). A zero amplitude would correspond to absolute, or infinite, attenuation, and any approach to such an ideal situation would be difficult to appreciate on a linear sketch. For example, if $|\mathbf{H}(j\omega)| = 0.001$ in Fig. 15.5(a) ($\frac{1}{10}$ of 1% of its peak value of 1), this would correspond to -60 dB (or 60 dB below its peak value of 0 dB). The latter figure means much more to a telephone engineer than the linear figure.

Even more often in practice it is of interest to consider the *attenuation* or *loss,* defined by

$$\alpha(\omega) = -20 \log_{10} |\mathbf{H}(j\omega)|$$

$$= -20 \log_{10} \left| \frac{\mathbf{V}_2(j\omega)}{\mathbf{V}_1(j\omega)} \right| \qquad (15.38)$$

$$= 20 \log_{10} \left| \frac{\mathbf{V}_1(j\omega)}{\mathbf{V}_2(j\omega)} \right| \text{ dB}$$

In this case, a frequency ω_1 passes if $\alpha(\omega_1)$ is relatively small and is attenuated if $\alpha(\omega_1)$ is relatively large. The decibel units enable us to tell with some standard precision the degree to which the frequency is attenuated.

EXAMPLE 15.8   As an example, suppose

$$\mathbf{H}(s) = \frac{1}{s^2 + \sqrt{2}s + 1}$$

As the reader may verify, the amplitude response is

$$|\mathbf{H}(j\omega)| = \frac{1}{\sqrt{1 + \omega^4}}$$

which is that of a low-pass filter with $\omega_c = 1$ rad/s. The linear sketch looks somewhat like that of Fig. 15.5(a). The attenuation is given in decibels by

$$\alpha(\omega) = 20 \log_{10} \sqrt{1 + \omega^4} \qquad (15.39)$$

$$= 10 \log_{10}(1 + \omega^4)$$

and is shown in Fig. 15.11 for $0 \le \omega \le 5$.

FIGURE 15.11 Attenuation of a low-pass filter

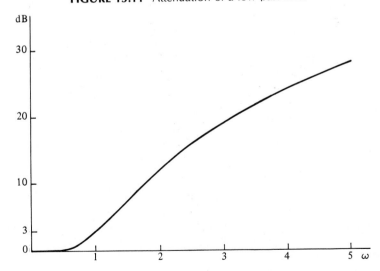

EXERCISES

15.7.1 Let the amplitude response $|\mathbf{H}(j\omega)|$ be such that $|\mathbf{H}(j\omega)|_{\max} = |\mathbf{H}(j0)| = K$, so that at cutoff $|\mathbf{H}(j\omega_c)| = K/\sqrt{2}$. Find the loss, given by (15.38), at $\omega = 0$ and at $\omega = \omega_c$. Note that ω_c corresponds to the "3-dB point," meaning that at $\omega = \omega_c$ the loss is approximately 3 dB more than at the point of minimum loss, $\omega = 0$ in this case.

Answer $\alpha(0) = -20 \log K$, $\alpha(\omega_c) = -20 \log K + 3$

15.7.2 Given the bandpass filter function

$$\mathbf{H}(s) = \frac{0.2s}{s^2 + 0.2s + 1}$$

find the loss in decibels at $\omega = 0.0001$, 0.5, 0.905, 1.0, 1.105, 10, and 100 rad/s.

Answer 94, 18, 3, 0, 3, 34, 54

15.8

FREQUENCY RESPONSE WITH SPICE

The frequency response of a network is easily obtained using SPICE by simply specifying a frequency range and the number of points desired within the range in the .AC solution control statement. The generation of a circuit file for a given frequency response is almost identical to procedures described in previous chapters for ac analysis. The .PRINT or .PLOT output control statements can be used for output to the console or the printer.

EXAMPLE 15.9 As a first example, let us plot the frequency response for a linear frequency sweep in the network of Fig. 15.12 from 10 to 500 kHz with v_g having an amplitude of 1 V.

FIGURE 15.12 *RLC* circuit with CCCS

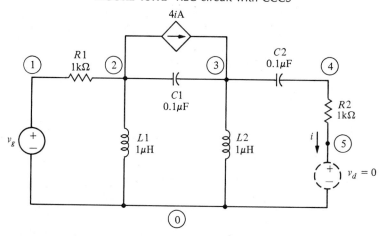

The source $v_d = 0$ (shown dashed) connected from node 5 to the reference node has been added for specifying the current i of the CCCS. A circuit file is

```
FREQUENCY RESPONSE FOR FIGURE 15.12
* DATA STATEMENTS
VG 0 1 AC 1 0
R1 1 2 1K
L1 2 0 1UH
C1 2 3 0.1UF
F  2 3 VD 4
L2 3 0 1UH
C2 3 4 0.1UF
R2 4 5 1K
VD 5 0 AC 0 0
* SOLUTION CONTROL STATEMENT
.AC LIN 25 10K 500K
* OUTPUT CONTROL STATEMENT
.PLOT AC IM(R2) IP(R2)
.END
```

The resulting plot for this example is shown in Fig. 15.13.

FIGURE 15.13 Frequency response for Fig. 15.12

```
LEGEND:

*: IM(R2)
+: IP(R2)

  FREQ         IM(R2)


(*)----------   1.0000E-11   1.0000E-09   1.0000E-07   1.0000E-05   1.0000E-03
(+)----------  -2.0000E+02  -1.0000E+02   0.0000E+00   1.0000E+02   2.0000E+02

 1.000E+04   2.452E-11 .  *   .             .             +         .
 3.042E+04   7.022E-10 .        *            .             +.        .
 5.083E+04   3.324E-09 .             *       .             +.        .
 7.125E+04   9.344E-09 .                *    .             +.        .
 9.167E+04   2.046E-08 .                   * .             +.        .
 1.121E+05   3.877E-08 .                    .*             +.        .
 1.325E+05   6.698E-08 .                    .*.            +.        .
 1.529E+05   1.088E-07 .                    . *            +.        .
 1.733E+05   1.693E-07 .                    .  *           +.        .
 1.938E+05   2.564E-07 .                    .   *          +.        .
 2.142E+05   3.820E-07 .                    .    *         +.        .
 2.346E+05   5.662E-07 .                    .      *       +.        .
 2.550E+05   8.452E-07 .                    .       *      +.        .
 2.754E+05   1.292E-06 .                    .         *    +.        .
 2.958E+05   2.078E-06 .                    .           *  +.        .
 3.163E+05   3.728E-06 .                    .             *+.        .
 3.367E+05   8.989E-06 .                    .             +*         .
 3.571E+05   1.186E-04 .           +        .             .        * .
 3.775E+05   1.065E-05 .            .+      .             *.        .
 3.979E+05   6.245E-06 .            .+      .             *.        .
 4.183E+05   4.756E-06 .            .+      .            *.         .
 4.388E+05   4.029E-06 .            .+      .            *.         .
 4.592E+05   3.613E-06 .            .+      .            *.         .
 4.796E+05   3.353E-06 .            .+      .           *.          .
 5.000E+05   3.184E-06 .            .+      .           *.          .
```

EXAMPLE 15.10 As a final example, consider the network of Fig. 15.4 having node assignments of 1 at v_1, 2 at the common node of the 2-ohm resistors, 3 at the op amp noninverting input, and 4 at its output. Let us determine the frequency response for a linear sweep from 0.001 to 1 Hz for v having a 1-V amplitude. A circuit file using the subcircuit file OPAMP.CKT of Chapter 4 is

```
FREQUENCY RESPONSE FOR LOW-PASS FILTER OF FIG. 15.4
* DATA STATEMENTS
V 1   1   0   AC   1   0
R 1   1   2   2
R 2   2   3   2
C 1   2   4   0.5
C 2   3   0   0.2 5
XO P AMP   4   3   4   OP AMP
* DEFINE SUBCIRCUIT FILE
.LIB OPAMP.CKT
* SOLUTION CONTROL STATEMENT
.AC LIN 25 0.001 1
* OUTPUT CONTROL STATEMENT
.PLOT AC VM(4) VP(4)
.END
```

The plot for this response is similar to that of Fig. 15.5.

EXERCISES

15.8.1 Using SPICE, find the frequency response from 10 Hz to 400 kHz for **V** in Fig. 15.10(a) if the capacitor and inductor values are 10 μF and 0.1 μH, respectively. Use a source current of 1 mA.

15.8.2 Using SPICE, find the frequency response for the network of Ex. 15.2.4 in the interval from 0.001 to 1 Hz.

PROBLEMS

15.1 For the circuit shown, $R = 1\ \Omega$, $L = \frac{1}{2}$ H, and $C = 0.02$ F. If the input and output are V_1 and V_2, respectively, find the network function and sketch the amplitude and phase responses. Show that the peak amplitude and zero phase occur at $\omega = 10$ rad/s.

15.2 Find $\mathbf{H}(s) = \mathbf{V}_2(s)/\mathbf{V}_1(s)$ and sketch the amplitude and phase responses. Show that the peak amplitude and zero phase occur at $\omega = 2$ rad/s.

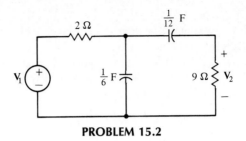

PROBLEM 15.2

15.3 For the circuit shown, $R_1 = R_2 = 0.5$ Ω and $R_3 = 0.01$ Ω. If the input and output are $\mathbf{V}_1$ and $\mathbf{V}_2$, respectively, find the network function and sketch the amplitude and phase responses. Show that the peak amplitude and zero phase occur at $\omega = 10$ rad/s. Compare these results with those of Prob. 15.1.

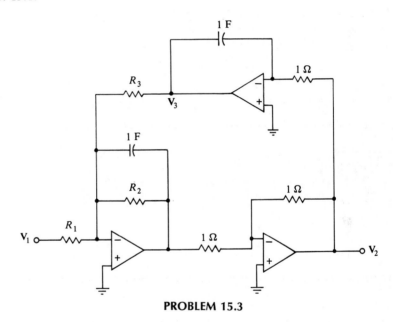

PROBLEM 15.3

15.4 For the circuit shown, find the network function, $\mathbf{H}(s) = \mathbf{V}_2(s)/\mathbf{V}_1(s)$, and sketch the amplitude and phase responses. Show that the peak amplitude and zero phase occur at $\omega = 0$.

PROBLEM 15.4

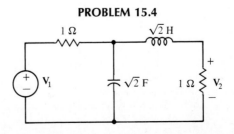

15.5 Show that in the general case the network function in the figure for Prob. 15.1 is given by

$$\mathbf{H}(s) = \frac{\mathbf{V}_2(s)}{\mathbf{V}_1(s)} = \frac{(R/L)s}{s^2 + (R/L)s + (1/LC)}$$

Thus by comparison with (15.22) and (15.25), show that the circuit is a bandpass filter with resonant, or center, frequency $\omega_0 = 1/\sqrt{LC}$, bandwidth $B = R/L$, and quality factor $Q = \omega_0/B = (1/R)\sqrt{L/C}$. Show also that the *gain G* of the filter, defined as its peak amplitude value, is given in this case by $G = 1$.

15.6 Show that in Prob. 15.5 if $L = Q$, $C = 1/Q$, and $R = 1$, then

$$\mathbf{H}(s) = \frac{(1/Q)s}{s^2 + (1/Q)s + 1}$$

which is a bandpass network function with quality factor Q, center frequency $\omega_0 = 1$ rad/s, bandwidth $B = 1/Q$ rad/s, and gain $G = 1$.

15.7 Show that in the general case of the figure for Prob. 15.3

$$\mathbf{H}(s) = \frac{\mathbf{V}_2(s)}{\mathbf{V}_1(s)} = \frac{(1/R_1)s}{s^2 + (1/R_2)s + (1/R_3)}$$

Thus show that the circuit is a bandpass filter with center frequency $1/\sqrt{R_3}$ rad/s, bandwidth $1/R_2$ rad/s, gain R_2/R_1, and quality factor $R_2/\sqrt{R_3}$. Let $R_1 = Q/G$, $R_2 = Q$, and $R_3 = 1$ to obtain the network function

$$\mathbf{H}(s) = \frac{\mathbf{V}_2(s)}{\mathbf{V}_1(s)} = \frac{(G/Q)s}{s^2 + (1/Q)s + 1}$$

15.8 Using the results of Prob. 15.7, obtain a bandpass filter having the form of the figure for Prob. 15.3 with $\omega_0 = 1$ rad/s, $Q = 10$, and $G = 2$. Find the approximate cutoff points and the bandwidth.

15.9 Show by finding $\mathbf{V}_2/\mathbf{V}_1$ that the circuit is a bandpass filter, and find the gain, the bandwidth, and the center frequency.

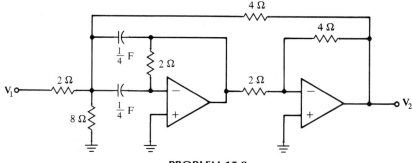

PROBLEM 15.9

15.10 Show that the network function $\mathbf{V}_3/\mathbf{V}_1$ in the figure for Prob. 15.1 is given by

$$\frac{\mathbf{V}_3}{\mathbf{V}_1} = \frac{1/LC}{s^2 + (R/L)s + (1/LC)}$$

Let $R = 1\ \Omega$, $L = 1/\sqrt{2}$ H, and $C = \sqrt{2}$ F, and show that the result is a low-pass filter with $\omega_c = 1$ rad/s by finding the amplitude response.

15.11 Show that the given circuit is a low-pass filter with $\omega_c = 1$ rad/s by finding $H(s) = V_2(s)/V_1(s)$ and the amplitude response.

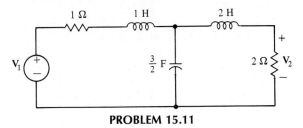

PROBLEM 15.11

15.12 One type of low-pass filter is the nth-order *Butterworth* filter, whose amplitude response is

$$|H(j\omega)| = \frac{K}{\sqrt{1 + \omega^{2n}}}, \qquad n = 1, 2, 3, \ldots$$

where K is a constant. Ideally, a filter would pass all frequencies in its passband equally well ($|H| > 0$ would be constant) and perfectly block all other frequencies ($|H|$ would be zero). As the accompanying figure shows, for $n = 2$, 3, and 8, the Butterworth filter improves (approaches ideal) as the order n increases. Show that $\omega_c = 1$ rad/s for any n and that the filters of Probs. 15.4 and 15.11 are Butterworth filters of second and third orders, respectively. Finally, sketch a fourth-order Butterworth response and compare it with the second, third, and eighth orders.

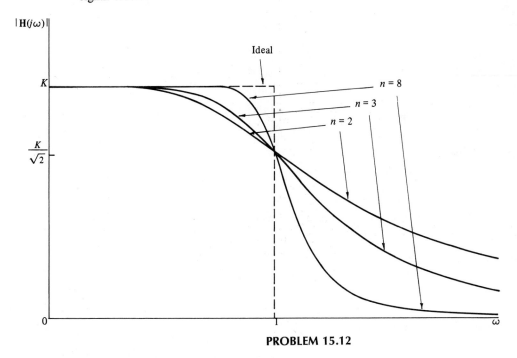

PROBLEM 15.12

15.13 Show that for the figure for Prob. 15.3

$$H(s) = \frac{V_3(s)}{V_1(s)} = -\frac{1/R_1}{s^2 + (1/R_2)s + (1/R_3)}$$

Choose values of the resistances so that the circuit is a low-pass Butterworth filter with $\omega_c = 1$ rad/s and $\mathbf{H}(0) = 2$. (*Suggestion:* By Prob. 15.10, the denominator of $\mathbf{H}(s)$ is required to be $s^2 + \sqrt{2}s + 1$.)

15.14 If the *gain* of a low-pass filter with the network function $\mathbf{H}(s)$ is defined to be $K = |\mathbf{H}(0)|$, compare the gains in the general cases of Probs. 15.10 and 15.13, which are, respectively, passive and active circuits that perform low-pass filtering. Note that gains higher than 1 are possible with active elements present. This feature and the absence of inductors, which are undesirable at lower frequencies, are advantages of active filters over passive filters.

15.15 Show that in the figure for Prob. 15.1

$$\frac{\mathbf{V}_4}{\mathbf{V}_1} = \frac{s^2}{s^2 + (R/L)s + (1/LC)}$$

Let $R = 1\ \Omega, L = 1/\sqrt{2}$ H, and $C = \sqrt{2}$ F, and show that the result is a high-pass filter with cutoff $\omega_c = 1$ rad/s.

15.16 Apply impedance and frequency scaling to the circuit of Prob. 15.1 to obtain a bandpass filter with a center frequency $f_0 = 10^5$ Hz, with a quality factor $Q = 5$, and using a capacitor of 1 nF. (*Suggestion:* See Prob. 15.6.)

15.17 Obtain an active bandpass filter using the configuration of Prob. 15.3 with $f_0 = 1000$ Hz, $Q = 10$, and $K = 2$, using capacitors of 0.01 μF. (*Suggestion:* See Prob. 15.7.)

15.18 Determine R_1 and R_2 so that the circuit is a first-order low-pass filter ($\mathbf{H} = \mathbf{V}_2/\mathbf{V}_1$) with $\omega_c = 1$ rad/s and a gain of 3. Scale the result to obtain $\omega_c = 10^5$ rad/s using a 1-nF capacitor.

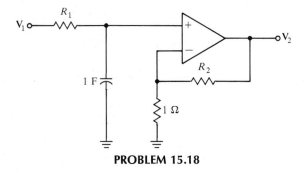

PROBLEM 15.18

15.19 Show by finding $\mathbf{H} = \mathbf{V}_2/\mathbf{V}_1$ that the given circuit is a bandpass filter, and find the center frequency and the bandwidth. Scale the circuit so that the center frequency is 20,000 rad/s using 0.01-μF capacitors.

PROBLEM 15.19

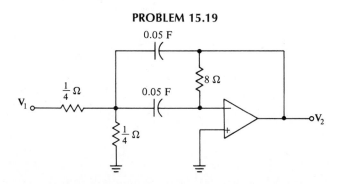

15.20 Show that the given circuit is a third-order low-pass Butterworth filter with $\omega_c = 1$ rad/s and a gain of 1. Scale the circuit so that the capacitances are 0.01 μF each and $\omega_c = 1000$ Hz.

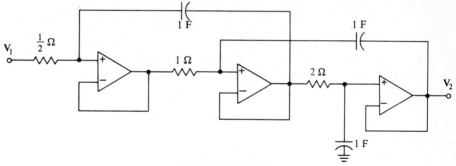

PROBLEM 15.20

15.21 Show that the network function of the given circuit is

$$\mathbf{H}(s) = \frac{\mathbf{V}_2}{\mathbf{V}_1} = \frac{K(s^2 + 1)}{s^2 + (1/Q)s + 1}$$

Thus the circuit is a band-reject filter with center frequency (rejected) $\omega_0 = 1$ rad/s. Also, as in the bandpass case, Q is the quality factor, and $B = \omega_0/Q = 1/Q$ is the bandwidth. Note that the *gain* is $\mathbf{H}(0) = K$, where $0 < K < 1$.

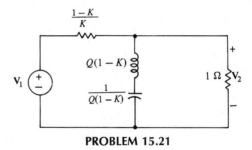

PROBLEM 15.21

15.22 Scale the network in the figure for Prob. 15.21 so that $\omega_0 = 10^5$ rad/s, $Q = 5$, $K = 0.5$, and the capacitance is 1 nF.

15.23 Show that the given circuit is a band-reject filter with a gain of 1, $Q = 1$, and a center frequency $\omega_0 = 1$ rad/s. Scale the network to obtain a center frequency $f_0 = 60$ Hz, using capacitances of 1 and 2 nF.

PROBLEM 15.23

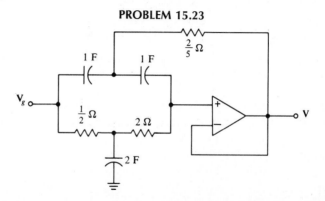

15.24 Show by finding $H = V_2/V_1$ and the amplitude and phase responses that the circuit is a second-order all-pass filter. (*Suggestion:* See Ex. 15.2.4.)

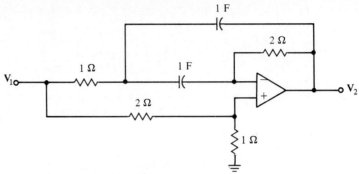

PROBLEM 15.24

15.25 The given circuit is a fourth-order bandpass filter (the denominator of the network function is fourth degree), with $\omega_0 = \sqrt{2}$ rad/s and $Q = 5$. Verify this by finding

$$H(s) = \frac{V_2}{V_1} = \frac{\frac{2}{25}s^2}{s^4 + \frac{2}{5}s^3 + \frac{102}{25}s^2 + \frac{4}{5}s + 4}$$

and so forth. [*Suggestion:* Find the amplitude response and verify that $\omega_0 = \sqrt{2}$ yields the peak point and that $\omega_{c1}, \omega_{c2} \approx \omega_0 \mp (\omega_0/10)$ satisfy the appropriate equation. Since $|H| \geq 0$, its maximum occurs when the maximum of $|H|^2$ occurs. Also $|H|$ and $|H|^2$ are functions of ω^2; thus the maximum occurs when

$$\frac{d}{d\omega^2}|H(j\omega)|^2 = 0$$

Since $(d/d\omega)|H|^2 = (d/d\omega^2)|H|^2 (d\omega^2/d\omega) = 2\omega(d/d\omega^2)|H|^2 = 0$, we must check $\omega = 0$ separately.]

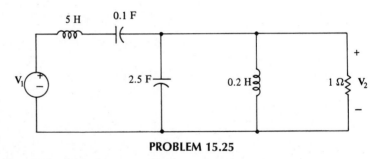

PROBLEM 15.25

COMPUTER APPLICATION PROBLEMS

15.26 Using SPICE, plot the frequency response of V_2 in Prob. 15.11 for $0.001 < f < 0.5$ Hz.

15.27 Repeat Prob. 15.19 to find the center frequency and bandwidth directly from a plot using SPICE for the network shown.

15.28 Using SPICE, plot the frequency response of the network of Prob. 15.25 for $0.001 < f < 0.5$ Hz.

500   Chapter 15   Frequency Response

15.29 Plot the frequency response of the network shown and determine the filter type (e.g., bandpass, etc.) and the characteristics (passband, cutoff frequency, etc.) using SPICE.

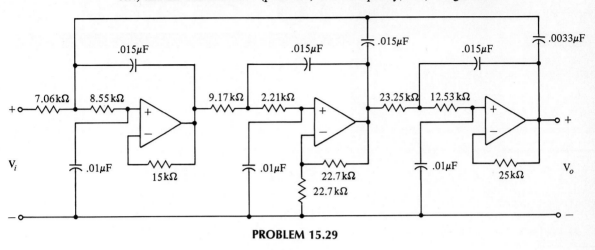

PROBLEM 15.29

16

George Westinghouse
1846–1914

Transformers

Westinghouse was one of the world's true noblemen . . . to whom humanity owes an immense debt of gratitude.
Nikola Tesla

In the battle of the currents of the 1880s, ac won over dc because of the fabulous inventions of Nikola Tesla, the availability of transformers to step up and step down the ac voltages, and the genius of George Westinghouse. Westinghouse had already made his fortune in 1869 with the invention of the air brake for railroad trains. He was shrewd enough to use his wealth to hire Tesla and to buy the patent from Lucien Gaulard and John D. Gibbs for their newly developed practical transformer.

Westinghouse was born in Central Bridge, New York, the son of a prosperous machine factory owner. He served in the Union army and navy during the Civil War and then attended Union College before striking out on his own. By the time he was 40 years old, he had formed the Westinghouse Air Brake Company, developed a system of pipes to conduct natural gas safely into homes, and invented the gas meter. He organized the Westinghouse Electric Company in 1886, which he used as a base to advocate successfully the ac system. Westinghouse was one of America's greatest inventors and one of the true giants of United States industry. ■

In our study of inductance in Chapter 7, we found that a changing current produces a changing magnetic flux which induces a voltage in a coil. In this chapter, we shall consider the effect of a changing magnetic flux that is common to two or more distinct coils. Neighboring coils which share a common magnetic flux are said to be *mutually coupled*. In mutually coupled circuits, a changing current in one coil winding produces an induced voltage in the remaining mutually coupled windings. The induced voltage is characterized by a mutual inductance which exists between the neighboring coils.

A system of mutually coupled coils which are wound on a composite form, or *core,* is commonly called a *transformer*. Transformers are available in a wide variety of sizes and shapes that are designed for numerous applications. Devices as small as an aspirin tablet, for instance, are common in radios, television sets, and stereos for connecting various amplifier stages of the system. On the other hand, transformers are designed for 60-Hz power applications which range in size from that of a ping-pong ball to those which are larger than an automobile.

We shall restrict ourselves to *linear* transformers (those whose constituent coils are linear) and begin by considering the properties of *self-* and *mutual* inductances. Energy-storage and impedance properties are then analyzed, and an important special case of the linear transformer, the ideal transformer, is introduced. We shall conclude the chapter with a discussion of equivalent circuits that are useful for representing linear transformers.

The circuit analyses presented in the chapter will include both time-domain and frequency-domain cases. Particular emphasis will be given to the frequency-domain case because the most important transformer applications occur in the ac steady state.

16.1

MUTUAL INDUCTANCE

In Sec. 7.4 we found that the inductance L of a linear inductor, as shown in Fig. 16.1, is related to the flux linkage λ by the expression

$$\lambda = N\phi = Li$$

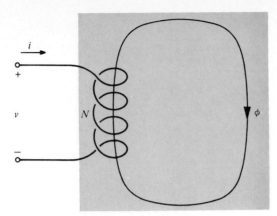

FIGURE 16.1 Simple inductor

From Faraday's law, the terminal voltage is given by

$$v = \frac{d\lambda}{dt} = L\frac{di}{dt}$$

It is evident from Faraday's law that a voltage is induced in a coil which contains a time-varying magnetic flux, regardless of the source of the flux. Let us therefore consider a second coil, having N_2 turns, positioned in the neighborhood of the first coil, having $N = N_1$ turns, as shown in Fig. 16.2(a). In this case, we have formed a simple transformer having two pairs of terminals in which coil 1 is referred to as the *primary winding* and coil 2 as the *secondary winding*.

To introduce several important inductive quantities, let us begin by examining the open-circuited secondary case of Fig. 16.2(a). The current i_1 produces a magnetic flux ϕ_{11} given by

$$\phi_{11} = \phi_{L1} + \phi_{21}$$

where ϕ_{L1} is the flux of i_1 that links coil 1 and not coil 2, called the *leakage flux*, and ϕ_{21} is the flux of i_1 that links coil 2 and coil 1, called the *mutual flux*.

We shall assume, as in the case of the linear inductor, that the flux in each coil links all turns of the coil. Since the secondary is an open circuit, no curent flows in coil 2, and the flux linkage of this coil is

$$\lambda_2 = N_2\phi_{21}$$

Therefore, the voltage v_2 is given by

$$v_2 = \frac{d\lambda_2}{dt} = N_2\frac{d\phi_{21}}{dt}$$

(We shall justify later that the polarity is as shown.)

In a linear transformer, the flux ϕ_{21} is proportional to i_1. Thus we may write

$$N_2\phi_{21} = M_{21}i_1 \tag{16.1}$$

where M_{21} is a *mutual inductance* in henrys (H). In terms of this mutual inductance, the open-circuit secondary voltage becomes

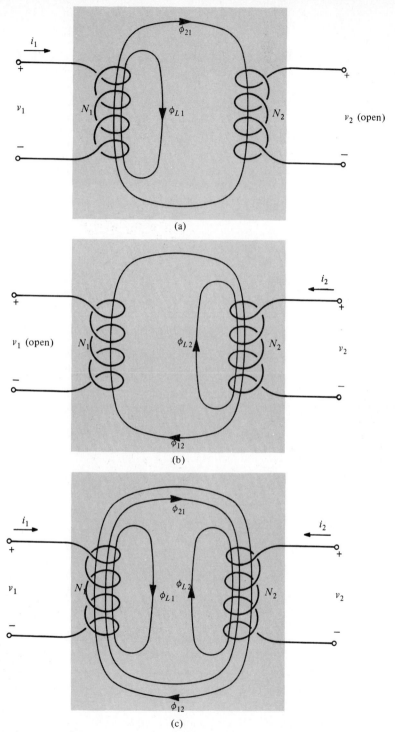

FIGURE 16.2 Mutually coupled coils in which (a) $i_2 = 0$; (b) $i_1 = 0$; (c) i_1 and i_2 are nonzero

$$v_2 = M_{21} \frac{di_1}{dt}$$

Let us now find the primary voltage v_1. We know that

$$v_1 = \frac{d\lambda_1}{dt}$$

where λ_1, the flux linkage of coil 1, is given by

$$\lambda_1 = N_1 \phi_{11}$$

As in the case of a single linear inductor, the inductance L_1 of the primary winding, sometimes called the *self-inductance* of the primary to distinguish it from the mutual inductance, is defined by

$$L_1 i_1 = N_1 \phi_{11} \qquad (16.2)$$

Hence we see that

$$v_1 = L_1 \frac{di_1}{dt}$$

in the case under consideration.

Next let us consider Fig. 16.2(b) in which the primary is an open circuit and a current i_2 flows in the secondary. Proceeding as before, we have

$$\phi_{22} = \phi_{L2} + \phi_{12}$$

where ϕ_{L2} is the leakage flux of i_2 that links coil 2 but not coil 1, and ϕ_{12} is the mutual flux of i_2 that links coil 1 and coil 2. The flux linkage of coil 1 is

$$\lambda_1 = N_1 \phi_{12}$$

Therefore, the primary voltage is given by

$$v_1 = \frac{d\lambda_1}{dt} = N_1 \frac{d\phi_{12}}{dt}$$

If we define

$$N_1 \phi_{12} = M_{12} i_2 \qquad (16.3)$$

where M_{12} is a mutual inductance, then the open-circuit primary voltage is

$$v_1 = M_{12} \frac{di_2}{dt}$$

In the next section we shall show that the mutual inductances M_{12} and M_{21} are equal; therefore we shall write

$$\boxed{M = M_{12} = M_{21} \qquad (16.4)}$$

and refer to M as the mutual inductance.

The secondary voltage is given by

$$v_2 = \frac{d\lambda_2}{dt}$$

where the flux linkage is

$$\lambda_2 = N_2\phi_{22}$$

We now define the relation

$$L_2 i_2 = N_2\phi_{22} \tag{16.5}$$

where L_2 is the self-inductance of coil 2 in henrys (H). Therefore,

$$v_2 = L_2\frac{di_2}{dt}$$

in the open-circuit primary case.

Let us now consider the general case of Fig. 16.2(c) in which both i_1 and i_2 are nonzero. The fluxes in coil 1 and coil 2, as shown in the figure, are

$$\phi_1 = \phi_{L1} + \phi_{21} + \phi_{12} = \phi_{11} + \phi_{12}$$
$$\phi_2 = \phi_{L2} + \phi_{12} + \phi_{21} = \phi_{21} + \phi_{22}$$

respectively. Therefore, the flux linkages of the primary and secondary coils are

$$\lambda_1 = N_1\phi_{11} + N_1\phi_{12}$$
$$\lambda_2 = N_2\phi_{21} + N_2\phi_{22}$$

Substituting from (16.1)–(16.5), we find upon differentiation that the primary and secondary voltages are

$$
\begin{aligned}
v_1 &= L_1\frac{di_1}{dt} + M\frac{di_2}{dt} \\[2mm]
v_2 &= M\frac{di_1}{dt} + L_2\frac{di_2}{dt}
\end{aligned}
\tag{16.6}
$$

It is clear that the voltages consist of self-induced voltages due to the inductances L_1 and L_2 and mutual voltages due to the mutual inductance M.

In the preceding discussion, the coil windings in Fig. 16.2 are such that the algebraic sign of the mutual voltage terms $M\,di_1/dt$ and $M\,di_2/dt$ are positive for the terminal voltage and current assignments as shown. In practice, it is, of course, undesirable to show a detailed sketch of the windings. This is avoided by the use of a *dot convention* which designates the polarity of the mutual voltage. Equivalent circuit symbols for the transformer of Fig. 16.2 are shown in Fig. 16.3. The polarity markings are assigned so that a positively increasing current flowing into a dotted (undotted) terminal in one winding induces a positive voltage at the dotted (undotted) terminal of the other winding.

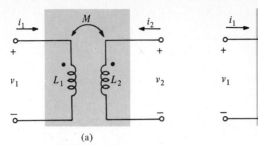

FIGURE 16.3 Circuit symbols for the transformer of Fig. 16.2: (a) dots on upper terminals; (b) dots on lower terminals

For the purpose of writing the describing equations, we may state the following rule:

A current i entering a dotted (undotted) terminal in one winding induces a voltage $M \, di/dt$ with positive polarity at the dotted (undotted) terminal of the other winding.

We note that with this rule it is unimportant whether the current i is increasing or not, since the sign of the induced voltage is accounted for by di/dt. That is, if i is increasing, the induced voltage $M \, di/dt$ is positive, and if i is decreasing, the induced voltage is negative. If i is a dc current, then, of course, the induced voltage is zero.

EXAMPLE 16.1 As an example, let us write the loop equations for Fig. 16.3(a). We see that i_2 enters a dotted terminal. Thus the mutual voltage $M \, di_2/dt$ has positive polarity at the dotted terminal of the primary. Similarly, since i_1 enters a dotted terminal, the mutual voltage $M \, di_1/dt$ has positive polarity at the dotted terminal of the secondary. Application of KVL around the primary and secondary circuits gives (16.6).

In a like manner, identical statements apply to the undotted terminals of Fig. 16.3(b), and KVL once again yields (16.6).

Let us now establish a method for placing polarity markings on a transformer. We begin by arbitrarily assigning a dot to a terminal, such as terminal a of Fig. 16.4(a). A current into this terminal produces a flux ϕ as shown. (The direction of ϕ is determined by the right-hand rule, which states that if the fingers of the right hand encircle a coil in the direction of the current, the thumb indicates the direction of the flux.) The dotted terminal in the remaining coil is the one which a current enters to produce a flux in the same direction as ϕ. (This statement, which may be used to obtain the polarity of the induced voltage, is a consequence of a rule known as *Lenz's law*. The reader may study this law, formulated in 1834 by the German scientist Heinrich F. E. Lenz, in a later course in electromagnetic field theory.) We note that a current flowing into b in Fig. 16.4(a) produces a flux in the same direction as ϕ. Thus terminal b receives the polarity dot. The circuit symbol for this transformer is shown in Fig. 16.4(b).

Evidently the polarity markings on a transformer are independent of the terminal voltage and current assignments. Consider, for instance, the assignment of Fig. 16.5(a). Applying KVL around the primary and secondary loops, we find

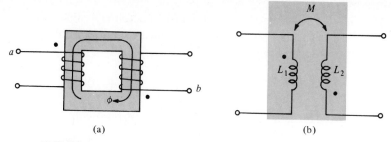

(a)   (b)

FIGURE 16.4 (a) Model of a transformer to determine polarity markings; (b) circuit symbol

$$v_1 = L_1 \frac{di_1}{dt} - M \frac{di_2}{dt}$$

$$v_2 = -M \frac{di_1}{dt} + L_2 \frac{di_2}{dt}$$

In Fig. 16.5(b), the voltage and current assignments have been changed. Application of KVL for these assignments yields

$$v_1 = -L_1 \frac{di_1}{dt} - M \frac{di_2}{dt}$$

$$v_2 = M \frac{di_1}{dt} + L_2 \frac{di_2}{dt}$$

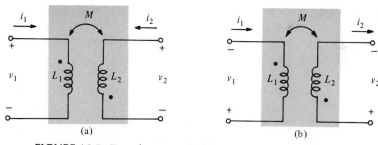

(a)   (b)

FIGURE 16.5 Transformer with different terminal voltage and current assignments

The above results can also be obtained by an alternative method for selecting the sign of the mutual voltages, as follows:

If both currents enter (or leave) the dotted terminals of the coils, the mutual and self-inductance terms for each terminal pair have the same sign; otherwise, they have opposite signs.

The student can easily verify the use of this method for the cases of Figs. 16.3 and 16.5.

EXAMPLE 16.2 As an example, let us find the open-circuit voltage v_2 in the circuit of Fig. 16.6, given that $i_1(0^-) = 0$. For $t > 0$, since no current flows in the secondary, applying KVL around the primary yields

$$\frac{di_1}{dt} + 10i_1 = 20$$

This is a first-order differential equation with a general solution of

$$i_1 = 2 + Ae^{-10t}$$

Since $i(0^+) = i(0^-) = 0$, $A = -2$ and

$$i_1 = 2(1 - e^{-10t}) \text{ A}$$

Therefore, in the secondary,

$$v_2 = -0.25\frac{di_1}{dt}$$

$$= -5e^{-10t} \text{ V}$$

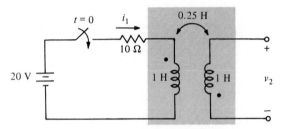

FIGURE 16.6 Circuit with an open-circuit secondary

Let us now consider the transformer of Fig. 16.3(a) in the case of an excitation having a complex frequency s. In this case, since differentiation in the time domain is equivalent to multiplication by s in the frequency domain, the phasor equations corresponding to (16.6) are

$$\mathbf{V}_1(s) = sL_1\mathbf{I}_1(s) + sM\mathbf{I}_2(s)$$
$$\mathbf{V}_2(s) = sM\mathbf{I}_1(s) + sL_2\mathbf{I}_2(s)$$

The phasor circuit for this network is shown in Fig. 16.7. In the case of a purely sinusoidal excitation, we simply replace s by $j\omega$.

EXAMPLE 16.3 For a final example, let us find the phasor voltage $\mathbf{V}_2(s)$ in the network of Fig. 16.8 due to the complex forcing function $\mathbf{V}_1(s)$. The loop equations in the primary and secondary are

$$\mathbf{V}_1 = (s + 2)\mathbf{I}_1 + s\mathbf{I}_2$$
$$0 = s\mathbf{I}_1 + (2s + 3)\mathbf{I}_2$$

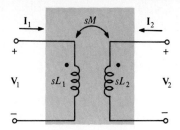

FIGURE 16.7 Phasor circuit for a transformer

Therefore,

$$\mathbf{I}_2 = \frac{\begin{vmatrix} s+2 & \mathbf{V}_1 \\ s & 0 \end{vmatrix}}{\begin{vmatrix} s+2 & s \\ s & 2s+3 \end{vmatrix}}$$

$$= \frac{-s\mathbf{V}_1}{s^2 + 7s + 6}$$

and if $\mathbf{H}(s)$ is the voltage ratio function, then

$$\mathbf{H}(s) = \frac{\mathbf{V}_2}{\mathbf{V}_1} = \frac{3\mathbf{I}_2}{\mathbf{V}_1} = \frac{-3s}{s^2 + 7s + 6}$$

$$= \frac{-3s}{(s+1)(s+6)}$$

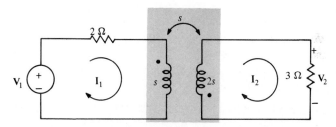

FIGURE 16.8 Circuit with a complex excitation

A pole-zero plot and a sketch of $|\mathbf{H}(j\omega)|$ are shown in Fig. 16.9. We may show that the maximum ac steady-state response occurs for $\omega = \sqrt{6}$ rad/s. Suppose, as an example, that $v_1 = 100 \cos 10t$ V. Then

$$\mathbf{V}_1 = 100 \text{ V}, \qquad s = j10 \text{ rad/s}$$

and

$$\mathbf{V}_2 = \mathbf{H}\mathbf{V}_1 = \frac{(-j30)(100)}{-100 + j70 + 6}$$

which simplifies to

$$\mathbf{V}_2 = 25.6\underline{/126.7°} \text{ V}$$

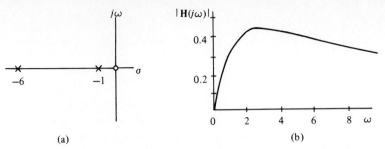

(a)   (b)

FIGURE 16.9 (a) Pole-zero and (b) $|H(j\omega)|$ plot for the network of Fig. 16.8

Thus the ac steady-state response is

$$v_2(t) = 25.6 \cos (10t + 126.7°) \text{ V}$$

EXERCISES

16.1.1   In the circuit of Fig. 16.3(a), $L_1 = 2$ H, $L_2 = 8$ H and $M = 3$ H. Find v_1 and v_2 at the time when the rates of change of the currents are

$$\frac{di_1}{dt} = 20 \text{ A}/s, \qquad \frac{di_2}{dt} = -6 \text{ A}/s$$

Answer 22, 12 V

16.1.2   In the circuit of Fig. 16.3(a), $L_1 = L_2 = 0.1$ H and $M = 10$ mH. Find v_1 and v_2 if (a) $i_1 = 0.1 \cos t$ A and $i_2 = 0.3 \sin t$ A, and (b) $i_1 = 10 \sin 100t$ mA and $i_2 = 0$.
  Answer (a) $3 \cos t - 10 \sin t$, $30 \cos t - \sin t$ mV;   (b) $100 \cos 100t$, $10 \cos 100t$ mV

16.1.3   Verify that the coil windings of Fig. 16.2 are consistent with the polarity markings of Fig. 16.3.

16.1.4   Find the phasor currents $\mathbf{I}_1$ and $\mathbf{I}_2$.
  Answer $4 - j4$, $j2$ A

EXERCISE 16.1.4

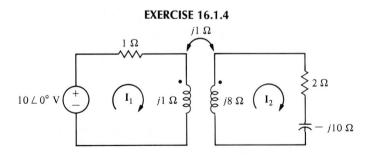

16.2

ENERGY STORAGE

We have previously shown that the energy stored in an inductor at time t is

$$w(t) = \frac{1}{2}Li^2(t)$$

Evidently, for a given inductance L, the energy is completely specified in terms of $i(t)$. Let us now determine the energy stored in a pair of mutually coupled inductors, such as those of Figs. 16.2–16.4. In so doing, we shall show that $M_{12} = M_{21} = M$ and establish limits for the magnitude of M.

In Fig. 16.3, the stored energy is the sum of the energies supplied to the primary and secondary terminals. The instantaneous powers delivered to these terminals, from (16.6), are

$$p_1 = v_1 i_1 = \left(L_1\frac{di_1}{dt} + M_{12}\frac{di_2}{dt}\right)i_1$$
$$p_2 = v_2 i_2 = \left(M_{21}\frac{di_1}{dt} + L_2\frac{di_2}{dt}\right)i_2$$

(16.7)

where M has been replaced by M_{12} and M_{21} in the appropriate terms.

Let us now perform a simple experiment. Suppose we start at time t_0 with $i_1(t_0) = i_2(t_0) = 0$. Since the magnetic flux is zero, no energy is stored in the magnetic field; that is, $w(t_0) = 0$. Next, assume, beginning at time t_0, that we maintain $i_2 = 0$ and increase i_1 until, at time t_1, $i_1(t_1) = I_1$ and $i_2(t_1) = 0$. During this interval, $i_2 = 0$ and $di_2/dt = 0$. Thus the energy accumulated during this time is

$$w_1 = \int_{t_0}^{t_1} (p_1 + p_2)\, dt = \int_{t_0}^{t_1} L_1 i_1 \frac{di_1}{dt}\, dt$$

$$= \int_0^{I_1} L_1 i_1\, di_1 = \frac{1}{2}L_1 I_1^2$$

As a final step, let us maintain $i_1 = I_1$ and increase i_2 until, at time t_2, $i_2(t_2) = I_2$. During this time interval, $di_1/dt = 0$, and the energy accumulated, using (16.7), is

$$w_2 = \int_{t_1}^{t_2} \left(M_{12}I_1\frac{di_2}{dt} + L_2 i_2 \frac{di_2}{dt}\right) dt$$

$$= \int_0^{I_2} (M_{12}I_1 + L_2 i_2)\, di_2$$

$$= M_{12}I_1 I_2 + \frac{1}{2}L_2 I_2^2$$

Thus the energy stored in the transformer at time t_2 is

$$w(t_2) = w(t_0) + w_1 + w_2$$

$$= \frac{1}{2}L_1I_1^2 + M_{12}I_1I_2 + \frac{1}{2}L_2I_2^2$$

Let us now repeat our experiment but reverse the order in which we increase i_1 and i_2. That is, in the interval from t_0 to t_1, we increase i_2 so that $i_2(t_1) = I_2$ while holding $i_1 = 0$. Finally, we maintain $i_2 = I_2$ while increasing i_1 so that $i_1(t_2) = I_1$. Using the same steps as before, we find in this case that

$$w(t_2) = \frac{1}{2}L_1I_1^2 + M_{21}I_1I_2 + \frac{1}{2}L_2I_2^2$$

Since $i_1(t_2) = I_1$ and $i_2(t_2) = I_2$ in both experiments, then $w(t_2)$ should be the same in both cases. Comparing our results, we see that this requires

$$M_{12} = M_{21} = M$$

If we repeat our experiment with the polarity markings of Fig. 16.4, the sign of the mutual voltage is negative, which causes the mutual term in the energy MI_1I_2 to be negative. Thus a general expression for the energy at any time t is given by

$$w(t) = \frac{1}{2}L_1i_1^2 \pm Mi_1i_2 + \frac{1}{2}L_2i_2^2 \qquad (16.8)$$

where the sign in the mutual term is positive if both currents enter dotted (or undotted) terminals; otherwise, it is negative.

The *coefficient of coupling* between the inductors indicates the amount of coupling and is defined by

$$k = \frac{M}{\sqrt{L_1L_2}} \qquad (16.9)$$

Clearly, $k = 0$ if no coupling exists between the coils, since $M = 0$. We can establish the upper limit of k by substituting from (16.1)–(16.5), which yields

$$k = \frac{M}{\sqrt{L_1L_2}} = \frac{\sqrt{M_{12}M_{21}}}{\sqrt{L_1L_2}} = \sqrt{\frac{\phi_{12}}{\phi_{11}}}\sqrt{\frac{\phi_{21}}{\phi_{22}}} \le 1$$

since $\phi_{12}/\phi_{11} \le 1$ and $\phi_{21}/\phi_{22} \le 1$. Thus we must have

$$0 \le k \le 1$$

or, equivalently,

$$0 \le M \le \sqrt{L_1L_2}$$

If $k = 1$, all of the flux links all of the turns of both windings, which is a *unity-coupled* transformer.

The value of k (and hence M) depends on the physical dimensions and number of turns of each coil, their relative positions to one another, and the magnetic properties of the core on which they are wound. Coils are said to be *loosely coupled* if $k < 0.5$, whereas those for which $k > 0.5$ are *tightly coupled*. Most air-core transformers are loosely coupled, in contrast to iron-core devices for which k can approach 1.

Let us now examine the values of i_1 and i_2 in (16.8) for which $w(t)$ is zero. From the quadratic formula, we may write

$$i_1 = \mp \frac{M i_2}{L_1} \pm \frac{i_2}{L_1} \sqrt{M^2 - L_1 L_2}$$

For real values of i_1 and i_2, we see that for $w(t) = 0$ we must have

$$
\begin{aligned}
i_1 = i_2 = 0 & \quad M < \sqrt{L_1 L_2} \\
i_1 = \pm \frac{M}{L_1} i_2, & \quad M = \sqrt{L_1 L_2}
\end{aligned}
\tag{16.10}
$$

where the sign of the mutual term is negative if both currents enter dotted (or undotted) terminals; otherwise, it is positive. The second equation of (16.10) shows that $w(t)$ can be zero in a unity-coupled ($k = 1$) transformer even though i_1 and i_2 are nonzero.

EXERCISES

16.2.1 Find the coefficient of coupling if $L_1 = 0.02$ H, $L_2 = 0.125$ H, and $M = 0.04$ H.
Answer 0.8

16.2.2 Find M if $L_1 = 0.4$ H, $L_2 = 0.9$ H, and (a) $k = 1$, (b) $k = 0.5$, and (c) $k = 0.01$.
Answer (a) 0.6 H; (b) 0.3 H; (c) 6 mH

16.2.3 Determine the energy stored at $t = 0$ in each case of Ex. 16.1.2.
Answer (a) 0.5 mJ, (b) 0

16.3

CIRCUITS WITH LINEAR TRANSFORMERS

A two-winding transformer is in general a four-terminal device in which the reference potential in the primary can be different from that of the secondary without altering the values of v_1, v_2, i_1, or i_2.

EXAMPLE 16.4 Consider, for example, the circuit of Fig. 16.10. KVL around the primary and secondary circuits gives

$$v_1(t) = R_1 i_1 + L_1 \frac{di_1}{dt} - M \frac{di_2}{dt}$$

$$0 = -M \frac{di_1}{dt} + R_2 i_2 + L_2 \frac{di_2}{dt}$$

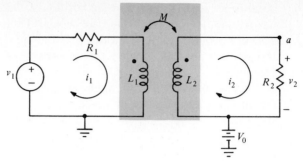

FIGURE 16.10 Circuit showing different reference potentials in primary and secondary

Evidently, the voltages and currents are not affected by V_0. For this reason, the secondary of the transformer is said to have *dc isolation* from the primary. Point a, of course, is at an absolute potential of $V_0 + i_2R_2$ volts with respect to the ground reference. If we now let $V_0 = 0$, it is seen that the bottom terminals of the transformer are connected and that the primary and secondary circuits have a common reference point. The transformer, in this case, is a three-terminal device.

EXAMPLE 16.5 As a second example, let us find the complete response i_2 for $t > 0$ in Fig. 16.11, given $M = 1/\sqrt{2}$ H and $i_1(0^-) = i_2(0^-) = 0$. Since the forced response for i_1 is a dc current (by inspection), no steady-state voltage is produced in the secondary; thus $i_{2f} = 0$. To find the natural response, let us first obtain the network function for $t > 0$. Application of KVL for a complex excitation $\mathbf{V}_1(s)$ yields

$$\mathbf{V}_1(s) = \left(\frac{3}{2}s + 2\right)\mathbf{I}_1 - \frac{1}{\sqrt{2}}s\mathbf{I}_2$$

$$0 = -\frac{1}{\sqrt{2}}s\mathbf{I}_1 + (s + 2)\mathbf{I}_2$$

from which we find

$$\mathbf{H}(s) = \frac{\mathbf{I}_2}{\mathbf{V}_1} = \frac{1}{\sqrt{2}}\frac{s}{(s + 1)(s + 4)}$$

The poles of $\mathbf{H}(s)$ are the natural frequencies, -1, -4, of the natural response.

FIGURE 16.11 Switching circuit containing a transformer

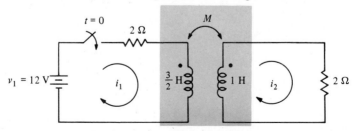

Thus, for $t > 0$,

$$i_2 = i_{2f} + i_{2n} = 0 + A_1 e^{-t} + A_2 e^{-4t}$$

From (16.8), $w(0^-) = 0$; therefore since the energy cannot change instantaneously in the absence of infinite forcing functions, $w(0^+) = 0$. From (16.10), noting that $M < \sqrt{L_1 L_2}$, we see that $i_1(0^+) = i_2(0^+) = 0$. To obtain a second initial condition for finding A_1 and A_2, we apply KVL around the primary and secondary at $t = 0^+$, which yields

$$12 = \frac{3}{2} \frac{di_1(0^+)}{dt} - \frac{1}{\sqrt{2}} \frac{di_2(0^+)}{dt}$$

$$0 = -\frac{1}{\sqrt{2}} \frac{di_1(0^+)}{dt} + \frac{di_2(0^+)}{dt}$$

Solving, we find $di_2(0^+)/dt = 6\sqrt{2}$ A/s. Evaluating A_1 and A_2, using $i_2(0^+)$ and $di_2(0^+)/dt$, the solution becomes

$$i_2 = 2\sqrt{2}(e^{-t} - e^{-4t}) \text{ A}$$

EXAMPLE 16.6 Suppose we now repeat this example for $M = \sqrt{L_1 L_2} = \sqrt{3}/2$. In this case the network function for $t > 0$ is

$$\mathbf{H}(s) = \frac{\sqrt{3/2}s}{5s + 4}$$

and the complete response is

$$i_2 = A e^{-0.8t}$$

We know that $i_1(0^-) = i_2(0^-) = 0$; however, if we take $i_2(0^+) = i_2(0^-)$ as before, then the unreasonable solution $i_2 = 0$ results. Also, applying KVL gives

$$12 = 2i_1 + \frac{3}{2} \frac{di_1}{dt} - \sqrt{\frac{3}{2}} \frac{di_2}{dt}$$

$$0 = -\sqrt{\frac{3}{2}} \frac{di_1}{dt} + 2i_2 + \frac{di_2}{dt}$$

Multiplying the latter equation by $\sqrt{3/2}$ and adding it to the former, we have

$$12 = 2i_1 + \sqrt{6}i_2 \qquad (16.11)$$

This result contradicts $i_1(0^+) = i_2(0^+) = 0$.

Recalling (16.10), we see that the energy is zero in the unity-coupled case when

$$i_1(0^+) = \frac{M}{L_1} i_2(0^+) = \sqrt{\frac{2}{3}} i_2(0^+)$$

Combining the last result and (16.11) gives $i_2(0^+) = 2.94$ A, and therefore

$$i_2 = 2.94 e^{-0.8t} \text{ A}$$

Thus the current in a unity-coupled transformer can change instantaneously as a result of the application of a finite forcing function.

EXAMPLE 16.7   As another example, let us find the steady-state response v_2 in Fig. 16.12. Applying KVL for a complex frequency s, we have

$$\mathbf{V}_1 = \left(s + 3 + \frac{1}{s}\right)\mathbf{I}_1 - \left(s + \frac{1}{s}\right)\mathbf{I}_2$$

$$0 = -\left(s + \frac{1}{s}\right)\mathbf{I}_1 + \left(2s + 1 + \frac{1}{s}\right)\mathbf{I}_2$$

for which

$$\mathbf{H}(s) = \frac{\mathbf{V}_2}{\mathbf{V}_1} = \frac{\mathbf{I}_2}{\mathbf{V}_1} = \frac{s^2 + 1}{s^3 + 7s^2 + 4s + 4}$$

Substituting $s = j2$ rad/s and $\mathbf{V}_1 = 16$ V, we find $\mathbf{V}_2 = 2\underline{/0°}$ V. Thus

$$v_2 = 2 \cos 2t \text{ V}$$

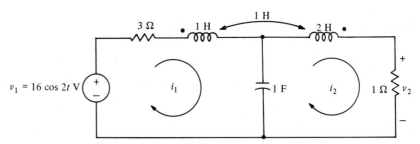

FIGURE 16.12 Circuit containing a linear transformer

EXAMPLE 16.8   As a final example, let us find the network function $\mathbf{V}_2/\mathbf{V}_1$ for the phasor circuit of Fig. 16.13. Let us first examine the voltage $\mathbf{V}_A$ which appears across winding A in loop 1. We see that the voltage due to $\mathbf{I}_1$ is the self-inductance term $(2s\,\mathbf{I}_1)$ and the mutual term of winding $C\,(2s\,\mathbf{I}_1)$. The voltage due to $\mathbf{I}_2$ arises from the mutual terms of winding $B\,(-s\,\mathbf{I}_2)$ and winding $C\,(-2s\,\mathbf{I}_2)$. Thus

$$\mathbf{V}_A = (2s + 2s)\mathbf{I}_1 - (s + 2s)\mathbf{I}_2$$

Similarly, we find

$$\mathbf{V}_B = -(s + 3s)\mathbf{I}_1 + (3s + 3s)\mathbf{I}_2$$

$$\mathbf{V}_C = (4s + 2s)\mathbf{I}_1 - (4s + 3s)\mathbf{I}_2$$

Applying KVL in loops 1 and 2 gives

$$\mathbf{V}_1 = 3\mathbf{I}_1 + \mathbf{V}_A + \mathbf{V}_C - 2\mathbf{I}_2$$

$$0 = -2\mathbf{I}_1 - \mathbf{V}_C + \mathbf{V}_B + 5\mathbf{I}_2$$

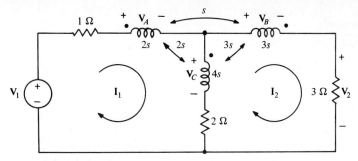

FIGURE 16.13 Circuit containing a three-winding transformer

or

$$\mathbf{V_1} = (10s + 3)\mathbf{I_1} - (10s + 2)\mathbf{I_2}$$

$$0 = -(10s + 2)\mathbf{I_1} + (13s + 5)\mathbf{I_2}$$

Solving these equations for the ratio $\mathbf{I_2}/\mathbf{V_1}$, we find the network function to be

$$\mathbf{H}(s) = \frac{\mathbf{V_2}}{\mathbf{V_1}} = \frac{3\mathbf{I_2}}{\mathbf{V_1}} = \frac{3(10s + 2)}{30s^2 + 49s + 11}$$

EXERCISES

16.3.1 Determine i_1 for $t > 0$ in the network of Fig. 16.11, given $M = 1/\sqrt{2}$ H. Assume the circuit is in steady state at $t = 0^-$.
Answer $6 - 4e^{-t} - 2e^{-4t}$ A

16.3.2 Find the forced response v_2 if $v_1 = 8e^{-4t} \cos 2t$ V.
Answer $\sqrt{2}e^{-4t} \cos (2t + 45°)$ V

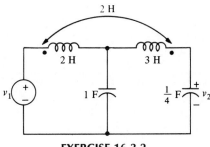

EXERCISE 16.3.2

16.3.3 Determine the network function $\mathbf{V_2}/\mathbf{V_1}$ for Fig. 16.13 if the polarity dot of coil B is placed on the other terminal.
Answer $6(s + 1)/(6s^2 + 45s + 11)$

REFLECTED IMPEDANCE

In this section we shall develop several important impedance relationships for the ac steady-state case. Let us begin by considering the phasor circuit of Fig. 16.14 having a practical source $\mathbf{V}_g$ with an impedance $\mathbf{Z}_2$ connected in the secondary.

Applying KVL at the primary terminals of the transformer we find

$$\mathbf{V}_1 = j\omega L_1 \mathbf{I}_1 - j\omega M \mathbf{I}_2$$
$$0 = -j\omega M \mathbf{I}_1 + (\mathbf{Z}_2 + j\omega L_2)\mathbf{I}_2 \tag{16.12}$$

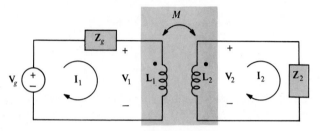

FIGURE 16.14 Circuit for deriving impedance relationships

Eliminating $\mathbf{I}_2$ from these equations we have

$$\mathbf{V}_1 = \left[j\omega L_1 - \frac{j\omega M\,(j\omega M)}{\mathbf{Z}_2 + j\omega L_2} \right] \mathbf{I}_1$$

so that the input impedance seen at the primary terminals of the transformer is

$$\mathbf{Z}_1 = \frac{\mathbf{V}_1}{\mathbf{I}_1} = j\omega L_1 + \frac{\omega^2 M^2}{\mathbf{Z}_2 + j\omega L_2} \tag{16.13}$$

The first part, $j\omega L_1$, depends entirely on the reactance of the primary. The second part is due to the mutual coupling, and it is called the *reflected impedance,* given by

$$\mathbf{Z}_R = \frac{\omega^2 M^2}{\mathbf{Z}_2 + j\omega L_2}$$

It may be thought of as the impedance inserted into, or *reflected* into, the primary by the secondary.

The input impedance as seen by the source $\mathbf{V}_g$ is evidently

$$\mathbf{Z}_{in} = \mathbf{Z}_g + \mathbf{Z}_1$$

Also the secondary-to-primary current ratio $\mathbf{I}_2/\mathbf{I}_1$ may be found from the second equation of (16.12), and the voltage ratio $\mathbf{V}_2/\mathbf{V}_1$ may be found from

$$\frac{\mathbf{V}_2}{\mathbf{V}_1} = \frac{\mathbf{Z}_2 \mathbf{I}_2}{\mathbf{V}_1} = \mathbf{Z}_2 \left(\frac{\mathbf{I}_2}{\mathbf{I}_1}\right)\left(\frac{\mathbf{I}_1}{\mathbf{V}_1}\right)$$

using (16.13). The results are

$$\frac{\mathbf{I}_2}{\mathbf{I}_1} = \frac{j\omega M}{\mathbf{Z}_2 + j\omega L_2}$$

$$\frac{\mathbf{V}_2}{\mathbf{V}_1} = \frac{j\omega M \mathbf{Z}_2}{j\omega L_1(\mathbf{Z}_2 + j\omega L_2) + \omega^2 M^2} \tag{16.14}$$

It is interesting to note that $\mathbf{Z}_R$ is independent of the dot locations on the transformer. If either dot in Fig. 16.14 is placed on the opposite terminal, the sign of the mutual term in each equation of (16.14) changes, which is equivalent to replacing M by $-M$. Since $\mathbf{Z}_R$ varies as M^2, its sign is unchanged. A second important property is illustrated by rationalizing $\mathbf{Z}_R$, which gives

$$\mathbf{Z}_R = \frac{\omega^2 M^2}{R_2^2 + (X_2 + \omega L_2)^2}[R_2 - j(X_2 + \omega L_2)]$$

where we have used the relation $\mathbf{Z}_2 = R_2 + jX_2$ for the load impedance. It is seen that the sign of the imaginary part of $\mathbf{Z}_R$ is minus. Therefore the reflected reactance is opposite that of the net reactance $X_2 + \omega L_2$ of the secondary. In particular, if X_2 is a capacitive reactance whose magnitude is less than ωL_2 or if it is an inductive reactance, then the reflected reactance is capacitive. Otherwise, the reflected reactance is either inductive or it is zero. In the latter case, X_2 must be a capacitive reactance $-1/\omega C$ with a resonant frequency $f_0 = \omega_0/2\pi = 1/2\pi\sqrt{L_2 C}$. In this case, for $\omega = \omega_0$,

$$\mathbf{Z}_R = \frac{\omega_0^2 M^2}{R_2}$$

and the reflected impedance is purely real.

Inspecting (16.13) and (16.14), we see that if the polarity dot on either winding of Fig. 16.14 occurs on the opposite terminal, then the current and voltage ratios require a sign change, whereas the impedance relations are unaffected.

EXERCISES

16.4.1 Given: In Fig. 16.14, $\mathbf{V}_g = 100\underline{/0^\circ}$ V, $\mathbf{Z}_g = 40\ \Omega$, $L_1 = 0.5$ H, $L_2 = 0.1$ H, $M = 0.1$ H, and $\omega = 100$ rad/s. If $\mathbf{Z}_2 = 10 - (j1000/\omega)\ \Omega$, find (a) $\mathbf{Z}_{in}$, (b) $\mathbf{I}_1$, (c) $\mathbf{I}_2$, (d) $\mathbf{V}_1$, and (e) $\mathbf{V}_2$.
  Answer (a) $50 + j50\ \Omega$; (b) $1 - j1$ A; (c) $1 + j1$ A; (d) $60 + j40$ V; (e) 20 V

16.4.2 Repeat Ex. 16.4.1 if the polarity dot is on the lower terminal of the secondary.
  Answer (a) $50 + j50\ \Omega$; (b) $1 - j1$A; (c) $-1 - j1$ A; (d) $60 + j40$ V; (e) -20 V

16.4.3 Find the frequency for which the reflected impedance in Fig. 16.14 is real if $L_2 = 4$ H and $\mathbf{Z}_2$ is a 6-Ω resistor in series with a 1/16-F capacitor.
  Answer 2 rad/s

16.5

THE IDEAL TRANSFORMER

An *ideal transformer* is a lossless unity-coupled transformer in which the self-inductances of the primary and secondary are infinite but their ratio is finite. Physical transformers which approximate this ideal case are the previously mentioned iron-core transformers. The primary and secondary coils are wound on a laminated iron-core structure such that nearly all of the flux links all of the turns of both coils. The reactances of the primary and secondary self-inductances are very large compared to moderate load impedances, and the coupling coefficient is nearly unity over the frequency range for which the device is designed. The ideal transformer is thus an approximate model for well-constructed iron-core transformers.

An important parameter that is necessary in describing the characteristics of an ideal transformer is the *turns ratio n*, defined by

$$n = \frac{N_2}{N_1} \qquad (16.15)$$

where N_1 and N_2 are the number of turns on the primary and secondary, respectively.

The flux produced in a winding of a transformer due to a current in the winding is proportional to the product of the current and the number of turns on the winding. Thus, in the primary and secondary windings,

$$\phi_{11} = \alpha N_1 i_1$$

$$\phi_{22} = \alpha N_2 i_2$$

where α is a constant of proportionality which depends on the physical properties of the transformer. (The constant α is the same in each case because we are assuming that there is no leakage flux and thus both flux paths are identical.) Substituting these relations into (16.2) and (16.5), we see that

$$L_1 = \alpha N_1^2, \qquad L_2 = \alpha N_2^2$$

Therefore,

$$\frac{L_2}{L_1} = \left(\frac{N_2}{N_1}\right)^2 = n^2 \qquad (16.16)$$

In the case of unity coupling, $M = \sqrt{L_1 L_2}$, the second of (16.14) becomes

$$\frac{\mathbf{V}_2}{\mathbf{V}_1} = \frac{j\omega \mathbf{Z}_2 \sqrt{L_1 L_2}}{j\omega L_1 (\mathbf{Z}_2 + j\omega L_2) + \omega^2 L_1 L_2}$$

$$= \sqrt{\frac{L_2}{L_1}}$$

$$= n$$

For the ideal transformer, in addition to unity coupling, the inductances L_1 and L_2 tend to infinity in such a way that the ratio of (16.16) is the constant n^2. From the first equation of (16.14) we see that in this case we have

$$\frac{\mathbf{I}_2}{\mathbf{I}_1} = \lim_{L_1, L_2 \to \infty} \frac{j\omega\sqrt{L_1 L_2}}{\mathbf{Z}_2 + j\omega L_2}$$

$$= \lim_{L_1, L_2 \to \infty} \frac{j\omega\sqrt{L_1/L_2}}{j\omega + (\mathbf{Z}_2/L_2)}$$

$$= \lim_{L_1, L_2 \to \infty} \frac{j\omega(1/n)}{j\omega + (\mathbf{Z}_2/L_2)}$$

$$= \frac{1}{n}$$

Thus the primary and secondary voltages and currents of an ideal transformer are related simply to the turns ratio by

$$\frac{\mathbf{V}_2}{\mathbf{V}_1} = n$$

$$\frac{\mathbf{I}_2}{\mathbf{I}_1} = \frac{1}{n} \tag{16.17}$$

Replacing n by N_2/N_1, we have

$$\frac{\mathbf{V}_2}{\mathbf{V}_1} = \frac{N_2}{N_1}$$

$$N_1\mathbf{I}_1 = N_2\mathbf{I}_2$$

Therefore, the voltages are in the same ratio as the turns, and the *ampere turns* (*NI*) are the same for both primary and secondary.

The symbol for an ideal transformer is shown in Fig. 16.15(a) with polarities such that (16.17) holds. The vertical lines are used to symbolize the iron core, and $1:n$ denotes the turns ratio. If one or the other, but not both, of the polarity dots is placed on the opposite terminal, then n is replaced by $-n$ in (16.17).

FIGURE 16.15 (a) Ideal transformer symbol; (b) circuit containing an ideal transformer

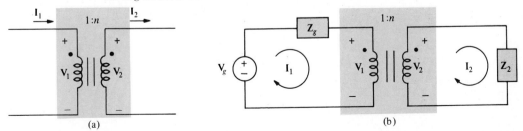

Figure 16.15(b) shows an ideal transformer connected to a secondary load Z_2 and a source V_g with internal impedance Z_g. The primary impedance Z_1 of (16.13), in the case of the ideal transformer, is given by

$$Z_1 = \frac{V_1}{I_1} = \frac{V_2/n}{nI_2} = \frac{V_2/I_2}{n^2}$$

or

$$Z_1 = \frac{Z_2}{n^2}, \qquad \frac{Z_2}{Z_1} = n^2 \qquad\qquad (16.18)$$

Thus the input impedance viewed from the terminals of the voltage source is

$$Z_{in} = Z_g + Z_1 = Z_g + \frac{Z_2}{n^2} \qquad\qquad (16.19)$$

The lossless property of an ideal transformer is easily demonstrated using (16.17). Taking the complex conjugate of the current ratio, we have, since n is real,

$$I_1^* = nI_2^*$$

from which

$$I_1^* = \frac{V_2 I_2^*}{V_1}$$

or

$$\frac{V_1 I_1^*}{2} = \frac{V_2 I_2^*}{2}$$

Thus the complex power applied to the primary is delivered to the load Z_2, and hence the transformer absorbs zero power.

In analyzing networks containing ideal transformers, it is often convenient to replace the transformer by an equivalent circuit before performing the analysis. Let us consider, for example, replacing the transformer and load impedance Z_2 of Fig. 16.15(b). Clearly, the input impedance seen by the generator V_g is Z_{in} given by (16.19), so that an equivalent circuit insofar as V_g is concerned is shown in Fig. 16.16.

FIGURE 16.16 Equivalent circuit for Fig. 16.15(b) obtained by replacing the secondary

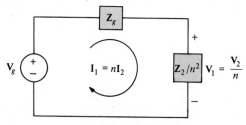

The voltages and currents can now be easily determined from the single-loop circuit.

Let us next replace the primary circuit and the transformer of Fig. 16.15(b) by its Thevenin equivalent. By (16.17) we have

$$\mathbf{I}_1 = n\mathbf{I}_2, \qquad \mathbf{V}_2 = n\mathbf{V}_1$$

so that for $\mathbf{V}_{oc}$, $\mathbf{I}_2 = 0$ and thus $\mathbf{I}_1 = 0$. Therefore

$$\mathbf{V}_{oc} = \mathbf{V}_2 = n\mathbf{V}_1 = n\mathbf{V}_g$$

For $\mathbf{I}_{sc}$, we have $\mathbf{V}_2 = 0$ and thus $\mathbf{V}_1 = 0$. Therefore

$$\mathbf{I}_{sc} = \mathbf{I}_2 = \frac{\mathbf{I}_1}{n} = \frac{\mathbf{V}_g}{n\mathbf{Z}_g}$$

and

$$\mathbf{Z}_{th} = \frac{\mathbf{V}_{oc}}{\mathbf{I}_{sc}} = n^2\mathbf{Z}_g$$

The resulting equivalent circuit is shown in Fig. 16.17.

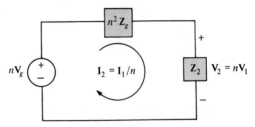

FIGURE 16.17 Thevenin equivalent circuit of Fig. 16.15(b)

Inspecting the results for the Thevenin equivalent circuit, we see that each primary voltage is multiplied by n, each primary current is multiplied by $1/n$, and each primary impedance is multiplied by n^2 when we replace the primary circuit and the transformer. It may be shown that this statement holds in general whether the result is the Thevenin circuit or not. On the other hand, if we replace the secondary circuit and transformer by an equivalent circuit, as in Fig. 16.16, we simply multiply each secondary voltage, current, and impedance by $1/n$, n, and $1/n^2$, respectively. If either dot on the transformer is reversed, we simply replace n by $-n$.

In applying the above-described procedure, the student is cautioned that the technique is valid if the transformer divides the circuit into two parts. When external connections exist between the windings, the method in general cannot be used. The equivalent circuits to be discussed in the next section are often useful for networks of this type.

EXAMPLE 16.9   As an example, let us find $\mathbf{V}_2$ in the circuit of Fig. 16.18(a), which contains an ideal transformer and a voltage-controlled current source. In Fig. 16.18(b) an equivalent circuit for the primary circuit and the transformer is shown. It should be noted that a minus sign has been used on primary voltages and currents to account for the polarity

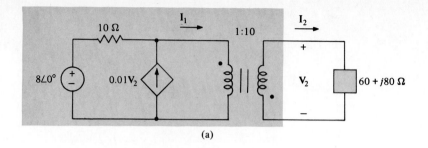

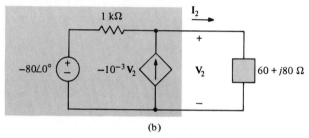

FIGURE 16.18 (a) Example circuit; (b) equivalent circuit

dots shown on the transformer. The nodal equation for V_2 gives

$$\frac{V_2 + 80\underline{/0^\circ}}{10^3} + 10^{-3}V_2 + \frac{V_2}{60 + j80} = 0$$

from which

$$V_2 = 5\sqrt{2}\underline{/-135^\circ} \text{ V}$$

So far we have considered only the ideal transformer in the ac steady-state case. In the general case, we see from (16.6) that

$$v_1 = L_1\frac{di_1}{dt} + M\left(\frac{v_2}{L_2} - \frac{M}{L_2}\frac{di_1}{dt}\right)$$

$$= \left(L_1 - \frac{M^2}{L_2}\right)\frac{di_1}{dt} + \frac{M}{L_2}v_2$$

Thus since $M^2 = L_1L_2$ and $\sqrt{L_2/L_1} = n$, we have

$$v_1 = \frac{v_2}{n}$$

Next, let us rearrange the first equation of (16.6) in the form

$$\frac{v_1}{L_1} = \frac{di_1}{dt} + \frac{M}{L_1}\frac{di_2}{dt}$$

$$= \frac{di_1}{dt} + n\frac{di_2}{dt}$$

Taking the limit as L_1 becomes infinite, we have

$$\frac{di_1}{dt} = -n\frac{di_2}{dt}$$

Integrating both sides, we have

$$i_1 = -ni_2 + C_1$$

where C_1 is a constant of integration. Since dc currents produce no time-varying magnetic flux, they do not contribute to the induced voltages or currents in the ideal transformer. Therefore if we neglect the constant C_1, then

$$i_1 = -ni_2$$

where the minus sign arises due to the direction assigned i_2 in Fig. 16.3(a). Thus the same current and voltage relationships are valid in the time domain as were found in the frequency domain if we neglect any dc currents.

EXERCISES

16.5.1 Find $\mathbf{V}_1$, $\mathbf{V}_2$, $\mathbf{I}_1$, and $\mathbf{I}_2$.
Answer $10\underline{/-36.9°}$ V, $50\underline{/143.1°}$ V, $2\underline{/0°}$ A, $0.4\underline{/180°}$ A

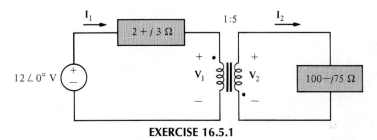

EXERCISE 16.5.1

16.5.2 In Fig. 16.15(b), $\mathbf{V}_g = 100\underline{/0°}$ V, $\mathbf{Z}_g = 10$ Ω, and $\mathbf{Z}_2 = 1$ kΩ. Find n such that $\mathbf{Z}_1 = \mathbf{Z}_g$, and then find the power delivered to $\mathbf{Z}_2$.
Answer 10, 250 W

16.5.3 Using Norton's theorem, show that the primary circuit and transformer of Fig. 16.15(b) is equivalent to a constant current source of $\mathbf{V}_g/n\mathbf{Z}_g$ A in parallel with an impedance of $n^2\mathbf{Z}_g$ Ω.

16.5.4 Find $\mathbf{V}_2$ in Fig. 16.18(a) by first replacing the secondary circuit and the transformer by an equivalent circuit.

16.6

EQUIVALENT CIRCUITS

Equivalent circuits for linear transformers are easily developed by considering the equations for primary and secondary currents and voltages. In Fig. 16.19(a) we have

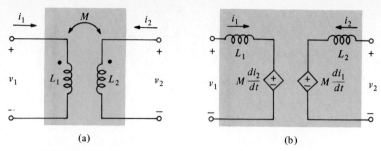

FIGURE 16.19 (a) Linear transformer; (b) equivalent circuit

$$v_1 = L_1 \frac{di_1}{dt} + M \frac{di_2}{dt}$$

$$v_2 = M \frac{di_1}{dt} + L_2 \frac{di_2}{dt}$$

It is evident that the circuit of Fig. 16.19(b) satisfies these equations. The dependent voltage sources, however, are controlled by the time derivatives of the primary and secondary currents. In the frequency domain, these sources can be considered as current-controlled voltage sources.

Let us now rearrange the equations in the form

$$v_1 = (L_1 - M) \frac{di_1}{dt} + M \left(\frac{di_1}{dt} + \frac{di_2}{dt} \right)$$

$$v_2 = M \left(\frac{di_1}{dt} + \frac{di_2}{dt} \right) + (L_2 - M) \frac{di_2}{dt}$$

These equations are satisfied by the T network of Fig. 16.20(b). Since this circuit is a three-terminal network, it is equivalent to the transformer connection of Fig. 16.20(a). If either polarity dot is changed to another terminal, we must replace M by $-M$ in the equivalent circuits.

FIGURE 16.20 (a) Linear transformer with common terminals; (b) equivalent T network

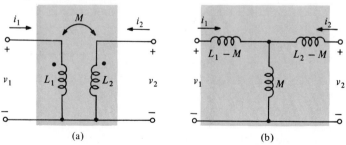

In the case of an ideal transformer as shown in Fig. 16.21(a), the currents and voltages are given by

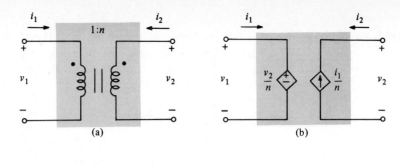

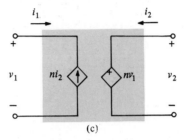

FIGURE 16.21 (a) Ideal transformer; (b) and (c) equivalent circuits

$$i_2 = \frac{-i_1}{n}, \qquad v_2 = nv_1$$

Clearly, the circuits of Figs. 16.21(b) and (c) satisfy these relations. If either of the dots are reversed, of course, we must replace n by $-n$ in the equivalent circuits.

EXERCISES

16.6.1 Employing the T equivalent circuit for the linear transformer in Fig. 16.12, determine the phasor current $\mathbf{I}_1$. (Note that an inductor with $L = 0$ is a short circuit.)
Answer $4.85\underline{/-15.9°}$ A

16.6.2 Find $\mathbf{V}_2$ in Fig. 16.18(a) using the equivalent circuit in Fig. 16.21(c).
Answer $5\sqrt{2}\underline{/-135°}$ V

16.7

SPICE ANALYSIS FOR TRANSFORMERS

SPICE can be used for analyzing circuits that contain linear transformers in the same manner as performed in previous chapters by employing the K data statement for defining the transformer. A description of the K statement is given in Appendix E. The K statement expresses the mutual coupling of the transformer in terms of the coefficient of coupling given in (16.9).

EXAMPLE 16.10 Consider as a first example the circuit of Fig. 16.11, redrawn in Fig. 16.22 for applying SPICE. It should be noted that the primary and secondary have been connected together at the reference node to avoid having the nodes of the secondary elements dangling (being disconnected) for SPICE analysis. A circuit file for the transient response for $0 < t < 1.5$ s is

```
SOLUTION OF FIG. 16.11 USING EQ. 16.9 FOR FINDING k
* DATA STATEMENTS
VIN 1 0 DC 12V
R1 1 2 2
L1 2 0 1.5 IC=0
L2 3 0 1 IC=0
R2 3 0 2
K L1 L2 0.577
* SOLUTION CONTROL STATEMENT
.TRAN 0.1 1.5 UIC
* OUTPUT CONTROL STATEMENT
.PLOT TRAN I(R2)
.END
```

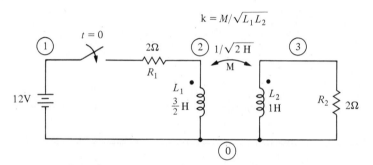

FIGURE 16.22 Circuit of Fig. 16.11 redrawn for SPICE

A plot of the output of this program is shown in Fig. 16.23.

FIGURE 16.23 Response for the circuit of Fig. 16.22

```
TIME         I(R2)
(*)----------      0.0000E+00    5.0000E-01    1.0000E+00    1.5000E+00    2.0000E+00
                   - - - - - - - - - - - - - - - - - - - - - - - - - - - - - - - - -
0.000E+00  3.174E-07 *                   .             .             .             .
1.000E-01  6.602E-01 .                   .      *      .             .             .
2.000E-01  1.044E+00 .                   .             . *           .             .
3.000E-01  1.242E+00 .                   .             .       *     .             .
4.000E-01  1.324E+00 .                   .             .          *  .             .
5.000E-01  1.333E+00 .                   .             .           * .             .
6.000E-01  1.295E+00 .                   .             .          *  .             .
7.000E-01  1.232E+00 .                   .             .       *     .             .
8.000E-01  1.155E+00 .                   .             .   *         .             .
9.000E-01  1.072E+00 .                   .             .*            .             .
1.000E+00  9.885E-01 .                   .           * .             .             .
1.100E+00  9.065E-01 .                   .        *    .             .             .
1.200E+00  8.284E-01 .                   .     *       .             .             .
1.300E+00  7.550E-01 .                   .   *         .             .             .
1.400E+00  6.867E-01 .                   . *           .             .             .
1.500E+00  6.238E-01 .                *  .             .             .             .
                   - - - - - - - - - - - - - - - - - - - - - - - - - - - - - - - - -
```

EXAMPLE 16.11   As a second example, let us find the solution for the network of Fig. 16.12, which has a coupling coefficient of 0.707. A circuit file for nodes being numbered sequentially clockwise, with node 1 being at the top of v_1, is

```
SPICE SOLUTION FOR FIG. 16.12.
* DATA STATEMENTS
V1 1 0 AC 16V
R1 1 2 3
L1 2 3 1H
C 3 0 1
L2 4 3 2H
R2 4 0 1
K L1 L2 0.707
* SOLUTION AND OUTPUT CONTROL STATEMENTS
.AC LIN 1 0.3183 0.3183
.PRINT AC VM(R2) VP(R2)
.END
```

which yields the solution

FREQ	VM(R2)	VP(R2)
3.183E−01	2.000E+00	−3.139E−03

In the case of an ideal transformer, solutions can be found using the equivalent circuits of Fig. 16.21(b) and (c) employing the procedures discussed in previous chapters.

EXERCISE

16.7.1   Using SPICE, find V_2 in Fig. 16.13 if $V_1 = 120 \underline{/0°}$ V, $f = 60$ Hz, and the 3-ohm resistor is replaced with 300 ohms.
Answer $30.75 \underline{/-75.1°}$ V.

PROBLEMS

16.1   In Fig. 16.2(a), $N_2 = 1000$ turns and $\phi_{21} = 50 \ \mu$Wb when $i_1 = 1$ A. Determine v_2 if $i_1 = 10 \cos 100t$ A.

16.2   If the inductance measured between terminals a and d is 0.2 H when terminals b and c are connected, and the inductance measured between terminals a and c is 0.6 H when terminals b and d are connected, find the mutual inductance M between the two coils, and the position of the dots.

PROBLEM 16.2

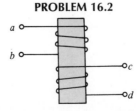

16.3   (a) Find v_1 and v_2 if $L_1 = 2$ H, $L_2 = 5$ H, $M = 3$ H, $i_1 = 5 \sin 2t$ A, and $i_2 = -3 \cos 4t$ A.
(b) Find v_2 if the secondary is open.

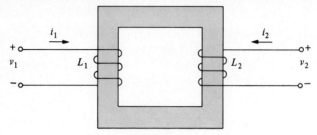

PROBLEM 16.3

16.4   Find v_1 and v_2 in Fig. 16.3 if $L_1 = 2$ H, $L_2 = 5$ H, $M = 3$ H, and the currents i_1 and i_2 are changing at the rates -10 A/s and -2 A/s, respectively.

16.5   Find v for $t > 0$ if $v_g = 4u(t)$ V.

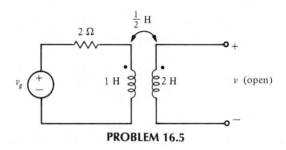

PROBLEM 16.5

16.6   Find the steady-state currents i_1 and i_2.

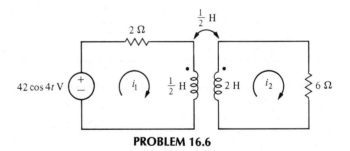

PROBLEM 16.6

16.7   Find the steady-state value of v.

PROBLEM 16.7

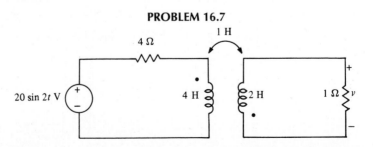

16.8   Find the energy stored at $t = 0$ if $\omega = 2$ rad/s. (Take the transient currents to be zero.)

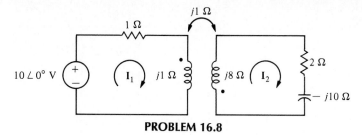

PROBLEM 16.8

16.9   (a) Find the energy stored in the transformer of Fig. 16.3(a) at a time when $i_1 = 2$ A and $i_2 = 4$ A if $L_1 = 2$ H, $L_2 = 10$ H, and $M = 4$ H. (b) Repeat part (a) if one of the dots is moved to the other terminal.

16.10 Show from (16.8) that a real transformer $(0 \le M \le \sqrt{L_1 L_2})$ satisfies the passivity condition $w(t) \ge 0$.

16.11 Find the steady-state current i_2.

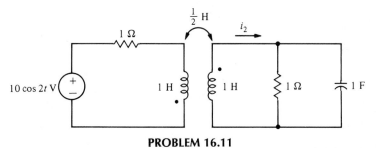

PROBLEM 16.11

16.12 Find i for $t > 0$ if $i(0) = 0$ and $v(0) = 4$ V.

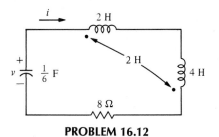

PROBLEM 16.12

16.13 Find the steady-state current i.

PROBLEM 16.13

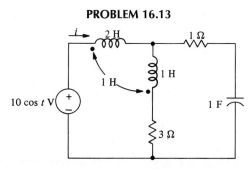

16.14 Find the natural frequencies present in $i(t)$.

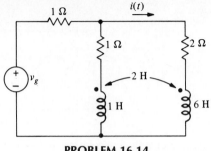

PROBLEM 16.14

16.15 Find the steady-state current i_1 in Prob. 16.6 using reflected impedance.

16.16 Find the steady-state current i_2 in Prob. 16.6 by replacing everything in the corresponding phasor circuit except the 6-Ω resistor by its Thevenin equivalent circuit.

16.17 Find the network function, $\mathbf{H}(s) = \mathbf{V}_2(s)/\mathbf{V}_1(s)$.

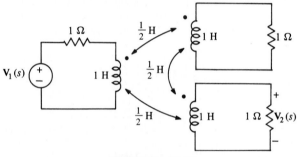

PROBLEM 16.17

16.18 Find $i(t)$ for $t > 0$ if $i(0^+) = 0$.

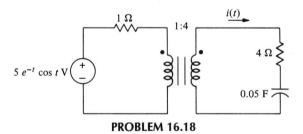

PROBLEM 16.18

16.19 Find the power delivered to the $(300 - j500)$-Ω load using reflected impedance.

PROBLEM 16.19

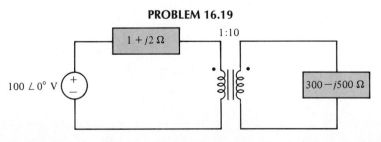

16.20 Find the average power delivered to the 4-Ω resistor.

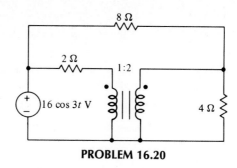

PROBLEM 16.20

16.21 Solve Prob. 16.19 by first finding the secondary phasor current by the method of Fig. 16.17.

16.22 The stepdown *autotransformer* has the secondary terminal 2 *tapped* to the primary winding at node 2, as shown. (a) If the secondary winding has N_2 turns and the primary winding has N_1 turns, find the voltage and current ratios. (b) Find $\mathbf{I}_2$ and $\mathbf{V}_2$ if $\mathbf{V}_1 = 100\underline{/0°}$ V, $\mathbf{I}_1 = 4\underline{/30°}$ A, $N_1 = 1000$ turns, and $N_2 = 400$ turns.

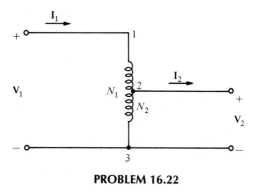

PROBLEM 16.22

16.23 Find the turns ratio n so that the maximum power is delivered to the 20-kΩ resistor.

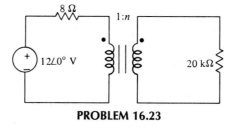

PROBLEM 16.23

16.24 Find the z- and y-parameters of the transformer of Fig. 16.20(a) in terms of s.

16.25 Find the steady-state voltage v using the equivalent T circuit for the transformer.

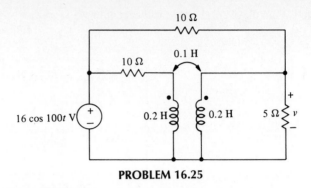

PROBLEM 16.25

COMPUTER APPLICATION PROBLEMS

16.26 Using SPICE, plot i for $0 < t < 5$ s in the circuit of Prob. 16.12

16.27 Using SPICE, find V_2 in Prob. 16.17 if $V_1 = 120 \underline{/0°}$ V and $f = 60$ Hz.

16.28 Using the equivalent circuit of Fig. 16.21(b) with SPICE, find the voltage of the 8-ohm resistor in Prob. 16.20.

16.29 In the circuit of Prob. 16.25, replace the 10-ohm resistor bridging the transformer with a 0.1-F capacitor. Use SPICE to plot the frequency response of the resulting circuit in the interval $1 < f < 50$ Hz.

17

Jean Baptiste Joseph Fourier
1768–1830

Fourier Series

A great mathematical poem.
    Lord Kelvin on the Fourier Series

In 1822 a greatly influential pioneer work on the mathematical theory of heat conduction was published by the great French mathematician, Egyptologist, and administrator Jean Baptiste Joseph Fourier. It was a masterpiece not only because of the new field of heat conduction that it explored, but also because of the infinite series of sinusoids that it developed; the latter became famous as the Fourier series. With the Fourier series, we are no longer restricted in the shortcut phasor methods to circuits whose inputs are sinusoids.

Fourier was born in Auxerre, France, the son of a tailor. He attended a local military school conducted by Benedictine monks and showed such proficiency in mathematics that he later became a mathematics teacher in the school. Like most Frenchmen his age he was swept into the politics of the French Revolution and its aftermath and more than once came near to losing his life. He was one of the first teachers in the newly formed Ecole Polytechnique and later became its professor of mathematical analysis. At age 30 Fourier was appointed scientific advisor by Napoleon on an expedition to Egypt and for 4 years was secretary of the Institute d'Egypte, the work of which marked Egyptology as a separate discipline. He was prefect of the department of Isere from 1801 to 1814, where he wrote his famous treatise on heat conduction. He completed a book on algebraic equations just before his death in 1830. ∎

Thus far, with few exceptions, we have considered excitations which were exponentials, sinusoids, or damped sinusoids, and the reader may have been impressed by the relative ease with which we can handle such functions. In particular, the concept of phasors and impedance allowed us to treat circuits with these excitations as if they were resistive circuits and find both their forced and natural responses.

There are many input functions, however, that are important in engineering that are not exponentials or sinusoids and for which the phasor methods do not apply directly. A few examples are square waves, sawtooth waves, and triangular pulses. Indeed, a function may be represented by a set of data points and have no analytical representation given at all. In this chapter we shall consider these additional functions and show how we may represent them in terms of the familiar sinusoidal functions and thus use phasors as before. The techniques we shall use were first given in 1822 by the great French mathematician and physicist Jean Baptiste Joseph Fourier, who lived from 1768 to 1830.

17.1

THE TRIGONOMETRIC FOURIER SERIES

The limitations of the phasor method may be dramatically seen by considering the functions of Fig. 17.1, which are highly useful circuit inputs, but none of which is sinusoidal or exponential. The sawtooth wave of Fig. 17.1(a) is the sweep generator signal that controls the electron beam of a cathode-ray oscilloscope and causes the image on the screen to quickly retrace itself every T seconds, giving the illusion of a stationary picture. The functions of Figs. 17.1(b) and (c) are, respectively, a *full-wave rectified* sine wave and a *half-wave rectified* sine wave, used in converting ac to pulsating dc. The full-wave rectified sine wave is a sine wave with its negative values changed to positive values, and the half-wave rectified sine wave is a sine wave with its negative values changed to zero. The signal of Fig. 17.1(d) is a *square wave,* which can be used as a *clock* to actuate a circuit at periodic times for small intervals when its value is positive. All these signals are very common, but since they are not

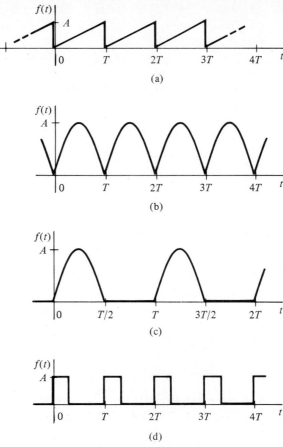

FIGURE 17.1 (a) Sawtooth wave; (b) full-wave rectified sine wave; (c) half-wave rectified sine wave; (d) square wave

exponentials or sinusoids, the phasor method cannot be applied directly to circuits for which they are inputs.

The waveforms of Fig. 17.1 have one thing in common, however: each is a function $f(t)$ that is periodic of period T. That is,

$$f(t) = f(t + T)$$

As Fourier showed, if such a function $f(t)$ satisfies a set of rather general conditions, it may be represented by the infinite series of sinusoids

$$f(t) = \frac{a_0}{2} + a_1 \cos \omega_0 t + a_2 \cos 2\omega_0 t + \ldots$$

$$+ b_1 \sin \omega_0 t + b_2 \sin 2\omega_0 t + \ldots$$

or, more compactly,

$$f(t) = \frac{a_0}{2} + \sum_{n=1}^{\infty} (a_n \cos n\omega_0 t + b_n \sin n\omega_0 t) \qquad (17.1)$$

where $\omega_0 = 2\pi/T$. This series is called the *trigonometric Fourier series,* or simply the *Fourier series,* of $f(t)$. The a's and b's are called the *Fourier coefficients* and depend, of course, on $f(t)$.

We see, therefore, that a nonsinusoidal wave, which has no phasor representation, may be represented by a series of sinusoids, each of which does have a phasor representation. Also, we see from (17.1) that a nonsinusoidal function may contain not just one frequency, like a sine wave, but an infinite number of frequencies, 0, ω_0, $2\omega_0$, ...

The Fourier coefficients may be determined rather easily by the use of Table 12.1. Let us begin by obtaining a_0, which may be done by integrating both sides of (17.1) over a full period; that is,

$$\int_0^T f(t)\, dt = \int_0^T \frac{a_0}{2}\, dt$$

$$+ \sum_{n=1}^{\infty} \int_0^T (a_n \cos n\omega_0 t + b_n \sin n\omega_0 t)\, dt$$

Since $T = 2\pi/\omega_0$, every term in the summation is zero by entry 2 in Table 12.1 ($\alpha = 0$), and therefore we have

$$a_0 = \frac{2}{T} \int_0^T f(t)\, dt \tag{17.2}$$

Next, let us multiply (17.1) through by $\cos m\omega_0 t$, where m is an integer, and integrate. This yields

$$\int_0^T f(t) \cos m\omega_0 t\, dt = \int_0^T \frac{a_0}{2} \cos m\omega_0 t\, dt$$

$$+ \sum_{n=1}^{\infty} a_n \int_0^T \cos m\omega_0 t \cos n\omega_0 t\, dt$$

$$+ \sum_{n=1}^{\infty} b_n \int_0^T \cos m\omega_0 t \sin n\omega_0 t\, dt$$

By entries 2, 4, and 5 of Table 12.1 (for $\alpha = \beta = 0$), every term in the right member is zero except the term where $n = m$ in the first summation. This term is given by

$$a_m \int_0^T \cos^2 m\omega_0 t\, dt = \frac{\pi}{\omega_0} a_m = \frac{T}{2} a_m$$

so that

$$a_m = \frac{2}{T} \int_0^T f(t) \cos m\omega_0 t\, dt, \qquad m = 1, 2, 3, \ldots \tag{17.3}$$

Finally, multiplying (17.1) by $\sin m\omega_0 t$, integrating, and applying Table 12.1, we have

$$b_m = \frac{2}{T} \int_0^T f(t) \sin m\omega_0 t\, dt, \qquad m = 1, 2, 3, \ldots \tag{17.4}$$

We note that (17.2) is the special case, $m = 0$, of (17.3) (which is why we used $a_0/2$ instead of a_0 for the constant term). Also, since $f(t)$ and $\cos n\omega_0 t$ are both periodic of period T, their product is periodic of period T. Similarly, the product of $f(t)$ and $\sin n\omega_0 t$ is periodic of period T. Thus we may integrate over any interval of length T, such as t_0 to $t_0 + T$, for arbitrary t_0, and the results will be the same. Therefore, we may summarize by giving the Fourier coefficients in the form

$$
\begin{aligned}
a_n &= \frac{2}{T} \int_{t_0}^{t_0+T} f(t) \cos n\omega_0 t \, dt, & n &= 0, 1, 2, \ldots \\
b_n &= \frac{2}{T} \int_{t_0}^{t_0+T} f(t) \sin n\omega_0 t \, dt, & n &= 1, 2, 3, \ldots
\end{aligned}
\tag{17.5}
$$

We have replaced the subscript m by n to correspond to the notation of (17.1).

The term $a_n \cos n\omega_0 t + b_n \sin n\omega_0 t$ in (17.1) is sometimes called the nth *harmonic*, by analogy with music theory. The case $n = 1$ is the first harmonic, or *fundamental*, with fundamental frequency ω_0. The case $n = 2$ is the second harmonic with frequency $2\omega_0$, and so forth. The term $a_0/2$ is the constant, or dc, component, and by (17.2) may be seen to be the average value of $f(t)$ over a period. It may be found quite often by inspection of the graph of $f(t)$.

The conditions that (17.1) is the Fourier series representing $f(t)$, where the Fourier coefficients are given by (17.5), are, as we have said, quite general and hold for almost any function we are likely to encounter in engineering. For the reader's information they are called the *Dirichlet* conditions and are as follows:

In any finite interval $f(t)$ has at most a finite number of finite discontinuities, a finite number of maxima and minima, and

$$
\int_0^T |f(t) \, dt| < \infty
\tag{17.6}
$$

The series converges to $\frac{1}{2}[f(t^+) + f(t^-)]$, which is $f(t)$ at the points where $f(t)$ is continuous.

EXAMPLE 17.1   As an example, let us find the Fourier series for the sawtooth wave of Fig. 17.2, given by

$$
\begin{aligned}
f(t) &= t, & -\pi < t < \pi \\
f(t + 2\pi) &= f(t)
\end{aligned}
\tag{17.7}
$$

Since $T = 2\pi$, we have $\omega_0 = 2\pi/T = 1$. If we choose $t_0 = -\pi$, then the first equation of (17.5) for $n = 0$ yields

$$
a_0 = \frac{1}{\pi} \int_{-\pi}^{\pi} t \, dt = 0
$$

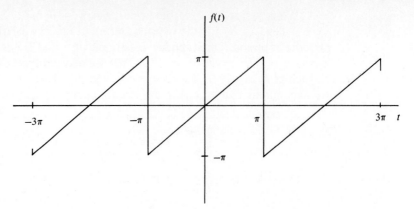

FIGURE 17.2 Sawtooth wave

For $n = 1, 2, 3, \ldots$, we have

$$a_n = \frac{1}{\pi} \int_{-\pi}^{\pi} t \cos nt \, dt$$

$$= \frac{1}{n^2\pi}(\cos nt + nt \sin nt) \Big|_{-\pi}^{\pi}$$

$$= 0$$

and

$$b_n = \frac{1}{\pi} \int_{-\pi}^{\pi} t \sin nt \, dt$$

$$= \frac{1}{n^2\pi}(\sin nt - nt \cos nt) \Big|_{-\pi}^{\pi}$$

$$= -\frac{2 \cos n\pi}{n}$$

$$= \frac{2(-1)^{n+1}}{n}$$

The case $n = 0$ had to be considered separately because of the appearance of n^2 in the denominator in the general case. Also, since $a_0/2$ is the average value of the sawtooth wave over a period, by inspection of Fig. 17.2 we see that $a_0 = 0$.

From our results the Fourier series for the sawtooth wave is

$$f(t) = 2\left(\frac{\sin t}{1} - \frac{\sin 2t}{2} + \frac{\sin 3t}{3} - \cdots\right) \qquad (17.8)$$

The fundamental and the second, third, and fifth harmonics are sketched for one period in Fig. 17.3. If a sufficient number of terms in (17.8) are taken, the series can be made to approximate $f(t)$ very nearly. For example, the first 10 harmonics are added and the result sketched in Fig. 17.4.

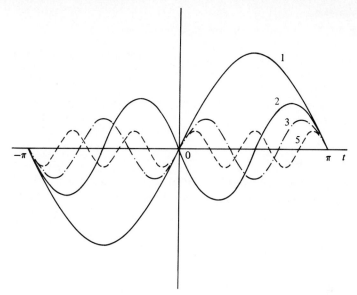

FIGURE 17.3 Four harmonics of (17.8)

FIGURE 17.4 Sum of the first 10 harmonics of (17.8)

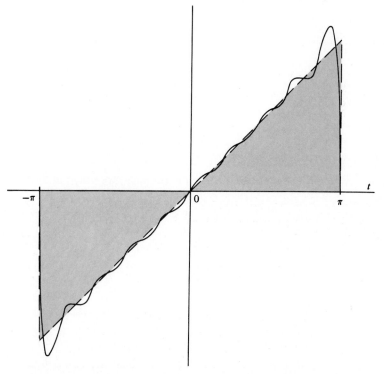

EXAMPLE 17.2 As another example, suppose we have

$$f(t) = 0, \qquad -2 < t < -1$$
$$= 6, \qquad -1 < t < 1$$
$$= 0, \qquad 1 < t < 2$$
$$f(t + 4) = f(t)$$

one period of which is shown in Fig. 17.5. Evidently, $T = 4$ and $\omega_0 = 2\pi/T = \pi/2$. If we take $t_0 = 0$ in (17.5), we must break each integral into three parts since on the interval from 0 to 4 $f(t)$ has 0, 6, and 0 values. If $t_0 = -1$, we only have to divide the integral into two parts since $f(t) = 6$ on -1 to 1 and $f(t) = 0$ on 1 to 3. Therefore let us choose this value of t_0 and obtain

$$a_0 = \frac{2}{4} \int_{-1}^{1} 6 \, dt + \frac{2}{4} \int_{1}^{3} 0 \, dt = 6$$

Also we have

$$a_n = \frac{2}{4} \int_{-1}^{1} 6 \cos \frac{n\pi t}{2} \, dt + \frac{2}{4} \int_{1}^{3} 0 \cos \frac{n\pi t}{2} \, dt$$

$$= \frac{12}{n\pi} \sin \frac{n\pi}{2}$$

and, finally,

$$b_n = \frac{2}{4} \int_{-1}^{1} 6 \sin \frac{n\pi t}{2} \, dt + \frac{2}{4} \int_{1}^{3} 0 \sin \frac{n\pi t}{2} \, dt$$

$$= 0$$

Thus the Fourier series is

$$f(t) = 3 + \frac{12}{\pi} \left(\cos \frac{\pi t}{2} - \frac{1}{3} \cos \frac{3\pi t}{2} + \frac{1}{5} \cos \frac{5\pi t}{2} - \cdots \right)$$

Since there are no even harmonics, we may put the result in the more compact form

$$f(t) = 3 + \frac{12}{\pi} \sum_{n=1}^{\infty} \frac{(-1)^{n+1} \cos \{[(2n - 1)\pi t]/2\}}{2n - 1}$$

FIGURE 17.5 One period ($-2 < t < 2$) of a periodic function

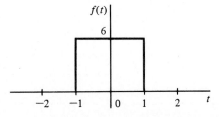

Chapter 17 Fourier Series

Anyone doubting the importance of selecting a good value of t_0 in (17.5) may wish to take $t_0 = 0$. The results will be the same, but not the effort.

EXERCISES

17.1.1   Find the Fourier series representation of the square wave

$$f(t) = 4, \qquad 0 < t < 1$$
$$= -4, \qquad 1 < t < 2$$
$$f(t + 2) = f(t)$$

Answer $\dfrac{16}{\pi} \displaystyle\sum_{n=1}^{\infty} \dfrac{\sin(2n - 1)\pi t}{2n - 1}$

17.1.2   Find the Fourier coefficients for

$$f(t) = 3, \qquad 0 < t < 1$$
$$= -1, \qquad 1 < t < 4$$
$$f(t + 4) = f(t)$$

Answer $a_0 = 0$, $a_n = \dfrac{4}{n\pi} \sin \dfrac{n\pi}{2}$, $b_n = \dfrac{4}{n\pi}\left(1 - \cos \dfrac{n\pi}{2}\right)$, $n = 1, 2, 3, \ldots$

17.1.3   Find the Fourier series for

$$f(t) = 2, \qquad 0 < t < \dfrac{\pi}{2}$$
$$= 0, \qquad \dfrac{\pi}{2} < t < \pi$$
$$f(t + \pi) = f(t)$$

Answer $1 + \dfrac{4}{\pi} \displaystyle\sum_{n=1}^{\infty} \dfrac{\sin 2(2n - 1)t}{2n - 1}$

17.1.4   Find the Fourier series for

$$f(t) = 0, \qquad -1 < t < -\dfrac{1}{2}$$
$$= 4, \qquad -\dfrac{1}{2} < t < \dfrac{1}{2}$$
$$= 0, \qquad \dfrac{1}{2} < t < 1$$
$$f(t + 2) = f(t)$$

Answer $2 + \dfrac{8}{\pi} \displaystyle\sum_{n=1}^{\infty} \dfrac{(-1)^{n+1} \cos (2n - 1)\pi t}{2n - 1}$

17.2
SYMMETRY PROPERTIES

If a function has symmetry about the vertical axis or the origin, then the computation of the Fourier coefficients may be greatly facilitated. A function $f(t)$ which is symmetrical about the vertical axis is said to be an *even* function and has the property

$$f(t) = f(-t) \tag{17.9}$$

for all t. That is, we may replace t by $-t$ without changing the function. Examples of even functions are t^2 and $\cos t$, and a typical even function is shown in Fig. 17.6(a). Evidently we could fold the figure along the vertical axis, and the two portions of the graph would coincide.

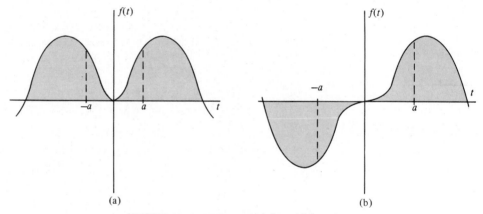

(a)   (b)

FIGURE 17.6 (a) Even and (b) odd function

An *odd* function $f(t)$ has the property

$$f(t) = -f(-t) \tag{17.10}$$

In other words, replacing t by $-t$ changes only the sign of the function. A typical odd function is shown in Fig. 17.6(b), and other examples are t and $\sin t$. Evidently we can fold the right half of the figure of an odd function and then rotate it about the t-axis so that it will coincide with the left half. Also, if a line is drawn from a point $[t, f(t)]$ through the origin, it will intercept the curve at $[-t, -f(t)]$. For this reason, an odd function is said to be symmetrical about the origin.

   Now let us see how symmetry properties can help us in determining the Fourier coefficients. In Fig. 17.6(a) we see that for $f(t)$ even,

$$\int_{-a}^{a} f(t)\, dt = 2 \int_{0}^{a} f(t)\, dt \tag{17.11}$$

546

This is true because the area from $-a$ to 0 is identical to that from 0 to a. This result may also be established analytically by writing

$$\int_{-a}^{a} f(t)\, dt = \int_{-a}^{0} f(t)\, dt + \int_{0}^{a} f(t)\, dt$$

$$= -\int_{a}^{0} f(-\tau)\, d\tau + \int_{0}^{a} f(t)\, dt$$

$$= \int_{0}^{a} f(\tau)\, d\tau + \int_{0}^{a} f(t)\, dt$$

$$= 2\int_{0}^{a} f(t)\, dt$$

In case $f(t)$ is odd, it is clear from Fig. 17.6(b) that

$$\int_{-a}^{a} f(t)\, dt = 0 \tag{17.12}$$

This is true because the integral from $-a$ to 0 is precisely the negative of that from 0 to a. This result may also be obtained analytically, as was done for even functions.

Products of two even functions or of two odd functions are even, whereas products of an even and an odd function are odd. For example, suppose $G(t)$ and $H(t)$ are both odd and that

$$F(t) = G(t)H(t)$$

Then we have

$$F(-t) = G(-t)H(-t)$$

$$= -G(t)[-H(t)]$$

$$= G(t)H(t)$$

$$= F(t)$$

and thus $F(t)$ is even. The other statements may be established similarly.

In the case of the Fourier coefficients we need to integrate the functions

$$g(t) = f(t) \cos n\omega_0 t \tag{17.13}$$

and

$$h(t) = f(t) \sin n\omega_0 t \tag{17.14}$$

If $f(t)$ is even, then since $\cos n\omega_0 t$ is even and $\sin n\omega_0 t$ is odd, $g(t)$ is even and $h(t)$ is odd. Therefore taking $t_0 = -T/2$ in (17.5), we have, for $f(t)$ even,

$$a_n = \frac{4}{T} \int_{0}^{T/2} f(t) \cos n\omega_0 t\, dt, \qquad n = 0, 1, 2, \ldots$$
$$b_n = 0, \qquad\qquad\qquad\qquad n = 1, 2, 3, \ldots \tag{17.15}$$

and for $f(t)$ odd,

$$a_n = 0, \qquad\qquad\qquad n = 0, 1, 2, \ldots$$
$$b_n = \frac{4}{T} \int_0^{T/2} f(t) \sin n\omega_0 t\, dt \qquad n = 1, 2, 3, \ldots$$

(17.16)

In either case one entire set of coefficients is zero, and the other set is obtained by taking twice the integral over half the period, as described by (17.15) and (17.16).

In summary, an even function has no sine terms, and an odd function has no constant or cosine terms in its Fourier series. As examples, the functions given earlier in (17.7) and Ex. 17.1.1 are odd functions, and their Fourier series have only sine terms. The function of Ex. 17.1.4 is even so that its series contains only cosine terms (including the dc case, $n = 0$). The function of Ex. 17.1.2 is neither even nor odd, and consequently both types of terms are present.

EXAMPLE 17.3   To illustrate the use of symmetry, let us find the Fourier coefficients of the square wave of Fig. 17.7, given by

$$f(t) = \quad 4, \qquad 0 < t < 1$$
$$= -4, \qquad 1 < t < 2$$
$$f(t + 2) = f(t)$$

which was given earlier in Ex. 17.1.1. Evidently $T = 2$, and thus $\omega_0 = \pi$. The function is odd, and, therefore, by (17.16) we have

$$a_n = 0, \qquad n = 0, 1, 2, \ldots$$

$$b_n = \frac{4}{2} \int_0^1 4 \sin n\pi t\, dt$$

$$= \frac{8}{n\pi}[1 - (-1)^n]$$

Therefore

$$b_n = 0, \qquad n \text{ even}$$

$$= \frac{16}{n\pi}, \qquad n \text{ odd}$$

which checks Ex. 17.1.1.

Figure 17.8 illustrates how well the Fourier series approximates the square wave. Only one term, the fundamental, is shown in (a), the sum of the first 9

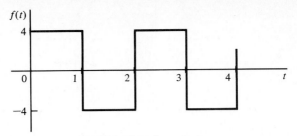

FIGURE 17.7 Square wave

harmonics is shown in (b), a much closer fit, and the sum of the first 19 harmonics, shown in (c), is a still closer fit. We note that in all three cases the Fourier series exhibits an *overshoot* (above or below the actual square-wave value shown dashed) at the points of discontinuity, $t = 0, 1, 2, \ldots$. This overshoot is present for any discontinuous function for any finite number of terms of the series, a fact known as the *Gibbs phenomenon.*

Another type of symmetry that leads to simplification in the calculation of the Fourier coefficients is *half-wave symmetry,* which a periodic function $f(t)$ possesses if

$$f\left(t + \frac{T}{2}\right) = -f(t) \tag{17.17}$$

where T is the period of $f(t)$. Such a function may be described as one whose graph from $T/2$ to T is its graph from 0 to $T/2$ shifted $T/2$ units to the right and inverted. An example of one period is shown in Fig. 17.9, where the dashed line represents the shifted half-period, which is then inverted.

As the reader is asked to show in Ex. 17.2.4, the Fourier coefficients of a function $f(t)$ with half-wave symmetry are given by

$$
\begin{aligned}
a_n &= 0, && n \text{ even} \\
&= \frac{4}{T} \int_0^{T/2} f(t) \cos n\omega_0 t \, dt, && n \text{ odd}
\end{aligned}
\tag{17.18}
$$

and

$$
\begin{aligned}
b_n &= 0, && n \text{ even} \\
&= \frac{4}{T} \int_0^{T/2} f(t) \sin n\omega_0 t \, dt, && n \text{ odd}
\end{aligned}
\tag{17.19}
$$

Thus the Fourier series contains only odd harmonics. For this reason $f(t)$ is sometimes referred to as an *odd harmonic* function. Also, half-wave symmetry is sometimes called *rotational symmetry,* being a form of symmetry produced by rotating electrical machinery.

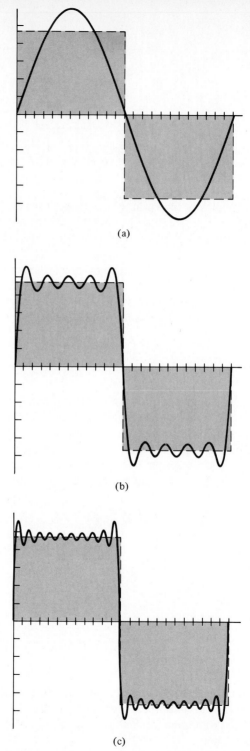

FIGURE 17.8 Approximations to the square wave using (a) the fundamental, (b) the first 9 harmonics, and (c) the first 19 harmonics

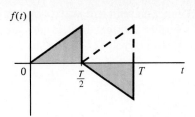

FIGURE 17.9 One period of a function with half-wave symmetry

EXAMPLE 17.4 As an example, let us find the Fourier series of $f(t)$ given in Fig. 17.10, which clearly has half-wave symmetry. If $T = 2\pi$ and $A = \pi$, then $\omega_0 = 2\pi/T = 1$, and

$$f(t) = t, \qquad 0 < t < \pi$$

Therefore by (17.18) we have, for n odd,

$$a_n = \frac{2}{\pi} \int_0^\pi t \cos nt \, dt,$$

$$= \frac{2}{\pi n^2} [\cos nt + nt \sin nt]_0^\pi$$

$$= -\frac{4}{n^2 \pi}$$

and by (17.19), for n odd,

$$b_n = \frac{2}{\pi} \int_0^\pi t \sin nt \, dt$$

$$= \frac{2}{\pi n^2} [\sin nt - nt \cos nt]_0^\pi$$

$$= \frac{2}{n}$$

The even-harmonic coefficients are zero, and thus the Fourier series is

$$f(t) = 2 \sum_{n=1}^\infty \frac{-2}{(2n-1)^2 \pi} \cos (2n-1)t + \frac{1}{2n-1} \sin (2n-1)t$$

FIGURE 17.10 Function with half-wave symmetry

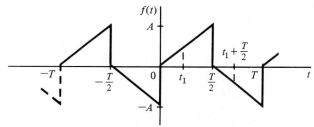

A function with half-wave symmetry may also be even or odd, in which case the work is further shortened. Examples are given in Probs. 17.4 and 17.5. The general case is given in Prob. 17.6.

EXERCISES

17.2.1 Find the Fourier series for the sawtooth wave of (17.7) using symmetry properties.

17.2.2 Find the Fourier series for the half-wave rectified sinusoid

$$f(t) = 0, \qquad -\frac{\pi}{2} < t < -\frac{\pi}{4}$$

$$= 4 \cos 2t, \qquad -\frac{\pi}{4} < t < \frac{\pi}{4}$$

$$= 0, \qquad \frac{\pi}{4} < t < \frac{\pi}{2}$$

$$f(t + \pi) = f(t)$$

shown in the accompanying figure.

Answer $a_0 = \dfrac{8}{\pi}$; $a_1 = 2$; $a_n = \dfrac{8}{\pi(1 - n^2)} \cos \dfrac{n\pi}{2}$, $n = 2, 3, 4, \ldots$; $b_n = 0$, $n = 1, 2, 3, \ldots$

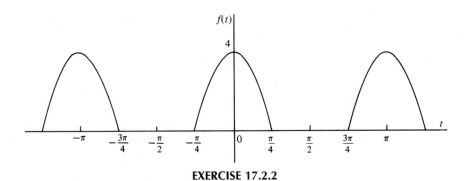

EXERCISE 17.2.2

17.2.3 Find the Fourier series for the full-wave rectified sinusoid

$$f(t) = |4 \sin 2t|$$

shown in the accompanying figure.

Answer $\dfrac{16}{\pi} \left(\dfrac{1}{2} + \sum_{n=1}^{\infty} \dfrac{1}{1 - 4n^2} \cos 4nt \right)$

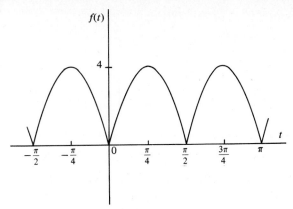

EXERCISE 17.2.3

17.2.4 Show that if $f(t + T/2) = -f(t)$, and

$$g_1(t) = f(t) \cos n\omega_0 t$$

$$g_2(t) = f(t) \sin n\omega_0 t$$

then

$$g_1\left(t + \frac{T}{2}\right) = (-1)^{n+1}g_1(t)$$

$$g_2\left(t + \frac{T}{2}\right) = (-1)^{n+1}g_2(t)$$

Use the result to derive (17.18) and (17.19).

17.2.5 Find the Fourier series for the function shown. Use the fact that it has half-wave symmetry.

Answer $\dfrac{16}{\pi} \displaystyle\sum_{n=1}^{\infty} \dfrac{(-1)^{n+1}}{2n - 1} \cos (2n - 1)\pi t$

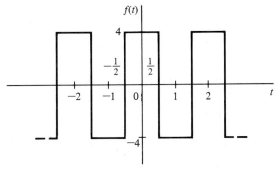

EXERCISE 17.2.5

17.2.6 The function shown in (a) of the accompanying figure is neither odd nor even, but

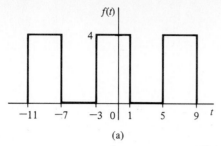

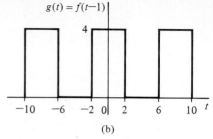

(a)

(b)

EXERCISE 17.2.6

when it is translated one unit to the right (t is replaced by $t - 1$), a new even function $g(t) = f(t - 1)$ is obtained, as shown in (b). Find the Fourier series for $f(t)$ by finding it for $g(t)$, taking advantage of symmetry, and then writing $f(t) = g(t + 1)$.

Answer $\dfrac{a_0}{2} = 2; \quad a_n = \dfrac{8}{n\pi} \sin \dfrac{n\pi}{2} \cos \dfrac{n\pi}{4}, \quad b_n = \dfrac{-8}{n\pi} \sin \dfrac{n\pi}{2} \sin \dfrac{n\pi}{4}$

17.3

RESPONSE TO PERIODIC EXCITATIONS

We are now in a position to find the forced response of a circuit to any periodic excitation that has a Fourier series. The excitation is expressed as a Fourier series, and the response to each term of the series is found by the phasor method. Then the total response is, by superposition, the sum of the individual responses.

EXAMPLE 17.5 Before proceeding to the general case, let us find the forced response $i(t)$ of the circuit of Fig. 17.11(a) excited by the square-wave voltage of Fig. 17.11(b). The source voltage is given by

FIGURE 17.11 (a) *RL* circuit; (b) its square-wave input

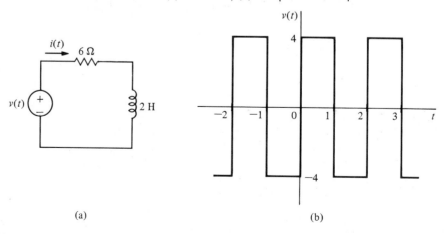

(a)

(b)

$$v(t) = \quad 4 \text{ V}, \qquad 0 < t < 1 \text{ s}$$
$$= -4 \text{ V}, \qquad 1 < t < 2 \text{ s}$$
$$v(t + 2) = v(t)$$

and its Fourier series was found earlier in Ex. 17.1.1 to be

$$v(t) = \frac{16}{\pi} \sum_{n=1}^{\infty} \frac{\sin (2n - 1)\pi t}{2n - 1} \text{ V}$$

Therefore, we may write

$$v(t) = \sum_{n=1}^{\infty} v_n$$

where

$$v_n = \frac{16}{(2n - 1)\pi} \sin (2n - 1)\pi t$$

$$= \frac{16}{(2n - 1)\pi} \cos [(2n - 1)\pi t - 90°]$$

is the $(2n - 1)$th harmonic of $v(t)$, with frequency $\omega_n = (2n - 1)\pi$ rad/s.
The phasor voltage is

$$\mathbf{V}_n = \frac{16}{(2n - 1)\pi} \underline{/-90°}$$

and the impedance seen at the terminals of the source at ω_n is

$$\mathbf{Z}(j\omega_n) = 6 + j2(2n - 1)\pi$$

$$= 2\sqrt{9 + \pi^2(2n - 1)^2} \underline{/ \tan^{-1} \frac{(2n - 1)\pi}{3}}$$

The phasor current is, therefore,

$$\mathbf{I}_n = |\mathbf{I}_n| \underline{/-90° - \theta_n}$$

where

$$|\mathbf{I}_n| = \frac{8}{(2n - 1)\pi \sqrt{9 + \pi^2(2n - 1)^2}} \tag{17.20}$$

and

$$\theta_n = \tan^{-1} \frac{(2n - 1)\pi}{3} \tag{17.21}$$

The time-domain current is given by

$$i_n(t) = |\mathbf{I}_n| \cos (\omega_n t - 90° - \theta_n)$$
$$= |\mathbf{I}_n| \sin (\omega_n t - \theta_n) \tag{17.22}$$

Therefore, by superposition, the forced component of the current is

$$i_f(t) = \sum_{n=1}^{\infty} |\mathbf{I}_n| \sin(\omega_n t - \theta_n) \qquad (17.23)$$

We note that we could shorten the procedure by basing the phasor on the sine function instead of the cosine function, in which case the phase angle of $-90°$ would not be added in at first and subtracted out at the end to obtain the second of (17.22).

We may obtain the natural response, and thus the complete response, as we have done in the past. A difficulty with the procedure is the necessity for finding the arbitrary constant in the natural response by summing an infinite series. This, of course, also presents difficulties in finding $i(t)$ for a given t. In this example, as we see by (17.20), the current amplitude decreases as n increases, and tends to zero as n becomes infinite. This is true in general for the Fourier coefficients, since we know the series converges. Therefore, the series may be truncated to a relatively few terms and still yield a good approximation.

The steady-state current, approximating (17.23) by its first six terms, is the solid curve of Fig. 17.12. Its peak of 0.57 A is reached at $t = 1$ s. The source voltage is shown dashed using a different scale.

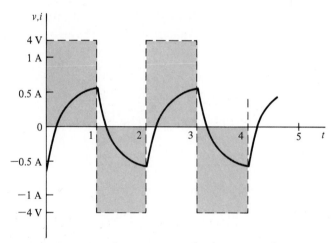

FIGURE 17.12 Steady-state current for the circuit of Fig. 17.11

To minimize the work involved in the phasor method, we need an alternative form of the Fourier series, using only cosine terms. To obtain such a form, we need only write the nth harmonic ($n \neq 0$) as a single sinusoid,

$$a_n \cos n\omega_0 t + b_n \sin n\omega_0 t = A_n \cos(n\omega_0 t + \phi_n) \qquad (17.24)$$

where by (10.9) and (10.10) the amplitude is

$$A_n = \sqrt{a_n^2 + b_n^2} \qquad (17.25)$$

and the phase is

$$\phi_n = -\tan^{-1}\frac{b_n}{a_n} \qquad (17.26)$$

The trigonometric Fourier series is then given by

$$f(t) = \frac{a_0}{2} + \sum_{n=1}^{\infty} A_n \cos\left(n\omega_0 t + \phi_n\right) \qquad (17.27)$$

As in the phasor case, care must be exercised in determining the quadrant of ϕ_n. The terminal side of the angle $\tan^{-1}(b_n/a_n)$ is in the quadrant in which the point (a_n, b_n) is located, which is determined by the signs of a_n and b_n.

If the nth harmonic is a voltage v_n, then its phasor representation is the voltage

$$\mathbf{V}_n = A_n \underline{/\phi_n} \,.$$

and the frequency is $\omega = n\omega_0$. The phasor representation of the dc component

$$v_0 = \frac{a_0}{2}$$

will be designated

$$\mathbf{V}_0 = A_0 \underline{/\phi_0}$$

where

$$A_0 = \left|\frac{a_0}{2}\right| \qquad (17.28)$$

and

$$\begin{aligned}\phi_0 &= 0, && a_0 > 0 \\ &= 180^\circ, && a_0 < 0\end{aligned} \qquad (17.29)$$

If the impedance seen by the voltage $\mathbf{V}_n$ is $\mathbf{Z}(jn\omega_0)$, which we designate

$$\mathbf{Z}(jn\omega_0) = \mathbf{Z}_n = |\mathbf{Z}_n| \underline{/\theta_n}$$

then the terminal phasor current is

$$\mathbf{I}_n = \frac{\mathbf{V}_n}{\mathbf{Z}_n} = \frac{A_n}{|\mathbf{Z}_n|} \underline{/\phi_n - \theta_n}$$

from which we may find the time-domain current i_n, as before, using phasors.

The response i_0 to the dc voltage v_0 is found as before, where capacitors are open circuited and inductors are short circuited. If $\mathbf{Z}(0)$ is the resistance of the circuit in this case, then the phasor of i_0 is

$$\mathbf{I}_0 = \frac{\mathbf{V}_0}{\mathbf{Z}(0)}$$

which determines i_0.

If the terminal voltage v is nonsinusoidal, but is periodic with a Fourier series

$$v = \sum_{n=0}^{\infty} v_n \qquad (17.30)$$

then by superposition the terminal current is

$$i = \sum_{n=0}^{\infty} i_n = \frac{a_0}{2\mathbf{Z}(0)} + \sum_{n=1}^{\infty} \frac{A_n}{|\mathbf{Z}_n|} \cos(n\omega_0 t + \phi_n - \theta_n) \qquad (17.31)$$

EXAMPLE 17.6 As an example, let us find the forced response i in the circuit of Fig. 17.13 if the source voltage is given by

$$
\begin{aligned}
v &= 4 \text{ V}, &\quad 0 < t < 1 \text{ s} \\
&= 0 \text{ ,} &\quad 1 < t < 4 \text{ s} \\
v(t + 4) &= v(t)
\end{aligned}
\qquad (17.32)
$$

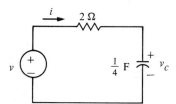

FIGURE 17.13 *RC* circuit

We have $\omega_0 = \pi/2$, and the Fourier coefficients of v may be shown to be

$$\frac{a_0}{2} = 1$$

$$a_n = \frac{4}{n\pi} \sin \frac{n\pi}{2}$$

$$b_n = \frac{4}{n\pi}\left(1 - \cos \frac{n\pi}{2}\right)$$

Therefore we have

$$
\begin{aligned}
A_n &= \sqrt{\left(\frac{4}{n\pi}\right)^2 \left[\sin^2 \frac{n\pi}{2} + \left(1 - \cos \frac{n\pi}{2}\right)^2\right]} \\
&= \frac{4}{n\pi}\sqrt{2\left(1 - \cos \frac{n\pi}{2}\right)} \\
&= \frac{8}{n\pi}\left|\sin \frac{n\pi}{4}\right|
\end{aligned}
$$

and

$$\phi_n = -\tan^{-1}\left[\frac{(4/n\pi)[1 - \cos{(n\pi/2)}]}{(4/n\pi)\sin{(n\pi/2)}}\right]$$

$$= -\frac{n\pi}{4}$$

The source voltage is therefore

$$v = 1 + \sum_{n=1}^{\infty} A_n \cos\left(\frac{n\pi t}{2} + \phi_n\right) \quad \text{V}$$

and the impedance is

$$\mathbf{Z}_n = 2 - j\frac{8}{n\pi} \quad \Omega$$

$$= \frac{2}{n\pi}\sqrt{n^2\pi^2 + 16} \bigg/ -\tan^{-1}\frac{4}{n\pi} \quad \Omega$$

Since the capacitor presents an infinite impedance to dc, the term $a_0/[2\mathbf{Z}(0)]$ is zero. Therefore, by (17.31) the current may be simplified to

$$i = 4 \sum_{n=1}^{\infty} \frac{|\sin n\pi/4|}{\sqrt{16 + n^2\pi^2}} \cos\left(\frac{n\pi t}{2} - \frac{n\pi}{4} + \tan^{-1}\frac{4}{n\pi}\right)$$

EXAMPLE 17.7 As another example, let us find the forced response v_c in Fig. 17.13. In the phasor domain, by voltage division, we have, for $n \neq 0$,

$$\mathbf{V}_{cn} = \frac{8/jn\pi}{2 + (8/jn\pi)}\mathbf{V}_n = \frac{4A_n \, /\phi_n}{4 + jn\pi} \tag{17.33}$$

where $\mathbf{V}_{cn}$ is the phasor of v_{cn}, the nth harmonic of v_c. Substituting for A_n and ϕ_n and simplifying yields

$$\mathbf{V}_{cn} = \frac{32|\sin{(n\pi/4)}|}{n\pi\sqrt{n^2\pi^2 + 16}} \bigg/ -\frac{n\pi}{4} - \tan^{-1}\frac{n\pi}{4}$$

The dc component $\mathbf{V}_{c0}$ is simply $\mathbf{V}_0 = 1$, since the current is zero in the dc case. The time-domain voltage is therefore

$$v_c = 1 + \frac{32}{\pi}\sum_{n=1}^{\infty} \frac{|\sin{(n\pi/4)}|}{n\sqrt{n^2\pi^2 + 16}} \cos\left(\frac{n\pi t}{2} - \frac{n\pi}{4} - \tan^{-1}\frac{n\pi}{4}\right) \quad \text{V}$$

In the general case, the network function $\mathbf{H}(s)$, as defined in Sec. 14.5, may be used to determine the output, say y for a given periodic input x. If $\mathbf{X}_n$ is the phasor of the nth harmonic of x, then $\mathbf{Y}_n$, the phasor of the nth harmonic of y, is given by

$$\mathbf{Y}_n = \mathbf{H}(jn\omega_0)\mathbf{X}_n$$

from which we may find y_n, the nth component of y. Then by superposition,

$$y = \sum_{n=0}^{\infty} y_n$$

EXAMPLE 17.8 In the last example, the network function was

$$\mathbf{H}(s) = \frac{\mathbf{V}_c(s)}{\mathbf{V}(s)} = \frac{2}{s + 2}$$

where $s = jn\omega_0 = jn\pi/2$. Therefore

$$\mathbf{V}_{cn} = \frac{2}{(jn\pi/2) + 2}\mathbf{V}_n = \frac{4A_n \underline{/\phi_n}}{4 + jn\pi}$$

which checks (17.33).

EXERCISES

17.3.1 Find the forced component of the voltage across the inductor in the circuit of Fig. 17.11.

Answer $\displaystyle\sum_{n=1}^{\infty} \frac{16}{\sqrt{9 + \omega_n^2}} \cos\left(\omega_n t - \tan^{-1}\frac{\omega_n}{3}\right)$, where $\omega_n = (2n - 1)\pi$

17.3.2 Find the forced component of v in the circuit of Prob. 11.17 if v_g is the voltage defined in (17.32).

Answer $\displaystyle -1 - \frac{8}{\pi}\sum_{n=1}^{\infty}\frac{1}{n}\left|\sin\frac{n\pi}{4}\right|\cos\left(\frac{n\pi t}{2} - 2\tan^{-1}\frac{n\pi}{4} - \frac{n\pi}{4}\right)$

17.3.3 Find the Fourier series of type (17.27) of $f(t)$ shown in (a) of the accompanying figure by translating the vertical axis one unit to the right, finding the trigonometric Fourier series of the new function $g(t)$ obtained in (b), and translating back to $f(t) = g(t - 1)$.

Answer $\displaystyle 1 + \frac{4}{\pi}\sum_{n=1}^{\infty}\frac{1}{n}\sin\frac{n\pi}{2}\cos\left(\frac{n\pi t}{2} - \frac{n\pi}{2}\right)$

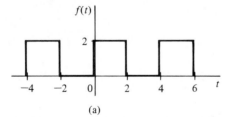

(a)

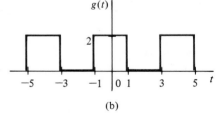

(b)

EXERCISE 17.3.3

17.3.4 Apply the idea of Ex. 17.3.3 to the example of (17.32).

17.4

AVERAGE POWER AND RMS VALUES

As we have seen, a periodic input signal that is represented by a Fourier series yields a periodic output expressed as a Fourier series. In particular, for the *RL* circuit of Fig.

17.11(a), an input voltage

$$v = \frac{16}{\pi} \sum_{n=1}^{\infty} \frac{\sin (2n - 1)\pi t}{2n - 1} \quad \text{V} \tag{17.34}$$

yielded a terminal current

$$i = \sum_{n=1}^{\infty} |\mathbf{I}_n| \sin[(2n - 1)\pi t - \theta_n] \quad \text{A} \tag{17.35}$$

where

$$|\mathbf{I}_n| = \frac{8}{(2n - 1)\pi\sqrt{9 + \pi^2(2n - 1)^2}}$$

and

$$\theta_n = \tan^{-1} \frac{(2n - 1)\pi}{3}$$

If we are interested in the average power delivered by the input, we have the problem of computing the average value of a product vi of two Fourier series. We will consider this problem in this section.

Let us begin by considering two Fourier series, which we designate

$$f = A_{\text{dc}} + \sum_{n=1}^{\infty} A_n \cos (n\omega_0 t + \phi_n) \tag{17.36}$$

and

$$g = B_{\text{dc}} + \sum_{m=1}^{\infty} B_m \cos (m\omega_0 t + \theta_m) \tag{17.37}$$

If we multiply these series together and take the average value, we have

$$\frac{1}{T}\int_0^T fg\, dt = \frac{1}{T}\int_0^T A_{\text{dc}} B_{\text{dc}} dt + \sum_{n=1}^{\infty} \frac{A_n B_{\text{dc}}}{T} \int_0^T \cos(n\omega_0 t + \phi_n)\, dt$$

$$+ \sum_{m=1}^{\infty} \frac{A_{\text{dc}} B_m}{T} \int_0^T \cos(m\omega_0 t + \theta_m)\, dt$$

$$+ \sum_{n=1}^{\infty} \sum_{m=1}^{\infty} \frac{A_n B_m}{T} \int_0^T \cos(m\omega_0 t + \theta_m) \cos(n\omega_0 t + \phi_n)\, dt$$

By entries 1 and 5 of Table 12.1, all the terms are zero in the first two series on the right and in the third series when $m \neq n$. Integrating the first term and applying entry 5 for the third series when $m = n$, we have

$$\frac{1}{T}\int_0^T fg\, dt = A_{\text{dc}} B_{\text{dc}} + \sum_{n=1}^{\infty} \frac{A_n B_n}{2} \cos(\phi_n - \theta_n) \tag{17.38}$$

If $f = v$, a terminal voltage, and $g = i$, the corresponding terminal current, where $A_{\text{dc}} = V_{\text{dc}}$, $A_n = V_n$, $B_{\text{dc}} = I_{\text{dc}}$, and $B_n = I_n$, then (17.38) is the average power delivered to the circuit, given by

$$P = \frac{1}{T} \int_0^T vi \, dt = V_{dc} I_{dc} + \sum_{n=1}^{\infty} \frac{V_n I_n}{2} \cos(\phi_n - \theta_n) \qquad (17.39)$$

Since V_n and I_n are the amplitudes of the nth harmonic voltage and current, ϕ_n and θ_n are the phases, and V_{dc} and I_{dc} are the dc components, this result is superposition of power,

$$P = P_{dc} + \sum_{n=1}^{\infty} P_n \qquad (17.40)$$

where

$$P_{dc} = V_{dc} I_{dc}$$

is the power delivered by the dc components, and

$$P_n = \frac{V_n I_n}{2} \cos(\phi_n - \theta_n)$$

is the power delivered by the nth harmonics v_n and i_n acting alone. This is, of course, a generalization of (12.15).

EXAMPLE 17.9   In the *RL* example with terminal voltage and current given by (17.34) and (17.35), the sine functions may be converted to cosines by subtracting 90° from their phase angles. Since these will cancel in (17.39), we have

$$P = \sum_{n=1}^{\infty} \frac{16}{2(2n-1)\pi} \left[\frac{8}{(2n-1)\pi\sqrt{9 + \pi^2(2n-1)^2}} \right] \cos \left[\tan^{-1} \frac{(2n-1)\pi}{3} \right]$$

which simplifies to

$$P = \frac{192}{\pi^2} \sum_{n=1}^{\infty} \frac{1}{(2n-1)^2[9 + \pi^2(2n-1)^2]} \quad \text{W} \qquad (17.41)$$

If $g = f$ in (17.38), then $B_{dc} = A_{dc}$, $B_n = A_n$, and $\theta_n = \phi_n$, so that we have

$$\frac{1}{T} \int_0^T f^2(t) \, dt = A_{dc}^2 + \sum_{n=1}^{\infty} \frac{A_n^2}{2} \qquad (17.42)$$

which holds for the Fourier series (17.36). This result is called *Parseval's equation*. If f is a current i (or a voltage v), then the left member of (17.42) is the square of its rms value, so that

$$I_{rms}^2 = I_{dc}^2 + \sum_{n=1}^{\infty} \frac{I_n^2}{2}$$

Since I_n is the amplitude of the nth harmonic, $I_n/\sqrt{2}$ is its rms value I_{nrms}, so we have

$$I_{rms} = \sqrt{I_{dc}^2 + \sum_{n=1}^{\infty} I_{nrms}^2} \qquad (17.43)$$

This is a generalization of (12.20), where the number of currents is infinite. A similar result may be stated for voltages.

EXAMPLE 17.10 As an example, let us find the rms value of v given by (17.34). We see that $V_{dc} = 0$ and

$$V_n = \frac{16}{\pi(2n - 1)} = \sqrt{2}\, V_{nrms}$$

(The rms value is the same for a sine or a cosine function, since they differ only by a phase angle.) Therefore we have

$$V_{rms} = \frac{8\sqrt{2}}{\pi} \sqrt{\sum_{n=1}^{\infty} \frac{1}{(2n - 1)^2}} \quad V \qquad (17.44)$$

EXERCISES

17.4.1 Find the average power delivered to the circuit of Fig. 17.13 by the source of (17.32).

Answer $16 \displaystyle\sum_{n=1}^{\infty} \frac{\sin^2 (n\pi/4)}{n^2 \pi^2 + 16}$

17.4.2 Find the rms value of the current in (17.35), and use the result to show that the power absorbed by the 6-Ω resistor of the circuit is the same as (17.41).

Answer $I_{rms} = \dfrac{4\sqrt{2}}{\pi} \sqrt{\displaystyle\sum_{n=1}^{\infty} \frac{1}{(2n - 1)^2[9 + \pi^2(2n - 1)^2]}}$

17.4.3 Apply Parseval's equation to the Fourier series (17.34) of the function

$$\begin{aligned} v = \quad &4, \qquad 0 < t < 1 \\ = \,&{-4}, \qquad 1 < t < 2 \end{aligned}$$

$$v(t + 2) = v(t)$$

to show that

$$\sum_{n=1}^{\infty} \frac{1}{(2n - 1)^2} = \frac{\pi^2}{8}$$

Use this result to find V_{rms} of (17.44). Check the answer by the definition of rms.
Answer 4 V

17.5

THE EXPONENTIAL FOURIER SERIES

We may obtain yet another form of the Fourier series by replacing the sine and cosine functions by their exponential equivalents, using Euler's formula. This form is called the *exponential* Fourier series and is extremely useful, particularly in considering frequency responses, one of the most important applications of Fourier series. In this section we will consider the exponential Fourier series, and in the following section we will investigate the frequency responses or frequency spectra.

Replacing the sinusoidal terms in the trigonometric Fourier series by their exponential equivalents,

$$\cos n\omega_0 t = \frac{1}{2}(e^{jn\omega_0 t} + e^{-jn\omega_0 t})$$

and

$$\sin n\omega_0 t = \frac{1}{j2}(e^{jn\omega_0 t} - e^{-jn\omega_0 t})$$

(see Appendix D), and collecting terms, we have

$$f(t) = \frac{a_0}{2} + \sum_{n=1}^{\infty}\left[\left(\frac{a_n - jb_n}{2}\right)e^{jn\omega_0 t} + \left(\frac{a_n + jb_n}{2}\right)e^{-jn\omega_0 t}\right] \tag{17.45}$$

If we define a new coefficient c_n by

$$c_n = \frac{a_n - jb_n}{2} \tag{17.46}$$

and substitute for a_n and b_n from (17.5), with $t_0 = -T/2$, we have

$$c_n = \frac{1}{T}\int_{-T/2}^{T/2} f(t)(\cos n\omega_0 t - j\sin n\omega_0 t)\,dt$$

or, by Euler's formula,

$$c_n = \frac{1}{T}\int_{-T/2}^{T/2} f(t)e^{-jn\omega_0 t}\,dt \tag{17.47}$$

We observe also that c_n^* (the conjugate of c_n) is given by

$$c_n^* = \frac{a_n + jb_n}{2}$$

$$= \frac{1}{T}\int_{-T/2}^{T/2} f(t)(\cos n\omega_0 t + j\sin n\omega_0 t)\,dt$$

which is evidently c_{-n} (c_n with n replaced by $-n$). That is,

$$c_{-n} = \frac{a_n + jb_n}{2} \tag{17.48}$$

Finally, let us observe that

$$\frac{a_0}{2} = \frac{1}{T}\int_{-T/2}^{T/2} f(t)\,dt$$

which by (17.47) is

$$\frac{a_0}{2} = c_0 \qquad (17.49)$$

Summing up, (17.46), (17.48), and (17.49) enable us to write (17.45) in the form

$$f(t) = c_0 + \sum_{n=1}^{\infty} c_n e^{jn\omega_0 t} + \sum_{n=1}^{\infty} c_{-n} e^{-jn\omega_0 t}$$

$$= \sum_{n=0}^{\infty} c_n e^{jn\omega_0 t} + \sum_{n=-1}^{-\infty} c_n e^{jn\omega_0 t}$$

We have combined c_0 with the first summation and replaced the dummy summation index n by $-n$ in the second summation. The result is more compactly written as

$$f(t) = \sum_{n=-\infty}^{\infty} c_n e^{jn\omega_0 t} \qquad (17.50)$$

where c_n is given by (17.47). This version of the Fourier series is, of course, called the exponential Fourier series.

EXAMPLE 17.11 As an example, let us find the exponential series for the rectangular wave of Ex. 17.1.1, given by

$$f(t) = \quad 4, \qquad 0 < t < 1$$

$$= -4, \qquad 1 < t < 2$$

with $T = 2$. We have $\omega_0 = 2\pi/T = \pi$, and thus by (17.47)

$$c_n = \frac{1}{2} \int_{-1}^{1} f(t) e^{-jn\pi t} \, dt$$

For $n \neq 0$ this is

$$c_n = \frac{1}{2} \int_{-1}^{0} (-4) e^{-jn\pi t} \, dt + \frac{1}{2} \int_{0}^{1} 4 e^{-jn\pi t} \, dt$$

$$= \frac{4}{jn\pi} [1 - (-1)^n]$$

Also we have

$$c_0 = \frac{1}{2} \int_{-1}^{1} f(t) \, dt$$

$$= \frac{1}{2} \int_{-1}^{0} 4 \, dt - \frac{1}{2} \int_{0}^{1} 4 \, dt$$

$$= 0$$

Since $c_n = 0$ for n even and $c_n = 8/jn\pi$ for n odd, we may write the exponential series in the form

$$f(t) = \frac{8}{j\pi} \sum_{n=-\infty}^{\infty} \frac{1}{2n-1} e^{j(2n-1)\pi t} \tag{17.51}$$

The reader may verify that this result is equivalent to that obtained in Ex. 17.1.1.

We may also find the forced response y to a periodic input x using the exponential Fourier series. That is, if x is given by

$$x = \sum_{n=-\infty}^{\infty} c_n e^{jn\omega_0 t}$$

a linear combination of complex excitations,

$$x_n = c_n e^{jn\omega_0 t}$$

we may use superposition to find the response y. We note that x_n has precisely the form of the complex excitations considered in Sec. 10.3, and therefore c_n is a phasor. The complex output y_n, due to x_n, is therefore

$$y_n = \mathbf{H}(jn\omega_0)c_n e^{jn\omega_0 t}$$

where $\mathbf{H}(s)$ is the network function. The response is therefore

$$y = \sum_{n=-\infty}^{\infty} y_n = \sum_{n=-\infty}^{\infty} \mathbf{H}(jn\omega_0)c_n e^{jn\omega_0 t}$$

which is the exponential Fourier series of y.

EXERCISES

17.5.1 Find the exponential Fourier series for

$$f(t) = 1 \qquad -1 < t < 1$$
$$= 0, \qquad 1 < |t| < 2$$
$$f(t+4) = f(t)$$

Answer $\dfrac{1}{2} + \dfrac{1}{\pi} \displaystyle\sum_{\substack{n=-\infty \\ n\neq 0}}^{\infty} \dfrac{\sin(n\pi/2)}{n} e^{jn\pi t/2}$

17.5.2 The function

$$\mathrm{Sa}(x) = \frac{\sin x}{x}$$

called the *sampling function*, occurs frequently in communication theory. [A closely related function is the *sinc function*, defined by sinc $x = (\sin \pi x)/\pi x = \mathrm{Sa}(\pi x)$.] Show that if we define

$$c_0 = \lim_{n \to 0} c_n$$

then the result of Ex. 17.5.1 may be given by

$$f(t) = \frac{1}{2} \sum_{n=-\infty}^{\infty} \mathrm{Sa}\left(\frac{n\pi}{2}\right) e^{jn\pi t/2}$$

17.5.3 Generalize the results of Ex. 17.5.2 by finding the exponential series of a train of pulses of width $\delta < T$, described by

$$f(t) = 1, \qquad -\frac{\delta}{2} < t < \frac{\delta}{2}$$

$$= 0, \qquad \frac{\delta}{2} < |t| < \frac{T}{2}$$

$$f(t + T) = f(t)$$

Answer $\dfrac{\delta}{T} \displaystyle\sum_{n=-\infty}^{\infty} \mathrm{Sa}\!\left(\dfrac{n\pi\delta}{T}\right) e^{j2n\pi t/T}$

17.5.4 Find the forced component of the current in Fig. 17.11 using the exponential Fourier series. Show that the exponential Fourier series for the current reduces to the trigonometric series (17.23). [*Suggestion:* The exponential Fourier series for the input $v(t)$ is given in (17.51).]

Answer $\dfrac{4}{j\pi} \displaystyle\sum_{n=-\infty}^{\infty} \dfrac{e^{j(2n-1)\pi t}}{(2n-1)[3 + j(2n-1)\pi]}$

17.5.5 Show that, if $f(t)$ is even, the exponential Fourier coefficients are

$$c_n = \frac{a_n}{2} = \frac{2}{T}\int_0^{T/2} f(t)\cos n\omega_0 t\, dt$$

and if $f(t)$ is odd, they are

$$c_n = \frac{b_n}{2j} = \frac{2}{jT}\int_0^{T/2} f(t)\sin n\omega_0 t\, dt$$

Use the results to obtain (17.51).

17.5.6 Show that, if $f(t)$ has half-wave symmetry, the exponential Fourier coefficients are

$$c_n = 0, \qquad\qquad n \text{ even}$$

$$= \frac{2}{T}\int_0^{T/2} f(t)e^{-jn\omega_0 t}\, dt, \qquad n \text{ odd}$$

17.6

FREQUENCY SPECTRA

The periodic nonsinusoidal inputs and outputs of this chapter contain not just one frequency, as was the case for sinusoidal functions, but an infinite number of frequencies, as we know from their Fourier series representations. Since the Fourier series displays the amplitude, as well as the phase, corresponding to each frequency, it is a simple matter to tell which frequencies are playing an important role in the shape of the output and which are not. This, of course, is one of the important applications of Fourier series and will be considered in more detail in this section.

An output function $f(t)$ represented in the time domain by the Fourier series

$$f(t) = \frac{a_0}{2} + \sum_{n=1}^{\infty} A_n \cos(n\omega_0 t + \phi_n)$$

may also be determined from frequency-domain information. That is, if we know the phasor quantities A_n and ϕ_n for each component, we can construct the function. The dc component follows from the case $n = 0$. If we wish to know if the nth harmonic is dominant or not, we need only look at A_n. This information is readily obtained from a plot of A_n versus frequency. Since the frequencies are discrete values such as 0, ω_0, $2\omega_0$, and so on, the plot will be a set of lines whose lengths are proportional to A_n.

If we use Euler's formula to write

$$\cos \omega t = \frac{e^{j\omega t} + e^{-j\omega t}}{2}$$

we see that both complex frequencies $j\omega$ and $-j\omega$ are present. Consequently $f(t)$ contains all the frequencies $\pm jn\omega_0$, for $n = 0, 1, 2, \ldots$. This leads us to consider the exponential Fourier series

$$f(t) = \sum_{n=-\infty}^{\infty} c_n e^{jn\omega_0 t}$$

where, as noted earlier,

$$c_n = \frac{a_n - jb_n}{2}$$

or, in polar form,

$$c_n = \frac{\sqrt{a_n^2 + b_n^2}}{2} \underline{/-\tan^{-1} \frac{b_n}{a_n}}$$

In terms of A_n and ϕ_n we have

$$c_n = \frac{A_n}{2} \underline{/\phi_n}$$

$$c_{-n} = \frac{A_n}{2} \underline{/-\phi_n}$$

(17.52)

We also have $c_0 = a_0/2$, so that

$$c_0 = A_0 \underline{/\phi_0}$$

(17.53)

Since from (17.52) and (17.53) we have

$$|c_n| = |c_{-n}| = \frac{A_n}{2}, \qquad n = 1, 2, 3, \ldots$$

$$|c_0| = A_0$$

the plot of $|c_n|$ versus frequency contains the same information as that of A_n versus frequency, and the plot of $|c_n|$ includes negative as well as positive values of n. This latter plot is called a *discrete amplitude spectrum* or a *line spectrum* and is analogous to the amplitude response considered in Chapter 15 for the continuous case. A similar line plot of ϕ_n is a *discrete phase spectrum* and is analogous to the phase response in the continuous case. The discrete amplitude and phase spectra contain all the information needed to construct the Fourier series.

EXAMPLE 17.12 As an example, the rectangular pulse of Ex. 17.5.1 had Fourier coefficients given by

$$c_0 = \frac{1}{2}$$

$$c_n = \frac{1}{n\pi} \sin \frac{n\pi}{2}, \qquad n = \pm 1, \pm 2, \pm 3, \ldots$$

Also, from Ex. 17.5.2, we may write this result in terms of the sampling function as

$$A_n = 2|c_n| = \left| \frac{\sin (n\pi/2)}{n\pi/2} \right|$$

$$= \left| \mathrm{Sa}\left(\frac{n\pi}{2}\right) \right|$$

Evidently, $A_n = 0$, n even, and $|2/n\pi|$, n odd, while $\phi_n = 0$ for $n = 1, 5, 9, \ldots$, and $\phi_n = \pm 180°$ for $n = 3, 7, 11, \ldots$. The amplitude spectrum is shown in Fig. 17.14 by the vertical solid lines. The dashed curve is the absolute value of the sampling function, which is also the *envelope* of the spectrum.

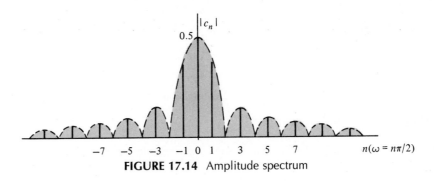

FIGURE 17.14 Amplitude spectrum

From Fig. 17.14 we see that the amplitude spectrum is symmetric about the vertical axis, which is to say it is an even function. This is true in general, as we see from (17.52), where replacing n by $-n$ results in the same amplitude A_n. Similarly, the phase spectrum is odd, since in (17.52) replacing n by $-n$ changes ϕ_n to $-\phi_n$. This latter fact can be used to advantage in sketching the phase spectrum, shown in Fig. 17.15. We have chosen $-180°$ as the phase for positive n and, to preserve the odd symmetry, $180°$ for negative n.

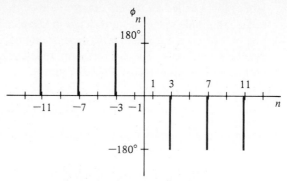

FIGURE 17.15 Phase spectrum

EXERCISES

17.6.1 Find the amplitude and phase spectra $|c_n|$ and ϕ_n of the sawtooth wave of (17.7).

Answer $|c_n| = |1/n|$, $n = \pm 1, \pm 2, \pm 3, \ldots$, $c_0 = 0$

$$\phi_n = -90°, \quad n = 1, 3, 5, \ldots, -2, -4, -6, \ldots$$

$$= 90°, \quad n = 2, 4, 6, \ldots, -1, -3, -5, \ldots$$

17.6.2 Find the amplitude spectrum of

$$f(t) = 1, \qquad -a < t < a$$

$$= 0, \qquad a < |t| < T/2$$

$$f(t + T) = f(t)$$

where $T/2 > a$. Note that this function is a generalization of that of Ex. 17.5.1 and that its spectrum envelope reaches zero at $1/a, 2/a, 3/a, \ldots$, which are independent of T. That is, the "width" of the envelope is independent of the pulse period.

Answer $|c_n| = (2a/T)|\,\text{Sa}\,(2\pi a n/T)|$

17.7

FOURIER SERIES AND SPICE

The Fourier coefficients can be found using SPICE for a trigonometric series of the form

$$f(t) = d_0 + \sum_{n=1}^{\infty} D_n \sin(n\omega_0 t + \psi_n) \tag{17.54}$$

Comparing (17.54) and (17.27), we see that

$$d_0 = \frac{a_0}{2}$$

$$D_n = A_n$$

$$\psi_n = \phi_n - \frac{\pi}{2}$$

The Fourier decomposition is performed in conjunction with a transient analysis using the previously described .TRAN and .FOUR commands (see Appendix E). The Fourier components of (17.54) can be found by a suitable definition of the transient interval used in the .TRAN command. In general, we can assign the duration of the transient response beginning at $t = 0$ s to the end of the desired response interval to be the period of the fundamental frequency

$$T = \frac{1}{f_0} = \frac{2\pi}{\omega_0}$$

This gives a series representation for a periodic function of period T whose waveform is that of the transient response.

EXAMPLE 17.13 As an example, let us find the Fourier coefficients of the waveform of Fig. 17.2. To perform a transient analysis, we need to define a circuit to which the Fourier decomposition can be applied to the response. Suppose we use the simple circuit composed of a current source with an output flowing from the reference node to node 1 equal to that of Fig. 17.2 connected to a 1-ohm resistor. The Fourier coefficients for this voltage can then be calculated using the .FOUR command. A circuit file that will produce this result using the PWL (piecewise linear waveform transient specification in I) is

```
EXAMPLE FOR THE FOURIER SERIES OF FIG. 17.2.
* DATA STATEMENTS
I 0 1 PWL(0 0A 3.1416 3.1416A 3.1417 −3.1416A 6.2832 0A)
R 1 0 1
.TRAN 0.1 6.2832
.FOUR 0.159 V(1)
.END
```

In this program, the waveform period is defined in the interval from 0 to 6.2832 s, the transient response interval. Therefore, a fundamental frequency of 0.159 is used in the .FOUR command. The resulting solution for the decomposition is

FOURIER COMPONENTS OF TRANSIENT RESPONSE V(1)

DC COMPONENT = −2.497780E−05

HARMONIC NO	FREQUENCY (HZ)	FOURIER COMPONENT	NORMALIZED COMPONENT	PHASE (DEG)	NORMALIZED PHASE (DEG)
1	1.590E−01	2.000E+00	1.000E+00	5.805E−04	0.000E+00
2	3.180E−01	1.001E+00	5.002E−01	−1.800E+02	−1.800E+02
3	4.770E−01	6.676E−01	3.338E−01	1.936E−04	−3.869E−04
4	6.360E−01	5.013E−01	2.506E−01	1.800E+02	1.800E+02
5	7.950E−01	4.016E−01	2.008E−01	1.156E−04	−4.649E−04
6	9.540E−01	3.353E−01	1.676E−01	1.800E+02	1.800E+02
7	1.113E+00	2.880E−01	1.440E−01	8.192E−05	−4.986E−04
8	1.272E+00	2.526E−01	1.263E−01	−1.800E+02	−1.800E+02
9	1.431E+00	2.252E−01	1.126E−01	6.606E−05	−5.144E−04

Once the Fourier coefficients are determined, the series can be written using (17.54) to describe the original function. The response of a network to a periodic function, as discussed in Sec. 17.3, can now be found by applying the .DC command for the dc component d_0 to the circuit and the .AC command for each harmonic component (amplitude D_n, frequency nf_0 and phase ψ_n). The solution, by superposition, is then the sum of the individual components in the time domain.

EXERCISES

17.7.1 Using SPICE, find the Fourier coefficients for the waveform of Fig. 17.7.

17.7.2 Repeat Exercise 17.7.1 for Fig. 17.10 if $T = 10$ μs and $A = 1$.

PROBLEMS

17.1 Find the trigonometric Fourier series for the functions
(a) $f(t) = 1, \, 0 < t < \pi$
$\qquad = 2, \, \pi < t < 2\pi$
$\quad f(t + 2\pi) = f(t)$.
(b) $f(t) = e^t, \, -\pi < t < \pi$
$\quad f(t + 2\pi) = f(t)$
(c) $f(t) = t^2, \, -1 < t < 1$
$\quad f(t + 2) = f(t)$.

17.2 Find the trigonometric Fourier series for the functions
(a) $f(t) = t + 1, \qquad 0 < t < \pi$
$\qquad = t - 1, \qquad -\pi < t < 0$
$\quad f(t + 2\pi) = f(t)$.
(b) $f(t) = |t|, \qquad -1 < t < 1$
$\quad f(t + 2) = f(t)$.
(c) $f(t) = \dfrac{A}{T}t, \qquad 0 < t < T$
$\quad f(t + T) = f(t)$
(d) $f(t) = 1 - |t|, \qquad -1 < t < 1$
$\quad f(t + 2) = f(t)$.

17.3 Show that the trigonometric Fourier series for the functions shown are as given. (In all cases, $\omega_0 = 2\pi/T$.)

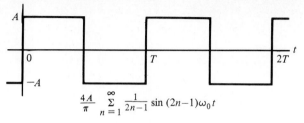

$$\frac{4A}{\pi} \sum_{n=1}^{\infty} \frac{1}{2n-1} \sin (2n-1)\omega_0 t$$

(a) Square wave

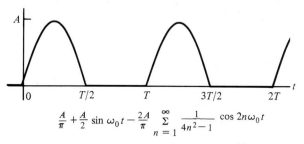

$$\frac{A}{\pi} + \frac{A}{2} \sin \omega_0 t - \frac{2A}{\pi} \sum_{n=1}^{\infty} \frac{1}{4n^2-1} \cos 2n\omega_0 t$$

(b) Half-wave rectified sine

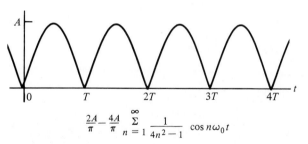

$$\frac{2A}{\pi} - \frac{4A}{\pi} \sum_{n=1}^{\infty} \frac{1}{4n^2-1} \cos n\omega_0 t$$

(c) Full-wave rectified sine

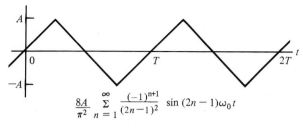

$$\frac{8A}{\pi^2} \sum_{n=1}^{\infty} \frac{(-1)^{n+1}}{(2n-1)^2} \sin (2n-1)\omega_0 t$$

(d) Triangular wave

PROBLEM 17.3

17.4 Find the trigonometric Fourier series for the function $f(t)$, one period of which is shown. (Note that this function is even and has half-wave symmetry.)

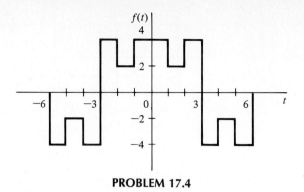

PROBLEM 17.4

17.5 Find the trigonometric Fourier series for the function $f(t)$, one period of which is shown. (Note that this function is odd and has half-wave symmetry.)

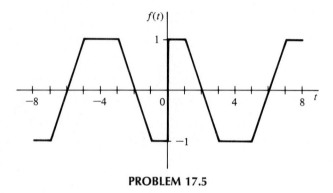

PROBLEM 17.5

17.6 Show that for $f(t)$ with half-wave symmetry, if $f(t)$ is even, all the Fourier coefficients are zero except

$$a_{2n-1} = \frac{8}{T} \int_0^{T/4} f(t) \cos(2n-1)\omega_0 t \, dt$$

and if $f(t)$ is odd, all the coefficients are zero except

$$b_{2n-1} = \frac{8}{T} \int_0^{T/4} f(t) \sin(2n-1)\omega_0 t \, dt$$

17.7 If the t, $f(t)$ axes are translated to the new τ, $g(\tau)$ axes shown so that the origin for the new axes is the point (t_0, f_0), the relations between the variables are $g = f - f_0$ and $\tau = t - t_0$. Show that

$$f(t) = f_0 + g(t - t_0)$$

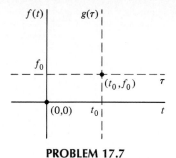

PROBLEM 17.7

17.8 Find the trignometric Fourier series for the function $f(t)$ shown by first translating the axes to move the origin to the point $(1, -1)$, so that symmetry can be exploited. [*Note:* Use the results of Probs. 17.7 and 17.3(a).]

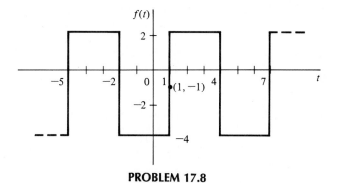

PROBLEM 17.8

17.9 Find the trigonometric Fourier series of the form (17.27) for the function

$$f(t) = 4 \sin (2t + \pi/3), \qquad -\frac{\pi}{6} < t < \frac{\pi}{3}$$

$$= 0, \qquad \frac{\pi}{3} < t < \frac{5\pi}{6}$$

$$f(t + \pi) = f(t)$$

by first translating the vertical axis to the point where the results of Prob. 17.3(b) may be used.

17.10 Find the trigonometric Fourier series of the form (17.27) for the function

$$f(t) = 4 |\cos 2t|$$

by first translating the vertical axis to the point where the results of Prob. 17.3(c) may be used.

17.11 Obtain the trigonometric series of $f(t)$ defined by

$$f(t) = 1, \qquad -1 < t < 1$$

$$= 0, \qquad 1 < |t| < 2$$

$$f(t + 4) = f(t)$$

17.12 Find the first three terms of the forced response $i(t)$ if $v_g(t)$ is the function of Prob. 17.11

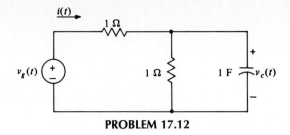

PROBLEM 17.12

17.13 Find the forced response $v(t)$ if v_g is the function of Ex. 17.2.3.

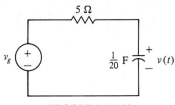

PROBLEM 17.13

17.14 Find the forced response $i(t)$ in Prob. 17.12 if $v_g(t)$ is the function of Prob. 17.1(c).

17.15 Find the forced response v_2 in the circuit of Fig. 15.4 if v_1 is the function of Prob. 17.2(c) with $A = 4$ and $T = 2$.

17.16 Repeat Prob. 17.15 if v_1 is the function of Ex. 17.2.3.

17.17 Given the function

$$f(t) = e^t, \qquad -\pi < t < \pi$$
$$f(t + 2\pi) = f(t)$$

show that its Fourier series is

$$f(t) = \frac{2 \sinh \pi}{\pi} \left[\frac{1}{2} + \sum_{n=1}^{\infty} \frac{(-1)^n}{n^2 + 1} (\cos nt - n \sin nt) \right]$$

17.18 Use the result of Prob. 17.17 to obtain the sum

$$\sum_{n=1}^{\infty} \frac{1}{n^2 + 1} = \frac{\pi \coth \pi - 1}{2}$$

Suggestion: Let $t = \pi$ and recall that the series converges to $\frac{1}{2}[f(\pi^+) + f(\pi^-)]$.

17.19 Find the Fourier series of

$$f(t) = t^2, \qquad -1 < t < 1$$
$$f(t + 2) = f(t)$$

and use the results at $t = 0$ and $t = 1$ to establish, respectively, the formulas

$$\sum_{n=1}^{\infty} \frac{(-1)^{n+1}}{n^2} = \frac{\pi^2}{12}$$

and

$$\sum_{n=1}^{\infty} \frac{1}{n^2} = \frac{\pi^2}{6}$$

17.20 Use Parseval's equation for the series of Prob. 17.19 to establish the formula

$$\sum_{n=1}^{\infty} \frac{1}{n^4} = \frac{\pi^4}{90}$$

17.21 Find the power delivered by the source in the circuit of Prob. 17.12 if v_g is the function of Ex. 17.1.1.

17.22 Find the rms value of $v_c(t)$ in the circuit of Prob. 17.12 if v_g is the function of Ex. 17.2.3.

17.23 Find the forced responses $i(t)$ and $v_c(t)$ and the power delivered to each of the 1-Ω resistors in the circuit of Prob. 17.12 if v_g is the function of Ex. 17.1.1.

17.24 Find the exponential Fourier series for the functions
(a) $f(t) = e^{-t}, \ -1 < t < 1$
$\quad f(t + 2) = f(t)$.
(b) $f(t) = t, \quad \quad -1 < t < 1$
$\quad \quad \ = 2 - t, \ 1 < t < 3$
$\quad f(t + 4) = f(t)$.
(c) $f(t) = t, \ 0 < t < 1$
$\quad f(t + 1) = f(t)$

17.25 Obtain the exponential series of the full-wave rectified sinusoid of Ex. 17.2.3 from its trigonometric series.

17.26 Obtain the trignometric series of $f(t)$ in Prob. 17.11 from the exponential series given in Ex. 17.5.1.

17.27 Find the discrete amplitude and phase spectra of the functions of Prob. 17.1.

17.28 Find the discrete amplitude and phase spectra of the function of (a) Ex. 17.1.1, (b) Ex. 17.1.2, (c) Ex. 17.1.3, and (d) Ex. 17.1.4.

17.29 If the exponential series (17.50) of $f(t)$ is the voltage across, or the current through, a 1-Ω resistor, then the instantaneous power is $f^2(t)$, and the average power is

$$P = \frac{1}{T} \int_0^T f^2(t) \, dt$$

Show that the average power is also given by

$$P = \frac{1}{T} \int_0^T f^2(t) \, dt = \sum_{n=-\infty}^{\infty} |c_n|^2$$

where c_n is the exponential Fourier coefficient. This result is known as *Parseval's theorem*. [*Suggestion:* Write $f^2(t)$ in the form

$$f^2(t) = \sum_{n=-\infty}^{\infty} \sum_{m=-\infty}^{\infty} c_n c_m e^{jn\omega_0 t} e^{jm\omega_0 t}$$

and integrate, noting that

$$\int_0^T e^{jn\omega_0 t} e^{jm\omega_0 t} \, dt = 0, \quad m, n = 0, \pm 1, \pm 2, \ldots$$

unless $m = -n$.]

17.30   From Prob. 17.29 the quantity $|c_n|^2$ is the average power associated with the frequency $n\omega_0$. The plot of $|c_n|^2$ versus frequency is a line spectrum known as the discrete power spectrum of $f(t)$. Plot this function in the cases of Prob. 17.24.

COMPUTER APPLICATION PROBLEMS

17.31   Using SPICE, repeat Prob. 17.3 for the first nine harmonics with $T = 1$ s.

17.32   Using SPICE, repeat Prob. 17.5 for the first nine harmonics.

17.33   For the function $f(t) = 10e^{-t}$, $0 < t < 1$ s, use SPICE to find the Fourier series for the first nine harmonics.

17.34   Using SPICE, repeat Prob. 17.12 if $v_g(t)$ is the function of Prob. 17.33.

Heinrich Rudolf Hertz
1857–1894

18

Fourier Transforms

That was no mean performance. [On his verification of Maxwell's theories]
Heinrich Hertz

The way for the development of radio, television, and radar was opened by the German physicist Heinrich Rudolf Hertz with his discovery in 1886–1888 of electromagnetic waves. His work confirmed the 1864 theory of the great English physicist James Clerk Maxwell that such waves existed.

Hertz was born in Hamburg, the oldest of five children in a prominent and prosperous family. After graduation from high school he spent a year with an engineering firm in Frankfurt, a year of volunteer military service in Berlin, and then a year at the University of Munich. Finally he entered the University of Berlin as a student of the great physicist Hermann von Helmholtz. Later Hertz received his doctorate and was a professor at Karlsruhe when he began his quest for electromagnetic waves. It was there that he met Elizabeth Doll, the daughter of one of his fellow professors, and after a 3-month courtship they were married. Only a few years after his famous discovery, Hertz died on New Year's Day in 1894 of a bone malignancy at the young age of 37. His researches ushered in the modern communication age, and in his honor the unit of frequency (cycles per second) was named the *hertz*. ∎

The Fourier series of the previous chapter enables us to go from the time domain to the frequency domain by calculating the Fourier coefficients for periodic time-domain functions. The Fourier coefficients are functions of integral multiples $n\omega_0$ of a basic frequency ω_0, and are therefore discrete quantities corresponding to the integer n. If the function under consideration is not periodic, we cannot find a Fourier series. In this case, as will be seen in this chapter, we may be able to find a *Fourier transform,* a function of the continuous frequency ω that corresponds to the time-domain function. We will develop the properties of the transform and consider a number of its applications. In particular, we will see that the transform plays the role for nonperiodic functions that phasors play for sinusoids, and that network functions may be obtained for transforms that are identical to those obtained for phasors.

18.1

THE FOURIER INTEGRAL

The Fourier transform is an excellent tool for considering frequency characteristics of the network signals and is used extensively in communication theory and signal processing. As will be seen in this chapter, Fourier transforms extend the phasor concept to nonperiodic functions that are more general than sinusoids and whose amplitude and phase spectra are continuous rather than discrete. These frequency spectra are extremely useful in determining dominant ranges of frequencies in exactly the same manner as the frequency responses of the transfer functions of Chapter 15. In the case of the Fourier transform, however, we are not restricted to sinusoids.

The continuous amplitude spectrum of Fig. 18.1 will provide us with some motivation for considering the Fourier transform. It is the amplitude response of an actual low-pass filter displayed on the screen of an oscilloscope. The horizontal axis is frequency, which starts at zero and increases to the right, and the vertical axis is the amplitude. Clearly, the dominant frequencies are the low ones, corresponding to large amplitudes, and the higher frequencies are suppressed, or filtered out, corresponding to very low amplitudes.

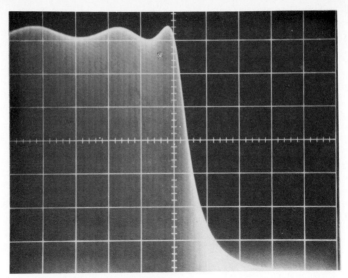

FIGURE 18.1 Amplitude response of a low-pass filter

We begin the development of the Fourier transform by considering a function $f(t)$ defined on an infinite interval, but which is not periodic and therefore cannot be represented by a Fourier series. It may be possible, however, to consider the function to be periodic with an *infinite* period and extend our previous results to include this case. The development we will give is nonrigorous, but the results may be obtained rigorously if $f(t)$ satisfies the Dirichlet conditions of Sec. 17.1 with (17.6) replaced by

$$\int_{-\infty}^{\infty} |f(t)| \, dt < \infty \tag{18.1}$$

Our strategy is to define a function $f_T(t)$ to be $f(t)$ on the interval $-T/2 < t < T/2$ and periodic of period T. That is,

$$f_T(t) = f(t), \qquad \frac{-T}{2} < t < \frac{T}{2}$$

$$f_T(t + T) = f(t)$$

Thus if $f(t)$ is the nonperiodic function shown in Fig. 18.2(a), then $f_T(t)$ is identical to $f(t)$ on $(-T/2, T/2)$ and is periodic of period T, as shown in Fig. 18.2(b). [Such

FIGURE 18.2 (a) A function $f(t)$; (b) its periodic extension $f_T(t)$

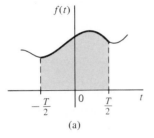

(a)

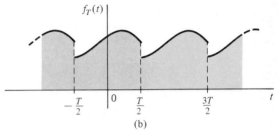

(b)

a function $f_T(t)$ is said to be the *periodic extension* of $f(t)$.] The exponential Fourier series for $f_T(t)$ is

$$f_T(t) = \sum_{n=-\infty}^{\infty} c_n e^{j2\pi n t/T} \tag{18.2}$$

where

$$c_n = \frac{1}{T} \int_{-T/2}^{T/2} f_T(x) e^{-j2\pi n x/T}\, dx \tag{18.3}$$

We have replaced ω_0 by $2\pi/T$ and are using the dummy variable x instead of t in the coefficient expression. Our intention is to let $T \to \infty$, in which case $f_T(t) \to f(t)$. We will then have extended the Fourier series concept to the nonperiodic function $f(t)$ by considering it to be periodic with an infinite period.

Since the limiting process requires that $\omega_0 = 2\pi/T \to 0$, for emphasis we replace $2\pi/T$ by $\Delta\omega$. Therefore substituting (18.3) into (18.2), we have

$$\begin{aligned}
f_T(t) &= \sum_{n=-\infty}^{\infty} \left[\frac{\Delta\omega}{2\pi} \int_{-T/2}^{T/2} f_T(x) e^{-jxn\Delta\omega}\, dx \right] e^{jtn\Delta\omega} \\
&= \sum_{n=-\infty}^{\infty} \left[\frac{1}{2\pi} \int_{-T/2}^{T/2} f_T(x) e^{-j(x-t)n\Delta\omega}\, dx \right] \Delta\omega
\end{aligned} \tag{18.4}$$

If we define the function

$$g(\omega, t) = \frac{1}{2\pi} \int_{-T/2}^{T/2} f_T(x) e^{-j\omega(x-t)}\, dx \tag{18.5}$$

then clearly the limit of (18.4) is given by

$$f(t) = \lim_{\substack{T \to \infty \\ (\Delta\omega \to 0)}} \sum_{n=-\infty}^{\infty} g(n\Delta\omega, t)\Delta\omega \tag{18.6}$$

By the fundamental theorem of integral calculus the last result appears to be

$$f(t) = \int_{-\infty}^{\infty} g(\omega, t)\, d\omega \tag{18.7}$$

But in the limit, $f_T \to f$ and $T \to \infty$ in (18.5) so that what appears to be $g(\omega, t)$ in (18.7) is really its limit, which by (18.5) is

$$\lim_{T \to \infty} g(\omega, t) = \frac{1}{2\pi} \int_{-\infty}^{\infty} f(x) e^{-j\omega(x-t)}\, dx$$

Therefore, (18.6) is actually

$$f(t) = \frac{1}{2\pi} \int_{-\infty}^{\infty} \left[\int_{-\infty}^{\infty} f(x) e^{-j\omega(x-t)}\, dx \right] d\omega \tag{18.8}$$

The expression on the right is called the *Fourier integral,* and it plays the same role for nonperiodic functions that the Fourier series plays for periodic functions. We will carry the analogy further in the next section when we separate the analogous expression for the Fourier coefficients from the Fourier integral, leading to the Fourier transform, a direct analogy to the Fourier series.

18.2

DEVELOPMENT OF THE FOURIER TRANSFORM

In this section we will use the Fourier integral to obtain the Fourier transform. We begin by rewriting (18.8) in the form

$$f(t) = \frac{1}{2\pi} \int_{-\infty}^{\infty} \left[\int_{-\infty}^{\infty} f(x) e^{-j\omega x} \, dx \right] e^{j\omega t} \, d\omega \qquad (18.9)$$

Now, let us define the expression in brackets to be the function

$$\mathbf{F}(j\omega) = \int_{-\infty}^{\infty} f(t) e^{-j\omega t} \, dt \qquad (18.10)$$

where we have changed the variable of integration from x to t. Then (18.9) becomes

$$f(t) = \frac{1}{2\pi} \int_{-\infty}^{\infty} \mathbf{F}(j\omega) e^{j\omega t} \, d\omega \qquad (18.11)$$

The function $\mathbf{F}(j\omega)$ is called the *Fourier transform* of $f(t)$, and $f(t)$ is called the *inverse Fourier transform* of $\mathbf{F}(j\omega)$. These facts are often stated symbolically as

$$\mathbf{F}(j\omega) = \mathscr{F}[f(t)]$$
$$f(t) = \mathscr{F}^{-1}[\mathbf{F}(j\omega)] \qquad (18.12)$$

where $\mathscr{F}$ denotes the operation of taking the Fourier transform. Also, (18.10) and (18.11) are collectively called the *Fourier transform pair,* the symbolism for which is

$$f(t) \longleftrightarrow \mathbf{F}(j\omega) \qquad (18.13)$$

Equation (18.11), the inverse transform, is a direct analogy to the Fourier series. The Fourier transform (18.10) corresponds to the Fourier coefficients in the previous chapter. Another description for these two analogies is to say that the Fourier transform is a *continuous* representation (ω being a continuous variable), whereas the Fourier coefficient is a *discrete* representation ($n\omega_0$, for n an integer, being a discrete variable). In other words, the Fourier transform is a *transformation* of the function $f(t)$ from the time domain to the frequency domain.

EXAMPLE 18.1   As an example, let us find the transform of

$$f(t) = e^{-at}u(t)$$

where $a > 0$. By definition (18.10) we have

$$\mathcal{F}[e^{-at}u(t)] = \int_{-\infty}^{\infty} e^{-at}u(t)e^{-j\omega t}\, dt$$

$$= \int_{0}^{\infty} e^{-(a+j\omega)t}\, dt$$

or

$$\mathcal{F}[e^{-at}u(t)] = \frac{1}{-(a+j\omega)}\, e^{-(a+j\omega)t}\,\Big|_{0}^{\infty}$$

Because $a > 0$, the upper limit is

$$\lim_{t\to\infty} e^{-at}(\cos\,\omega t - j\,\sin\,\omega t) = 0$$

since the expression in parentheses is bounded while the exponential goes to zero. Thus we have

$$\mathcal{F}[e^{-at}u(t)] = \frac{1}{a + j\omega} \qquad\qquad (18.14)$$

or

$$e^{-at}u(t) \longleftrightarrow \frac{1}{a + j\omega}, \qquad a > 0$$

EXAMPLE 18.2   As another example, let us find the transform of the single rectangular pulse

$$f(t) = A, \qquad -\frac{\delta}{2} < t < \frac{\delta}{2}$$

$$= 0, \quad |t| > \frac{\delta}{2} \qquad\qquad (18.15)$$

which is shown in Fig. 18.3. By definition we have

$$\mathbf{F}(j\omega) = \int_{-\infty}^{\infty} f(t)e^{-j\omega t}\, dt$$

$$= \int_{-\delta/2}^{\delta/2} Ae^{-j\omega t}\, dt$$

$$= \frac{2A}{\omega}\left(\frac{e^{j\omega\delta/2} - e^{-j\omega\delta/2}}{j2}\right)$$

or

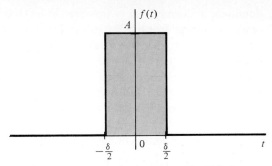

FIGURE 18.3 Finite pulse of width δ.

$$\mathbf{F}(j\omega) = \frac{2A}{\omega} \sin \frac{\omega\delta}{2} \qquad (18.16)$$

Alternatively, we may write

$$\mathbf{F}(j\omega) = A\delta \, \mathrm{Sa}\!\left(\frac{\omega\delta}{2}\right)$$

where Sa $(\omega\delta/2)$ is the sampling function defined in Ex. 17.5.2.

The Fourier transform is usually complex, but in this case, as we see, it is a real function; therefore, it can be sketched in the real plane, as shown in Fig.18.4(a). The frequencies where the curve crosses the horizontal axis are $\pm\omega_1,\ \pm\omega_2,\ \pm\omega_3,\ \dots$, where $\omega_1 = 2\pi/\delta$, $\omega_2 = 4\pi/\delta$, $\omega_3 = 6\pi/\delta$, $\dots$. From the graph we see that the

FIGURE 18.4 (a) Fourier transform of the rectangular pulse; (b) its absolute value

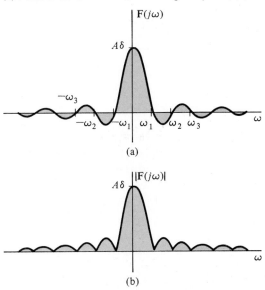

(a)

(b)

dominant frequency is $\omega = 0$, where $\mathbf{F}(j\omega)$ has its maximum value $A\delta$, and the width of the dominant frequency band is $2\omega_1 = 4\pi/\delta$. If δ, the width of the rectangular pulse, increases, then the peak $A\delta$ increases and the width $4\pi/\delta$ decreases, thus making the frequency $\omega = 0$ more dominant. As δ becomes large, the pulse approaches the constant function A, a dc value, and the width of the dominant frequency band shrinks to zero. The only frequency present is then $\omega = 0$, which is consistent with a dc function.

Figure 18.4(b) is a sketch of the absolute value of $\mathbf{F}(j\omega)$, a continuous spectrum that is like the envelope of the corresponding discrete spectrum of Fig. 17.14. In the discrete case the separation between spectral lines is $\omega_0 = 2\pi/T$, which tends to zero as $T \to \infty$, resulting in the continuous spectrum.

EXERCISES

18.2.1 Find $\mathscr{F}[e^{-a|t|}]$, where $a > 0$.

Answer $2a/(\omega^2 + a^2)$

18.2.2 Find $\mathbf{F}(j\omega)$ if

$$f(t) = t + 1, \qquad -1 < t < 0$$
$$= -t + 1, \qquad 0 < t < 1$$
$$= 0, \qquad \text{elsewhere}$$

Answer $[2(1 - \cos \omega)]/\omega^2 = \mathrm{Sa}^2(\omega/2)$

18.2.3 Find $f(t)$ if

$$\mathbf{F}(j\omega) = 1, \qquad -1 < \omega < 1$$
$$= 0, \qquad |\omega| > 1$$

Answer $(1/\pi)\,\mathrm{Sa}(t)$

18.3

FOURIER TRANSFORM PROPERTIES

In this section we will consider the general nature of the Fourier transform and develop some of its more important properties. We begin by using Euler's formula to replace the exponential function in (18.10) by its trigonometric equivalent, resulting in

$$\mathbf{F}(j\omega) = \int_{-\infty}^{\infty} f(t)(\cos \omega t - j \sin \omega t)\, dt$$

$$= \int_{-\infty}^{\infty} f(t) \cos \omega t\, dt - j \int_{-\infty}^{\infty} f(t) \sin \omega t\, dt$$

Since $f(t)$ is real, the two integrals are real, and the transform has been decomposed into its real and imaginary parts. That is, for a given ω, the transform is a complex number, given by

$$\mathbf{F}(j\omega) = \text{Re } \mathbf{F}(j\omega) + j\text{Im } \mathbf{F}(j\omega) \qquad (18.17)$$

where the real part is

$$\text{Re } \mathbf{F}(j\omega) = \int_{-\infty}^{\infty} f(t) \cos \omega t \, dt \qquad (18.18)$$

and the imaginary part is

$$\text{Im } \mathbf{F}(j\omega) = -\int_{-\infty}^{\infty} f(t) \sin \omega t \, dt \qquad (18.19)$$

The representation of (18.17) is the rectangular form of the transform, which we may readily put in the polar form

$$\mathbf{F}(j\omega) = |\mathbf{F}(j\omega)| e^{j\phi(\omega)} \qquad (18.20)$$

where the magnitude is

$$|\mathbf{F}(j\omega)| = \sqrt{\text{Re}^2\mathbf{F}(j\omega) + \text{Im}^2\mathbf{F}(j\omega)} \qquad (18.21)$$

and the phase is

$$\phi(\omega) = \tan^{-1}\frac{\text{Im } \mathbf{F}(j\omega)}{\text{Re } \mathbf{F}(j\omega)} \qquad (18.22)$$

These properties, as the reader may have noted, are identical in form to those of the network function $\mathbf{H}(j\omega)$ of Sec. 15.1.

We note from (18.18) and (18.19) that Re $\mathbf{F}(j\omega)$ is an even function and Im $\mathbf{F}(j\omega)$ is an odd function, since $\cos \omega t$ is even and $\sin \omega t$ is odd in ω. The magnitude $|\mathbf{F}(j\omega)|$ is therefore even, since squaring an even function and squaring an odd function results in an even function in each case. The phase $\phi(\omega)$ is odd, since the argument of the arctangent is a ratio of an odd to an even function and is therefore odd. We will leave the proof of these statemetns to Ex. 18.3.1.

EXAMPLE 18.3   As an example, the transform of $e^{-at}u(t)$, $a > 0$, given in (18.14), is

$$\mathbf{F}(j\omega) = \frac{1}{a + j\omega}$$

$$= \frac{a - j\omega}{a^2 + \omega^2}$$

$$= \frac{a}{a^2 + \omega^2} + j\frac{-\omega}{a^2 + \omega^2}$$

Thus we have

$$\text{Re } \mathbf{F}(j\omega) = \frac{a}{a^2 + \omega^2}$$

which is even, and

$$\text{Im } \mathbf{F}(j\omega) = \frac{-\omega}{a^2 + \omega^2}$$

which is odd. The magnitude is

$$|\mathbf{F}(j\omega)| = \frac{1}{\sqrt{a^2 + \omega^2}}$$

which is even, and the phase is

$$\phi(\omega) = -\tan^{-1}\frac{\omega}{a}$$

which is odd.

If $f(t)$ is even, then from (18.18) and (18.19) we see that Im $\mathbf{F}(j\omega) = 0$ and

$$\mathbf{F}(j\omega) = 2 \int_0^\infty f(t) \cos \omega t \, dt \qquad (18.23)$$

which is both real and even. If $f(t)$ is odd, then Re $\mathbf{F}(j\omega) = 0$ and

$$\mathbf{F}(j\omega) = -2j \int_0^\infty f(t) \sin \omega t \, dt \qquad (18.24)$$

which is pure imaginary and its imaginary part is odd.

EXAMPLE 18.4   As an example, the rectangular pulse centered about the origin, given in (18.15), is an even function, and its transform, given in (18.16), is

$$\mathbf{F}(j\omega) = \frac{2A}{\omega} \sin \frac{\omega\delta}{2}$$

which is both real and even. An example of an odd function is

$$\begin{aligned} f(t) &= -e^t, & t < 0 \\ &= e^{-t}, & t > 0 \end{aligned} \qquad (18.25)$$

and, as the reader is asked to show in Ex. 18.3.2, its transform is

$$F(j\omega) = \frac{-2j\omega}{1 + \omega^2} \tag{18.26}$$

which is pure imaginary with an odd imaginary part.

From (18.17), since Re $F(j\omega)$ is even and Im $F(j\omega)$ is odd, we may replace ω by $-\omega$ and obtain

$$F(-j\omega) = \text{Re } F(j\omega) - j\text{Im } F(j\omega)$$

the right member of which is the conjugate $F^*(j\omega)$. Therefore we have

$$F^*(j\omega) = F(-j\omega)$$

and thus

$$|F(j\omega)|^2 = F(j\omega)F(-j\omega)$$

EXAMPLE 18.5 For example, for the transform

$$F(j\omega) = \frac{1}{a + j\omega}$$

we have

$$|F(j\omega)|^2 = \frac{1}{a + j\omega} \cdot \frac{1}{a - j\omega} = \frac{1}{a^2 + \omega^2}$$

as before.

EXERCISES

18.3.1 Prove using the definition of even and odd functions that Re $F(j\omega)$ and $|F(j\omega)|$ are even and that Im $F(j\omega)$ and $\phi(\omega)$ are odd.

18.3.2 Show that the transform of the function of (18.25) is given by (18.26).

18.3.3 Solve Ex. 18.2.1 using (18.23).

18.4

FOURIER TRANSFORM OPERATIONS

There are a number of operations that may be performed on Fourier transforms in general, which we will find extremely useful. We will consider a number of these and tabulate them at the end of the section. One of the most often used properties is linearity, which is a consequence of the fact that the Fourier transform is an integral transform, and thus it is a *linear* operation. That is, the transform of the combination

$$v(t) = c_1 f_1(t) + c_2 f_2(t)$$

is the combination of the transforms

$$V(j\omega) = c_1 F_1(j\omega) + c_2 F_2(j\omega)$$

where V, F_1, and F_2 are, respectively, the Fourier transforms of v, f_1, and f_2, and c_1 and c_2 are constants. This concept of linearity enables us to readily find transforms of relatively complicated functions from transforms of simpler functions.

EXAMPLE 18.6 For example, the transform of the function

$$f(t) = 2(e^{-2t} - e^{-3t})u(t) \tag{18.27}$$

is, by linearity and (18.14),

$$
\begin{aligned}
\mathbf{F}(j\omega) &= \frac{2}{2 + j\omega} - \frac{2}{3 + j\omega} \\
&= \frac{2}{(2 + j\omega)(3 + j\omega)}
\end{aligned}
\tag{18.28}
$$

Another operation involving Fourier transforms that we shall find useful is that of time differentiation. Suppose we wish to find the Fourier transform of the derivative of a function $f(t)$. By definition, if

$$f(t) \longleftrightarrow \mathbf{F}(j\omega)$$

then

$$f(t) = \frac{1}{2\pi} \int_{-\infty}^{\infty} \mathbf{F}(j\omega)e^{j\omega t}\, d\omega \tag{18.29}$$

from which we obtain

$$
\begin{aligned}
\frac{df(t)}{dt} &= \frac{1}{2\pi} \int_{-\infty}^{\infty} \frac{d}{dt}[\mathbf{F}(j\omega)e^{j\omega t}]\, d\omega \\
&= \frac{1}{2\pi} \int_{-\infty}^{\infty} [j\omega \mathbf{F}(j\omega)]e^{j\omega t}\, d\omega
\end{aligned}
$$

Therefore, we have

$$\frac{df(t)}{dt} \longleftrightarrow j\omega \mathbf{F}(j\omega) \tag{18.30}$$

That is, the transform of the derivative of f is found by simply multiplying the transform of f by $j\omega$. This result may be extended readily to the general case

$$\frac{d^n f(t)}{dt^n} \longleftrightarrow (j\omega)^n \mathbf{F}(j\omega) \tag{18.31}$$

where $n = 0, 1, 2, \ldots$. We are assuming that the derivatives involved exist and that the interchange of operations of differentiation and integration is valid. The conditions under which this is true are quite general and hold for almost any function we are likely to encounter.

EXAMPLE 18.7 As another example, let us find the transform of $f(t - \tau)$, where τ is a constant. Replacing t by $t - \tau$ in (18.29), we have

$$f(t - \tau) = \frac{1}{2\pi} \int_{-\infty}^{\infty} \mathbf{F}(j\omega) e^{j\omega(t-\tau)} \, d\omega$$

$$= \frac{1}{2\pi} \int_{-\infty}^{\infty} [\mathbf{F}(j\omega) e^{-j\omega\tau}] e^{j\omega t} \, d\omega$$

from which we conclude, by (18.29), that

$$f(t - \tau) \longleftrightarrow \mathbf{F}(j\omega) e^{-j\omega\tau} \qquad (18.32)$$

The physical significance of this result is that a *delay* in the time domain [the function $f(t - \tau)$ is $f(t)$ delayed τ seconds] corresponds to a phase shift (an addition of $-\omega\tau$ to the phase) in the frequency domain.

EXAMPLE 18.8 To illustrate the use of (18.32), let us transform the pulse $p(t)$ of Fig. 18.5. This is the finite pulse $f(t)$ of Fig. 18.3 shifted to the right $\delta/2$ units. Since we have

$$p(t) = f\left(t - \frac{\delta}{2}\right)$$

and, by (18.16),

$$\mathbf{F}(j\omega) = \frac{2A}{\omega} \sin \frac{\omega\delta}{2}$$

the transform of $p(t)$ by (18.32) is

$$\mathbf{P}(j\omega) = \left(\frac{2A}{\omega} \sin \frac{\omega\delta}{2}\right) e^{-j\omega\delta/2}$$

Next, let us transform the function $e^{j\omega_0 t} f(t)$. In this case we will illustrate the procedure with the transform definition (18.10), which we rewrite as

FIGURE 18.5 Finite pulse shifted

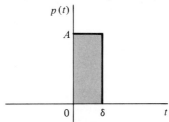

591

$$\mathbf{F}(j\omega) = \int_{-\infty}^{\infty} f(t)e^{-j\omega t}\, dt \qquad (18.33)$$

Replacing ω by $\omega - \omega_0$ results in

$$\mathbf{F}(j\omega - j\omega_0) = \int_{-\infty}^{\infty} f(t)e^{-j(\omega - \omega_0)t}\, dt$$

$$= \int_{-\infty}^{\infty} \left[e^{j\omega_0 t}f(t) \right] e^{-j\omega t}\, dt$$

from which we have by (18.33)

$$\boxed{e^{j\omega_0 t}f(t) \longleftrightarrow \mathbf{F}(j\omega - j\omega_0) \qquad (18.34)}$$

Comparing the last two results, we see that in (18.32) translating the time function adds a phase to the frequency function, whereas in (18.34) translating the frequency function adds a phase to the time function.

EXAMPLE 18.9 To illustrate (18.34), let us transform

$$g(t) = (\cos \omega_0 t)f(t)$$

$$= \frac{1}{2}\left[e^{j\omega_0 t}f(t) + e^{-j\omega_0 t}f(t) \right]$$

By using linearity and applying (18.34) to each term, we have

$$(\cos \omega_0 t)f(t) \longleftrightarrow \frac{1}{2}[\mathbf{F}(j\omega - j\omega_0) + \mathbf{F}(j\omega + j\omega_0)] \qquad (18.35)$$

This result is important in *modulation,* where frequency components of a signal are translated to a different position in the frequency spectrum.

The similarity of the integrals for $f(t)$ and its transform in (18.11) and (18.10) suggests that interchanging t and ω in some way would lead to a new set of transforms. Indeed, if in (18.11) we replace ω by x and t by $-\omega$, and multiply both sides of the equation by 2π, we have

$$2\pi f(-\omega) = \int_{-\infty}^{\infty} \mathbf{F}(jx)e^{-jx\omega}\, dx$$

The right member, as we see from (18.10), is the Fourier transform of $F(jt)$. Thus we have

$$\mathbf{F}(jt) \longleftrightarrow 2\pi f(-\omega) \qquad (18.36)$$

EXAMPLE 18.10 As examples, if we apply (18.36) to (18.14) and (18.16), respectively, we have

$$\frac{1}{a + jt} \longleftrightarrow 2\pi e^{a\omega}u(-\omega), \qquad a > 0$$

and

$$\frac{2A}{t}\sin\frac{t\delta}{2} \longleftrightarrow 2\pi A, \qquad |\omega| < \frac{\delta}{2}$$

$$\longleftrightarrow 0, \qquad |\omega| > \frac{\delta}{2}$$

We have tabulated the Fourier transform operations derived in this section, along with others, in Table 18.1. The reader is asked to derive the others in the exercises.

TABLE 18.1 Fourier Transform Operations

$f(t)$	$\mathbf{F}(j\omega)$
1. $c_1 f_1(t) + c_2 f_2(t)$	$c_1\mathbf{F}_1(j\omega) + c_2\mathbf{F}_2(j\omega)$
2. $f(t/a)$	$|a|\mathbf{F}(j\omega a)$
3. $f(t - \tau)$	$e^{-j\omega\tau}\mathbf{F}(j\omega)$
4. $e^{j\omega_0 t}f(t)$	$\mathbf{F}(j\omega - j\omega_0)$
5. $\mathbf{F}(jt)$	$2\pi f(-\omega)$
6. $f(-t)$	$\mathbf{F}(-j\omega)$
7. $\dfrac{d^n}{dt^n}f(t)$	$(j\omega)^n\mathbf{F}(j\omega), \ n = 0, 1, 2, \ldots$
8. $t^n f(t)$	$(-1)^n \dfrac{d^n}{d(j\omega)^n}\mathbf{F}(j\omega), \ n = 0, 1, 2, \ldots$

EXERCISES

18.4.1 Derive operations 2, 6, and 8 of Table 18.1.

18.4.2 Using the result of Ex. 18.2.1 and entry 3 of Table 18.1, find the inverse transform of $2e^{-j\omega}/(\omega^2 + 1)$.
Answer $e^{-|t - 1|}$

18.4.3 Use the result of Ex. 18.2.1 and entry 5 of Table 18.1 to find the transform of $1/(t^2 + 1)$.
Answer $\pi e^{-|\omega|}$

18.5

NETWORK FUNCTIONS

In this section we will use Fourier transforms to extend the concept of network functions, considered in Chapter 14, to circuits with nonsinusoidal excitations. As we will see, the network functions are exactly the same as in the case of circuits with sinusoidal excitations. Therefore, all the properties of transfer functions that we have considered are valid in this more general case. The only difference is that here the inputs and outputs are Fourier transforms rather than phasors.

We begin by considering the general circuit of Fig. 18.6, where, to be specific, we have taken the input and output as the voltages $v_i(t)$ and $v_o(t)$, respectively. We could as well let one or both of these functions be currents. We will take the describing

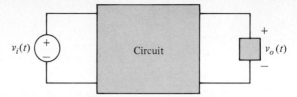

FIGURE 18.6 General circuit

equation of the circuit to be (14.22), which we repeat as

$$a_n\frac{d^n v_o}{dt^n} + a_{n-1}\frac{d^{n-1} v_o}{dt^{n-1}} + \cdots + a_1\frac{dv_o}{dt} + a_0 v_o$$

$$= b_m\frac{d^m v_i}{dt^m} + b_{m-1}\frac{d^{m-1} v_i}{dt^{m-1}} + \cdots + b_1\frac{dv_i}{dt} + b_0 v_i$$

Taking the Fourier transform of both sides, making use of (18.31) and linearity, we have

$$[a_n(j\omega)^n + a_{n-1}(j\omega)^{n-1} + \cdots + a_1 j\omega + a_0]\mathbf{V}_o(j\omega)$$

$$= [b_m(j\omega)^m + b_{m-1}(j\omega)^{m-1} + \cdots + b_1 j\omega + b_0]\mathbf{V}_i(j\omega)$$

The functions $\mathbf{V}_o$ and $\mathbf{V}_i$ are the Fourier transforms of the output and input functions, v_o and v_i. From this result we may write

$$\frac{\mathbf{V}_o(j\omega)}{\mathbf{V}_i(j\omega)} = \mathbf{H}(j\omega)$$

$$= \frac{b_m(j\omega)^m + b_{m-1}(j\omega)^{m-1} + \cdots + b_0}{a_n(j\omega)^n + a_{n-1}(j\omega)^{n-1} + \cdots + a_0} \tag{18.37}$$

where $\mathbf{H}(j\omega)$ is the identical network function of Chapter 14, evaluated at $s = j\omega$.

This development indicates that the network function, defined previously as the ratio of the output phasor to the input phasor, is precisely the same as the ratio of the output transform to the input transform, and in the latter case the functions involved do not have to be sinusoids. This generalization may be used also to find the time-domain responses.

EXAMPLE 18.11 For example, suppose the input $v_i(t)$ is given by

$$v_i(t) = e^{-3t}u(t) \tag{18.38}$$

and is related to the output $v_o(t)$ by the equation

$$\frac{dv_o}{dt} + 2v_o = 2v_i$$

Transforming the equation, we have

$$(j\omega + 2)V_o(j\omega) = 2V_i(j\omega)$$

from which the transfer function is

$$\mathbf{H}(j\omega) = \frac{\mathbf{V}_o(j\omega)}{\mathbf{V}_i(j\omega)} = \frac{2}{j\omega + 2} \tag{18.39}$$

Also by (18.14) we have

$$\mathbf{V}_i(j\omega) = \frac{1}{3 + j\omega} \tag{18.40}$$

so that the transform of the output is

$$\mathbf{V}_o(j\omega) = \mathbf{H}(j\omega)\mathbf{V}_i(j\omega)$$

$$= \frac{2}{(2 + j\omega)(3 + j\omega)} \tag{18.41}$$

By (18.28) this is

$$\mathbf{V}_o(j\omega) = 2\frac{1}{(2 + j\omega)} - 2\frac{1}{(3 + j\omega)}$$

so that by linearity, or by (18.27), we have

$$v_o(t) = 2(e^{-2t} - e^{-3t})u(t) \tag{18.42}$$

Since the network function concept implies only one input, the time-domain responses obtained in this manner are responses of initially *relaxed* circuits. (No initial energy is stored.) We could develop general methods for using Fourier transforms to find time-domain solutions, such as (18.42), but the Fourier transform is better suited for other applications. We prefer, therefore, to consider another transform, the *Laplace transform*, which unlike the Fourier transform takes into account nonzero initial conditions. We will discuss the Laplace transform in detail in the next two chapters.

Since the transfer functions of this chapter are identical to those considered earlier, we will not go into the details of their frequency responses, poles and zeros, natural frequencies, and so on. These topics have all been considered in Chapters 14 and 15. We will conclude our treatment of Fourier transforms in the next section with a consideration of *energy of a signal*, a topic for which the Fourier transform is well suited.

EXERCISES

18.5.1 If $x(t)$ is the input and $y(t)$ is the output, find the network function, using Fourier transforms, where

$$y'' + 4y' + 3y = 4x$$

Answer $4/(3 - \omega^2 + j4\omega)$

18.5.2 If in Ex. 18.5.1

$$x = e^{-2t}u(t)$$

find $\mathcal{F}[y(t)]$.

Answer $4/[(j\omega + 1)(j\omega + 2)(j\omega + 3)]$

18.5.3   Verify in Ex. 18.5.2 that

$$\mathscr{F}[y(t)] = \frac{2}{j\omega + 1} - \frac{4}{j\omega + 2} + \frac{2}{j\omega + 3}$$

and find $y(t)$ for the case $y(0) = y'(0) = 0$.
Answer $(2e^{-t} - 4e^{-2t} + 2e^{-3t})u(t)$

18.6

PARSEVAL'S EQUATION FOR FOURIER TRANSFORMS

As there is for periodic functions, there is also a Parseval's equation for nonperiodic functions, given by

$$\int_{-\infty}^{\infty} f^2(\tau)\, d\tau = \frac{1}{2\pi} \int_{-\infty}^{\infty} |\mathbf{F}(j\omega)|^2\, d\omega \qquad (18.43)$$

This result, which is analogous to (17.42), will be obtained in this section, where we will see that it is related to the energy dissipated by a current or voltage $f(t)$.

We begin by multiplying both sides of the inverse transform (18.29) by a function $g(t)$, resulting in

$$f(t)g(t) = g(t)\left(\frac{1}{2\pi}\right)\int_{-\infty}^{\infty} \mathbf{F}(j\omega)e^{j\omega t}\, d\omega$$

Integrating both sides with respect to t yields

$$\int_{-\infty}^{\infty} f(\tau)g(\tau)\, d\tau = \frac{1}{2\pi} \int_{-\infty}^{\infty}\int_{-\infty}^{\infty} \mathbf{F}(j\omega)g(\tau)e^{j\omega\tau}\, d\omega\, d\tau$$

where t has been replaced by τ as the variable of integration. Interchanging the order of integration in the right member results in

$$\int_{-\infty}^{\infty} f(\tau)g(\tau)\, d\tau = \frac{1}{2\pi} \int_{-\infty}^{\infty} \mathbf{F}(j\omega)\left[\int_{-\infty}^{\infty} g(\tau)e^{j\omega\tau}\, d\tau\right] d\omega \qquad (18.44)$$

The expression in brackets, by definition of the Fourier transform in (18.10), is simply $\mathbf{G}(-j\omega)$, the transform of $g(t)$ with ω replaced by $-\omega$. Since $\mathbf{G}(-j\omega)$ is the conjugate $\mathbf{G}^*(j\omega)$, we may write (18.44) as

$$\int_{-\infty}^{\infty} f(\tau)g(\tau)\, d\tau = \frac{1}{2\pi} \int_{-\infty}^{\infty} \mathbf{F}(j\omega)\mathbf{G}^*(j\omega)\, d\omega \qquad (18.45)$$

This is a generalized form of Parseval's equation, and it reduces to (18.43) when $g(t) = f(t)$.

Since the magnitude is an even function, we may write Parseval's equation in the form

$$\int_{-\infty}^{\infty} f^2(\tau) \, d\tau = \frac{1}{\pi} \int_0^{\infty} |\mathbf{F}(j\omega)|^2 \, d\omega \qquad (18.46)$$

A further simplification is possible if $f(t)$ is zero for t negative. Such a function is called a *causal* function, and for this case Parseval's equation is

$$\int_0^{\infty} f^2(\tau) \, d\tau = \frac{1}{\pi} \int_0^{\infty} |\mathbf{F}(j\omega)|^2 \, d\omega \qquad (18.47)$$

If $f(t)$ is the voltage (or the current) associated with a 1-Ω resistor, the left member of Parseval's equation (18.43) is the total energy delivered, which is sometimes called the 1-Ω, or normalized, energy of the signal $f(t)$. The right member is thus the total energy computed using the transform of $f(t)$. More generally, if $f(t)$ and $g(t)$ are the current and voltage, respectively, of any element, then the left member of (18.45) is the total energy delivered to the element using time-domain variables, and the right member is the total energy using the transformed variables.

Parseval's equation can be used to give us a better interpretation of the Fourier transform by relating $|\mathbf{F}(j\omega)|$ to energy. Since the left member of Parseval's equation (18.43) is energy, measured in joules, and $d\omega$ is in radians, the units of $|\mathbf{F}(j\omega)|^2/2\pi$ are joules/radian. Also ω is 2π times the frequency (Hz), so the units of 2π are rad/Hz. Therefore the units of $|\mathbf{F}(j\omega)|^2$ are (J/rad)(rad/Hz) or J/Hz. Thus $|\mathbf{F}(j\omega)|^2$, being energy per unit of frequency, is *energy density*. The energy of the signal $f(t)$ between two frequencies (Hz) therefore will be the area under the squared amplitude curve between the two frequencies.

EXAMPLE 18.12 As an illustration of Parseval's equation, let us consider the circuit of Fig. 18.7 with $v_o(0) = 0$ and input voltage

$$v_i(t) = e^{-3t}u(t) \quad \text{V}$$

The energy w_i of the input signal is

$$w_i = \int_{-\infty}^{\infty} v_i^2 \, dt = \int_{-\infty}^{\infty} [e^{-3t}u(t)]^2 \, dt$$

$$= \int_0^{\infty} e^{-6t} \, dt = \frac{1}{6} \quad \text{J}$$

The transform of $v_i(t)$ is

$$\mathbf{V}_i(j\omega) = \frac{1}{3 + j\omega}$$

FIGURE 18.7 *RC* circuit

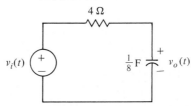

so that by Parseval's equation (18.47) the input signal energy is also given by

$$w_i = \frac{1}{\pi} \int_0^\infty |V_i(j\omega)|^2 \, d\omega$$

$$= \frac{1}{\pi} \int_0^\infty \frac{d\omega}{9 + \omega^2}$$

$$= \frac{1}{3\pi} \tan^{-1} \frac{\omega}{3} \bigg|_0^\infty$$

$$= \frac{1}{6} \ \text{J}$$

If we are interested in the energy w_o of the output voltage $v_o(t)$, the network function by voltage division is

$$\mathbf{H}(j\omega) = \frac{\mathbf{V}_o(j\omega)}{\mathbf{V}_i(j\omega)} = \frac{8/j\omega}{4 + (8/j\omega)} = \frac{2}{2 + j\omega}$$

so that

$$w_o = \frac{1}{\pi} \int_0^\infty |\mathbf{V}_o(j\omega)|^2 \, d\omega = \frac{1}{\pi} \int_0^\infty |\mathbf{H}(j\omega)|^2 |\mathbf{V}_i(j\omega)|^2 \, d\omega$$

$$= \frac{1}{\pi} \int_0^\infty \left| \frac{2}{2 + j\omega} \right|^2 \left| \frac{1}{3 + j\omega} \right|^2 d\omega$$

$$= \frac{4}{\pi} \int_0^\infty \frac{d\omega}{(4 + \omega^2)(9 + \omega^2)}$$

$$= \frac{4}{5\pi} \int_0^\infty \left[\frac{1}{4 + \omega^2} - \frac{1}{9 + \omega^2} \right] d\omega$$

$$= \frac{1}{15} \ \text{J} \tag{18.48}$$

The difference, $\frac{1}{6} - \frac{1}{15} = \frac{1}{10}$ J, between the energy of the input and output signals is the 1-Ω energy dissipated in the circuit.

Since $|\mathbf{F}(j\omega)|^2$ is the energy density associated with a function $f(t) = \mathscr{F}^{-1}[\mathbf{F}(j\omega)]$, a plot of $|\mathbf{F}(j\omega)|^2$ versus frequency (Hz) is an energy spectrum. The same information conveyed by the energy spectrum can be obtained from the amplitude response, $|\mathbf{F}(j\omega)|$ versus ω, which gives us some physical insight into the amplitude spectra we considered in detail in Chapter 15.

EXERCISES

18.6.1 Obtain the energy given in (18.48) by using the time-domain voltage v_o for the circuit of Fig. 18.7. Also find the resistor voltage and show that the total energy dissipated in the resistor is the $\frac{1}{10}$ J difference between w_i and w_o.

Answer $v = 3e^{-3t} - 2e^{-2t}$ V

18.6.2 An ideal low-pass filter has a magnitude response

$$|\mathbf{H}(j\omega)| = 1, \qquad -1 < \omega < 1$$
$$= 0, \qquad \text{elsewhere}$$

and an input $v_i = 4e^{-2t}u(t)$ V. Find the 1-Ω energy of the input signal and of the output signal.

Answer $w_i = 4$ J, $w_o = \dfrac{8}{\pi} \tan^{-1} 0.5 \approx 1.18$ J

18.6.3 Find the 1-Ω energy of the output signal i if $v_g = 4e^{-4t}u(t)$ V.

Answer $1/6$ J

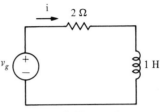

EXERCISE 18.6.3

PROBLEMS

18.1 Find the Fourier transforms of the following functions:
(a) $f(t) = u(t) - u(t - 1)$.
(b) $f(t) = e^{-at} \cos bt \, u(t), \, a > 0$.
(c) $f(t) = e^{-at}[u(t) - u(t - 1)]$.

18.2 Find the Fourier transforms of the following functions:
(a) $f(t) = e^{at}u(-t), \, a > 0$.
(b) $f(t) = e^{-at} \sin bt \, u(t), \, a > 0$.
(c) $f(t) = e^{at} \sin bt \, u(-t), \, a > 0$.
(d) $f(t) = e^{-at}u(t) - e^{at}u(-t), \, a > 0$.

18.3 Show that $f(t)$ in Prob. 18.2(d) is odd and find its transform using (18.24).

18.4 Find the real and imaginary parts of the transforms of Prob. 18.2(a), (b), and (c).

18.5 Find the magnitude and phase of the transforms of Prob. 18.2.

18.6 Use the linearity property, (18.14), and the result of Prob. 18.2(a) to solve Prob. 18.2(d).

18.7 Use linearity and a trigonometric identity to find the Fourier transform of

$$f(t) = e^{-t} \cos\left(2t + \frac{\pi}{4}\right)u(t)$$

18.8 Use the operations of Table 18.1 to find the Fourier transforms of the following functions:
(a) $te^{-2t}u(t)$, (b) $te^{-2t} \cos t \, u(t)$, (c) $t^2e^{-2t}u(t)$, (d) $te^{2t}u(-t)$,
(e) $e^{2t} \sin t \, u(-t)$, (f) $t^n e^{-2|t|}$, for $n = 1, 2$.

18.9 Use operations 1 and 4 of Table 18.1 to find the transforms of
(a) $e^{-2t} \sin 3t \, u(t)$, and (b) $e^{-2t} \cos 3t \, u(t)$.

599

18.10 Find the Fourier transform of (a) $e^{-t}u(t-1)$ and (b) $e^{-2t}\cos{(\pi t/4)}u(t-2)$ using operation 3 of Table 18.1.

18.11 Use operations 1 and 3 of Table 18.1 to find the Fourier transform of

$$f(t) = 2, \qquad -4 < t < -2 \text{ and } 1 < t < 3$$
$$= 0, \qquad \text{elsewhere}$$

18.12 Find the Fourier transforms of

(a) $f(t) = \dfrac{1}{a^2 + t^2}$.

(b) $f(t) = \dfrac{1}{a + jt}$, $a > 0$.

(c) $f(t) = \text{Sa}\,(t) = \dfrac{\sin t}{t}$.

(d) $f(t) = [\text{Sa}\,(t)]^2$.

18.13 Find the inverse transforms of (a) $\mathbf{F}(j\omega)$, (b) $j\omega\mathbf{F}(j\omega)$, and (c) $e^{-2j\omega}\mathbf{F}(j\omega)$, where $\mathbf{F}(j\omega) = e^{-a|\omega|}$, $a > 0$.

18.14 A function $f(t)$ has a Fourier transform given by

$$\mathbf{F}(j\omega) = \frac{1 + j\omega}{6 - \omega^2 + 5j\omega} = \frac{2}{j\omega + 3} - \frac{1}{j\omega + 2}$$

Without finding $f(t)$, obtain the transforms of (a) $f(t-2)$, (b) $e^{-t}f(t)$, (c) $f(2t)$, and (d) $f(-t)$. Check the results by finding $f(t)$ and computing the various transforms.

18.15 Find the inverse transforms of (a) $1/(j\omega - 1)$ and (b) $1/(j\omega - 1)^2$.

18.16 If the output is $i(t)$ and the input is $v_g(t)$ in the circuit of Prob. 17.12, find the network function.

18.17 If in the circuit of Prob. 17.12 the output is $v_c(t)$ and the input is $v_g(t)$, find (a) the network function $\mathbf{H}(j\omega)$ and (b) $v_c(t)$ if $v_c(0) = 0$ and $v_g(t) = e^{-t}u(t)$ V. *Suggestion:* Note that

$$\frac{1}{(x + 1)(x + 2)} = \frac{1}{x + 1} - \frac{1}{x + 2}$$

18.18 Find the amplitude and phase response of a network having input $x(t)$ and output $y(t)$ related by

$$y'' + 5y' + 4y = 2x'$$

18.19 If an initially relaxed circuit has an input $x(t) = (e^{-t} - e^{-2t})u(t)$ and an output $y = te^{-t}u(t)$, find the network function $\mathbf{H}(j\omega)$ and the describing differential equation.

18.20 If the output transform is

$$\mathbf{F}(j\omega) = \frac{1}{1 + j\omega}$$

find the energy of the output in $-1 < \omega < 1$ and in $-\infty < \omega < \infty$.

18.21 An ideal high-pass filter has a magnitude response

$$|\mathbf{H}(j\omega)| = 0, \qquad -1 < \omega < 1$$
$$= 1, \qquad \text{elsewhere}$$

and an input $v_i(t) = e^{-t}u(t)$ V. Find the energy of the output signal.

18.22 An ideal bandpass filter has a magnitude response

$$|\mathbf{H}(j\omega)| = 1, \qquad -\sqrt{3} < \omega < -1 \text{ and } 1 < \omega < \sqrt{3}$$
$$= 0, \qquad \text{elsewhere}$$

and an input $v_i(t) = e^{-t}u(t)$ V. Find the energy of the output function.

18.23   An ideal band-reject filter has a magnitude response

$$|\mathbf{H}(j\omega)| = 0, \qquad -\sqrt{3} < \omega < -1/\sqrt{3} \text{ and } 1/\sqrt{3} < \omega < \sqrt{3}$$
$$= 1, \qquad \text{elsewhere}$$

and an input $v_i(t) = e^{-t}u(t)$ V. Find the energy of the output function.

18.24   Find the network function if the output is $y(t)$, the input is $x(t)$, and they are related by other variables q_1, q_2, and q_3 by

$$\frac{dq_1}{dt} = -q_1 - q_3 + x$$

$$\frac{dq_2}{dt} = -q_2 + q_3$$

$$\frac{dq_3}{dt} = \frac{1}{2}(q_1 - q_2)$$

$$y = q_2$$

19

Pierre Simon Laplace
1749–1827

Laplace Transforms

There was no need for God in my hypothesis [to Napoleon, who had to ask].

Pierre Simon Laplace

Pierre Simon, Marquis de Laplace, the famous French astronomer and mathematician, is credited with the transform that bears his name and allows us to further generalize the generalized phasor method to analyze circuits with nonsinusoidal inputs. Laplace was better known, however, for *Celestial Mechanics,* his master work, which summarized the achievements in astronomy from the time of Newton.

Laplace was born in Beaumont-en-Auge, in Normandy, France. Little is known of his early life other than that his father was a farmer because the snobbish Laplace, after he became famous, did not like to speak of his humble origins. Rich neighbors, it is said, recognized his talent and helped finance his education, first at Caen and later at the military school in Beaumont. Through the efforts of the famous physicist d'Alembert, who was impressed by his abilities and his effrontery, Laplace became a professor of mathematics in Paris at age 20. He was an opportunist, shifting his political allegiance as required so that his career successfully spanned three regimes in revolutionary France—the republic, the empire of Napoleon, and the Bourbon restoration. Napoleon made him a count and Louis XVIII made him a marquis. His mathematical abilities, however, were genuine, inspiring the great mathematician Siméon Poisson to label him the Isaac Newton of France. ∎

As we have seen in the previous chapter, the Fourier transform may be used to extend the concept of network functions to cases where the excitation is nonsinusoidal. It may also be used in certain applications to find the forced response to an input function which possesses a Fourier transform. With the addition of the Fourier transform to our arsenal we are able to add networks with certain nonsinusoidal, nonperiodic excitations to our set of solvable circuits. We have thus made considerable progress since we first solved resistive circuits with dc excitations in Chapter 2. Along the way we have developed differential equation techniques for circuits containing storage elements, phasor techniques for sinusoidal excitations, generalized phasors for damped sinusoids, trigonometric and exponential Fourier series methods for periodic excitations, and now Fourier transforms for certain nonperiodic excitations.

In this chapter it remains for us to consider a most elegant procedure, the *Laplace transform* method, which can be applied to a much wider variety of excitation functions than the Fourier transform method and which gives at once both the natural and forced responses for a given set of initial conditions. In the process we shall define a special and highly useful function, known as the *impulse function,* and, in the succeeding chapter, use its properties to obtain general time-domain solutions to any linear circuit whose excitation possesses a Laplace transform.

The Laplace transform may be obtained as a generalization of the Fourier transform and is developed in this way by many authors. However, to make this chapter self-contained and independent of the previous chapter, we shall give a completely independent treatment of the Laplace transform. A treatment based on the Fourier transform will be left to the problems to give the reader an optional alternative development.

19.1

DEFINITION

The *Laplace transform,* named for the French astronomer and mathematician Pierre Simon, Marquis de Laplace (1749–1827), is defined by

$$\mathcal{L}[f(t)] = \mathbf{F}(s) = \int_0^\infty f(t)e^{-st}\,dt \tag{19.1}$$

The function $\mathbf{F}(s)$ is the Laplace transform of $f(t)$ and is a function of the generalized frequency, $s = \sigma + j\omega$, considered earlier in Chapter 14. We note that our definition is sometimes referred to as the *one-sided* or *unilateral* transform, as distinguished from the *two-sided* or *bilateral* transform. The latter is defined by (19.1) with the lower limit 0 replaced by $-\infty$. However, we shall not need this generalization since for the circuits we consider the function $f(t)$ is of interest only for $t \geq 0$. The behavior of the circuit for $t < 0$ is past history and is accounted for by the initial conditions.

If $f(t)$ has an infinite discontinuity at $t = 0$, we will understand that the lower limit of the Laplace transform integral is 0^-. This is because it is desirable to include the effects of *impulse* functions, which have infinite discontinuities at the origin, as we will see.

Comparing (19.1) with the definition of the Fourier transform of the previous chapter, we see that

$$\mathcal{F}[e^{-\sigma t}f(t)u(t)] = \mathcal{L}[f(t)]$$

The presence of the factors $e^{-\sigma t}$ and $u(t)$ accounts for the greater versatility of the Laplace transform. For example, the Fourier transform of a function $f(t)$ may not exist because the condition

$$\int_{-\infty}^\infty |f(t)|\,dt < \infty \tag{19.2}$$

fails to hold. It is possible, however, that for the same function $f(t)$ a value of σ may be found so that

$$\int_{-\infty}^\infty |e^{-\sigma t}f(t)u(t)|\,dt = \int_0^\infty e^{-\sigma t}|f(t)|\,dt < \infty \tag{19.3}$$

As a general rule, if $f(t) = 0$ for $t < 0$ and $\mathcal{F}[f(t)]$ exists, then the factor $u(t)$ in (19.3) is redundant, and the factor $e^{-\sigma t}$ is not necessary for the existence of the integral. In this case, the Laplace transform with $s = j\omega$ is the Fourier transform.

Typical functions $f(t)$ encountered in circuit theory will have Laplace transforms, which is to say that the integral of (19.1) will exist. The mathematical conditions sufficient for its existence are that $f(t)$ be *sectionally continuous* in every finite interval of $t \geq 0$ and that $f(t)$ be of *exponential order* $e^{\alpha t}$ as $t \to \infty$. The first of these means that every interval of $t \geq 0$ can be subdivided into a finite number of parts in each of which $f(t)$ is continuous with finite limits as t approaches either end point from the interior. The second requires that constants M and α exist such that, for all t sufficiently large,

$$|f(t)| < Me^{\alpha t} \tag{19.4}$$

in which case Re $s = \sigma > \alpha$ will ensure that (19.3) holds. We will not belabor these two conditions, since the first is sufficient for ordinary integrals to exist, and the second is an added requirement for the improper Laplace transform integral to converge. To fail these conditions requires an unusual function like e^{t^2}, which we are unlikely to see in circuit analysis.

EXAMPLE 19.1 To illustrate the computation of a Laplace transform, let us consider

$$f(t) = e^{-at}u(t)$$

where $a > 0$. Then by (19.1) the transform is

$$\mathbf{F}(s) = \int_0^\infty e^{-at}e^{-st}\, dt$$

$$= -\frac{1}{s+a}e^{-(s+a)t}\,\Big|_0^\infty$$

If Re $s = \sigma > -a$, then the value at the upper limit is zero, and we have

$$\mathscr{L}[e^{-at}u(t)] = \frac{1}{s+a} \tag{19.5}$$

In general, we shall define $f(t)$ in (19.1) as the *inverse* Laplace transform of $\mathbf{F}(s)$, with the symbolism

$$f(t) = \mathscr{L}^{-1}[\mathbf{F}(s)] \tag{19.6}$$

The inverse transform may be expressed explicitly as an integral, as was true of the Fourier transform (see Prob. 19.18). However, for our purposes we prefer to obtain the inverse transform by a simpler procedure, which we shall outline in Sec. 19.6.

EXAMPLE 19.2 As an example, (19.5) may be written in the form

$$\mathscr{L}^{-1}\left[\frac{1}{s+a}\right] = e^{-at}u(t) \tag{19.7}$$

EXAMPLE 19.3 As another, simpler, but highly useful example, the transform of the step function $u(t)$ is given by

$$\mathscr{L}[u(t)] = \int_0^\infty e^{-st}\, dt = \frac{1}{s} \tag{19.8}$$

provided Re $s > 0$. An alternative statement is, of course,

$$\mathscr{L}^{-1}\left[\frac{1}{s}\right] = u(t)$$

This function further illustrates the versatility of the Laplace transform. The step function does not have a Fourier transform since for it (19.2) does not hold. (That is, there is no Fourier transform in the *conventional* sense. It is possible by the use of generalized functions to derive a Fourier transform of the step function, expressed in terms of an *impulse* function, to be defined in Sec. 19.5. However, we shall not consider transforms of this type here.)

As mentioned earlier, the Laplace transform can be used to take into account the initial conditions prevailing in the circuit. To gain some insight into how this is accomplished, let us consider the transform of the derivative $f'(t)$, given by (19.1) as

$$\mathcal{L}[f'(t)] = \int_0^\infty f'(t)e^{-st}\, dt$$

Integrating by parts, we have

$$\mathcal{L}[f'(t)] = f(t)e^{-st}\bigg|_0^\infty + s\int_0^\infty f(t)e^{-st}\, dt$$

If $f(t)$ and s are such that the integrated part vanishes at the upper limit, as will usually be the case, then in view of (19.1), this result becomes

$$\mathcal{L}\left[\frac{df(t)}{dt}\right] = s\mathbf{F}(s) - f(0) \qquad (19.9)$$

[We are considering the case $f(t)$ to be a continuous function on $0 \le t \le \infty$.] Thus the initial condition $f(0)$ is automatically built into the transform of the derivative. This was not the case for the Fourier transform.

EXAMPLE 19.4 As a last example in this section, let us demonstrate how (19.9) may be used to obtain transforms. Consider the ramp function $f(t) = tu(t)$, or

$$f(t) = t, \qquad t \ge 0$$
$$= 0, \qquad t < 0$$

for which $f(0) = 0$ and

$$f'(t) = 1, \qquad t \ge 0$$
$$= 0, \qquad t < 0$$

or $f'(t) = u(t)$. Substituting these values into (19.9) yields

$$\mathcal{L}[u(t)] = s\mathbf{F}(s) - 0 = \frac{1}{s}$$

from which

$$\mathcal{L}[tu(t)] = \mathbf{F}(s) = \frac{1}{s^2}$$

EXERCISES

19.1.1 Find the Laplace transform of (a) $\sin kt\, u(t)$, (b) $\cos kt\, u(t)$, and (c) $(1 + 3e^{-2t})u(t)$.

Answer (a) $\dfrac{k}{s^2 + k^2}$; (b) $\dfrac{s}{s^2 + k^2}$; (c) $\dfrac{1}{s} + \dfrac{3}{s + 2}$

19.1.2 Find $\mathcal{L}[\sin t\, u(t)]$ using (19.9) and the result of Ex. 19.1.1(b).

Answer $\dfrac{1}{s^2 + 1}$

19.1.3 (a) Use the definition of the Laplace transform and integration by parts to show that, for n a nonnegative integer,

$$\mathcal{L}[t^n u(t)] = \frac{n}{s}\mathcal{L}[t^{n-1} u(t)]$$

(b) Replace n successively in this result by $n-1$, $n-2$, $\ldots$, $n-k$ to obtain

$$\mathcal{L}[t^{n-1} u(t)] = \frac{n-1}{s}\mathcal{L}[t^{n-2} u(t)]$$

$$\mathcal{L}[t^{n-2} u(t)] = \frac{n-2}{s}\mathcal{L}[t^{n-3} u(t)]$$

$$\vdots$$

$$\mathcal{L}[t^{n-k} u(t)] = \frac{n-k}{s}\mathcal{L}[t^{n-k-1} u(t)]$$

19.1.4 (a) Show from the results of Ex. 19.1.3 that

$$\mathcal{L}[t^n u(t)] = \frac{n(n-1)(n-2)\cdots(n-k)}{s^{k+1}}\mathcal{L}[t^{n-k-1} u(t)]$$

(b) Let $k = n - 1$ and show that

$$\mathcal{L}[t^n u(t)] = \frac{n!}{s^{n+1}}, \qquad n = 0, 1, 2, \ldots$$

19.2

SOME SPECIAL RESULTS USING LINEARITY

As was the case for the Fourier transform, the Laplace transform is an integral and is therefore a linear operation. Thus we may say that

$$\mathcal{L}[c_1 f_1(t) + c_2 f_2(t)] = c_1 F_1(s) + c_2 F_2(s) \qquad (19.10)$$

where F_1 and F_2 are the Laplace transforms of f_1 and f_2 and c_1 and c_2 are constants. Thus we may obtain a variety of transforms from two known ones. For example, the function of Ex. 19.1.1.(c) may be easily transformed by writing

$$\mathcal{L}[u(t) + 3e^{-2t}u(t)] = \mathcal{L}[u(t)] + 3\mathcal{L}[e^{-2t}u(t)]$$

$$= \frac{1}{s} + \frac{3}{s + 2}$$

EXAMPLE 19.5   As another example, since we have

$$\mathcal{L}[e^{jkt}u(t)] = \frac{1}{s - jk}$$

and

$$\mathcal{L}[e^{-jkt}u(t)] = \frac{1}{s + jk}$$

we may write, using (19.10),

$$\mathcal{L}[\cos kt \, u(t)] = \mathcal{L}\left[\frac{e^{jkt} + e^{-jkt}}{2}u(t)\right]$$

$$= \frac{1}{2}\mathcal{L}[e^{jkt}u(t)] + \frac{1}{2}\mathcal{L}[e^{-jkt}u(t)]$$

$$= \frac{1}{2}\left[\frac{1}{s - jk} + \frac{1}{s + jk}\right]$$

$$= \frac{s}{s^2 + k^2} \tag{19.11}$$

In a similar manner we may show that

$$\mathcal{L}[\sin kt \, u(t)] = \frac{k}{s^2 + k^2} \tag{19.12}$$

Both of these results were previously given in Ex. 19.1.1.

Two special cases of (19.10) are of importance: the case $c_2 = 0$ and the case $c_1 = c_2 = 1$, which yield respectively

$$\mathcal{L}[c_1 f_1(t)] = c_1 \mathbf{F}_1(s)$$

and

$$\mathcal{L}[f_1(t) + f_2(t)] = \mathbf{F}_1(s) + \mathbf{F}_2(s)$$

In other words, the transform of a constant times a function is the constant times the transform of the function, and the transform of a sum is the sum of the transforms.

The inverse transform is also a linear operation, which is seen by writing (19.10) in the form

$$c_1 f_1(t) + c_2 f_2(t) = \mathcal{L}^{-1}[c_1 \mathbf{F}_1(s) + c_2 \mathbf{F}_2(s)]$$

(In other words, we have taken the inverse transform of both sides of the equation.) Since f_1 and f_2 are inverse transforms of $\mathbf{F}_1$ and $\mathbf{F}_2$, this result may be written

$$\mathcal{L}^{-1}[c_1 \mathbf{F}_1(s) + c_2 \mathbf{F}_2(s)] = c_1 \mathcal{L}^{-1}[\mathbf{F}_1(s)] + c_2 \mathcal{L}^{-1}[\mathbf{F}_2(s)] \tag{19.13}$$

which is a linear operation.

EXAMPLE 19.6 The linearity of the inverse transform allows us to invert a variety of transforms. For example, suppose we are required to find $f(t)$ if

$$\mathbf{F}(s) = \frac{4}{s} - \frac{3}{s+2} \tag{19.14}$$

We may write, by (19.13),

$$f(t) = \mathscr{L}^{-1}\left[\frac{4}{s} + \frac{-3}{s+2}\right]$$

$$= 4\mathscr{L}^{-1}\left[\frac{1}{s}\right] - 3\mathscr{L}^{-1}\left[\frac{1}{s+2}\right]$$

From (19.8) and (19.7) with $a = 2$, we have

$$f(t) = (4 - 3e^{-2t})u(t) \tag{19.15}$$

The result may be summarized as

$$\mathscr{L}^{-1}\left[\frac{s+8}{s(s+2)}\right] = (4 - 3e^{-2t})u(t) \tag{19.16}$$

where we have obtained a common denominator in (19.14).

The foregoing example suggests a general method of obtaining inverse transforms. Transforms such as (19.16) are expanded in *partial fractions* such as (19.14), each fraction is inverted, and the inverse transform is the sum of the inverted fractions. We will consider the method of partial fractions in Sec. 19.6.

EXAMPLE 19.7 As a last example, let us use the Laplace transform method to find $i(t)$, $t > 0$, in the circuit of Fig. 19.1, given that $v_g = 24u(t)$ V and $i(0) = 1$ A. The mesh equation is

$$3\frac{di}{dt} + 6i = 24u(t)$$

which may be solved by taking the Laplace transform of both sides and solving the result for $\mathscr{L}[i(t)]$, which we designate $\mathbf{I}(s)$. The process yields

FIGURE 19.1 *RL* circuit

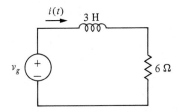

$$\mathscr{L}\left[3\frac{di}{dt}\right] + \mathscr{L}[6i] = \mathscr{L}[24u(t)]$$

which, using (19.9), becomes

$$3[s\mathbf{I}(s) - 1] + 6\mathbf{I}(s) = \frac{24}{s}$$

Solving for the transform of the current results in

$$\mathbf{I}(s) = \frac{(24/s) + 3}{3s + 6} = \frac{s + 8}{s(s + 2)}$$

which, as we have seen in (19.16), inverts to

$$i(t) = (4 - 3e^{-2t})u(t) \quad \text{A}$$

or, for $t > 0$,

$$i(t) = 4 - 3e^{-2t} \quad \text{A}$$

EXERCISES

19.2.1 Find the Laplace transform of (a) $(2 + 3e^{-t})u(t)$, (b) $(3t - 5e^{6t})u(t)$, and (c) $\cosh kt\, u(t)$.

Answer (a) $\dfrac{2}{s} + \dfrac{3}{s + 1}$; (b) $\dfrac{3}{s^2} - \dfrac{5}{s - 6}$; (c) $\dfrac{s}{s^2 - k^2}$

19.2.2 Find the inverse transform of (a) $\dfrac{6}{s^2} - \dfrac{5s}{s^2 + 4}$, (b) $\dfrac{3s + 4}{s^2 + 16}$, and (c) $\dfrac{s + 4}{s(s + 2)} = \dfrac{2}{s} - \dfrac{1}{s + 2}$.

Answer (a) $(6t - 5\cos 2t)u(t)$; (b) $(3\cos 4t + \sin 4t)u(t)$; (c) $(2 - e^{-2t})u(t)$

19.2.3 Use Laplace transforms to find $i(t)$, $t > 0$, in Fig. 19.1 if $v_g = 12u(t)$ V and (a) $i(0) = 1$ A and (b) $i(0) = 2$ A. [*Suggestion:* For (a), see Ex. 19.2.2(c).]

Answer (a) $2 - e^{-2t}$ A; (b) 2 A

19.3

TRANSLATION THEOREMS

There are a number of theorems that facilitate the use of Laplace transforms. We will derive many of these in the remainder of this chapter, beginning in this section, where we consider two useful translation results. In one, s is replaced by $s + a$ (frequency translation), and in the other, t is replaced by $t - \tau$ (time translation). To obtain the first, let us consider the transform

$$\mathscr{L}[e^{-at}f(t)] = \int_0^{\infty} e^{-at}f(t)e^{-st}\, dt$$

$$= \int_0^\infty f(t)e^{-(s+a)t}\, dt$$

By comparing this result with (19.1), we see that

$$\boxed{\mathcal{L}[e^{-at}f(t)] = \mathbf{F}(s + a) \qquad (19.17)}$$

where $\mathbf{F}(s)$ is the transform of $f(t)$.

EXAMPLE 19.8   As an example, let us find the transform of $te^{-3t}u(t)$. We let $f(t) = tu(t)$ so that $\mathbf{F}(s) = \dfrac{1}{s^2}$, and from (19.17), with $a = 3$, we have

$$\mathcal{L}[te^{-3t}u(t)] = \mathbf{F}(s + 3) = \frac{1}{(s + 3)^2}$$

Two other useful examples are the transforms of the damped sinusoids

$$\mathcal{L}[e^{-at} \cos bt\, u(t)] = \frac{s + a}{(s + a)^2 + b^2} \qquad (19.18)$$

and

$$\mathcal{L}[e^{-at} \sin bt\, u(t)] = \frac{b}{(s + a)^2 + b^2} \qquad (19.19)$$

as the reader may verify using (19.11) and (19.12). It should be noted that since a and b are real, the denominators above are irreducible quadratics.

EXAMPLE 19.9   To illustrate these last two results, suppose it is required to find $f(t)$, given that its transform is

$$\mathbf{F}(s) = \frac{6s}{s^2 + 2s + 5}$$

We may complete the square in the denominator, obtaining

$$\mathbf{F}(s) = \frac{6s}{(s + 1)^2 + 2^2}$$

which we may write as

$$\mathbf{F}(s) = \frac{6(s + 1) - 6}{(s + 1)^2 + 2^2}$$

$$= 6\left[\frac{s + 1}{(s + 1)^2 + 2^2}\right] - 3\left[\frac{2}{(s + 1)^2 + 2^2}\right]$$

Finally, by (19.18) and (19.19), we have

$$f(t) = (6e^{-t} \cos 2t - 3e^{-t} \sin 2t)u(t) \qquad (19.20)$$

To obtain the time translation theorem, let us consider the function $f(t - \tau)u(t - \tau)$, which is the function $f(t)$ made zero for $t < 0$ and delayed τ units, as shown in Fig. 19.2. The transform of this function is

$$\mathcal{L}[f(t - \tau)u(t - \tau)] = \int_0^\infty f(t - \tau)u(t - \tau)e^{-st}\, dt$$

$$= \int_\tau^\infty f(t - \tau)e^{-st}\, dt$$

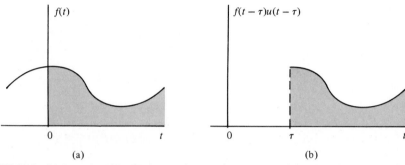

FIGURE 19.2   (a) Function $f(t)$; (b) $f(t)$ made zero for $t < 0$ and translated τ units to the right

Making the change of variable $t = \tau + x$ yields

$$\mathcal{L}[f(t - \tau)u(t - \tau)] = \int_0^\infty f(x)e^{-s(\tau+x)}\, dx$$

$$= e^{-s\tau}\int_0^\infty f(x)e^{-sx}\, dx$$

or the time translation theorem,

$$\boxed{\mathcal{L}[f(t - \tau)u(t - \tau)] = e^{-s\tau}\mathbf{F}(s) \qquad\qquad (19.21)}$$

where, again, $\mathbf{F}(s)$ is the transform of $f(t)$. Comparing this result with the frequency translation theorem (19.17) we see that in the latter case, translation in the frequency domain (replacing s by $s + a$) corresponds to multiplication by an exponential in the time domain, whereas in (19.21), translation in the time domain corresponds to multiplication by an exponential in the frequency domain.

EXAMPLE 19.10   To illustrate the time translation result, let us find the transform of $f(t) = e^{-3t}u(t - 2)$, which we write in the form

$$e^{-3t}u(t - 2) = e^{-3(t-2)-6}u(t - 2)$$

$$= e^{-6}e^{-3(t-2)}u(t - 2)$$

Since $\mathcal{L}[e^{-3t}u(t)] = \mathbf{F}(s) = 1/(s + 3)$, we have, by (19.21),

$$\mathscr{L}[e^{-3t}u(t-2)] = e^{-6}\left(\frac{e^{-2s}}{s+3}\right)$$

Another useful result, which can be used to obtain a generalization of (19.17), is the *scale change property*

$$\mathscr{L}[f(ct)] = \frac{1}{c}\mathbf{F}\left(\frac{s}{c}\right), \qquad c > 0 \qquad\qquad (19.22)$$

To obtain this result, we note from the definition of the transform that

$$\mathscr{L}[f(ct)] = \int_0^\infty f(cx)e^{-sx}\,dx$$

where we have used x as the variable of integration. Replacing cx by t in the integral results in

$$\mathscr{L}[f(ct)] = \int_0^\infty f(t)e^{-st/c}\,\frac{dt}{c}$$

$$= \frac{1}{c}\int_0^\infty f(t)e^{-(s/c)t}\,dt$$

which is (19.22).

Replacing c by $1/a$ in (19.22) results in

$$\mathscr{L}\left[\frac{1}{a}f\left(\frac{t}{a}\right)\right] = \mathbf{F}(as)$$

Finally, we may replace s by $s + (b/a)$ and use (19.17) to obtain the generalization of the frequency translation property:

$$\mathscr{L}\left[\frac{1}{a}e^{-bt/a}f\left(\frac{t}{a}\right)\right] = \mathbf{F}(as + b) \qquad\qquad (19.23)$$

EXERCISES

19.3.1 Find the Laplace transform of (a) $te^{-2t}u(t)$, (b) $u(t-3)$, and (c) $f(t) = 1$, $0 < t < 2$, and $f(t) = 0$, elsewhere.

Answer (a) $\dfrac{1}{(s+2)^2}$; (b) $\dfrac{e^{-3s}}{s}$; (c) $\dfrac{1}{s}(1 - e^{-2s})$

19.3.2 Find the inverse transform of (a) $\dfrac{e^{-2s}}{s+3}$, (b) $\dfrac{s+1}{s^2}e^{-s}$, and (c) $\dfrac{2s}{s^2+2s+5}$.

Answer (a) $e^{-3(t-2)}u(t-2)$; (b) $tu(t-1)$; (c) $(2\cos 2t - \sin 2t)e^{-t}u(t)$

19.3.3 Use the scale change property (19.22) to obtain the transform of $\cos kt$ given that $\mathscr{L}[\cos t] = s/(s^2+1)$.

Answer $\dfrac{s}{s^2+k^2}$

19.3.4 Use the frequency translation theorem and the transform of $t^n u(t)$ of Ex. 19.1.4 to derive the transform

$$\mathcal{L}[e^{-at}t^n u(t)] = \frac{n!}{(s + a)^{n+1}}; \qquad n = 0, 1, 2, \ldots$$

19.4

CONVOLUTION

Because of linearity, the transform of a sum is the sum of the transforms. Another combination that occurs often in circuit theory, as well as in general system theory, is the product of two transforms,

$$\mathbf{Y}(s) = \mathbf{F}(s)\mathbf{G}(s)$$

and it is extremely important to have a general method of obtaining the inverse

$$y(t) = \mathcal{L}^{-1}[\mathbf{F}(s)\mathbf{G}(s)]$$

in terms of $f(t)$ and $g(t)$, the inverse transforms of $\mathbf{F}(s)$ and $\mathbf{G}(s)$. This may be done in general, as will be seen in this section, and the resulting value of $y(t)$ is called the *convolution* of $f(t)$ and $g(t)$.

We begin with the definition

$$\mathbf{F}(s) = \int_0^\infty f(\tau)e^{-s\tau} \, d\tau$$

which when multiplied through by $\mathbf{G}(s)$ results in

$$\mathbf{F}(s)\mathbf{G}(s) = \int_0^\infty f(\tau)[\mathbf{G}(s)e^{-s\tau}] \, d\tau$$

By (19.21) this becomes

$$\mathbf{F}(s)\mathbf{G}(s) = \int_0^\infty f(\tau)\mathcal{L}[g(t - \tau)u(t - \tau)] \, d\tau$$

$$= \int_0^\infty f(\tau)\left[\int_0^\infty g(t - \tau)u(t - \tau)e^{-st} \, dt\right] d\tau$$

Interchanging the order of integration and noting that $u(t - \tau) = 0$ for $\tau > t$, we have

$$\mathbf{F}(s)\mathbf{G}(s) = \int_0^\infty e^{-st}\left[\int_0^t f(\tau)g(t - \tau) \, d\tau\right] dt$$

But by definition of the transform this is

$$\mathbf{F}(s)\mathbf{G}(s) = \mathcal{L}\left[\int_0^t f(\tau)g(t - \tau) \, d\tau\right]$$

Thus the function $y(t)$ is given by

$$\mathcal{L}^{-1}[\mathbf{F}(s)\mathbf{G}(s)] = \int_0^t f(\tau)g(t - \tau) \, d\tau \tag{19.24}$$

This result is known as the *convolution theorem,* and the integral involved is the convolution of f and g, symbolized by

$$f(t) * g(t) = \int_0^t f(\tau)g(t - \tau)\, d\tau \qquad (19.25)$$

The order of the functions f and g is immaterial, as the reader is asked to show in Ex. 19.4.1.

EXAMPLE 19.11 As an example, let us calculate the convolution of e^{-t} and e^{-2t}. By (19.25) we have

$$e^{-t} * e^{-2t} = \int_0^t e^{-\tau}e^{-2(t - \tau)}\, d\tau$$

$$= e^{-2t}\int_0^t e^{\tau}\, d\tau$$

$$= e^{-t} - e^{-2t}$$

EXAMPLE 19.12 The convolution integral may be made clearer by interpreting it graphically as the area under the curve $f(\tau)g(t - \tau)$ between $\tau = 0$ and $\tau = t$. For example, suppose $f(\tau)$ and $g(\tau)$ are as shown in Figs. 19.3(a) and (b). The function $g(-\tau)$ is $g(\tau)$ *folded* as in Fig. 19.3(c), and $g(t - \tau)$ is the folded function translated t units, shown in Fig. 19.3(d). To consider the product $f(\tau)g(t - \tau)$ for various values of t, we have sketched both functions on the same set of axes, shown in Fig. 19.4. In (a) for $t < 0$, the product is zero, as is the area. For $0 < t < 1$, shown in (b), the product is 2 over $0 < \tau < t$ and zero elsewhere; thus the area is $2t$, as noted. In (c) for $1 < t < 2$, the product is 2 on $t - 1 < \tau < 1$ and zero elsewhere; therefore, the area is $2(2 - t)$, as shown. Finally, in (d) for $t > 2$, the product and the area are zero. The convolution is therefore

$$f(t) * g(t) = 0, \qquad t < 0$$

$$= 2t, \qquad 0 < t < 1$$

$$= 2(2 - t), \qquad 1 < t < 2$$

$$= 0, \qquad t > 2$$

FIGURE 19.3 (a) $f(\tau)$; (b) $g(\tau)$; (c) $g(-\tau)$; (d) $g(t - \tau)$

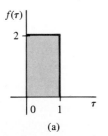
(a)

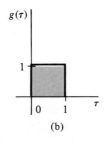
(b)

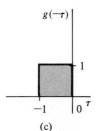
(c)

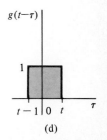
(d)

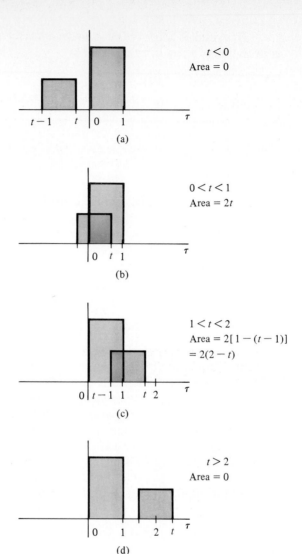

FIGURE 19.4 $f(\tau)$ and $g(t - \tau)$ for various values of t

EXAMPLE 19.13 As another example, let $g(t) = u(t)$ in (19.24). Then $\mathbf{G}(s) = 1/s$ and, on the interval $0 < \tau < t$, $g(t - \tau) = u(t - \tau) = 1$, so (19.24) becomes

$$\mathcal{L}^{-1}\left[\frac{\mathbf{F}(s)}{s}\right] = \int_0^t f(\tau)\, d\tau$$

or

$$\mathcal{L}\left[\int_0^t f(\tau)\, d\tau\right] = \frac{\mathbf{F}(s)}{s} \tag{19.26}$$

Thus the transform of an integral of this type is the transform of its integrand divided by s. This contrasts with the transform of a derivative, which, we will recall, involves the transform of the function multiplied by s.

EXAMPLE 19.14 As a final example, let us use Laplace transforms to find the current $i(t)$ for $t > 0$ in Fig. 19.5, if $v(0) = 2$ V. The loop equation is

$$4i(t) + 8\int_0^t i(\tau)\,d\tau + 2 = 14$$

or

$$i(t) + 2\int_0^t i(\tau)\,d\tau = 3$$

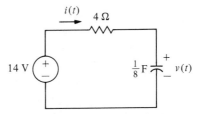

FIGURE 19.5 *RC* circuit

Transforming, we have

$$\mathbf{I}(s) + \frac{2}{s}\mathbf{I}(s) = \frac{3}{s}$$

where $\mathbf{I}(s)$ is the transform of $i(t)$. [We note that we are treating 3 like $3u(t)$ since $t > 0$.] Solving for the transform, we have

$$\mathbf{I}(s) = \frac{3/s}{1 + (2/s)} = \frac{3}{s + 2}$$

which yields, for $t > 0$,

$$i(t) = 3e^{-2t} \quad \text{A}$$

EXERCISES

19.4.1 Show that the operation of convolution is commutative; that is,

$$f(t) * g(t) = g(t) * f(t)$$

[*Suggestion:* Make a change of variable, $\tau = t - x$, in (19.25).]

19.4.2 Find $\mathcal{L}^{-1}[1/(s^2 + 1)^2]$. [*Suggestion:* Use the convolution integral, where $\mathbf{F}(s) = \mathbf{G}(s) = 1/(s^2 + 1)$.]
Answer $\frac{1}{2}(\sin t - t \cos t)u(t)$

19.4.3 Use Laplace transforms to find $i(t)$ for $t > 0$ if $i(0) = 2$ A and $v(0) = 2$ V.
Answer $2e^{-2t}(\cos 3t + \sin 3t)$ A

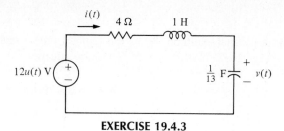

EXERCISE 19.4.3

19.4.4 Solve for $y(t)$, for $t > 0$:

$$y(t) = 1 + \int_0^t 2y(t - \tau)e^{-2\tau}\, d\tau$$

[*Suggestion:* The integral is the convolution of $2y(t)$ and e^{-2t}.]
Answer $1 + 2t$

19.4.5 Solve for x, for $t > 0$, if

$$x'(t) + x(t) + \int_0^t x(\tau)e^{\tau - t}\, d\tau = 0, \qquad x(0) = 1$$

Answer $e^{-t} \cos t$

19.4.6 Show that

$$\mathcal{L}\left[\int_{-\infty}^t f(\tau)\, d\tau\right] = \frac{\mathbf{F}(s)}{s} + \frac{\displaystyle\int_{-\infty}^0 f(\tau)\, d\tau}{s}$$

$$\left[\textit{Suggestion:} \int_{-\infty}^t f(\tau)\, d\tau = \int_{-\infty}^0 f(\tau)\, d\tau + \int_0^t f(\tau)\, d\tau.\right]$$

19.5

THE IMPULSE FUNCTION

Thus far we have obtained a number of Laplace transforms, ranging in complexity from $1/s$, the transform of the unit step function, to some very complicated functions arising from the translation theorems and the convolution theorem. The reader may have noticed, however, that one very simple transform, $\mathbf{F}(s) = 1$, is missing. In this section we will consider this transform, and we will subsequently see that its corresponding time-domain function is one of the most useful functions in circuit and system theory.

The function whose transform is 1 is called the *impulse function* and is denoted by $\delta(t)$. That is,

$$\mathcal{L}[\delta(t)] = 1 \tag{19.27}$$

The impulse function is not a function in the conventional sense, and indeed it was originally called an "improper function" by the great British physicist Paul A. M. Dirac (1902–   ), who is noted for his prediction of the existence of the positive electron, or positron, but who was also a pioneer in the development of the impulse function. The impulse function can be defined and its properties established rigorously using generalized or distribution function theory. We will give a plausibility argument for its existence and use it quite successfully as an ordinary function.

EXAMPLE 19.15 Let us begin by considering the finite pulse of Fig. 19.6, centered about the origin, of width a and height $1/a$, defined by

$$f(t) = \frac{1}{a}, \qquad -\frac{a}{2} < t < \frac{a}{2}$$

$$= 0 \qquad \text{elsewhere}$$

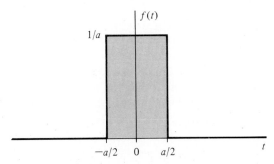

FIGURE 19.6 Finite pulse

The area under the pulse is 1 and remains fixed regardless of the value of a. If a is made smaller, then the base of the pulse shrinks and its height increases, maintaining the constant area of 1. In the limit as a tends toward zero, the pulse approaches an infinite pulse occurring over zero time but still associated with an area of 1. This is the impulse function, sometimes called the *unit* impulse function to indicate its relationship to the unit area. Graphically, it is represented by an arrow erected at $t = 0$, as shown in Fig. 19.7(a). A more general function is $\delta(t - \tau)$, which is an impulse occurring at $t = \tau$, as indicated in Fig. 19.7(b).

In Fig. 19.7, the impulse functions are of *strength* 1 because of their association with the area of 1 in Fig. 19.6. An impulse of strength k is denoted by $k\delta(t)$ and is represented by an arrow as in Fig. 19.7(a), marked (k).

Mathematically, the impulse function is defined by

$$\delta(t) = 0, \qquad t \neq 0$$

$$\int_{-\infty}^{\infty} \delta(t)\, dt = 1 \tag{19.28}$$

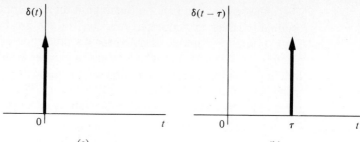

FIGURE 19.7 Impulses (a) $\delta(t)$ and (b) $\delta(t - \tau)$

or more generally by

$$\delta(t - \tau) = 0, \qquad t \neq \tau$$

$$\int_{-\infty}^{\infty} \delta(t - \tau) \, dt = 1 \qquad\qquad (19.29)$$

Thus the impulse is zero everywhere except at its point of discontinuity, at which it has an area of 1 concentrated. Physically, the impulse describes very well a force of very large magnitude exerted for an extremely short time, like that of a hammer suddenly striking a mass on a spring to put it into motion.

An important property associated with the impulse function is the *sampling*, or *sifting*, property, described by

$$\int_{a}^{b} f(t)\delta(t - \tau) \, dt = f(\tau) \qquad\qquad (19.30)$$

where $a < \tau < b$ and $f(t)$ is continuous at $t = \tau$. This property may be made plausible by noting that since $\delta(t - \tau)$ is zero except at $t = \tau$, we may write

$$\int_{a}^{b} f(t)\delta(t - \tau) \, dt = \int_{\tau-\epsilon}^{\tau+\epsilon} f(t)\delta(t - \tau) \, dt$$

Thus if ϵ is sufficiently small, since $f(t)$ is continous at τ, then $f(t)$ is approximately $f(\tau)$ between $\tau - \epsilon$ and $\tau + \epsilon$. Thus we may factor $f(\tau)$ out of the integral, and by the nature of the impulse function we have

$$\int_{a}^{b} f(t)\delta(t - \tau) \, dt = f(\tau)\int_{\tau-\epsilon}^{\tau+\epsilon} \delta(t - \tau) \, dt$$

$$= f(\tau)$$

The sampling property is very useful in determining the Laplace transform of the impulse function, given by

$$\mathcal{L}[\delta(t)] = \int_{0^-}^{\infty} e^{-st}\delta(t) \, dt$$

(Recall that in the event of an infinite discontinuity of the integrand at 0, we are using

0^- as the lower limit.) By the sampling property, with $\tau = 0$, we have

$$\mathcal{L}[\delta(t)] = e^{-st}\Big|_{t=0} = 1 \qquad (19.31)$$

which is (19.27). The shifted impulse has the more general transform,

$$\mathcal{L}[\delta(t - \tau)] = \int_{0^-}^{\infty} e^{-st}\delta(t - \tau)\,dt$$

$$= e^{-s\tau}$$

which reduces to 1 for $\tau = 0$.

EXAMPLE 19.16 To illustrate how an impulse function may arise, let us invert the transform

$$\mathbf{F}(s) = \frac{s}{s + 2}$$

This is an improper fraction (the numerator is of the same degree as the denominator), which has not been encountered thus far. However, by long division it may be written

$$\mathbf{F}(s) = 1 - \frac{2}{s + 2}$$

and by linearity its inverse is

$$f(t) = \delta(t) - 2e^{-2t}u(t)$$

At this point let us relate the impulse function to another well-known function by recalling the expression for the transform of a derivative,

$$\mathcal{L}[f'(t)] = s\mathbf{F}(s) - f(0^-) \qquad (19.32)$$

where 0 has been replaced by 0^- for the case of a discontinuity at $t = 0$. If we let $f(t) = u(t)$, this relation becomes

$$\mathcal{L}[u'(t)] = s\left(\frac{1}{s}\right) - 0 = 1$$

Thus we would like to conclude that

$$\frac{du(t)}{dt} = \delta(t) \qquad (19.33)$$

since they have the same transform. Unfortunately, this does not follow from rigorous mathematics, treating $u(t)$ and $\delta(t)$ as conventional functions. However, (19.33) may be established using generalized function theory and is a good practical tool used extensively in systems and circuit analysis. A plausibility argument for its validity is given in Ex. 19.5.2.

If we formally replace $f(t)$ in (19.32) by $\delta(t)$, we have

$$\mathcal{L}[\delta'(t)] = s\mathcal{L}[\delta(t)] - 0 = s \qquad (19.34)$$

Thus $\delta'(t)$, called a *doublet*, has Laplace transform s, and thus is a useful function, though not a conventional one. The functions $u(t)$, $\delta(t) = u'(t)$, and $\delta'(t) = u''(t)$ are members of the family of *singular functions,* and may be shown formally, but not conventionally, by means of (19.32) to have the transforms

$$\mathscr{L}\left[\frac{d^n u(t)}{dt^n}\right] = s^{n-1}; \qquad n = 0, 1, 2, \ldots \qquad (19.35)$$

The step and impulse are the most often used members of the family, but the doublet may arise occasionally and, in any case, it is useful in our later discussion of stability. Like the step and impulse functions, the doublet has a graph that can be established by the following plausibility argument.

Consider the triangular pulse $p(t)$ of Fig. 19.8(a). Its area is 1, as was the case for the rectangular pulse of Fig. 19.6. If $h \to 0$ in Fig. 19.8(a), the area of 1 is preserved, and the legs of the triangle shrink toward zero while the altitude tends toward infinity. Thus the shape in the limiting position of Fig. 19.8(a) is precisely $\delta(t)$, a pulse of infinite magnitude occurring over an interval of zero time. It is possible then to take

$$\delta(t) = \lim_{h \to 0} p(t)$$

and thus

$$\delta'(t) = \frac{d}{dt}\left[\lim_{h \to 0} p(t)\right]$$

If we could interchange the derivative and limit operations in this last expression, we would have

$$\delta'(t) = \lim_{h \to 0} p'(t)$$

which by Fig. 19.8(b) is seen to be two pulses tending toward zero bases and infinite heights in opposite directions. Thus we are led to represent the doublet by the double arrows of Fig. 19.9.

FIGURE 19.8 (a) A triangular pulse and (b) its derivative

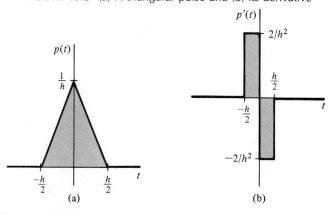

(a)

(b)

Another useful result concerning the impulse function is

$$f(t)\delta(t) = f(0)\delta(t) \tag{19.36}$$

where $f(t)$ is continuous at $t = 0$. This is plausible since $\delta(t) = 0$ for $t \neq 0$. It arises, for example, when we differentiate functions that contain $u(t)$ as a factor, formally using the rule for differentiation of a product.

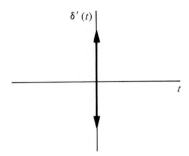

FIGURE 19.9 Graph of the doublet

EXAMPLE 19.17 As a last example, let us find the current $i(t)$ in the initially relaxed circuit $[i(0^-) = 0]$ of Fig. 19.10 if the input is $v_g(t) = \delta(t)$ V. The current in this case is the *impulse response* (response to a unit impulse), which will be considered in detail in Sec. 20.4. The loop equation is

$$\frac{di(t)}{dt} + 2i(t) = \delta(t)$$

which transforms to

$$s\mathbf{I}(s) + 2\mathbf{I}(s) = 1$$

The current transform is

$$\mathbf{I}(s) = \frac{1}{s + 2}$$

and the current is

$$i(t) = e^{-2t}u(t) \quad \text{A} \tag{19.37}$$

FIGURE 19.10 *RL* circuit

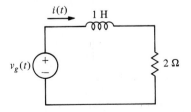

To check the answer, we note that the initial condition $i(0^-) = 0$ holds. To see if the differential equation is satisfied, we need

$$\frac{di}{dt} = e^{-2t}\delta(t) - 2e^{-2t}u(t)$$

which by (19.36) is

$$\frac{di}{dt} = \delta(t) - 2e^{-2t}u(t)$$

Substituting this expression and (19.37) into the differential equation, we see that it is formally satisfied.

EXERCISES

19.5.1 Evaluate the integral

$$\int_a^3 (t^2 + 3 \cos 2t)\delta(t) \, dt$$

where (a) $a = -1$ and (b) $a = 1$.
Answer (a) 3; (b) 0

19.5.2 Show that the pulse of Fig. 19.6 is given by

$$f(t) = \frac{u[t + (a/2)] - u[t - (a/2)]}{a}$$

so that

$$\lim_{a \to 0} f(t) = \frac{du(t)}{dt}$$

This is a plausibility argument for establishing

$$\delta(t) = \frac{du(t)}{dt}$$

19.5.3 Use the results of Ex. 19.5.2 to obtain formally

$$\frac{d}{dt}[f(t)u(t)] = f'(t)u(t) + f(t)\delta(t)$$

$$= f'(t)u(t) + f(0)\delta(t)$$

[If $f(0) = 0$, this result becomes

$$\frac{d}{dt}f(t)u(t) = f'(t)u(t)]$$

19.5.4 Use the result of Ex. 19.5.3 and (19.9), which, because of the discontinuity, we write as

$$\mathcal{L}[f'(t)] = s\mathcal{L}[f(t)] - f(0^-)$$

to derive $\mathcal{L}[\delta(t)] = 1$, where $f(t) = e^{-at}u(t)$.

19.6

THE INVERSE TRANSFORM

From the first sections of this chapter, it is clear that we may compile a lengthy table of functions and their Laplace transforms by repeated applications of the definition of the transform and the theorems. We may then obtain a wide variety of inverse transforms by matching entries in the table. This will be our objective in this section, and for this purpose we will use Table 19.1, which contains most of the common functions $f(t)$ and their transforms $\mathbf{F}(s)$, which we have derived previously. (Entry 10 follows directly from Ex. 19.3.4 with n replaced by $n - 1$.) For ready reference, a more complete table of transforms is given inside the back cover of the book.

TABLE 19.1   Short Table of Laplace Transforms

	$f(t)$	$\mathbf{F}(s)$
1.	$\delta(t)$	1
2.	$u(t)$	$\dfrac{1}{s}$
3.	e^{-at}	$\dfrac{1}{s + a}$
4.	$\sin kt$	$\dfrac{k}{s^2 + k^2}$
5.	$\cos kt$	$\dfrac{s}{s^2 + k^2}$
6.	$e^{-at} \sin bt$	$\dfrac{b}{(s + a)^2 + b^2}$
7.	$e^{-at} \cos bt$	$\dfrac{s + a}{(s + a)^2 + b^2}$
8.	t	$\dfrac{1}{s^2}$
9.	te^{-at}	$\dfrac{1}{(s + a)^2}$
10.	$\dfrac{t^{n-1}e^{-at}}{(n - 1)!}$	$\dfrac{1}{(s + a)^n};\qquad n = 1, 2, 3, \ldots$

Since we are interested only in functions defined for $t > 0$, we will omit the factor $u(t)$ except where it is necessary, as in the time translation formula (19.21). We have retained $u(t)$ in entry 2 of Table 19.1, which for $t > 0$ will be used as

$$\mathcal{L}[1] = \frac{1}{s}$$

We have omitted the factor $u(t)$ in the other entries of the table.

EXAMPLE 19.18 To illustrate the use of the table, suppose we are required to find the inverse of the transform

$$\mathbf{F}(s) = \frac{6}{s + 4} + \frac{2}{s^2 + 9} - \frac{3}{s}$$

Since the inverse transform is a linear operation, we have

$$f(t) = 6\mathcal{L}^{-1}\left[\frac{1}{s + 4}\right] + \frac{2}{3}\mathcal{L}^{-1}\left[\frac{3}{s^2 + 3^2}\right] - 3\mathcal{L}^{-1}\left[\frac{1}{s}\right]$$

which by entries 3, 4, and 2 of the table yields

$$f(t) = 6e^{-4t} + \frac{2}{3}\sin 3t - 3$$

EXAMPLE 19.19 As another example, let us invert the transform

$$\mathbf{F}(s) = \frac{2(s + 10)}{(s + 1)(s + 4)}$$

There is no direct entry in Table 19.1 which we can use to obtain $f(t)$ in this case; however, we may obtain a *partial fraction expansion* of $\mathbf{F}(s)$ and apply the table to each term in the expansion.

Obtaining a partial fraction expansion is the opposite operation of getting a common denominator. That is, we ask ourselves what simple fractions add together to yield $\mathbf{F}(s)$. Since in this case the transform is a proper fraction (the numerator is of lower degree than the denominator), the partial fractions will be proper fractions and must therefore be of the form

$$\mathbf{F}(s) = \frac{2(s + 10)}{(s + 1)(s + 4)} = \frac{A}{s + 1} + \frac{B}{s + 4}$$

The constants A and B are determined so as to make the second and third members an identity in s.

The simplest means of determining A and B is to note that

$$(s + 1)\mathbf{F}(s) = \frac{2(s + 10)}{s + 4} = A + \frac{B(s + 1)}{s + 4}$$

and

$$(s + 4)\mathbf{F}(s) = \frac{2(s + 10)}{s + 1} = \frac{A(s + 4)}{s + 1} + B$$

Since these must be identities for all s, let us evaluate the first at $s = -1$, which eliminates B, and the second at $s = -4$, which eliminates A. The results are

$$A = (s + 1)\mathbf{F}(s)\Big|_{s=-1} = \frac{2(9)}{3} = 6$$

$$B = (s + 4)\mathbf{F}(s)\Big|_{s=-4} = \frac{2(6)}{-3} = -4$$

Therefore, we have

$$\mathbf{F}(s) = \frac{6}{s + 1} - \frac{4}{s + 4}$$

and, by Table 19.1,

$$f(t) = 6e^{-t} - 4e^{-4t}$$

We should note that the formal computation of A and B above is a shorthand notation for what we are actually doing, which is

$$A = \lim_{s \to -1} (s + 1)\mathbf{F}(s)$$

and so forth.

The foregoing example, as the reader may recall from partial fraction expansions in calculus, is the relatively simple case of distinct linear factors, such as $s - p$ in the denominator of $\mathbf{F}(s)$. The pole $s = p$ is in this case a *simple* pole, or a pole of *order* one, and in general for each such simple pole the partial fraction expansion contains a term $A/(s - p)$. In the case of a denominator factor $(s - p)^n$, where $n = 2, 3, 4, \ldots$, the pole $s = p$ is a *multiple* pole, or a pole of order n. The factors comprising $(s - p)^n$ are thus not distinct, and the partial fraction expansion, of course, must be modified. We will consider multiple poles later.

EXAMPLE 19.20 As another example involving only simple poles, the expansion of

$$\mathbf{F}(s) = \frac{2s}{(s + 1)(s + 2)(s + 3)}$$

is given by

$$\frac{2s}{(s + 1)(s + 2)(s + 3)} = \frac{A}{s + 1} + \frac{B}{s + 2} + \frac{C}{s + 3}$$

where

$$A = \frac{2s}{(s + 2)(s + 3)}\Big|_{s=-1} = -1$$

$$B = \frac{2s}{(s + 1)(s + 3)}\Big|_{s=-2} = 4$$

$$C = \frac{2s}{(s + 1)(s + 2)}\Big|_{s=-3} = -3$$

The inverse transform is therefore

$$f(t) = -e^{-t} + 4e^{-2t} - 3e^{-3t}$$

Complex poles occur in conjugate pairs and their corresponding coefficients in the expansion are complex conjugates. For example, if $\mathbf{F}(s)$ has simple poles $s = \alpha \pm j\beta$ and an expansion

$$\mathbf{F}(s) = \frac{A}{s - \alpha - j\beta} + \frac{B}{s - \alpha + j\beta}$$

then

$$A = (s - \alpha - j\beta)\mathbf{F}(s)\Big|_{s=\alpha+j\beta}$$

and

$$B = (s - \alpha + j\beta)\mathbf{F}(s)\Big|_{s=\alpha-j\beta}$$

From this we see that $B = A^*$, because $\mathbf{F}(s)$ is a ratio of polynomials in s with real coefficients. The inverse transform is

$$f(t) = Ae^{(\alpha+j\beta)t} + A^*e^{(\alpha-j\beta)t}$$

which is the sum of a complex number and its conjugate. Therefore

$$f(t) = 2 \operatorname{Re}[Ae^{(\alpha+j\beta)t}]$$

If $A = |A|e^{j\theta}$, then we have

$$f(t) = 2 \operatorname{Re}[|A|e^{\alpha t}e^{j(\beta t + \theta)}]$$
$$= 2|A|e^{\alpha t} \cos(\beta t + \theta)$$

EXAMPLE 19.21 As an example, the transform

$$\mathbf{F}(s) = \frac{s}{(s + 1)(s^2 + 2s + 2)} \tag{19.38}$$

has the partial fraction expansion

$$\mathbf{F}(s) = \frac{s}{(s + 1)(s + 1 - j1)(s + 1 + j1)}$$

$$= \frac{A}{s + 1} + \frac{B}{s + 1 - j1} + \frac{B^*}{s + 1 + j1}$$

where

$$A = \frac{s}{s^2 + 2s + 2}\Big|_{s=-1} = -1$$

and

$$B = \left. \frac{s}{(s + 1)(s + 1 + j1)} \right|_{s=-1+j1} = \frac{1 - j1}{2} = \frac{1}{\sqrt{2}} \underline{/-45^\circ}$$

Thus we have

$$f(t) = Ae^{-t} + 2\,\text{Re}[Be^{(-1+j1)t}]$$

$$= -e^{-t} + 2\,\text{Re}\left[\frac{1}{\sqrt{2}}e^{-t}\underline{/1t - 45^\circ}\right]$$

$$= -e^{-t} + \sqrt{2}e^{-t}\cos(t - 45^\circ)$$

An alternative form is

$$f(t) = -e^{-t} + \sqrt{2}e^{-t}(\cos t \cos 45^\circ + \sin t \sin 45^\circ)$$

$$= -e^{-t} + e^{-t}(\cos t + \sin t) \tag{19.39}$$

EXAMPLE 19.22 If it is desirable to keep the quadratic factor intact, the expansion of (19.38) is

$$\mathbf{F}(s) = \frac{s}{(s + 1)(s^2 + 2s + 2)} = \frac{A}{s + 1} + \frac{Cs + D}{s^2 + 2s + 2} \tag{19.40}$$

where $Cs + D$ is the most general numerator for the quadratic term. As before $A = -1$, and to find C and D we clear (19.40) of fractions, obtaining

$$s = -(s^2 + 2s + 2) + (Cs + D)(s + 1)$$

which is an identity in s. Equating coefficients of s^2 gives

$$0 = -1 + C$$

from which $C = 1$. Equating coefficients of s^0 yields

$$0 = -2 + D$$

or $D = 2$. The transform is therefore

$$\mathbf{F}(s) = \frac{-1}{s + 1} + \frac{s + 2}{s^2 + 2s + 2}$$

$$= \frac{-1}{s + 1} + \frac{s + 1}{(s + 1)^2 + 1} + \frac{1}{(s + 1)^2 + 1}$$

The inverse transform is the function of (19.39).

Multiple poles from denominator factors $(s + a)^n$ require a partial fraction expansion

$$\mathbf{F}(s) = \frac{A_n}{(s + a)^n} + \frac{A_{n-1}}{(s + a)^{n-1}} + \cdots + \frac{A_k}{(s + a)^k} + \cdots + \frac{A_1}{s + a} + \mathbf{F}_1(s)$$

$$\tag{19.41}$$

where $\mathbf{F}_1(s)$ corresponds to the remaining poles of $\mathbf{F}(s)$

EXAMPLE 19.23 For example,

$$F(s) = \frac{4}{(s + 1)^2(s + 2)}$$

$$= \frac{A}{(s + 1)^2} + \frac{B}{s + 1} + \frac{C}{s + 2} \qquad (19.42)$$

Coefficients A and C may be obtained easily, as

$$A = (s + 1)^2 \mathbf{F}(s)\bigg|_{s=-1} = \frac{4}{s + 2}\bigg|_{s=-1} = 4$$

and

$$C = (s + 2)\mathbf{F}(s)\bigg|_{s=-2} = \frac{4}{(s + 1)^2}\bigg|_{s=-2} = 4$$

Thus C is obtained as before, and A, the coefficient corresponding to the highest degree of the repeated linear factor, may be found in an identical manner. To obtain B, we may clear (19.42) of fractions, obtaining

$$4 = 4(s + 2) + B(s + 1)(s + 2) + 4(s + 1)^2$$

Equating coefficients of s^2 yields

$$0 = B + 4$$

so that $B = -4$.

The transform of (19.42) is therefore

$$\mathbf{F}(s) = \frac{4}{(s + 1)^2} - \frac{4}{s + 1} + \frac{4}{s + 2}$$

so that

$$f(t) = 4te^{-t} - 4e^{-t} + 4e^{-2t}$$

The first term on the right is obtained using entry 9 in Table 19.1.

We may also find B in (19.42) by giving s any value other than -1 or -2. If $s = 0$, for example, (19.42) becomes

$$\frac{4}{1(2)} = \frac{4}{1} + \frac{B}{1} + \frac{4}{2}$$

from which $B = -4$.

We will next consider a general method of finding all the coefficients corresponding to multiple poles. We begin by multiplying (19.41) through by $(s + a)^n$, resulting in

$$(s + a)^n \mathbf{F}(s) = A_n + A_{n-1}(s + a) + \cdots + A_k(s + a)^{n-k}$$

$$+ \cdots + A_1(s + a)^{n-1} + (s + a)^n \mathbf{F}_1(s)$$

Differentiating $n - k$ times gives

$$\frac{d^{n-k}}{ds^{n-k}}[(s + a)^n F(s)] = (n - k)! A_k + (s + a)G(s) \qquad (19.43)$$

where

$$(s + a)G(s) = \frac{d^{n-k}}{ds^{n-k}}[A_{k+1}(s + a)^{n-k+1} + \cdots + A_1(s + a)^{n-1}$$
$$+ (s + a)^n F_1(s)]$$

Evaluating (19.43) at $s = -a$ yields, for $k = 1, 2, 3, \ldots, n$,

$$\boxed{A_k = \frac{1}{(n - k)!} \frac{d^{n-k}}{ds^{n-k}}[(s + a)^n F(s)]\Big|_{s=-a} \qquad (19.44)}$$

EXAMPLE 19.24 To illustrate this result, we find B in (19.42) to be

$$B = \frac{1}{(2 - 1)!} \frac{d^{2-1}}{ds^{2-1}}\left[\frac{4}{s + 2}\right]\Big|_{s=-1}$$

$$= \frac{-4}{(s + 2)^2}\Big|_{s=-1} = -4$$

which checks the previous answer.

EXAMPLE 19.25 Multiple complex poles are handled exactly like multiple real poles, the only difference being that complex numbers are involved. For example, the transform

$$F(s) = \frac{4}{(s + 1)(s^2 + 2s + 2)^2}$$

$$= \frac{A}{s + 1} + \frac{B}{(s + 1 - j1)^2} + \frac{B^*}{(s + 1 + j1)^2} + \frac{C}{s + 1 - j1} + \frac{C^*}{s + 1 + j1}$$
$$(19.45)$$

has coefficients given by

$$A = \frac{4}{(s^2 + 2s + 2)^2}\Big|_{s=-1} = 4$$

$$B = \frac{4}{(s + 1)(s + 1 + j1)^2}\Big|_{s=-1+j1} = j1 = 1\underline{/90°}$$

and

$$C = \frac{1}{1!} \frac{d}{ds} \left[(s + 1 - j1)^2 \frac{4}{(s + 1)(s^2 + 2s + 2)^2} \right]_{s=-1+j1}$$

$$= \frac{d}{ds} \left[\frac{4}{(s + 1)(s + 1 + j1)^2} \right]_{s=-1+j1}$$

$$= -4 \left[\frac{(s + 1)(2) + s + 1 + j1}{(s + 1)^2(s + 1 + j1)^3} \right]_{s=-1+j1} = -2$$

Therefore the time-domain function is

$$f(t) = 4e^{-t} + 2\,\mathrm{Re}[(1\,\underline{/90^\circ})te^{(-1+j1)t}] + 2\,\mathrm{Re}[-2e^{(-1+j1)t}]$$

$$= 4e^{-t} + 2te^{-t} \cos(t + 90^\circ) - 4e^{-t} \cos t$$

$$= 4e^{-t} - 2te^{-t} \sin t - 4e^{-t} \cos t$$

EXAMPLE 19.26 As a last example, let us invert

$$\mathbf{F}(s) = \frac{9s^3}{(s + 1)(s^2 + 2s + 10)}$$

We note that the degree of the numerator is the same as that of the denominator, so that long division is required. It is not necessary to actually perform the division since it is evident that it will result in 9 plus a proper fraction. Therefore, let us write

$$\mathbf{F}(s) = \frac{9s^3}{(s + 1)(s^2 + 2s + 10)}$$

$$= 9 + \frac{A}{s + 1} + \frac{Bs + C}{s^2 + 2s + 10} \tag{19.46}$$

We find A as before, from

$$A = (s + 1)\mathbf{F}(s) \Big|_{s=-1} = \frac{-9}{1 - 2 + 10} = -1$$

To obtain B and C, let us multiply (19.46) by the denominator of $\mathbf{F}(s)$, obtaining

$$9s^3 = 9(s + 1)(s^2 + 2s + 10) - (s^2 + 2s + 10) + (Bs + C)(s + 1)$$

The coefficients of s^2 are

$$0 = 18 + 9 - 1 + B$$

from which $B = -26$. The coefficients of s^0 are

$$0 = 90 - 10 + C$$

yielding $C = -80$. Therefore, we have

$$\mathbf{F}(s) = 9 - \frac{1}{s + 1} - \frac{26s + 80}{s^2 + 2s + 10}$$

which we rearrange as

$$F(s) = 9 - \frac{1}{s + 1} - 26\left[\frac{s + 1}{(s + 1)^2 + 3^2}\right] - 18\left[\frac{3}{(s + 1)^2 + 3^2}\right]$$

Therefore, by Table 19.1, we have

$$f(t) = 9\delta(t) - e^{-t}(1 + 26 \cos 3t + 18 \sin 3t)$$

As observed earlier, an explicit relation for the inverse transform is given in Prob. 19.18, but for our purposes we may generally use partial fractions and Table 19.1 to obtain the inverse transforms.

EXERCISES

19.6.1   Find the inverse transform of

(a) $\dfrac{2}{(s + 1)(s + 2)(s + 3)}$,

(b) $\dfrac{s}{(s^2 + 1)(s^2 + 4)}$,

(c) $\dfrac{s^4 + 5s^3 + 21s^2 + 47s + 78}{(s^2 + 9)(s^2 + 2s + 5)}$, and

(d) $\dfrac{2s^3 + 2s^2 - 4s + 8}{s^2(s^2 + 4)}$.

Answer (a) $e^{-t} - 2e^{-2t} + e^{-3t}$; (b) $\frac{1}{3}(\cos t - \cos 2t)$;
(c) $\delta(t) + \cos 3t + \sin 3t + 2e^{-t} \cos 2t$;
(d) $2t - 1 + 3 \cos 2t$

19.6.2   Find the inverse transform of $\dfrac{2s^3 + 5s^2 + 2s - 2}{s(s + 1)}$.
Answer $2\delta'(t) + 3\delta(t) - (2 - e^{-t})u(t)$

19.6.3   Find B and C in (19.46) by subtracting the first two terms in the right member from the left member. (Note that $A = -1$.)
Answer $B = -26$, $C = -80$

19.6.4   Given the transform

$$F(s) = \frac{3s^2 + 8s + 8}{(s + 1)^2(s + 2)}$$

find $f(t)$.
Answer $3te^{-t} - e^{-t} + 4e^{-2t}$

19.6.5   Find the inverse transform of

$$F(s) = \frac{2s^2 + s + 2}{s(s^2 + 1)^2}$$

Answer $2 - 2 \cos t + \frac{1}{2}(\sin t - t \cos t)$

19.7

DIFFERENTIATION THEOREMS

In this section we will complete our list of Laplace transforms and properties and summarize the latter in a table for easy reference. In particular, we will derive two results, using in one case differentiation in the time domain and, in the other, differentiation in the frequency domain.

In Sec. 19.1 we derived the transform of a derivative, which we repeat as

$$\mathcal{L}[f'(t)] = s\mathbf{F}(s) - f(0) \qquad (19.47)$$

If we formally replace f by f', we have

$$\mathcal{L}[f''(t)] = s\mathcal{L}[f'(t)] - f'(0)$$

or, by (19.47),

$$\mathcal{L}[f''(t)] = s^2\mathbf{F}(s) - sf(0) - f'(0) \qquad (19.48)$$

We may replace f by f' again in (19.48) to obtain $\mathcal{L}[f'''(t)]$, and so forth, with the general result being

$$\mathcal{L}[f^{(n)}(t)] = s^n\mathbf{F}(s) - s^{n-1}f(0) - s^{n-2}f'(0) - \cdots - f^{(n-1)}(0) \qquad (19.49)$$

where $f^{(n)}$ is the nth derivative. The functions $f, f', \ldots, f^{(n-1)}$ are assumed to be continuous on $(0, \infty)$, and $f^{(n)}$ is continuous except possibly for a finite number of finite discontinuities.

EXAMPLE 19.27 As an example, if in (19.49) we let $f(t) = t^n$, for n a nonnegative integer, then $f^{(n)}(t) = n!$ and $f(0) = f'(0) = \cdots = f^{(n-1)}(0) = 0$. Therefore we have

$$\mathcal{L}[n!] = s^n\mathcal{L}[t^n]$$

or

$$\mathcal{L}[t^n] = \frac{1}{s^n}\mathcal{L}[n!] = \frac{n!}{s^{n+1}}; \qquad n = 0, 1, 2, \ldots \qquad (19.50)$$

a result we obtained earlier in Ex. 19.1.4 by another method.

EXAMPLE 19.28 As another example, let us invert the transform

$$\mathbf{F}(s) = \frac{6}{s^4(s+1)}$$

which has the partial fraction expansion

$$\mathbf{F}(s) = \frac{A}{s^4} + \frac{B}{s^3} + \frac{C}{s^2} + \frac{D}{s} + \frac{E}{s+1}$$

The coefficients may be found by the method of the previous section; but instead we will illustrate an alternative method that is very effective for high-order real poles and is based simply on long division. We carry out a long division of 1 by $1 + s$ until a remainder is obtained of degree equal to the order of the multiple pole (in this case 4). The result is

$$\mathbf{F}(s) = \frac{6}{s^4}\left[\frac{1}{1+s}\right] = \frac{6}{s^4}\left[1 - s + s^2 - s^3 + \frac{s^4}{s+1}\right]$$

$$= \frac{6}{s^4} - \frac{6}{s^3} + \frac{6}{s^2} - \frac{6}{s} + \frac{6}{s+1}$$

or

$$\mathbf{F}(s) = \frac{3!}{s^4} - 3\frac{2!}{s^3} + 6\frac{1!}{s^2} - \frac{6}{s} + \frac{6}{s+1}$$

The first four terms in the right member are of the form (19.50), and thus the inverse transform is

$$f(t) = t^3 - 3t^2 + 6t - 6 + 6e^{-t}$$

To obtain the frequency-domain differentiation formulas, we will differentiate the Laplace transform with respect to s. That is,

$$\frac{d\mathbf{F}(s)}{ds} = \frac{d}{ds}\int_0^\infty f(t)e^{-st}\,dt$$

Assuming that we can interchange the operations of differentiation and integration, we have

$$\frac{d\mathbf{F}(s)}{ds} = \int_0^\infty \frac{d}{ds}[f(t)e^{-st}]\,dt$$

$$= \int_0^\infty [-tf(t)]e^{-st}\,dt$$

From the last integral we see that by definition the transform of $-tf(t)$ is $d\mathbf{F}(s)/ds$, or equivalently,

$$\mathcal{L}[tf(t)] = -\frac{d\mathbf{F}(s)}{ds}$$

where $\mathbf{F}(s) = \mathcal{L}[f(t)]$.

EXAMPLE 19.29 As an example, if $f(t) = u(t)$, then $\mathbf{F}(s) = 1/s$, and

$$\mathcal{L}[tu(t)] = -\frac{d}{ds}\left(\frac{1}{s}\right) = \frac{1}{s^2}$$

As another example, if $f(t) = \cos kt$, then $\mathbf{F}(s) = s/(s^2 + k^2)$, and we have

$$\mathcal{L}[t \cos kt] = -\frac{d}{ds}\left(\frac{s}{s^2 + k^2}\right) = \frac{s^2 - k^2}{(s^2 + k^2)^2}$$

We may repeatedly differentiate the transform to obtain the general case

$$\frac{d^n \mathbf{F}(s)}{ds^n} = \int_0^\infty [(-t)^n f(t)] e^{-st}\, dt$$

from which we conclude that

$$\mathcal{L}[t^n f(t)] = (-1)^n \frac{d^n \mathbf{F}(s)}{ds^n}; \qquad n = 0, 1, 2, \ldots \qquad (19.51)$$

EXAMPLE 19.30 As an example, if $f(t) = \sin t$, then $\mathbf{F}(s) = 1/(s^2 + 1)$, and we have (for $n = 2$)

$$\mathcal{L}[t^2 \sin t] = (-1)^2 \frac{d^2}{ds^2}\left(\frac{1}{s^2 + 1}\right)$$

$$= \frac{d}{ds}\left[\frac{-2s}{(s^2 + 1)^2}\right]$$

$$= \frac{2(3s^2 - 1)}{(s^2 + 1)^3}$$

The properties of the Laplace transform that we have obtained are listed in Table 19.2.

TABLE 19.2 Short List of Laplace Transform Properties

	$f(t)$	$\mathbf{F}(s)$
1.	$cf(t)$	$c\mathbf{F}(s)$
2.	$f_1(t) + f_2(t)$	$\mathbf{F}_1(s) + \mathbf{F}_2(s)$
3.	$\dfrac{df(t)}{dt}$	$s\mathbf{F}(s) - f(0)$
4.	$\dfrac{d^n f(t)}{dt^n}$	$s^n \mathbf{F}(s) - s^{n-1}f(0) - s^{n-2}f'(0)$ $-s^{n-3}f''(0) - \cdots - f^{(n-1)}(0)$
5.	$\displaystyle\int_0^t f(\tau)\, d\tau$	$\dfrac{\mathbf{F}(s)}{s}$
6.	$e^{-at}f(t)$	$\mathbf{F}(s + a)$
7.	$f(t - \tau)u(t - \tau)$	$e^{-s\tau}\mathbf{F}(s)$
8.	$f * g = \displaystyle\int_0^t f(\tau)g(t - \tau)\, d\tau$	$\mathbf{F}(s)\mathbf{G}(s)$
9.	$f(ct),\ c > 0$	$\dfrac{1}{c}\mathbf{F}\left(\dfrac{s}{c}\right)$
10.	$t^n f(t),\quad n = 0, 1, 2, \ldots$	$(-1)^n \mathbf{F}^{(n)}(s)$

EXERCISES

19.7.1 Find the Laplace transform of (a) t^5, (b) $t^4 e^{-2t}$, (c) $t \sin kt$, and (d) $t^2 \cos t$.

Answer (a) $\dfrac{120}{s^6}$; (b) $\dfrac{24}{(s+2)^5}$; (c) $\dfrac{2ks}{(s^2+k^2)^2}$; (d) $\dfrac{2s(s^2-3)}{(s^2+1)^3}$

19.7.2 Find the inverse transform of $\dfrac{s^4 - s^3 - 4s^2 - 6s - 2}{s^2(s+1)^3}$.

Answer $e^{-t}(t^2 - t + 1) - 2t$

19.7.3 Derive, for $n = 1, 2, 3, \ldots$, entry 10 of Table 19.1:

$$\mathscr{L}^{-1}\left[\frac{1}{(s+a)^n}\right] = \frac{t^{n-1} e^{-at}}{(n-1)!}$$

[*Suggestion:* Use (19.50) and property 6 of Table 19.2.]

19.7.4 Use the method of long division considered in this section to invert the transform

$$\mathbf{F}(s) = \frac{2}{s(s+1)^3}$$

(*Suggestion:* Change the variable from s to p by letting $s + 1 = p$, perform the long division, and change back to s.)

Answer $f(t) = 2 - 2e^{-t} - 2te^{-t} - t^2 e^{-t}$

19.7.5 Extend the formula (19.51) to $n = -1$ by integrating the transform, written in the form

$$\mathbf{F}(x) = \int_0^\infty f(t) e^{-xt}\, dt$$

between the limits $x = s$ to $x = \infty$ to obtain

$$\mathscr{L}\left[\frac{f(t)}{t}\right] = \int_s^\infty \mathbf{F}(x)\, dx$$

(*Suggestion:* Change the order of integration in the double integral.)

19.8

APPLICATIONS TO INTEGRODIFFERENTIAL EQUATIONS

The Laplace transform, like the Fourier transform, may be used to solve differential equations. However, the Laplace transform method has the advantage of yielding the complete solution with the initial conditions accounted for automatically, as we shall see in this section. This use of the Laplace transform is one of its most elegant applications.

Evidently, if we transform both members of a linear differential equation with constant coefficients, the result will be an algebraic equation in the transformed variable. We may see this from the general formula (19.49) for the transform of the derivatives. This formula also shows that the initial conditions are automatically taken into account. We may then solve for the transform of the unknown and invert it to obtain the time-domain answer. We have, in fact, already considered simple illustrations of this type earlier in the chapter.

EXAMPLE 19.31 As an example, let us find the solution $x(t)$, for $t > 0$, of the system of equations

$$x'' + 4x' + 3x = e^{-2t}$$

$$x(0) = 1, \qquad x'(0) = 2$$

Transforming, we have

$$s^2 X(s) - s - 2 + 4[sX(s) - 1] + 3X(s) = \frac{1}{s + 2}$$

from which

$$X(s) = \frac{s^2 + 8s + 13}{(s + 1)(s + 2)(s + 3)}$$

The partial fraction expansion is

$$X(s) = \frac{3}{s + 1} - \frac{1}{s + 2} - \frac{1}{s + 3}$$

so that the time-domain answer, for $t > 0$, is

$$x(t) = 3e^{-t} - e^{-2t} - e^{-3t}$$

As we have also seen, we may handle certain integrodifferential equations directly without differentiating to remove the integrals. We need only transform the integrals by means of

$$\mathcal{L}\left[\int_0^t f(\tau)\, d\tau \right] = \frac{F(s)}{s}$$

EXAMPLE 19.32 We will illustrate an integrodifferential equation with the circuit of Fig. 19.11, where we will find the current $i(t)$ if there is no initial stored energy. The equations are

$$\frac{di}{dt} + 2i + 5\int_0^t i\, dt = u(t)$$

$$i(0) = 0$$

FIGURE 19.11 *RLC* circuit

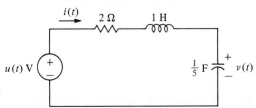

Transforming, we obtain

$$s\mathbf{I}(s) + 2\mathbf{I}(s) + \frac{5}{s}\mathbf{I}(s) = \frac{1}{s}$$

or

$$\mathbf{I}(s) = \frac{1}{s^2 + 2s + 5} = \frac{1}{2}\left[\frac{2}{(s + 1)^2 + 4}\right]$$

Therefore the response is

$$i(t) = \frac{1}{2}e^{-t}\sin 2t \text{ A}$$

EXAMPLE 19.33 As a last example, let us solve the simultaneous equations

$$\frac{dx}{dt} + x + 4y = 10$$

$$x - \frac{dy}{dt} - y = 0$$

for $x(t)$ and $y(t)$, given that $x(0) = 4$ and $y(0) = 3$. We could first eliminate y and solve the resulting second-order equation for x. We would then need a second condition on x, which we would have to find from the differential equations. It is easier with the Laplace transform method to transform both equations first and then solve for $\mathbf{X}(s)$, the transform of $x(t)$.

The transformed equations are

$$s\mathbf{X}(s) - 4 + \mathbf{X}(s) + 4\mathbf{Y}(s) = \frac{10}{s}$$

$$\mathbf{X}(s) - s\mathbf{Y}(s) + 3 - \mathbf{Y}(s) = 0$$

from which we have

$$\mathbf{X}(s) = \frac{\begin{vmatrix} \dfrac{10}{s} + 4 & 4 \\ -3 & -s - 1 \end{vmatrix}}{\begin{vmatrix} s + 1 & 4 \\ 1 & -s - 1 \end{vmatrix}}$$

After some simplification, this yields

$$\mathbf{X}(s) = \frac{4s^2 + 2s + 10}{s[(s + 1)^2 + 4]}$$

The partial fraction expansion may be shown to be

$$\mathbf{X}(s) = \frac{2}{s} + \frac{2(s + 1)}{(s + 1)^2 + 4} - \frac{4}{(s + 1)^2 + 4}$$

from which the time-domain solution is

$$x = 2 + 2e^{-t}(\cos 2t - \sin 2t)$$

In a similar manner we may show that the transform of y is

$$\mathbf{Y}(s) = \frac{2}{s} + \frac{s+1}{(s+1)^2 + 4} + \frac{2}{(s+1)^2 + 4}$$

and thus that

$$y = 2 + e^{-t}(\cos 2t + \sin 2t)$$

EXERCISES

19.8.1   Solve for $x(t)$, for $t > 0$, by using Laplace transforms:
(a) $x'' + 3x' + 2x = \delta(t)$
$x(0^-) = x'(0^-) = 0$.
(b) $x'' + 4x' + 3x = 4e^{-t}$
$x(0) = x'(0) = 4$.
Answer (a) $(e^{-1} - e^{-2t})u(t)$; (b) $e^{-t}(2t + 7) - 3e^{-3t}$

19.8.2   Find v for $t > 0$ using Laplace transforms if $v(0) = 8$ V.
Answer $6 + 2e^{-2t}$ V

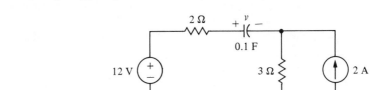

EXERCISE 19.8.2

19.8.3   Use Laplace transforms to find $x(t)$ and $y(t)$:

$$x' = x + 2y$$

$$y' = 2x + y$$

where $x(0) = 2$, $y(0) = 0$.
Answer $x = e^{3t} + e^{-t}$; $y = e^{3t} - e^{-t}$

PROBLEMS

19.1   Find the Laplace transform of the functions
(a) $f(t) = -1, \quad 0 \le t \le 3$
$= 1, \quad t > 3$
(b) $f(t) = 1, \quad 0 < a < t < b$
$= 0, \quad \text{otherwise}$

19.2 Evaluate the integral $\int te^{at}\,dt$ and use the result and the definition of the Laplace transform to find the transforms of the functions

(a) t,

(b) te^{-at},

(c) $f(t) = 0$, $\quad 0 < t < 2$,

$\quad\quad\ = t$, $\quad t > 2$

and (d) as shown.

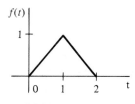

PROBLEM 19.2(d)

19.3 Find $\mathcal{L}[\cos t\, u(t)]$ using (19.9) and Ex. 19.1.2.

19.4 Using linearity, find the Laplace transform of the functions
(a) $\sinh kt\, u(t)$, (b) $(1 - e^{-2t})u(t)$, (c) $\sin^2 3t\, u(t)$, and (d) $\cos^2 t\, u(t)$.
[*Suggestion:* In (c) and (d), use a trigonometric identity.]

19.5 Find the inverse Laplace transform of

(a) $\dfrac{2}{s + 4} + \dfrac{3}{s^2}$,

(b) $\dfrac{1}{s} - \dfrac{s}{s^2 + 4}$,

(c) $\dfrac{2s + 6}{s^2 + 9}$,

(d) $\dfrac{1}{s + 1} + \dfrac{1}{s^2} - \dfrac{1}{s}$.

19.6 Find the Laplace transform of the functions
(a) $e^{-t}\cos 3t\, u(t)$, (b) $e^{-2t}\sin 4t\, u(t)$, (c) $e^{-3t}(2t + 1)\, u(t)$, and (d) $e^{-4t}\sinh t\, u(t)$.

19.7 Find the inverse transform of

(a) $\dfrac{6s}{s^2 + 2s + 10}$, (b) $\dfrac{2s + 10}{s^2 + 4s + 13}$, (c) $\dfrac{1}{(s + 2)^2}$, and (d) $\dfrac{2s - 6}{s^2 + 2s + 2}$.

19.8 Find (a) $\mathcal{L}[\sin 2t \sinh t\, u(t)]$ and (b) $\mathcal{L}[\cos t \cosh 2t\, u(t)]$.

19.9 Find the Laplace transform of (a) $u(t) - u(t - 1)$, (b) $e^{-2t}[u(t) - u(t - 2)]$, (c) $t\, u(t - 2)$, and (d) $e^{-t}u(t - 2)$.

19.10 Solve Problem 19.1 by expressing the functions in terms of step functions and using the time translation theorem.

19.11 Solve Problem 19.2(c) and (d) by expressing the functions as ramp and step functions and using the time translation theorem.

19.12 Find $\mathcal{L}[t \sin 2t\, u(t)]$. (*Suggestion:* Write $\sin 2t$ in terms of exponentials.)

19.13 Find the inverse transform of

(a) $\dfrac{1 + e^{-\pi s}}{s^2 + 4}$, (b) $\dfrac{e^{-4s} - e^{-7s}}{s^2}$, (c) $\dfrac{5s + 6}{s^2 + 9}e^{-\pi s}$, and (d) $\dfrac{2(1 + e^{-\pi s/2})}{s^2 + 2s + 5}$.

19.14 Show that $\mathcal{L}^{-1}\left[\dfrac{s^2}{(s^2 + k^2)^2}\right] = \dfrac{1}{2}\left(t \cos kt + \dfrac{1}{k} \sin kt\right).$

19.15 Use the results of Problem 19.14 and Exercise 19.4.2 to find $\mathcal{L}^{-1}\left[\dfrac{s^2 - 1}{(s^2 + 1)^2}\right].$

19.16 Use convolution to find the inverse transform of (a) $\dfrac{1}{s^2 + 5s + 6}$ and (b) $\dfrac{1}{s(s^2 + 1)}.$

19.17 Solve for $y(t)$, $t > 0$:

$$y(t) = \cos t + \int_0^t e^{-(t - \tau)}y(\tau)\, d\tau$$

19.18 In the Fourier integral (18.8), replace $f(t)$ by $e^{-\sigma t}f(t)$, and show that if $f(t) = 0$, for $t < 0$, then

$$f(t) = \frac{1}{2\pi}\int_{-\infty}^{\infty} e^{st}\left[\int_0^{\infty} f(x)e^{-sx}\, dx\right] d\omega$$

In the second integral, replace ω by $(s - \sigma)/j$ and obtain

$$f(t) = \frac{1}{2\pi j}\int_{\sigma - j\infty}^{\sigma + j\infty} e^{st}\, \mathbf{F}(s)ds$$

This is the inverse Laplace transform analogous to (18.11) in the Fourier transform case and is valid if σ is sufficiently large to ensure that (19.3) holds.

19.19 Find the inverse Laplace transform of

(a) $\dfrac{s}{(s + a)(s + b)}$, $b \neq a.$

(b) $\dfrac{s + 3}{(s + 1)(s + 2)}.$

(c) $\dfrac{2(s^2 - 6)}{s^3 + 4s^2 + 3s}.$

(d) $\dfrac{5s^3 - 3s^2 + 2s - 1}{s^4 + s^2}.$

19.20 Find the inverse transform of

(a) $\dfrac{3s^2 + 6}{(s^2 + 1)(s^2 + 4)}.$

(b) $\dfrac{4s^2}{s^4 - 1}.$

(c) $\dfrac{s^2 + 5s + 5}{(s + 1)(s + 2)^2}.$

(d) $\dfrac{s^3 - 1}{s^3 + s}.$

19.21 Find the inverse transform of

(a) $\dfrac{s^2 + 4s + 7}{(s + 1)(s^2 + 2s + 5)}.$

(b) $\dfrac{s + 2}{(s^2 + 2s + 2)(s + 1)^2}.$

(c) $\dfrac{4(s^3 - s^2 + 3s - 15)}{(s^2 + 9)(s^2 + 4s + 13)}$.

(d) $\dfrac{100}{(s + 2)(s^2 + 2s + 5)^2}$.

19.22 Show that if $f(t + T) = f(t)$, then

$$\mathcal{L}[f(t)] = \frac{\displaystyle\int_0^T e^{-st}f(t)\, dt}{1 - e^{-sT}}$$

[*Suggestion:* Write the transform as an infinite series of integrals over $(0, T)$, $(T, 2T)$, . . . , and sum.]

19.23 Use the result of Problem 19.22 to obtain the transforms of the functions

(a) $f(t) = 1, \quad 0 < t < 1$
$\qquad = 0, \quad 1 < t < 2$
$\quad f(t + 2) = f(t)$.

(b) $f(t) = |\sin t|$.

19.24 Note that if $p(t)$ is a pulse of some shape of finite duration, occurring on $0 < t < T\,[p(t) = 0$ elsewhere], and that if $f(t)$ is defined by

$$f(t) = p(t), \quad 0 < t < T$$

$$f(t + T) = f(t)$$

then $p(t) = f(t)[u(t) - u(t - T)]$, and the result of Prob. 19.22 is

$$\mathcal{L}[f(t)] = \frac{\mathcal{L}[p(t)]}{1 - e^{-sT}}$$

Use this fact to find the inverse transform of

$$F(s) = \frac{(1 - e^{-s})^2}{s(1 - e^{-2s})}$$

19.25 Find the Laplace transform of $\cos t$ using the results of Prob. 19.22.

19.26 Evaluate $\displaystyle\int_a^6 (t^2 + \sin 2t)[\delta(t) - 3\delta(t - 2)]\, dt$ for (a) $a = -2$, (b) $a = 1$, and (c) $a = 3$.

19.27 The square wave defined by

$$f(t) = 2, \quad 0 < t < 1$$

$$\qquad = -2, \quad 1 < t < 2$$

$$f(t + 2) = f(t)$$

has a Fourier series given by

$$f(t) = \sum_{n=1}^{\infty} b_n \sin n\pi t$$

Find the coefficients b_n by applying the sampling property to determine the coefficients of the Fourier series for $f'(t)$. [*Suggestion:* Note that $f'(t)$ is a train of impulse functions.]

19.28 Integrate the sampling property (19.30) by parts to obtain the corresponding doublet property,

$$\int_a^b f(t)\delta'(t - \tau)\, dt = -f'(\tau), \quad a < \tau < b$$
$$= 0, \quad \text{otherwise}$$

19.29 Use the idea of Prob. 19.27 and the sampling property of Prob. 19.28 to find the Fourier coefficients for $f(t)$ defined in (17.7). (*Suggestion:* Differentiate the Fourier series twice.)

19.30 Find the inverse transform of

(a) $\dfrac{27}{(s + 1)^3(s + 4)}$.

(b) $\dfrac{1}{s(s + 1)^4}$.

19.31 An *LC* series circuit with no initial stored energy is driven by a voltage source $v_g = 10 \sin 2t$ V. Find the current leaving the positive terminal of the source if $L = 5$ H and $C = 1/20$ F. (*Suggestion:* See Ex. 19.7.1)

19.32 Using Laplace transforms, solve the following for $t > 0$:
(a) $x'' + x = 0$, $x(0) = -1$, $x'(0) = 1$.
(b) $x'' + 2x' + 2x = 0$, $x(0) = 0$, $x'(0) = 1$.
(c) $x''' - 2x'' + 2x' = 0$, $x(0) = x'(0) = 1$, $x''(0) = 2$.
(d) $x'' + 4x' + 3x = 4 \sin t + 8 \cos t$, $x(0) = 3$, $x'(0) = -1$.
(e) $x'' + 4x' + 3x = 4e^{-3t}$, $x(0) = x'(0) = 0$.
(f) $x'' + 4x' + 3x = 4e^{-t} + 8e^{-3t}$, $x(0) = x'(0) = 0$.

19.33 Using Laplace transforms, solve the following for $t > 0$:

(a) $x' + 4x + 3\displaystyle\int_0^t x(\tau)\, d\tau = 5$, $x(0) = 1$.

(b) $x' + 4\displaystyle\int_0^t x(\tau)\, d\tau = 3 \sin t$, $x(0) = 4$.

19.34 Solve for x for $t > 0$:

$$x' + x + y' + y = 1$$
$$-2x + y' - y = 0$$
$$x(0) = 0, \qquad y(0) = 1.$$

19.35 Using Laplace transforms, solve Prob. 9.2.

19.36 Find v for $t > 0$ in the circuit of Prob. 14.14 if the initial stored energy is zero and $v_g = \delta(t)$ V.

Oliver Heaviside
1850–1925

20

Laplace Transform Applications

[Electromagnetics] has been said to be too complicated. This probably came from a simple-minded man.

Oliver Heaviside

The words inductance, capacitance, and impedance were given to us by the great English physicist and engineer, Oliver Heaviside, who also pioneered in the use of Laplace and Fourier transforms in the analysis of electric circuits, first suggested the existence of an ionized atmospheric layer (now called the ionosphere) that can reflect radio waves, and is said to have predicted the increase of mass of a charge moving at great speeds before Einstein formulated his theory of relativity. He not only coined the word impedance but introduced its concept to the solution of ac circuits.

Heaviside was born in London, the youngest of four sons of Thomas Heaviside, an engraver and watercolorist, and Rachel Elizabeth West, a sister-in-law of the famous physicist Sir Charles Wheatstone. Young Oliver's schooling ended when he was 16, but he trained himself at home in languages, mathematics, and the natural sciences. He became a telegraph operator in 1870, but in 1874 he was forced to retire because of increasing deafness. From then until his death he led a hermitlike existence, devoting himself to investigations of electrical phenomena and publishing such works as *Electrical Papers* in 1892 and a three-volume treatise, *Electromagnetic Theory* (1893–1912). His free and original use of mathematics was decades ahead of his time and evoked controversy with his contemporaries. Nevertheless, his fame spread, and because of his great store of knowledge and the scientific help he generously extended to all who sought it, his home became known as The Inexhaustible Cavity. He was elected Fellow of the Royal Society in 1891 and received an honorary doctorate from the University of Gottingen. To solve the problem of his sometimes being unable to pay his dues, the Institution of Electrical Engineers made him an honorary member, and shortly before his death awarded him its first Faraday Medal. ∎

As noted in the previous chapter, the Laplace transform is tailor-made for the solution of linear integrodifferential equations with constant coefficients, and so it is for the solution of linear electric circuits with lumped elements. In this chapter we will concentrate on applying the Laplace transform to the solution of electric circuits, first writing the describing equations and then transforming them. We will also see how we can bypass the time-domain equations, as we did in the phasor method, and go directly to a transformed circuit, which can be solved by algebraic means using the methods applied to resistive circuits.

We will see, too, how the transforms can be used to define the network functions and to obtain the step and impulse responses, the frequency responses, and the steady-state ac responses. The transforms, as·we will see, are very similar to the generalized phasors of Chapter 14 and are more versatile because they apply to a larger class of input functions.

20.1

APPLICATION TO ELECTRIC CIRCUITS

The Laplace transform method is an elegant procedure that can be used for solving electric circuits by transforming their describing integrodifferential equations into algebraic equations and applying the rules of algebra. In this section we will illustrate in depth how this is done by considering a variety of examples.

EXAMPLE 20.1   As a first example, let us find i for $t > 0$ in the circuit of Fig. 20.1, if $v(0) = 4$ V and $i(0) = 2$ A. The nodal equation at node a (whose node voltage is $12 - v$) is

$$\frac{dv}{dt} = \frac{12 - v}{1} + i \tag{20.1}$$

and the right mesh equation is

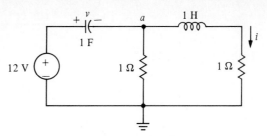

FIGURE 20.1 Second-order circuit

$$\frac{di}{dt} + i = 12 - v \tag{20.2}$$

We could eliminate v between these equations and solve the resulting second-order equation in i. However, with the Laplace transform method, we first transform each equation and then solve the resulting algebraic equations for the transform of v.

Transforming the equations, we have

$$s\mathbf{V}(s) - 4 = \frac{12}{s} - \mathbf{V}(s) + \mathbf{I}(s)$$

$$\tag{20.3}$$

$$s\mathbf{I}(s) - 2 + \mathbf{I}(s) = \frac{12}{s} - \mathbf{V}(s)$$

where $\mathbf{V}(s)$ and $\mathbf{I}(s)$ are, respectively, the transforms of v and i. Eliminating $\mathbf{V}(s)$ and simplifying yields

$$\mathbf{I}(s) = \frac{2(s + 5)}{s^2 + 2s + 2} = \frac{2(s + 1)}{(s + 1)^2 + 1} + \frac{8}{(s + 1)^2 + 1} \tag{20.4}$$

Therefore the current is

$$i = e^{-t}(2 \cos t + 8 \sin t) \quad \text{A} \tag{20.5}$$

EXAMPLE 20.2   As another example, let us find v for $t > 0$ in the circuit of Fig. 20.2, given that $v(0) = 0$. We note that the circuit has two sources, each of which has a different frequency. The generalized phasor method of Chapter 14 requires that we use super-position, but, as we will see, the Laplace transform method will handle both frequencies at once. The nodal equation at node a is

FIGURE 20.2 Circuit with two sources

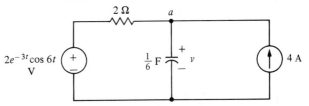

$$\frac{v - 2e^{-3t}\cos 6t}{2} + \frac{1}{6}\frac{dv}{dt} = 4$$

Transforming yields

$$3\mathbf{V}(s) - \frac{6(s + 3)}{(s + 3)^2 + 36} + s\mathbf{V}(s) = \frac{24}{s}$$

which may be written

$$(s + 3)\mathbf{V}(s) = \frac{24}{s} + \frac{6(s + 3)}{(s + 3)^2 + 36}$$

or

$$\mathbf{V}(s) = \frac{24}{s(s + 3)} + \frac{6}{(s + 3)^2 + 36} \tag{20.6}$$

Expanding the first term on the right into partial fractions yields

$$\mathbf{V}(s) = \frac{8}{s} - \frac{8}{s + 3} + \frac{6}{(s + 3)^2 + 6^2}$$

Inverting, we have the time-domain voltage, for $t > 0$,

$$v = 8 - 8e^{-3t} + e^{-3t}\sin 6t \quad \text{V}$$

EXAMPLE 20.3 As a final example, let us find the output voltage v, for $t > 0$, in the circuit of Fig. 20.3, where $v(0) = 2$ V and $v_c(0) = 2$ V. The circuit contains an op amp, which poses no difficulty for the Laplace transform method. Since the node voltage v_1 is the sum of v_c and v, we have $v_1(0) = 2 + 2 = 4$ V. The nodal equations at v_1 and v (of the $\frac{1}{5}$-farad capacitor) are, respectively,

$$\frac{v_1 - 6}{1} + \frac{dv_1}{dt} - \frac{dv}{dt} + \frac{v_1 - v}{1} = 0$$

and

$$\frac{1}{5}\frac{dv}{dt} + \frac{v - v_1}{1} = 0$$

FIGURE 20.3 Circuit with an op amp

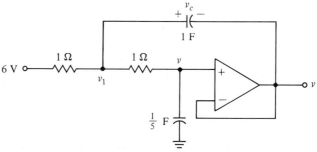

Transforming these equations results in

$$V_1(s) - \frac{6}{s} + sV_1(s) - 4 - sV(s) + 2 + V_1(s) - V(s) = 0$$

$$\frac{s}{5}V(s) - \frac{2}{5} + V(s) - V_1(s) = 0$$

Substituting $V_1(s)$ from the second equation in the first equation yields

$$(s + 2)\left[\frac{(s + 5)V(s) - 2}{5}\right] - sV(s) - V(s) = \frac{6}{s} + 2$$

Solving for $V(s)$ and simplifying gives

$$V(s) = \frac{2s^2 + 14s + 30}{s(s^2 + 2s + 5)}$$

The partial fraction expansion is given by

$$V(s) = \frac{6}{s} - \frac{4s - 2}{s^2 + 2s + 5}$$

which can be rearranged in the form

$$V(s) = \frac{6}{s} - 4\left[\frac{s + 1}{(s + 1)^2 + 2^2}\right] + 3\left[\frac{2}{(s + 1)^2 + 2^2}\right]$$

The inverse transform for $t > 0$ is therefore

$$v = 6 - 4e^{-t}\cos 2t + 3e^{-t}\sin 2t \quad V \tag{20.7}$$

EXERCISES

20.1.1   Show that the transform $I(s)$ for the given circuit is

$$I(s) = \frac{V_g(s) + i(0) - v_c(0)/s}{s + 4 + 1/Cs}$$

and determine $i(t)$, for $t > 0$, if $C = \frac{1}{3}$ F, $v_g(t) = 6$ V, $i(0) = 1$ A, and $v_c(0) = 1$ V.
Answer $2e^{-t} - e^{-3t}$ A

EXERCISE 20.1.1

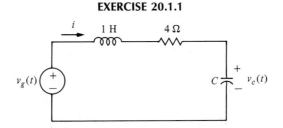

20.1.2 Find $i(t)$, for $t > 0$, in Ex. 20.1.1 if $C = \frac{1}{13}$ F, $v_g(t) = 6$ V, $i(0) = 2$ A, and $v_c(0) = 4$ V.

Answer $2e^{-2t} (\cos 3t - \frac{1}{3} \sin 3t)$ A

20.1.3 Find $i(t)$, for $t > 0$, in Ex. 20.1.1 if $C = \frac{1}{4}$ F, $v_g(t) = 6$ V, $i(0) = 1$ A, and $v_c(0) = 2$ V.

Answer $e^{-2t}(2t + 1)$ A

20.2

THE TRANSFORMED CIRCUIT

When dealing with phasors we were able to omit the steps of writing the differential equations by going directly to the phasor circuit. The circuit was analyzed by resistive circuit techniques, the phasor output was found, and from this the time-domain solution was obtained. We may perform a similar type of analysis with Laplace transforms by first obtaining a *transformed* circuit. The rest of the procedure is similar to that of the phasor method except that with Laplace transforms we are able to deal with more general functions and obtain the complete solution. That is, the initial conditions are satisfied as well.

To see how the transformed circuit is obtained, let us consider, for example, the time-domain expression of KVL, denoted by

$$v_1(t) + v_2(t) + \cdots + v_n(t) = 0 \tag{20.8}$$

Transforming this equation yields

$$\mathbf{V}_1(s) + \mathbf{V}_2(s) + \cdots + \mathbf{V}_n(s) = 0 \tag{20.9}$$

where $\mathbf{V}_i(s)$ is the transform of $v_i(t)$. The *transformed voltages* thus satisfy KVL as the phasors did. A similar development will show also that transformed currents satisfy KCL. Since the transformed voltages are transforms of voltages across inductors, resistors, and capacitors, as well as source transforms and terms representing initial conditions, they may be found for the case of interest and substituted directly into (20.9) without writing the time-domain equations at all. In fact, we may transform the time-domain voltage–current relations for the various circuit elements and use the result to obtain circuit element models in the *s*-domain. These may then be substituted into the circuit for the time-domain models to obtain the transformed circuit, which may then be used to write equations like (20.9) directly. This is, of course, exactly what was done to obtain the phasor circuits.

Let us first consider a resistance R, with current i_R and voltage v_R, for which

$$v_R = Ri_R$$

Transforming this equation, we have

$$\mathbf{V}_R(s) = R\mathbf{I}_R(s) \tag{20.10}$$

which is represented by the transformed resistor element of Fig. 20.4(a).

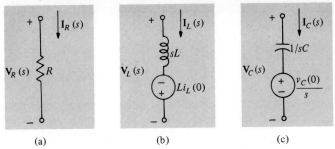

FIGURE 20.4 Transformed circuit elements

Next, let us consider an inductor L for which

$$v_L = L\frac{di_L}{dt}$$

Transforming, we have

$$\mathbf{V}_L(s) = sL\,\mathbf{I}_L(s) - Li_L(0) \qquad (20.11)$$

which may be represented by an inductor with impedance

$$\mathbf{Z}_L(s) = sL$$

in series with a source, $Li_L(0)$, with the proper polarity, as shown in Fig. 20.4(b). The included voltage source takes into account the initial condition $i_L(0)$. It is an *impulsive* source, since

$$\mathcal{L}[Li_L(0)\delta(t)] = Li_L(0)$$

In the case of a capacitor C we have

$$v_C = \frac{1}{C}\int_0^t i_C\,dt + v_C(0)$$

Transforming, we obtain

$$\mathbf{V}_C(s) = \frac{1}{sC}\mathbf{I}_C(s) + \frac{1}{s}v_C(0) \qquad (20.12)$$

which is represented in Fig. 20.4(c) as a capacitor with impedance

$$\mathbf{Z}_C(s) = \frac{1}{sC}$$

in series with a source, $v_C(0)/s$, accounting for the initial condition.

Independent sources are simply labeled with their transforms. Dependent sources are transformed in the same way as passive elements. For example, a controlled voltage source defined by

$$v_1 = Kv_2$$

transforms to

$$\mathbf{V}_1(s) = K\mathbf{V}_2(s)$$

which in the transformed circuit is a source controlled by a transformed variable.

    In summary, we obtain the transformed circuit from the original circuit by replacing the elements by their transformed elements and the currents and voltages by their transforms. The transformed elements of passive elements are those shown in Fig. 20.4, and the transformed sources are those of the original circuit labeled with the transforms of their time-domain values. The transformed circuit is then solved using resistive circuit methods for the transforms of the desired time-domain variables.

EXAMPLE 20.4   As an example, suppose we require $i(t)$, for $t > 0$, in Fig. 20.5(a), given that $i(0) = 4$ A and $v(0) = 8$ V. The transformed network is shown in Fig. 20.5(b), from which we have

$$\mathbf{I}(s) = \frac{[2/(s + 3)] + 4 - (8/s)}{3 + s + (2/s)}$$

This may be written

$$\mathbf{I}(s) = -\frac{13}{s + 1} + \frac{20}{s + 2} - \frac{3}{s + 3}$$

so that, for $t > 0$,

$$i(t) = -13e^{-t} + 20e^{-2t} - 3e^{-3t}$$

As a check, the time-domain equations for Fig. 20.5(a) are

$$\frac{di}{dt} + 3i + 2\int_0^t i\,dt + 8 = 2e^{-3t}$$

$$i(0) = 4$$

FIGURE 20.5 (a) Circuit; (b) its transformed counterpart

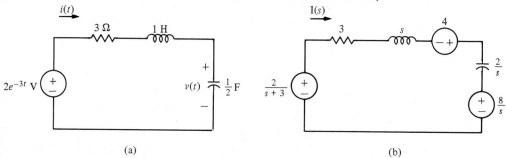

    (a)   (b)

Transforming, we have

$$sI(s) - 4 + 3I(s) + \frac{2}{s}I(s) + \frac{8}{s} = \frac{2}{s+3}$$

which is the loop equation of the circuit of Fig. 20.5(b), and which yields $I(s)$ as before.

We may also solve (20.10)–(20.12) explicitly for the currents. The results are

$$I_R(s) = \frac{V_R(s)}{R}$$

$$I_L(s) = \frac{1}{sL}V_L(s) + \frac{1}{s}i_L(0)$$

$$I_C(s) = sC V_C(s) - C v_C(0)$$

These results are often useful in nodal analysis. As the reader may verify, the transformed elements describing these equations are shown in Fig. 20.6. The labels on the passive elements are impedances, as in Fig. 20.4.

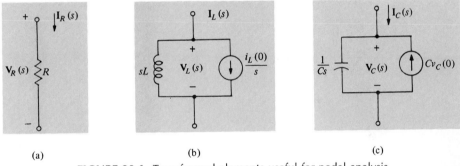

FIGURE 20.6 Transformed elements useful for nodal analysis

All the network theorems that apply to resistive and phasor circuits apply to the transformed circuit. An added advantage in the latter case is that initial conditions may be incorporated. As an example, the reader may verify that the elements of Fig. 20.6 are the Norton equivalents of those of Fig. 20.4, and can be obtained using source transformations.

EXAMPLE 20.5   As another example, let us find the voltage v in the circuit of Fig. 20.7(a) with initial conditions $i(0) = 1$ A and $v(0) = 4$ V. Using Fig. 20.6, we obtain the transformed circuit of Fig. 20.7(b). The nodal equation at the top node is given by

$$\frac{V}{4} + \frac{V}{3s} + \frac{sV}{24} = \frac{4}{24} - \frac{1}{s}$$

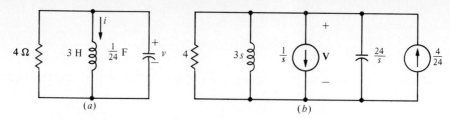

FIGURE 20.7 (a) Parallel *RLC* circuit; (b) its transformed counterpart

from which we have

$$\mathbf{V} = \frac{4s - 24}{(s + 2)(s + 4)} = \frac{-16}{s + 2} + \frac{20}{s + 4}$$

or

$$v = -16e^{-2t} + 20e^{-4t} \text{ V}$$

EXAMPLE 20.6   As an example illustrating op amp transformations, let us solve the circuit of Fig. 20.3 for v, using the transformed circuit method. The transformed circuit is shown in Fig. 20.8, where we note that the transformed element for an op amp is the op amp itself. This is because the currents into and the voltage across the input terminals are all zero, and the transform of zero is zero.

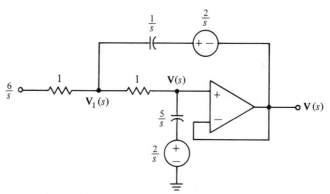

FIGURE 20.8 Transformed circuit for Fig. 20.3

At node $\mathbf{V}_1(s)$ we have

$$\frac{\mathbf{V}_1(s) - (6/s)}{1} + \frac{\mathbf{V}_1(s) - \mathbf{V}(s)}{1} + s\left\{\mathbf{V}_1(s) - \left[\mathbf{V}(s) + \frac{2}{s}\right]\right\} = 0$$

and at the noninverting input node we have

$$\frac{\mathbf{V}(s) - \mathbf{V}_1(s)}{1} + \frac{s}{5}\left[\mathbf{V}(s) - \frac{2}{s}\right] = 0$$

Solving for $\mathbf{V}(s)$ yields

$$V(s) = \frac{2s^2 + 14s + 30}{s(s^2 + 2s + 5)}$$

$$= \frac{6}{s} - 4\left[\frac{s + 1}{(s + 1)^2 + 4}\right] + 3\left[\frac{2}{(s + 1)^2 + 4}\right]$$

Inverting we have

$$v(t) = 6 - 4e^{-t}\cos 2t + 3e^{-t}\sin 2t \quad V$$

which checks (20.7).

EXAMPLE 20.7 Since the procedure using transformed circuits is identical to that using phasor circuits, we may obtain transformed Thevenin and Norton circuits exactly as we obtained their phasor circuits. To illustrate the process, let us replace everything to the right of the 4-Ω resistor of Fig. 20.7(a) by its Thevenin equivalent and use the result to find $v(t)$.

The transformed circuit to the right of the 4-Ω resistor is obtained from Fig. 20.7(b) and is shown with $\mathbf{V}_{oc}(s)$ across the open terminals in Fig. 20.9(a). (We note that, unlike the Thevenin phasor circuits, the transformed circuits have the initial conditions built in.) The nodal equation is

$$\frac{\mathbf{V}_{oc}(s)}{3s} + \frac{1}{s} + \frac{s}{24}\mathbf{V}_{oc}(s) = \frac{1}{6}$$

from which the transformed open-circuit voltage is

$$\mathbf{V}_{oc}(s) = \frac{4(s - 6)}{s^2 + 8}$$

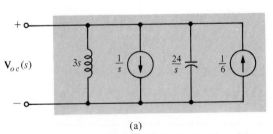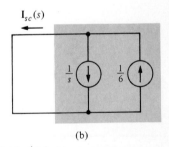

(a)   (b)

FIGURE 20.9 Circuit for obtaining (a) $\mathbf{V}_{oc}(s)$; (b) $\mathbf{I}_{sc}(s)$

If we short-circuit the input terminals of Fig. 20.9(a), we obtain Fig. 20.9(b), since the inductor and capacitor are shorted out. The nodal equation is

$$\mathbf{I}_{sc}(s) + \frac{1}{s} = \frac{1}{6}$$

from which the transformed short-circuit current is

$$\mathbf{I}_{sc}(s) = \frac{s - 6}{6s}$$

The Thevenin impedance is therefore

$$\mathbf{Z}_{th}(s) = \frac{\mathbf{V}_{oc}(s)}{\mathbf{I}_{sc}(s)} = \frac{\left[\dfrac{4(s-6)}{s^2+8}\right]}{\left[\dfrac{s-6}{6s}\right]} = \frac{24s}{s^2+8}$$

and the Thevenin equivalent circuit is shown in Fig. 20.10, with the 4-Ω resistor connected across its terminals. The voltage $\mathbf{V}(s)$ is the voltage across the resistor, as indicated.

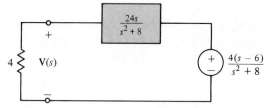

FIGURE 20.10 Thevenin equivalent circuit terminated in a 4-Ω resistor

The nodal equation for $\mathbf{V}(s)$ is

$$\frac{\mathbf{V}(s)}{4} + \frac{\mathbf{V}(s) - \dfrac{4(s-6)}{s^2+8}}{\dfrac{24s}{s^2+8}} = 0$$

from which, after simplification, we have

$$\mathbf{V}(s) = \frac{4(s-6)}{(s+2)(s+4)} = \frac{-16}{s+2} + \frac{20}{s+4}$$

The time-domain voltage for $t > 0$ is therefore

$$v(t) = -16e^{-2t} + 20e^{-4t} \quad \text{V}$$

as obtained before.

EXERCISES

20.2.1 Find v for $t > 0$ using the transformed circuit method if $v(0) = 10$ V.
Answer $10(2 - e^{-t}) + \frac{1}{2}e^{-t}\sin 2t$ V

EXERCISE 20.2.1

20.2.2 Find the voltage $v(t)$, for $t > 0$, across the terminals of the RLC parallel circuit, excited by a current source $i_g(t) = e^{-2t}$ A, using the transformed circuit method. The passive elements are $R = 4\ \Omega$, $L = 5$ H, and $C = \frac{1}{20}$ F; the initial inductor current is $i_L(0) = 1$ A; and $v(0) = 2$ V. (The current i_g is directed into the positive terminal, and the current i_L is directed out of the positive terminal of v.)
Answer $-14e^{-t} + 20e^{-2t} - 4e^{-4t}$ V

20.2.3 Find i and v in Fig. 20.1 using the transformed circuit.
Answer $v = 12 - e^{-t}(8 \cos t - 2 \sin t)$ V

20.2.4 Derive the transformed circuits of Fig. 20.6 by transforming the time-domain equations for the currents i_R, i_L, and i_C.

20.2.5 For the transformer of Fig. 16.3(a), shown here as (a), obtain the transformed circuit shown in (b). Use the transformed circuit to find v_2 if in (a) $L_1 = 4$ H, $L_2 = 1$ H, $M = 1$ H, v_1 is replaced by a practical voltage source of 7 V (with the positive polarity at the top) and a 4-Ω internal resistance, and v_2 is the voltage across a 1-Ω resistor. Take $i_1(0) = 3/2$ A and $i_2(0) = 0$.
Answer $\frac{1}{4}(e^{-2t/3} - e^{-2t})$ V

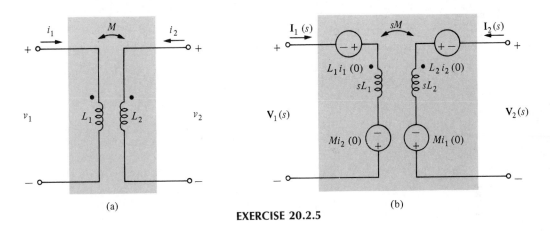

(a)

(b)

EXERCISE 20.2.5

20.2.6 Find $i(t)$ for $t > 0$ in the circuit of Fig. 20.5(a) by replacing everything except the 3-Ω resistor in its transformed circuit by its Norton equivalent circuit.

Answer $\mathbf{I}_{sc} = \dfrac{4s^2 + 6s - 24}{(s + 3)(s^2 + 2)}$ A; $Z_{th} = \dfrac{s^2 + 2}{s}\ \Omega$;

$$i(t) = -13e^{-t} + 20e^{-2t} - 3e^{-3t}\ \text{A}$$

20.3

NETWORK FUNCTIONS

In Sec. 14.4 we defined a network function in the generalized frequency domain to be the ratio $\mathbf{H}(s)$ of the output phasor to the input phasor of a circuit having a single excitation and response. In this section we will define the network function to be the ratio of the Laplace transform of the output to the Laplace transform of the input, and we will see that the two definitions are precisely the same. Again, we are assuming a single input and a single output so that all the initial circuit conditions are zero.

Suppose in the general case that the input v_i and output v_o are related by the differential equation

$$a_n \frac{d^n v_o}{dt^n} + a_{n-1} \frac{d^{n-1} v_o}{dt^{n-1}} + \cdots + a_1 \frac{dv_o}{dt} + a_0 v_o$$

$$= b_m \frac{d^m v_i}{dt^m} + b_{m-1} \frac{d^{m-1} v_i}{dt^{m-1}} + \cdots + b_1 \frac{dv_i}{dt} + b_0 v_i$$

and that the initial conditions are all zero; that is,

$$v_o(0) = \frac{dv_o(0)}{dt} = \cdots = \frac{d^{n-1} v_o(0)}{dt^{n-1}} = v_i(0) = \frac{dv_i(0)}{dt} = \cdots = \frac{d^{m-1} v_i(0)}{dt^{m-1}} = 0$$

Then, transforming, we have

$$(a_n s^n + a_{n-1} s^{n-1} + \cdots + a_1 s + a_0) V_o(s)$$

$$= (b_m s^m + b_{m-1} s^{m-1} + \cdots + b_1 s + b_0) V_i(s)$$

from which the network function, or transfer function, is given by

$$\mathbf{H}(s) = \frac{\mathbf{V}_o(s)}{\mathbf{V}_i(s)} = \frac{b_m s^m + b_{m-1} s^{m-1} + \cdots + b_1 s + b_0}{a_n s^n + a_{n-1} s^{n-1} + \cdots + a_1 s + a_0} \qquad (20.13)$$

Comparing this result with (18.37), obtained previously for the Fourier transform, we see that $\mathbf{H}(s)$ is the same network function, with $j\omega$ generalized to s. Also $\mathbf{H}(s)$, the ratio of the Laplace transforms of the output to the input, is precisely the network function as it was originally defined for the generalized phasors in (14.24). Moreover, since admittances and impedances are network functions, they are precisely the same in the Laplace transform domain as they were in the case of the generalized phasors. With Laplace transforms, however, the situation is more general. We may have any function that is Laplace transformable as the input or output, instead of being restricted to damped sinusoids.

EXAMPLE 20.8   In most cases an easier way to obtain the network function is from the transformed circuit. To illustrate, if the input of the RLC series circuit is $v_g(t)$ and the output is $i(t)$, then we have

$$\mathbf{H}(s) = \frac{\mathbf{I}(s)}{\mathbf{V}_g(s)} = \frac{1}{\mathbf{Z}(s)}$$

$$= \frac{1}{sL + R + (1/sC)}$$

If $R = 2 \, \Omega$, $L = 1$ H, and $C = 0.2$ F, this becomes

$$\mathbf{H}(s) = \frac{s}{s^2 + 2s + 5} \qquad (20.14)$$

For an initially relaxed circuit, the knowledge of the network function is sufficient to determine the external behavior of the circuit for *any* transformable input. In other words, in this case, having $\mathbf{H}(s)$ is equivalent to having the circuit itself. To see this, let us write (20.13) in the form

$$\mathbf{V}_o(s) = \mathbf{H}(s)\mathbf{V}_i(s) \qquad (20.15)$$

Thus if we know the input $v_i(t)$, we may find $\mathbf{V}_i(s)$ and subsequently $\mathbf{V}_o(s)$, from which the output $v_o(t)$ follows.

EXAMPLE 20.9 As an example, if the circuit with the transfer function (20.14) has an input

$$v_g(t) = u(t) \quad \text{V}$$

then

$$\mathbf{V}_g(s) = \frac{1}{s}$$

and we have

$$\mathbf{I}(s) = \mathbf{H}(s)\mathbf{V}_g(s) = \frac{\mathbf{H}(s)}{s}$$

$$= \frac{1}{s^2 + 2s + 5} \qquad (20.16)$$

Inverting, we have

$$i(t) = \frac{1}{2}e^{-t}\sin 2t \quad \text{A}$$

EXAMPLE 20.10 As another example, suppose the initially relaxed circuit with network function (20.14) has an input

$$v_g(t) = \sin t\, u(t) \quad \text{V}$$

Then the input transform is

$$\mathbf{V}_g(s) = \frac{1}{s^2 + 1}$$

and the output current transform is

$$\mathbf{I}(s) = \left[\frac{s}{s^2 + 2s + 5}\right]\left[\frac{1}{s^2 + 1}\right]$$

which may be expanded to

$$\mathbf{I}(s) = \frac{1}{5}\left(\frac{s}{s^2 + 1}\right) + \frac{1}{10}\left(\frac{1}{s^2 + 1}\right) - \frac{1}{5}\left[\frac{s + 1}{(s + 1)^2 + 4}\right] - \frac{3}{20}\left[\frac{2}{(s + 1)^2 + 4}\right]$$

Therefore the time-domain current is

$$i(t) = \frac{1}{10}(2 \cos t + \sin t) - \frac{1}{20}e^{-t}(4 \cos 2t + 3 \sin 2t) \quad A$$

EXAMPLE 20.11 As another example, let us find the transfer function for the circuit of Fig. 20.11(a) if the output is $i(t)$ and the input is $v_g(t)$. The transformed circuit is shown in Fig. 20.11(b), and since it is a ladder network, let us use the method of proportionality.

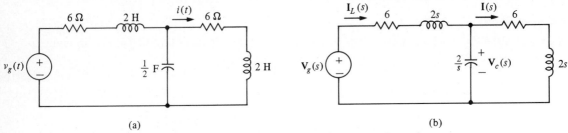

FIGURE 20.11 (a) Time-domain circuit; (b) its transformed circuit

Arbitrarily taking

$$\mathbf{I}(s) = 1$$

we have

$$\mathbf{V}_c(s) = (2s + 6)\mathbf{I}(s) = 2s + 6$$

$$\mathbf{I}_L(s) = \frac{\mathbf{V}_c(s)}{(2/s)} + \mathbf{I}(s) = \frac{s}{2}(2s + 6) + 1 = s^2 + 3s + 1$$

and

$$\mathbf{V}_g(s) = (2s + 6)\mathbf{I}_L(s) + \mathbf{V}_c(s)$$
$$= (2s + 6)(s^2 + 3s + 1) + 2s + 6$$
$$= (2s + 6)(s^2 + 3s + 2)$$

The transfer function is therefore

$$\mathbf{H}(s) = \frac{\mathbf{I}(s)}{\mathbf{V}_g(s)} = \frac{1}{(2s + 6)(s^2 + 3s + 2)}$$

$$= \frac{1/2}{(s + 1)(s + 2)(s + 3)}$$

EXAMPLE 20.12 As a final example, suppose the input of an initially relaxed circuit is

$$v_i(t) = 6e^{-t} \cos t \quad V$$

the output is

$$v_o(t) = 6e^{-t} \sin t \quad V$$

and we wish to find the output when the input is $u(t)$ V. The transfer function is

$$H(s) = \frac{V_o(s)}{V_i(s)} = \frac{\left[\dfrac{6}{(s+1)^2+1}\right]}{\left[\dfrac{6(s+1)}{(s+1)^2+1}\right]} = \frac{1}{s+1}$$

The output $v_o(t)$ when the input is $u(t)$ is therefore

$$v_o(t) = \mathcal{L}^{-1}\left[H(s)\frac{1}{s}\right]$$

$$= \mathcal{L}^{-1}\left[\frac{1}{s(s+1)}\right]$$

$$= \mathcal{L}^{-1}\left[\frac{1}{s} - \frac{1}{s+1}\right]$$

$$= 1 - e^{-t} \quad V, \qquad \text{for } t > 0$$

EXERCISES

20.3.1 Given the equation

$$y''(t) + 2y'(t) + 5y(t) = 10x(t)$$

where $y(t)$ and $x(t)$ are the output and input, respectively, of a circuit, find the transfer function and the step response.

Answer $\dfrac{10}{s^2 + 2s + 5}$; $(2 - 2e^{-t}\cos 2t - e^{-t}\sin 2t)u(t)$

20.3.2 If in the circuit of Prob. 9.10 the output is $i(t)$ and the input is the voltage source, find $H(s)$ for the three cases given.

Answer (a) $\dfrac{s}{2s^2 + 7s + 6}$; (b) $\dfrac{s}{s^2 + 4s + 4}$; (c) $\dfrac{s}{s^2 + 2s + 2}$

20.3.3 Find the step response in cases (b) and (c) of Ex. 20.3.2.
Answer (b) $te^{-2t}u(t)$ A; (c) $e^{-t}\sin t\, u(t)$ A

20.4

STEP AND IMPULSE RESPONSES

As we have seen in the previous section, if $Y(s)$ and $X(s)$ are the transformed output and input, respectively, of an initially relaxed circuit, then the network function is

$$H(s) = \frac{Y(s)}{X(s)}$$

and the output is given by

$$Y(s) = H(s)X(s) \qquad\qquad (20.17)$$

The step response, which we designate $r(t)$, with transform $R(s)$, was defined in Chapter 8 as the response of an initially relaxed circuit to the unit step input $u(t)$. It is particularly easy to get using the transfer function, since by (20.17) we have

$$\mathcal{L}[r(t)] = R(s) = H(s)\mathcal{L}[u(t)]$$

or

$$R(s) = \frac{H(s)}{s} \qquad\qquad (20.18)$$

EXAMPLE 20.13 As an example, the network function of the circuit of Fig. 20.12 was obtained in the previous section as

$$H(s) = \frac{I(s)}{V_g(s)} = \frac{s}{s^2 + 2s + 5}$$

and the step response was found to be

$$r(t) = i(t) = \mathcal{L}^{-1}\left[\frac{H(s)}{s}\right]$$

$$= \mathcal{L}^{-1}\left[\frac{1}{s^2 + 2s + 5}\right]$$

$$= \frac{1}{2}e^{-t}\sin 2t \quad \text{A}$$

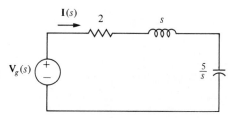

FIGURE 20.12 *RLC* transformed circuit

The *impulse response* is the response of an initially relaxed circuit to a unit impulse input $\delta(t)$. Since $\mathcal{L}[\delta(t)] = 1$, the impulse response, by (20.17), is

$$y(t) = \mathcal{L}^{-1}[H(s) \cdot 1] = \mathcal{L}^{-1}[H(s)]$$

Since it is the inverse of the transfer function $H(s)$, it is natural to denote the impulse response by $h(t)$. That is,

$$h(t) = \mathcal{L}^{-1}[H(s)] \qquad\qquad (20.19)$$

Therefore if we know the impulse response, we can find the transfer function,

$$\mathbf{H}(s) = \mathscr{L}[h(t)]$$

from which we can find the response to *any* input. For this reason, the impulse response is one of the most important output functions in circuit theory. Of course, we may also find the transfer function from the step response, but the process is not as elegant and direct, since the transform of the impulse response *is* the transfer function.

EXAMPLE 20.14 As an example, the impulse response of the circuit of Fig. 20.12 is

$$h(t) = \mathscr{L}^{-1}\left[\frac{s}{s^2 + 2s + 5}\right]$$

$$= e^{-t}\left(\cos 2t - \frac{1}{2}\sin 2t\right)u(t)$$

EXAMPLE 20.15 As another example, let us find the impulse response for the circuit of Fig. 20.13(a) if the input is $i_g(t)$ and the output is $v(t)$. From the transformed circuit of Fig. 20.13(b), we see that the network function is

$$\mathbf{H}(s) = \frac{\mathbf{V}(s)}{\mathbf{I}_g(s)} = \mathbf{Z}(s)$$

where $\mathbf{Z}(s)$ is the input impedance seen at the source terminals. Therefore we have

$$\mathbf{H}(s) = \frac{(2s + 8)(10/s)}{2s + 8 + (10/s)} = \frac{10(s + 4)}{s^2 + 4s + 5}$$

$$= \frac{10(s + 2)}{(s + 2)^2 + 1} + \frac{20}{(s + 2)^2 + 1}$$

The impulse response is therefore

$$h(t) = 10e^{-2t}(\cos t + 2 \sin t) \quad \text{V}$$

Another point should be made about the impulse response. Since, in general, $\mathbf{V}_o(s)$ is $\mathbf{H}(s)\mathbf{V}_i(s)$, a product of two transforms, we may write immediately, by means

FIGURE 20.13 (a) Time-domain and (b) corresponding transformed circuit

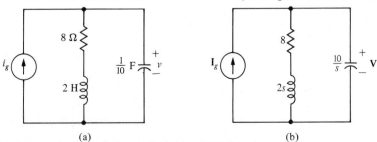

(a)   (b)

of the convolution theorem,

$$v_o(t) = h(t) * v_i(t) = \int_0^t h(\tau)v_i(t - \tau)d\tau \qquad (20.20)$$

Therefore, if we know the impulse response and the input, we may find the output of an initially relaxed network directly using the convolution theorem.

EXAMPLE 20.16 As an example, let us find the step response of the circuit of Fig. 20.12 using convolution. Since the input is $u(t)$, the step response $r(t) = i(t)$ is given by

$$r(t) = i(t) = h(t) * u(t)$$
$$= \int_0^t e^{-\tau}\left(\cos 2\tau - \frac{1}{2}\sin 2\tau\right)u(\tau)u(t - \tau)d\tau$$
$$= \int_0^t e^{-\tau}\left(\cos 2\tau - \frac{1}{2}\sin 2\tau\right)d\tau$$
$$= \frac{1}{2}e^{-t}\sin 2t$$

which checks the previous result.

In general, since the step response is

$$r(t) = \mathcal{L}^{-1}\left[\frac{\mathbf{H}(s)}{s}\right]$$

we have

$$r(t) = \int_0^t h(\tau)\, d\tau \qquad (20.21)$$

which shows that the step response may be obtained directly from the impulse response. The converse is true as well, since differentiating (20.21) yields

$$h(t) = \frac{dr(t)}{dt} \qquad (20.22)$$

Figure 20.14 may be used to sum up the various input–output relations for a linear, lumped-parameter circuit that is initially relaxed. The figures shown are *block diagrams* representing the circuit, and the arrows indicate the inputs and outputs. In (a) we have a frequency-domain block diagram, with $\mathbf{H}(s)$ representing the circuit. The other cases are time-domain diagrams, with $h(t)$ representing the circuit.

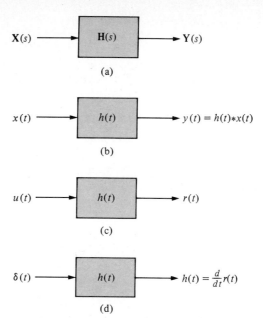

(a)

(b)

(c)

(d)

FIGURE 20.14 Block diagrams representing circuits

EXAMPLE 20.17 To illustrate Figs. 20.14(c) and (d), let us find the step and impulse responses $v(t)$ for the circuit of Fig. 20.15(a). The transformed circuit is shown in Fig. 20.15(b), from which we have, by voltage division,

$$\mathbf{H}(s) = \frac{\mathbf{V}(s)}{\mathbf{V}_g(s)} = \frac{\dfrac{7(7/2s)}{7 + (7/2s)}}{2s + 9 + \dfrac{7(7/2s)}{7 + (7/2s)}}$$

which simplifies to

$$\mathbf{H}(s) = \frac{7/4}{(s + 1)(s + 4)} \tag{20.23}$$

The step response is therefore

FIGURE 20.15 (a) Linear, lumped circuit; (b) its transformed circuit

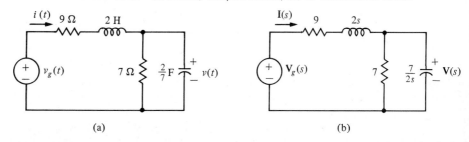

(a)

(b)

$$r(t) = \mathcal{L}^{-1}\left[\frac{\mathbf{H}(s)}{s}\right] = \mathcal{L}^{-1}\left[\frac{7/4}{s(s+1)(s+4)}\right]$$

$$= \mathcal{L}^{-1}\left[\frac{7}{48}\left(\frac{3}{s} - \frac{4}{s+1} + \frac{1}{s+4}\right)\right]$$

or

$$r(t) = \frac{7}{48}(3 - 4e^{-t} + e^{-4t})u(t) \quad \text{V}$$

We may obtain the impulse response as

$$h(t) = \mathcal{L}^{-1}[\mathbf{H}(s)]$$

but to illustrate Fig. 20.14(d), we will find it from

$$h(t) = \frac{dr(t)}{dt} = \frac{d}{dt}\left[\frac{7}{48}(3 - 4e^{-t} + e^{-4t})u(t)\right]$$

$$= \frac{7}{48}[(3 - 4e^{-t} + e^{-4t})\delta(t) + (4e^{-t} - 4e^{-4t})u(t)]$$

Since the expression in the first parentheses is zero for $t = 0$, we have

$$h(t) = \frac{7}{12}(e^{-t} - e^{-4t})u(t) \quad \text{V} \qquad (20.24)$$

We have defined the impulse response $h(t)$ by relating it to the transfer function $\mathbf{H}(s)$. This is unnecessary, of course, since we may find $h(t)$ from the circuit, as the output when the input is $\delta(t)$. In the example we have just considered, this procedure, using Laplace transforms, will lead to $h(t) = \mathcal{L}^{-1}[\mathbf{H}(s)]$, which as the reader may verify, is correct.

EXAMPLE 20.18 To completely separate the impulse response from the transfer function, however, we will solve the time-domain circuit of Fig. 20.15(a) for $v(t) = h(t)$ when $v_g(t) = \delta(t)$ and there is no initial stored energy. That is,

$$i(0^-) = v(0^-) = 0$$

The loop equation for $i(t)$ and the nodal equation for $v(t)$ are

$$2\frac{di}{dt} + 9i + v = v_g$$

$$\frac{v}{7} + \frac{2\,dv}{7\,dt} = i \qquad (20.25)$$

Substituting i from the second equation into the first equation and simplifying yields the describing equation,

$$\frac{d^2v}{dt^2} + 5\frac{dv}{dt} + 4v = \frac{7}{4}\delta(t) \tag{20.26}$$

Also, from the second of (20.25) and the initial conditions, we have

$$v(0^-) = 0, \qquad \frac{dv(0^-)}{dt} = 0$$

For $t < 0$, the describing equation is

$$\frac{d^2v}{dt^2} + 5\frac{dv}{dt} + 4v = 0 \tag{20.27}$$

since $\delta(t) = 0$ if $t \neq 0$. The general solution is

$$v(t) = Ae^{-t} + Be^{-4t} \tag{20.28}$$

but $A = B = 0$ because of the initial conditions. Therefore we have

$$v(t) = 0, \qquad t < 0 \tag{20.29}$$

To find $v(t)$ for $t > 0$, we have the same equation (20.27) and the same type of solution (20.28), but the initial conditions will be different. Evidently, since $v(t)$ is continuous, we have

$$v(0^+) = v(0^-) = 0$$

but there will be a jump in dv/dt at $t = 0$. To see this, we note that since the right-hand side of (20.26) contains an impulse, then the left-hand side must also contain an impulse. The impulse must be in the second derivative, because if it were in the first derivative, the second derivative would contain a doublet, and $v(t)$ would contain a step, which, of course, is not the case. If we integrate both sides of (20.26) between 0^- and 0^+, we have

$$\frac{dv(0^+)}{dt} - \frac{dv(0^-)}{dt} + \int_{0^-}^{0^+}\left(5\frac{dv}{dt} + v\right)dt = \frac{7}{4}\int_{0^-}^{0^+}\delta(t)\,dt = \frac{7}{4}$$

The integral in the left member is zero because v is continuous and dv/dt has a finite step in the interval of integration. Therefore, since $dv(0^-)/dt = 0$, we have

$$\frac{dv(0^+)}{dt} = \frac{7}{4}$$

Applying the initial conditions at $t = 0^+$ to (20.28) yields

$$A + B = 0$$

$$-A - 4B = \frac{7}{4}$$

from which $A = 7/12$ and $B = -7/12$. Therefore, by (20.28) and (20.29), we have the impulse response given earlier in (20.24).

20.4.1 Given the network function

$$\mathbf{H}(s) = \frac{2(s + 2)}{(s + 1)(s + 3)}$$

find (a) $h(t)$, (b) the step response, and (c) the forced response to an input e^{-2t}. In (c) take $y(0) = 0$ and $y'(0) = 2$.

Answer (a) $(e^{-t} + e^{-3t})u(t)$; (b) $\frac{1}{3}(4 - 3e^{-t} - e^{-3t})u(t)$; (c) $(e^{-t} - e^{-3t})u(t)$

20.4.2 Given the equation

$$y''(t) + 4y'(t) + 3y(t) = 6x(t)$$

where $x(t)$ is the input and $y(t)$ is the output, find (a) the network function, (b) the impulse response, and (c) the step response.

Answer (a) $\dfrac{6}{s^2 + 4s + 3}$; (b) $3(e^{-t} - e^{-3t})u(t)$; (c) $(2 - 3e^{-t} + e^{-3t})u(t)$

20.4.3 Find $h(t)$ from $r(t)$ in Ex. 20.4.1 using

$$\frac{dr(t)}{dt} = h(t)$$

20.5

STABILITY

An important concern in circuit theory is whether the output signal remains bounded or increases indefinitely following the application of an input signal. An unbounded output could seriously damage or even destroy the circuit, and therefore it is important to know before applying the input if the circuit can accommodate the expected output. We can answer this question by determining the *stability* of the circuit, which we will consider in this section.

We define a circuit to have *bounded input-bounded output* (BIBO) stability if any bounded input results in a bounded output. Such a circuit is also said to be *absolutely stable* or *unconditionally stable*. We can determine the stability in this sense by examining the poles of the network function, as we will see.

In the general case, as we know, the transform $\mathbf{V}_o(s)$ of the output function $v_o(t)$ is related to $\mathbf{V}_i(s)$, the transform of the input function $v_i(t)$, by the relation

$$\mathbf{V}_o(s) = \mathbf{H}(s)\mathbf{V}_i(s) \tag{20.30}$$

where $\mathbf{H}(s)$ is the network function of the circuit. If $s = p$ is any pole of $\mathbf{V}_o(s)$, then the denominator of $\mathbf{V}_o(s)$ contains a factor $(s - p)^n$, where n is a positive integer. if the pole p is of order n $(n > 1)$, the partial fraction expansion is of the form

$$\mathbf{V}_o(s) = \frac{K_n}{(s - p)^n} + \frac{K_{n-1}}{(s - p)^{n-1}} + \cdots + \frac{K_1}{s - p} + \mathbf{V}_i(s) \tag{20.31}$$

where $\mathbf{V}_1(s)$ contains the other poles of $\mathbf{V}_o(s)$. If $s = p$ is a simple pole, (20.31) reduces to

$$\mathbf{V}_o(s) = \frac{K}{s - p} + \mathbf{V}_1(s) \tag{20.32}$$

Inverting (20.31), using Ex. 19.7.3, we have

$$v_o(t) = A_n t^{n-1} e^{pt} + A_{n-1} t^{n-2} e^{pt} + \cdots + A_1 e^{pt} + v_1(t) \tag{20.33}$$

where $A_j = K_j/(j - 1)!$, $j = 1, 2, 3, \ldots, n$. This is also the inverse of (20.32) if all the A's are zero except A_1. In either case ($s = p$ a simple or a higher-order pole), if p is a real positive number or a complex number with a positive real part, $v_o(t)$ is unbounded because e^{pt} is a growing exponential. Therefore, for absolute stability there can be no pole of $\mathbf{V}_o(s)$ that is positive or has a positive real part. In other words, $\mathbf{V}_o(s)$ can have *no* pole in the right half of the s-plane.

Since $v_i(t)$ is bounded, $\mathbf{V}_i(s)$ has no poles in the right half-plane. Therefore, since the only poles of $\mathbf{V}_o(s)$ are those of $\mathbf{H}(s)$ and $\mathbf{V}_i(s)$, no pole of $\mathbf{H}(s)$ for an absolutely stable circuit can be in the right half-plane.

Let us now consider the possibility of $j\omega$-axis poles in $\mathbf{H}(s)$. Such poles occur when the numerator has a higher degree than the denominator (a pole at infinity, which we consider a $j\omega$-axis pole), or from denominator factors like s^m, $m > 0$ (a pole at $s = 0$), and $(s^2 + \omega_1^2)^n$, $n > 0$ (poles at $\pm j\omega_1$). Consider, for example, the network functions

$$\mathbf{H}_1(s) = s \tag{20.34}$$

$$\mathbf{H}_2(s) = \frac{1}{s} \tag{20.35}$$

and

$$\mathbf{H}_3(s) = \frac{1}{s^2 + 1} \tag{20.36}$$

all of which have $j\omega$-axis poles. $\mathbf{H}_1(s)$ has a pole at infinity, $\mathbf{H}_2(s)$ has a pole at zero, and $\mathbf{H}_3(s)$ has poles at $\pm j1$. If the input for (20.34) is $v_i(t) = u(t)$, a bounded input, then $\mathbf{V}_i(s) = 1/s$ and

$$\mathbf{V}_o(s) = s\left(\frac{1}{s}\right) = 1$$

Thus $v_o(t) = \delta(t)$, an unbounded output. In the case of (20.35), let $v_i(t) = u(t)$ again. Then

$$\mathbf{V}_o(s) = \left(\frac{1}{s}\right)\left(\frac{1}{s}\right) = 1/s^2 \tag{20.37}$$

and $v_o(t) = t$, an unbounded output. Finally, in the case of (20.36), let $v_i(t) = \sin t$, which is bounded. Then we have

$$\mathbf{V}_i(s) = \frac{1}{s^2 + 1}$$

and

$$V_o(s) = \frac{1}{(s^2 + 1)^2} \tag{20.38}$$

for which, by Ex. 19.4.2,

$$v_o(t) = \frac{1}{2}(\sin t - t \cos t) \tag{20.39}$$

which again is unbounded.

From these results we see that no type of $j\omega$-axis pole can be permitted in $\mathbf{H}(s)$ if the network is absolutely stable, because for each of the three types (infinity, zero, and finite nonzero poles), we can find a bounded input that results in an unbounded output. Evidently, from (20.33), left half-plane poles are acceptable, since for p negative, or complex with a negative real part, the terms in $v_o(t)$ are bounded. The condition for BIBO stability is therefore simple:

A network is absolutely stable if its network function has only left half-plane poles.

EXAMPLE 20.19 As an example, if V_o and V_i are the output and input, respectively, of the transformed circuit of Fig. 20.16(a), the network function, by voltage division, is

$$\mathbf{H}(s) = \frac{V_o(s)}{V_i(s)} = \frac{4}{s + 4 + (3/s)}$$

$$= \frac{4s}{(s + 1)(s + . 3)}$$

Both the poles, $s = -1$ and $s = -3$, are in the left half-plane, and therefore the network is absolutely stable. There is no bounded input that can produce an unbounded output.

EXAMPLE 20.20 As a second example, let us find the network function $\mathbf{H}(s) = \mathbf{I}(s)/\mathbf{V}_i(s)$ for the circuit of Fig. 20.16(b). The loop equation is

$$(s + 1)\mathbf{I} - 3\mathbf{I} = \mathbf{V}_i$$

from which

FIGURE 20.16 (a) Absolutely stable and (b) unstable network

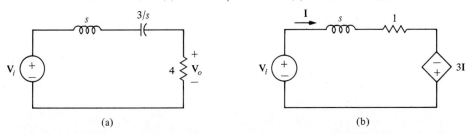

(a)   (b)

$$\mathbf{H}(s) = \frac{\mathbf{I}(s)}{\mathbf{V}_i(s)} = \frac{1}{s-2}$$

The pole $s = 2$ is a right half-plane pole and thus the circuit is unstable.

Absolutely stable circuits are, of course, desirable, but there are many useful circuits that are not absolutely stable. Consider, for example, the three commonplace circuits of Fig. 20.17. If the network functions are $\mathbf{V}(s)/\mathbf{I}(s)$ in (a) and (b), and $\mathbf{V}_o(s)/\mathbf{V}_i(s)$ in (c), they are $\mathbf{H}_1(s)$, $\mathbf{H}_2(s)$, and $\mathbf{H}_3(s)$ of (20.34), (20.35), and (20.36), respectively, and thus none of the circuits is absolutely stable. The difficulty that led to their unbounded outputs was due to the choice of inputs as much as to their network functions. The inputs used in (20.37) and (20.38), for example, caused the *double* $j\omega$-axis poles (of order 2) in the output transforms, which led to the unbounded outputs. If we place *conditions* on the choice of inputs, the outputs will be bounded. Of course, if the network functions themselves contain $j\omega$-axis poles of order higher than 1, the outputs will be unbounded. This leads us to define a second class of circuits:

A circuit is *conditionally stable* if its network function has no right half-plane poles or multiple $j\omega$-axis poles (those with order greater than 1).

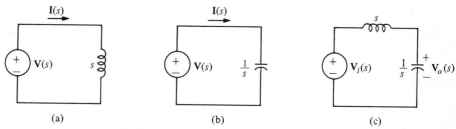

(a)   (b)   (c)

FIGURE 20.17 Conditionally stable circuits

We are now in position to define an *unstable* circuit as one that is neither absolutely nor conditionally stable. As an example, the circuit of Fig. 20.16(b), as noted earlier, is unstable.

EXERCISES

20.5.1 Determine if the following are network functions of absolutely stable, conditionally stable, or unstable circuits, and state the reasons.

(a) $\dfrac{4}{s^2 + s - 2}$; (b) $\dfrac{2s^2}{s^2 + 3s + 2}$; (c) $\dfrac{s^3 + 1}{s + 2}$; (d) $\dfrac{s + 1}{s^3 + 2s^2 + s + 2}$;

(e) $\dfrac{5}{(s^2 + 9)^2(s + 3)}$.

Answer (a) Unstable, pole at 1; (b) absolutely stable, poles at -1, -2; (c) unstable, double pole at infinity; (d) conditionally stable, poles at $\pm j1$, -2; (e) unstable, double pole at $\pm j3$

20.5.2 It may be shown that a circuit is absolutely stable if its impulse response $h(t)$ satisfies the condition

$$\int_0^\infty |h(\tau)|\, d\tau < \infty$$

Evaluate this integral for the network function

$$H(s) = \frac{3s + 4}{(s + 1)(s + 2)}$$

and show that it represents an absolutely stable circuit.
Answer Integral = 2

20.6

INITIAL- AND FINAL-VALUE THEOREMS

In this section we will consider two useful theorems of Laplace transforms known as the *initial-value theorem* and the *final-value theorem*. In particular, they enable us to evaluate the *initial* value of a function $f(t)$, the value of $f(t)$ as t approaches 0^+, and the *final* value, the limit as t becomes infinite, from the transform $F(s)$, without actually finding $f(t)$.

We begin with the initial-value theorem, which states that if $F(s)$ is a ratio of polynomials that is a proper fraction (the degree of the numerator is less than that of the denominator), then

$$f(0^+) = \lim_{s \to \infty} s F(s) \qquad (20.40)$$

To establish the theorem, let us begin with the transform of $f'(t)$, given by

$$\mathcal{L}[f'(t)] = \int_{0^-}^\infty f'(t)e^{-st}\, dt = s F(s) - f(0^-) \qquad (20.41)$$

First, considering the case $f(t)$ continuous at $t = 0$ $[f(0^+) = f(0^-)]$, and taking the limit of both sides of (20.41) as $s \to \infty$, we have

$$\lim_{s \to \infty} \int_{0^-}^\infty f'(t)e^{-st}\, dt = \int_{0^-}^\infty f'(t)\left[\lim_{s \to \infty} e^{-st}\right] dt = 0$$

$$= \lim_{s \to \infty} s F(s) - f(0^-)$$

From this we have

$$\lim_{s \to \infty} s F(s) = f(0^-) = f(0^+)$$

which is (20.40) for $f(t)$ continuous at $t = 0$.

Next let us consider the case where $f(t)$ has a finite discontinuity at $t = 0$. That is,

$$f(t) = g(t) + Au(t) \qquad (20.42)$$

where $g(t)$ is continuous at $t = 0$ $[g(0^+) = g(0^-)]$. The functions $f(t)$, $g(t)$, and $Au(t)$ are shown in Fig. 20.18, where we see that for $t < 0$, $f(t) = g(t)$, and for $t > 0$, $f(t) = g(t) + A$. Therefore

$$f(0^-) = g(0^-)$$
$$f(0^+) = g(0^+) + A \qquad (20.43)$$

or

$$A = f(0^+) - g(0^+)$$
$$= f(0^+) - g(0^-)$$
$$= f(0^+) - f(0^-)$$

Transforming (20.42) and multiplying the result by s yields

$$s\mathbf{F}(s) = s\mathbf{G}(s) + A \qquad (20.44)$$

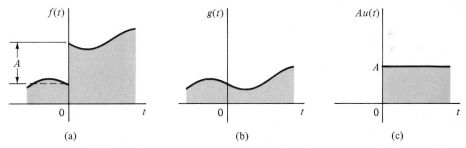

FIGURE 20.18 Discontinuous function $f(t)$ and its component parts

Since $g(t)$ is continuous at $t = 0$, (20.40) holds for it, as we have just shown. Therefore, taking the limit of (20.44) as $s \to \infty$, we have

$$\lim_{s \to \infty} s\mathbf{F}(s) = \lim_{s \to \infty} s\mathbf{G}(s) + A$$
$$= g(0^+) + A$$

By (20.43), this is

$$\lim_{s \to \infty} s\mathbf{F}(s) = f(0^+)$$

which is (20.40). Thus the initial-value theorem holds for $f(t)$ continuous or with a finite discontinuity at $t = 0$.

EXAMPLE 20.21 As an example, for the transform

$$\mathbf{F}(s) = \frac{4(s + 1)}{s^2 + 2s + 5}$$

we have

$$f(0^+) = \lim_{s \to \infty} s\mathbf{F}(s) = \lim_{s \to \infty} \frac{4s(s + 1)}{s^2 + 2s + 5} = 4$$

As a check, inverting $\mathbf{F}(s)$ yields

$$f(t) = 4e^{-t} \cos 2t$$

so that $f(0^+) = 4$, as before.

EXAMPLE 20.22 As another example, consider

$$f(t) = \delta(t) + 4e^{-t} \tag{20.45}$$

Its transform is

$$\mathbf{F}(s) = 1 + \frac{4}{s+1} = \frac{s+5}{s+4}$$

We note that the initial-value theorem does not apply, since $\mathbf{F}(s)$ is not a proper fraction. Also we have

$$\lim_{s\to\infty} s\mathbf{F}(s) = \lim_{s\to\infty} \frac{s(s+5)}{s+4} = \infty$$

whereas from (20.45) we have $f(0^+) = 4$.

The final-value theorem states that if $\mathbf{F}(s)$ has no right half-plane poles and no $j\omega$-axis poles except possibly at the origin, then

$$\lim_{t\to\infty} f(t) = \lim_{s\to 0} s\mathbf{F}(s) \tag{20.46}$$

To establish the theorem, let us take the limit as $s \to 0$ of the second and third members of (20.41), resulting in

$$\lim_{s\to 0} s\mathbf{F}(s) - f(0^-) = \int_{0^-}^{\infty} f'(t) \, dt$$

$$= \lim_{t\to\infty} f(t) - f(0^-)$$

from which (20.46) follows.

EXAMPLE 20.23 As an example, the transform

$$\mathbf{F}(s) = \frac{5s+2}{s(s+1)}$$

satisfies the conditions of the final-value theorem, and

$$\lim_{s\to 0} s\mathbf{F}(s) = 2$$

Inverting $\mathbf{F}(s)$, we have

$$f(t) = 2u(t) + 3e^{-t}$$

and thus

$$\lim_{t \to \infty} f(t) = 2$$

which checks.

EXAMPLE 20.24 The transform

$$\mathbf{F}(s) = \frac{1}{s - 1}$$

has a right half-plane pole, and therefore the final-value theorem is not applicable to it. Also, since $f(t) = e^t$, we see that

$$\lim_{s \to 0} s\mathbf{F}(s) = 0$$

but

$$\lim_{t \to \infty} f(t) = \infty$$

which, of course, do not agree.

EXAMPLE 20.25 The final-value theorem is useful in checking to see if an output function "settles down" as time increases. For example, if we have a low-pass filter with network function

$$\mathbf{H}(s) = \frac{2}{s^2 + 2s + 2}$$

the transform of the step response is

$$\mathbf{R}(s) = \frac{\mathbf{H}(s)}{s}$$

The final value of the step response $r(t)$ is then

$$\lim_{t \to \infty} r(t) = \lim_{s \to 0} s\mathbf{R}(s) = \lim_{s \to 0} \mathbf{H}(s) = 1$$

EXERCISES

20.6.1 Use the initial-value theorem, in those cases where it is applicable, to find the initial value $f(0^+)$ for the following functions $\mathbf{F}(s)$.

(a) $\dfrac{4s}{s^2 + 3s + 2}$; (b) $\dfrac{4s^2}{s^2 + 3s + 2}$; (c) $\dfrac{2(s^2 + 4)}{(s^2 + 2s + 4)(s + 2)}$; (d) $\dfrac{4}{s^2 + 3s + 2}$

Answer (a) 4; (b) not applicable; (c) 2; (d) 0

20.6.2 Use the final-value theorem, in those cases where it applies, to find the final value of $f(t)$ for the following functions $\mathbf{F}(s)$.

(a) $\dfrac{6}{s(s + 1)(s + 2)}$; (b) $\dfrac{2s + 3}{(s + 1)(s^2 + 1)}$; (c) $\dfrac{s}{s^2 + 2s + 5}$; (d) $\dfrac{4}{s(s - 2)}$

Answer (a) 3; (b) not applicable; (c) 0; (d) not applicable

20.6.3 Find $f(t)$ in Ex. 20.6.1(b) and show that the application of the initial-value theorem does not yield the correct answer.

Answer $f(t) = 4\delta(t) + 4e^{-t} - 16e^{-2t}$

20.6.4 Find $f(t)$ in Ex. 20.6.2(d) and show that the application of the final-value theorem does not yield the correct answer.

Answer $f(t) = -2 + 2e^{2t}$

20.7

STEADY-STATE SINUSOIDAL RESPONSE

It is possible, for a sinusoidal input, to extract the term from the partial fraction expansion of the output transform that leads to the steady-state sinusoidal response. This can be done, as we will see in this section, without performing the entire partial fraction expansion.

EXAMPLE 20.26 As an example, the network function $V(s)/V_i(s)$ of the circuit of Fig. 20.8 is

$$\mathbf{H}(s) = \frac{s^2 + 7s + 15}{3(s^2 + 2s + 5)}$$

so if the circuit is driven by an input

$$v_i(t) = \sin t \quad \text{V}$$

with transform

$$\mathbf{V}_i(s) = \frac{1}{s^2 + 1}$$

the output transform is

$$\mathbf{V}(s) = \frac{s^2 + 7s + 15}{3(s^2 + 1)(s^2 + 2s + 5)}$$

$$= \frac{As + B}{s^2 + 1} + \mathbf{V}_1(s) \tag{20.47}$$

where $\mathbf{V}_1(s)$ is the rest of the partial fraction expansion. The transform $\mathbf{V}_1(s)$ contains the poles from the quadratic $s^2 + 2s + 5$, which are in the left half-plane. Therefore, its inverse transform will be a transient term. In other words, the steady-state component $v_{ss}(t)$ of the output $v(t)$ will come from the first term of the right member of (20.47).

To find v_{ss} we therefore need A and B. Multiplying (20.47) through by $s^2 + 1$ we have

$$(s^2 + 1)\mathbf{V}(s) = \frac{s^2 + 7s + 15}{3(s^2 + 2s + 5)} = As + B + (s^2 + 1)\mathbf{V}_1(s)$$

Now letting $s = j$, we have

$$\frac{-1 + j7 + 15}{3(-1 + j2 + 5)} = jA + B$$

or

$$jA + B = \frac{7}{6}$$

so that, equating real parts and equating imaginary parts, we have $A = 0$ and $B = 7/6$. The steady-state response is therefore

$$v_{ss} = \mathcal{L}^{-1}\left[\frac{7/6}{s^2 + 1}\right] = \frac{7}{6}\sin t \quad \text{V}$$

In the general case, suppose the network is absolutely stable with network function $\mathbf{H}(s)$ and input $v_i = \cos \omega_0 t$. The output transform $\mathbf{F}(s)$ is then

$$\mathbf{F}(s) = \frac{s\mathbf{H}(s)}{s^2 + \omega_0^2} \tag{20.48}$$

where the factor $s^2 + \omega_0^2$ is not repeated in the denominator of $\mathbf{H}(s)$. (Otherwise the circuit is not absolutely stable.) The partial fraction expansion is of the form

$$\mathbf{F}(s) = \frac{s\mathbf{H}(s)}{s^2 + \omega_0^2} = \frac{As + B}{s^2 + \omega_0^2} + \mathbf{F}_1(s)$$

$$= \frac{K}{s - j\omega_0} + \frac{K^*}{s + j\omega_0} + \mathbf{F}_1(s) \tag{20.49}$$

where we have further expanded the sinusoidal steady-state term into two terms. The constant K is given by

$$K = \lim_{s \to j\omega_0} (s - j\omega_0)\mathbf{F}(s) = \frac{s\mathbf{H}(s)}{s + j\omega_0}\bigg|_{s=j\omega_0}$$

$$= \frac{\mathbf{H}(j\omega_0)}{2} = \frac{1}{2}|\mathbf{H}(j\omega_0)| \underline{/\phi_0} \tag{20.50}$$

where $\phi_0 = \phi(\omega_0)$ is the phase of $\mathbf{H}(j\omega_0)$.

Since K^* is the conjugate of K, (20.49) inverts to

$$f(t) = 2 \operatorname{Re} K e^{j\omega_0 t} + f_1(t)$$

where $f(t)$ is the output, and the first term on the right, denoted by

$$f_{ss}(t) = 2 \operatorname{Re} K e^{j\omega_0 t}$$

is the steady-state component of the output. By (20.50) it may be written

$$f_{ss}(t) = 2 \operatorname{Re}\left[\frac{1}{2}|\mathbf{H}(j\omega_0)|e^{j(\omega_0 t + \phi_0)}\right]$$

or

$$f_{ss}(t) = |\mathbf{H}(j\omega_0)| \cos (\omega_0 t + \phi_0) \tag{20.51}$$

This is exactly the same as the steady-state sinusoidal response obtained earlier using phasors.

If the network is conditionally stable and has other $j\omega$-axis poles different from $\pm j\omega_0$, the steady-state components due respectively to these poles are obtained in exactly the same way as (20.51).

If the input function has a phase angle θ_0, that is,

$$v_i = V_m \cos (\omega_0 t + \theta_0) \tag{20.52}$$

then it may be shown that (20.51) becomes

$$f_{ss}(t) = V_m |\mathbf{H}(j\omega_0)| \cos (\omega_0 t + \theta_0 + \phi_0) \tag{20.53}$$

The details are left to Exs. 20.7.1 and 20.7.2.

EXERCISES

20.7.1 Show that

$$\mathscr{L}[\cos (kt + \theta)] = \frac{s \cos \theta - k \sin \theta}{s^2 + k^2}$$

(*Suggestion:* Expand the cosine of the sum of two angles and then take the transform.)

20.7.2 Derive (20.53) using the result of Ex. 20.7.1.

20.7.3 Find the steady-state response to $v_i = 8 \cos (2t + 30°)$ if the network function is

$$\mathbf{H}(s) = \frac{4}{s^2 + 2s + 8}$$

Answer $4\sqrt{2} \cos (2t - 15°)$

20.8

BODE PLOTS

The magnitude and phase responses of the network functions obtained by Laplace transform methods are exactly like those of Chapter 15 obtained using phasors. This is because the network functions in the two cases are exactly the same, as we have seen. Thus there is no need for us to repeat the frequency responses here. Instead we will present a shortcut method of obtaining special cases of these responses known as *Bode plots,* named for the American engineer Hendrik W. Bode (1905–   ), who was a pioneer in the development of feedback systems. Bode plots are sketches of magnitude, expressed in decibels (dB), and phase versus the logarithm of frequency.

The factored network function is of the form of (15.30), which we write as

$$\mathbf{H}(s) = \frac{K_1(s + z_1)(s + z_2) \cdots (s + z_m)}{(s + p_1)(s + p_2) \cdots (s + p_n)}$$

where the z's and p's are the negatives of the zeros and poles. (Since we will be dealing with stable functions whose poles are in the left half-plane, we have used $-z_i$ and $-p_i$ for the zeros and poles.) For the Bode plot it is convenient to factor the product of the z's out of the numerator and the product of the p's out of the denominator, resulting in

$$\mathbf{H}(s) = \frac{K(1 + s/z_1)(1 + s/z_2) \cdots (1 + s/z_m)}{(1 + s/p_1)(1 + s/p_2) \cdots (1 + s/p_n)} \tag{20.54}$$

The new constant K contains K_1 and these products.

There may be repeated factors, of course, as well as factors like s^r (which we leave intact) and quadratic factors like $s^2 + (\omega_k/Q_k)s + \omega_k^2$ of (15.25), which becomes

$$M_k = 1 + \frac{1}{\omega_k Q_k} s + \frac{s^2}{\omega_k^2} \tag{20.55}$$

when ω_k^2 is factored out. If the poles (or zeros) are real, we will have only linear factors like those of (20.54) and factors like s^r. Irreducible quadratic factors like (20.55), corresponding to the product of two linear factors, will arise, however, when the poles (or zeros) are complex. We will consider only linear or linear repeated factors at first, and examine quadratic factors in the next section. Therefore, the factors of (20.54) in this section will be of the form

$$M_1 = s^r \tag{20.56}$$

which we will consider separately, and

$$M_2 = 1 + \frac{s}{a_k} \tag{20.57}$$

which may be repeated.

For $s = j\omega$, the network function becomes

$$\mathbf{H}(j\omega) = |\mathbf{H}(j\omega)| \,/\underline{\phi(\omega)} \tag{20.58}$$

where, for factors like (20.57), the magnitude is

$$|\mathbf{H}(j\omega)| = \frac{|K| \cdot |1 + j\omega/z_1| \cdot |1 + j\omega/z_2| \cdots}{|1 + j\omega/p_1| \cdot |1 + j\omega/p_2| \cdots} \tag{20.59}$$

and the phase is

$$\phi(\omega) = \text{phase of } K + \tan^{-1}\frac{\omega}{z_1} + \tan^{-1}\frac{\omega}{z_2} + \cdots$$
$$- \tan^{-1}\frac{\omega}{p_1} - \tan^{-1}\frac{\omega}{p_2} - \cdots \tag{20.60}$$

The phase of K is zero if $K > 0$ and $180°$ if $K < 0$.

We will consider the magnitude response first. Expressing the magnitude in decibels, as defined in Sec. 15.7, we have, from (20.59),

$$A_{dB} = 20 \log |\mathbf{H}(j\omega)| = 20 \log |K| + 20 \log \left| 1 + \frac{j\omega}{z_1} \right| + \cdots$$

$$- 20 \log \left| 1 + \frac{j\omega}{p_1} \right| - \cdots \tag{20.61}$$

Since the Bode plot is the plot of $|\mathbf{H}(j\omega)|$ in decibels (the log magnitude) versus $\log \omega$, we see that it is obtained by finding each term in (20.61) and adding and subtracting them graphically. All the terms except $\log |K|$ will come from one or the other of (20.56) and (20.57).

The logarithmic plot also has the virtue that plots of $\log (k\omega)^n$ are straight lines. That is,

$$y = 20 \log (k\omega)^n = 20n(\log \omega + \log k)$$

$$= 20nx + 20n \log k \tag{20.62}$$

which is the equation of a straight line in terms of $x = \log \omega$. To determine the slope and its units, let us consider two points on the line,

$$y_1 = 20n \log \omega_1 + 20n \log k$$

$$y_2 = 20n \log \omega_2 + 20n \log k$$

for which

$$y_2 - y_1 = 20n(\log \omega_2 - \log \omega_1)$$

$$= 20n \log \frac{\omega_2}{\omega_1} \tag{20.63}$$

If $\omega_2 = 2\omega_1$, the frequency interval between ω_1 and ω_2 is defined to be an *octave*. In this case,

$$y_2 - y_1 = 20n \log 2 \approx 6n$$

The change in y is therefore $6n$ dB, while the change in ω is one octave. Thus the slope is $6n$ dB/octave.

If $\omega_2 = 10\omega_1$, the frequency interval between ω_1 and ω_2 is a *decade*. In this case, (20.63) becomes

$$y_2 - y_1 = 20n \log 10 = 20n$$

so that the slope is $20n$ dB/decade. This is, of course, equivalent to $6n$ dB/octave.

The Bode plots are easy to draw if we use *semilog* graph paper, for which the vertical axis is linear, with equal distances separating 1 dB, 2 dB, and so on, and the horizontal axis is logarithmic, with equal distances separating $\log \omega_1$, $\log 10\omega_1$, $\log 100\omega_1$, and so on. For simplicity, we may label these equal distances ω_1, $10\omega_1$, $100\omega_1$, and so on, so that each is of length one decade. Thus we will actually have decibels plotted versus ω, although the equal horizontal distances are proportional to $\log \omega$. In the case of octaves, the equal distances are labeled ω_1, $2\omega_1$, $4\omega_1$, and so on, in which case the length of each interval is an octave.

EXAMPLE 20.27 As an example, let us plot the line of (20.62). If $k\omega = 1$, or $\omega = 1/k$, then $y = 0$. This point, together with the slope of $20n$ dB/decade, enables us to sketch the line in Fig. 20.19(a). The same line is shown in Fig. 20.19(b), where the horizontal axis is marked in octaves, and the slope is $6n$ dB/octave. We note that in each graph, the origin, $\omega = 0$, is not shown, since log 0 is not defined. We may extend the plot as far toward $\omega = 0$ as we wish by plotting points for $0.1/k$, $0.01/k$, and so on, or in the case of octaves, $k/2$, $k/4$, and so on.

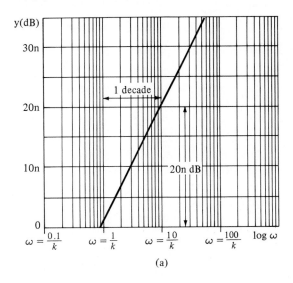

(a)

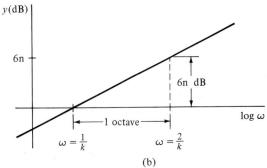

(b)

FIGURE 20.19 Plot of a straight line using logarithmic coordinates in (a) decades and (b) octaves

Let us now consider the magnitude and phase responses for each of the terms in $|\mathbf{H}(j\omega)|$ and $\phi(\omega)$. The term $20 \log |K|$ in (20.61) is a constant, positive if $|K| > 1$ and negative if $|K| < 1$. As we have noted, the phase of K is $0°$ if $K > 0$ and $180°$ if $K < 0$. Thus their plots are horizontal lines.

If there is a term s^r, it can be combined with the constant K either as $(\sqrt[r]{|K|}s)^r$, if s^r is in the numerator, or as $(s/\sqrt[r]{|K|})^r$, if s^r is in the denominator of $\mathbf{H}(s)$. (In the latter case, $r = 1$, or else the circuit is unstable.) In either case, the magnitude graph can be obtained from the graph of (20.62). If s^r is in the numerator, the magnitude graph is like that of Fig. 20.19. If s^r is in the denominator, the magnitude graph will be the negative of those of Fig. 20.19. The phase will be that of $(j\omega)^r$, which is $90r$ degrees for a numerator term and $-90r$ degrees for a denominator term.

The graph of the typical linear term in (20.61),

$$M = 20 \log \left| 1 + \frac{j\omega}{k} \right|$$

will be approximated by straight line segments. That is, for $\omega \ll k$ (*much* less than k),

$$M \approx 20 \log 1 = 0$$

and for $\omega \gg k$ (*much* greater than k),

$$M \approx 20 \log \frac{\omega}{k} = 20 \log \omega - 20 \log k$$

The first is a horizontal line coinciding with the horizontal axis, and the second is a line with slope 20 dB/decade. The *break point,* or *corner frequency,* is the point $\omega = k$, where the two lines intersect. If the term corresponding to M is in the numerator, the sum in (20.61) will contain M, which is sketched in Fig. 20.20(a). If the term is in the denominator, $-M$ appears in (20.61), as shown in Fig. 20.20(b).

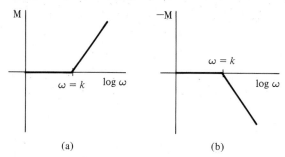

(a)   (b)

FIGURE 20.20 (a) Sketch of $| 1 + j\omega/k |$, and (b) of $1/| 1 + j\omega/k |$

The phase of the linear term $1 + j\omega/k$ is $\tan^{-1}(\omega/k)$ for a numerator term, and $-\tan^{-1}(\omega/k)$ for a denominator term. The phase may be plotted precisely as shown by the solid curve of Fig. 20.21, using the following values:

$$\phi = 5.7° \text{ at } \omega = \frac{k}{10}$$

$$= 26.6° \text{ at } \omega = \frac{k}{2}$$

$$= 45° \text{ at } \omega = k$$

$$= 63.4° \text{ at } \omega = 2k$$

$$= 84.3° \text{ at } \omega = 10k$$

These are for a numerator term. The signs are changed for a denominator term. An approximation, good enough in many cases, is the dashed-line plot of Fig. 20.21. It is $0°$ for ω less than one decade below the corner frequency k, $90°$ for ω greater than one decade above k, and a straight line connecting these, which passes through $45°$ at k.

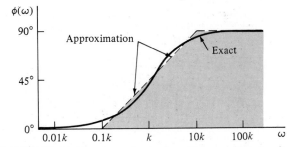

FIGURE 20.21 Phase responses for a linear numerator factor

EXAMPLE 20.28 As an example, let us find the Bode magnitude plot for the function

$$\mathbf{H}(s) = \frac{32(s+1)}{s(s+8)} = \frac{1+s}{(s/4)[1+(s/8)]}$$

The magnitude plots of each of the three factors are shown dashed (two ways) in Fig. 20.22. The plot for $1+s$ is labeled (1), that for $s/4$ (in the denominator) is labeled (2), and that for $1+s/8$ (in the denominator) is labeled (3). Their sum is the magnitude of $\mathbf{H}(j\omega)$ in decibels and is represented by the solid line of Fig. 20.22.

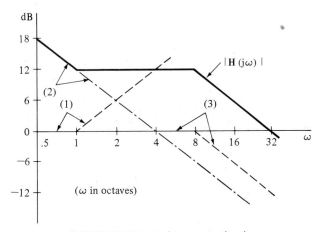

FIGURE 20.22 Bode magnitude plot

The phase response plots are shown in Fig. 20.23, with (1), (2), and (3) representing the same factors as in Fig. 20.22. Again, the phase response of $\mathbf{H}(j\omega)$ is their sum indicated by the solid line.

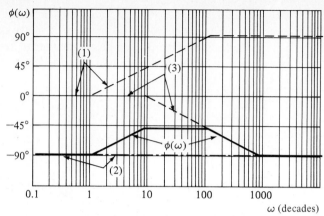

FIGURE 20.23 Bode phase plot

If we have a linear factor repeated, such as

$$M = 20 \log \left| 1 + \frac{j\omega}{k} \right|^r = 20r \log \left| 1 + \frac{j\omega}{k} \right|$$

the only change in the Bode magnitude plot is that the slope is $20r$ rather than 20 dB/decade, or $6r$ rather than 6 dB/octave. The phase will be r times the phase for the linear factor.

The straight line representations of the factors corresponding to $1 + j\omega/k$ are approximations, of course, and they may be adequate for most applications. They can be corrected, however, if necessary. The largest error occurs at the break point k, where for a numerator term the magnitude is 3 dB too small, as we will see. In general, the error is given by

$$
\begin{aligned}
\text{Error} &= 10 \log \left(1 + \frac{\omega^2}{k^2} \right), \qquad \omega \le k \\
&= 10 \log \left(1 + \frac{k^2}{\omega^2} \right), \qquad \omega > k
\end{aligned}
\tag{20.64}
$$

as the reader is asked to show in Ex. 20.8.1. If the factor is a numerator term, these amounts should be added to the plot, and if the factor is a denominator term, they should be subtracted. At $\omega = k$, both expressions yield 3 dB, as noted previously.

EXAMPLE 20.29 As an example, the approximate plot of $M = 20 \log |1 + j\omega/10|$ is shown dashed in Fig. 20.24. At the corner frequency $\omega = 10$, 3 dB has been added to produce the more accurate, corrected plot denoted by the solid line. If it is desired to correct the response at more points, we may show, using (20.64), that at $\omega = 2k$ or at $\omega = k/2$ (one octave above and below k) the error is approximately 0.97dB, and at $\omega = 10k$ or $0.1k$ (one decade above and below k) the error is approximately 0.04 dB. These values should be added to responses of numerator terms and subtracted from those of denominator terms.

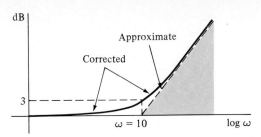

FIGURE 20.24 Magnitude plot corrected at the corner frequency

---

EXERCISES

20.8.1 Derive Eq. (20.64).

20.8.2 Find the equation of the approximate phase response of Fig. 20.21.

Answer $\phi(\omega) = 0, \qquad \omega < k/10$

$$= \frac{\pi}{4} \log \frac{10\omega}{k}, \qquad \frac{k}{10} \le \omega \le 10k$$

$$= \frac{\pi}{2}, \qquad \omega > 10k$$

20.8.3 (a) From the straight line approximations to the Bode magnitude plot of $\mathbf{H}(s) = [40(s + 4)]/(s + 16)$, find $|\mathbf{H}(j\omega)|$ in decibels at $\omega = 4$, 8, 16, and 20 rad/s. (b) Find the exact values at these points.

Answer (a) 20, 26, 32, 32; (b) 22.75, 26.02, 29.29, 30.06

20.9

QUADRATIC FACTORS

In this section we will consider the case of irreducible quadratic factors in the network function such as M_k given previously in (20.55). To facilitate the development, we will make the substitution

$$Q_k = \frac{1}{2\zeta_k} \tag{20.65}$$

so that (20.55) becomes

$$M_k = 1 + \frac{2\zeta_k}{\omega_k}s + \frac{s^2}{\omega_k^2} \tag{20.66}$$

The parameter ζ_k is called the *damping ratio*.

The motivation for the substitution (20.65) may be seen from Fig. 20.25, which shows the location of a typical pole or zero due to the irreducible factor. Setting $M_k = 0$ in (20.66) yields the pole or zero s_k given by

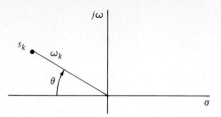

FIGURE 20.25 Location of a complex pole or zero

$$s_k = \omega_k[-\zeta_k + j\sqrt{1 - \zeta_k^2}] \tag{20.67}$$

and its conjugate. Since these are complex, we must have $0 < \zeta_k < 1$. Therefore, we may define

$$\zeta_k = \cos\theta$$

where θ is measured from the negative real axis, as shown in Fig. 20.25. Then (20.67) becomes

$$s_k = \omega_k(-\cos\theta + j\sin\theta)$$

Therefore s_k is as shown in Fig. 20.25, where $\cos^{-1}\zeta_k = \theta$ is its angle from the negative real axis, and ω_k is its magnitude. If $\zeta_k = 0$, we see by (20.67) that the pole is on the $j\omega$-axis, and if $\zeta_k = 1$, it is on the negative real axis. (We are not interested here in $\zeta_k \geq 1$, for in that case M_k is the product of two linear factors.)

For $s = j\omega$, the quadratic factor is given by

$$M_k = 1 + \frac{2\zeta_k}{\omega_k}j\omega - \frac{\omega^2}{\omega_k^2}$$

For $\omega \ll \omega_k$, we see that

$$|M_k| \approx 1$$

and for $\omega \gg \omega_k$,

$$|M_k| \approx \left| -\frac{\omega^2}{\omega_k^2} \right| = \frac{\omega^2}{\omega_k^2}$$

Therefore, for $\omega \ll \omega_k$,

$$20\log|M_k| \approx 0 \tag{20.68}$$

and for $\omega \gg \omega_k$,

$$20\log|M_k| \approx 20\log\frac{\omega^2}{\omega_k^2} = 40\log\frac{\omega}{\omega_k} \tag{20.69}$$

The asymptotic behavior (near zero and for large ω) is thus exactly like that of the linear factor except that the slope is 40 dB/decade instead of 20 dB/decade. If the quadratic factor is in the denominator, as it almost always is, the sign in (20.69) is changed, so the slope is -40 dB/decade.

The break point, as we see from equating (20.68) and (20.69), is $\omega = \omega_k$. This

Chapter 20 Laplace Transform Applications

is the same result as in the case of the linear factors, where we had $\omega_k = k$. Indeed, the straight line approximation for the quadratic factor looks exactly like that of the linear factor except for the slope of the high-frequency line. The error near the break point, however, can be quite large for the quadratic factor, especially if ζ_k is small. Graphs of the exact magnitudes of a denominator quadratic factor for various values of ζ_k (labeled as ζ) are shown in the top portion of Fig. 20.26. These may be used to make corrections at various points, such as the peak point. The lower portion of Fig. 20.26 is the phase response for various values of ζ_k (again labeled ζ). The abscissa is ω/ω_k, so that $\omega = \omega_k$ when $\omega/\omega_k = 1$.

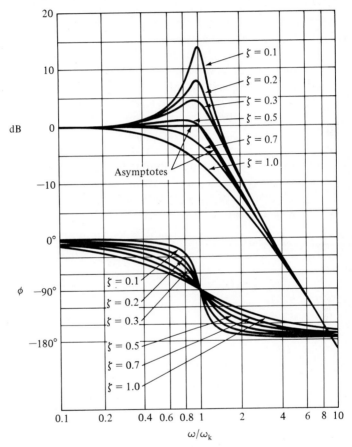

FIGURE 20.26 Log-magnitude and phase-angle curves of the irreducible quadratic factor (20.66) in the denominator

The straight line approximation to the magnitude response is labeled *asymptotes* in Fig. 20.26, and appears to be the dividing line between the cases with peaks and those without. From the exact magnitude response, it may be shown that the dividing line occurs for $\zeta_k = 1/\sqrt{2}$. That is, there is no peak in the curve for $1/\sqrt{2} \le \zeta_k < 1$, and for $0 < \zeta_k < 1/\sqrt{2}$ the peak occurs at

$$\omega = \omega_k \sqrt{1 - 2\zeta_k^2} = \omega_{\text{max}} \tag{20.70}$$

and the peak value is

$$M_{\text{max}} = \frac{1}{2\zeta_k \sqrt{1 - \zeta_k^2}} \tag{20.71}$$

The reader is asked to derive these results in Ex. 20.9.1.

From the exact expression for the phase, we may show that it varies from $0°$ to $-180°$ (for a denominator quadratic term), and it is $-90°$ at the corner frequency $\omega = \omega_k$. The graphs of Fig. 20.26 may be used to sketch the phase.

We are not allowing right half-plane poles in $\mathbf{H}(s)$ for reasons of stability. There are cases, however, such as in all-pass filter functions, where there are right half-plane zeros. The phase is said to be *minimum* phase if there are no right half-plane zeros in the network function, and it is *nonminimum* phase if there are right half-plane zeros. To see why the terms minimum and nonminimum are applicable to the phase, let us consider Fig. 20.27, which shows a left half-plane zero z_1 and its mirror image, a right half-plane zero z_2. Using the graphical method of Sec. 15.5, we see that each zero contributes the same magnitude $M_1 = M_2$ to the magnitude response, as shown by the arrows drawn to $j\omega$. However, z_1 contributes α_1 to the phase, whereas the right half-plane zero z_2 contributes the much larger α_2. Thus the first is minimum and the second is not.

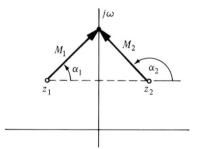

FIGURE 20.27 Vectors representing right half-plane and left half-plane zeros

In the case of a right half-plane zero, a factor like $1 - j\omega/k$ will be present in the numerator, which will contribute a term

$$\phi_1 = -\tan^{-1} \frac{\omega}{k}$$

to the phase $\phi(\omega)$. This term has the same sign as those contributed by the denominator terms, so the net effect is to increase the absolute value of the phase.

EXERCISES

20.9.1 Show that the magnitude response of a quadratic factor exhibits peaks if $0 < \zeta_k < 1/\sqrt{2}$, at the frequency given by (20.70) with peak value given by (20.71).

20.9.2 Given

$$F(s) = \frac{1}{s^2 + 2\zeta_k \omega_k s + \omega_k^2}$$

Find $f(t)$ and demonstrate that if $0 < \zeta_k < 1$, then the term *damping ratio* is appropriate for $\zeta_k = \dfrac{1}{2Q_k}$, since increasing ζ_k increases the damping.

Answer $f(t) = \dfrac{1}{b} e^{-at} \sin bt$, where $a = \zeta_k \omega_k$ and $b = \omega_k \sqrt{1 - \zeta_k^2}$

20.9.3 Given

$$F(s) = \frac{64}{s^2 + 8s + 64}$$

Find ζ_k, ω_k, $\omega_{\max}$, and $M_{\max}$. Compare the value of $20 \log M_{\max}$ with that obtained from Fig. 20.26.

Answer $\zeta_k = 0.5$, $\omega_k = 8$, $\omega_{\max} = 4\sqrt{2}$, $M_{\max} = 2/\sqrt{3}$

PROBLEMS

20.1 Use the describing equation and Laplace transforms to find v for $t > 0$ if (a) $i_g = 2u(t)$ A, and (b) $i_g = 2e^{-t}u(t)$ A.

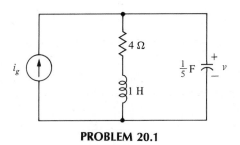

PROBLEM 20.1

20.2 Use the describing equation and Laplace transforms to find i for $t > 0$ if the circuit is in steady state at $t = 0^-$.

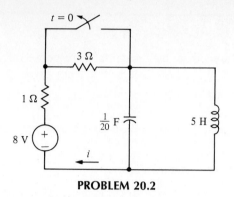

PROBLEM 20.2

20.3 Use the describing equation and Laplace transforms to find v for $t > 0$ if $i_1(0) = -1$ A and $i_2(0) = 0$.

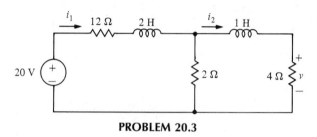

PROBLEM 20.3

20.4 Use the describing equation and Laplace transforms to find i for $t > 0$ if $i(0) = 2$ A and $v(0) = 6$ V.

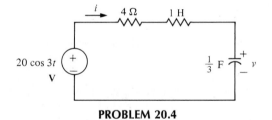

PROBLEM 20.4

20.5 Solve Prob. 9.7 using the transformed circuit.

20.6 Solve Prob. 9.9 using the transformed circuit.

20.7 Solve Prob. 9.10 using the transformed circuit.

20.8 Solve Prob. 9.14 using the transformed circuit.

20.9 Solve Prob. 9.18 using the transformed circuit.

20.10 Solve Prob. 9.19 using the transformed circuit.

20.11 Solve Prob. 9.20 using the transformed circuit.

20.12 Solve Prob. 9.21 using the transformed circuit.

20.13 Solve Prob. 9.24 using the transformed circuit.

20.14 Find i_2 in the circuit of Fig. 16.11 using the transformed circuit, if $M = 1/\sqrt{2}$ H.

20.15 Solve Prob. 16.12 using the transformed circuit.

Chapter 20   Laplace Transform Applications

20.16 Show by means of the convolution theorem that

$$\mathscr{L}^{-1}\left[\frac{s}{(s^2 + a^2)^2}\right] = \frac{t}{2a}\sin at\, u(t)$$

Use this result to find i for $t > 0$ in the initially relaxed circuit.

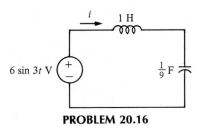

PROBLEM 20.16

20.17 If $x(t)$ is the input and $y(t)$ is the output, find the step response $r(t)$ and the impulse response $h(t)$ for the following:
(a) $y'' + 6y' + 5y = 20x$.
(b) $y'' + 4y' + 13y = 13x$.

20.18 Find i, for $t > 0$, using the transformed circuit, if $v(0) = 4$ V and $i(0) = 2$ A.

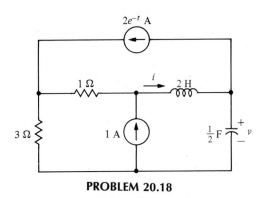

PROBLEM 20.18

20.19 If a network has an impulse response

$$h(t) = te^{-2t}u(t)$$

find the forced response to an input

$$f(t) = e^{-2t}\cos t\, u(t)$$

20.20 Solve for x and y, valid for $t > 0$:

$$2x' + 4x + y' + 7y = 0$$

$$x' + x + y' + 3y = \delta(t)$$

20.21 Find the network function and the impulse response if the output is $i(t)$.

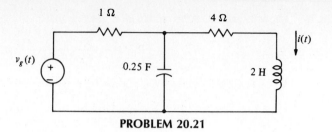

PROBLEM 20.21

20.22   Find the response $i(t)$ in Prob. 20.21 if the circuit is initially relaxed and $v_g(t) = u(t) - u(t - 1)$ V.

20.23   Find the response of a circuit with the network function

$$\mathbf{H}(s) = \frac{1}{s + 1}$$

and no initial stored energy to the input $f(t)$ of Prob. 19.23(a). *Suggestion:* Find the inverse transform involved by the procedure of Prob. 19.24, but note that if $p(t) \neq 0$, for $t > T$, then the inverse is not periodic but is

$$p(t) + p(t - T) + p(t - 2T) + \cdots$$

20.24   With the presence of impulse currents or voltages in a circuit it is possible, theoretically, to instantaneously change inductor currents and capacitor voltages. Demonstrate this by finding $v(t)$ and $i(t)$ for $-\infty < t < \infty$ in the given circuit. The switch is closed at $t = 0$ and $v(0^-) = 0$.

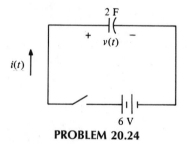

PROBLEM 20.24

20.25   The impulse response of a circuit is

$$h(t) = \sqrt{2}\, e^{-t/\sqrt{2}} \sin \frac{t}{\sqrt{2}}\, u(t)$$

Find the network function and the amplitude response, and thus show that the circuit is a second-order Butterworth low-pass filter with $\omega_c = 1$ rad/s.

20.26   Find the step response for the circuit of Prob. 9.22 using a transformed circuit.

20.27   Replace the 12-V source in Prob. 9.14 by $v_g(t)$, and use transformed circuits to find $\mathbf{H}(s) = \mathbf{I}(s)/\mathbf{V}_g(s)$. Also, in each case find $h(t)$ and $r(t)$.

20.28   Find $\mathbf{H}(s) = \mathbf{V}_o(s)/\mathbf{V}_i(s)$, $h(t)$, and $r(t)$ in Problem 20.28 on next page.

20.29   Find i for $t > 0$ in Prob. 9.10(b) by replacing everything to the left of resistor R by its transformed Thevenin equivalent circuit.

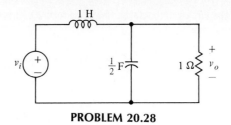

PROBLEM 20.28

20.30 In the transformed circuit for Prob. 9.18, replace everything except the $\frac{1}{2}$-F capacitor by its Thevenin equivalent circuit and find v.

20.31 Show that for the circuit described by

$$\mathbf{H}(s) = \frac{a}{(s^2 + 2as + 1)(s + a)}$$

where $0 < a < 1$, the step response is nondecreasing.

20.32 In Prob. 9.14, replace the 12-V source by v_g and obtain the network function $\mathbf{H}(s) = \mathbf{I}(s)/\mathbf{V}_g(s)$ for a general R and μ. If $R > 0$, find μ in terms of R so that the circuit is (a) absolutely stable, (b) conditionally stable, and (c) unstable.

20.33 If the network function of the circuit in the dashed box is $-\mathbf{G}(s) = \mathbf{V}_2(s)/[-\mathbf{V}_3(s)]$, show that

$$\mathbf{H}(s) = \frac{\mathbf{V}_2(s)}{\mathbf{V}_1(s)} = \frac{\mathbf{G}(s)}{1 + \mathbf{G}(s)}$$

(This is true as long as v_2 is the output of an op amp.) Show also that

$$\mathbf{G}(s) = \frac{1/R_1 R_2}{s(s + 1/R_1)}$$

and therefore

$$\mathbf{H}(s) = \frac{1/R_1 R_2}{s^2 + (1/R_1)s + (1/R_1 R_2)}$$

Finally, find R_1 and R_2 so that the poles of $\mathbf{H}(s)$ are at $s = -1 \pm j1$.

PROBLEM 20.33

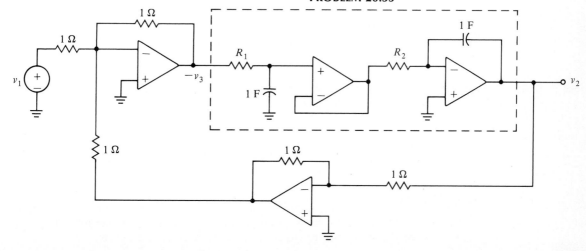

20.34 If the circuit in the dashed box of Prob. 20.33 is replaced by the circuit shown, find $-\mathbf{G}(s)$ and show that

$$\mathbf{H}(s) = \frac{1/C}{s^3 + 4s^2 + 3s + (1/C)}$$

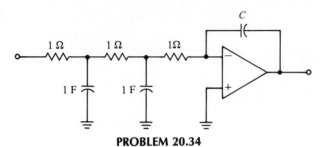

PROBLEM 20.34

20.35 A test that can be used to determine if a polynomial

$$Q(s) = a_0 s^n + a_1 s^{n-1} + \cdots + a_{n-1} s + a_n, \qquad a_0 > 0$$

has right half-plane zeros is *Routh's algorithm*, for which we construct a *Routh array*:

$$
\begin{array}{c|cccc}
s^n & a_0 & a_2 & a_4 & \cdots \\
s^{n-1} & a_1 & a_3 & a_5 & \cdots \\
s^{n-2} & b_1 & b_2 & b_3 & \cdots \\
s^{n-3} & c_1 & c_2 & c_3 & \cdots \\
s^{n-4} & d_1 & d_2 & d_3 & \cdots \\
s^{n-5} & e_1 & e_2 & e_3 & \cdots \\
\vdots & \vdots & \vdots & \vdots & \cdots \\
s^2 & f_1 & f_2 & & \\
s^1 & g_1 & & & \\
s^0 & h_1 & & &
\end{array}
$$

The first two rows are respectively the coefficients of s_n, s^{n-2}, s^{n-4}, . . . , and of s^{n-1}, s^{n-3}, s^{n-5}, ... The other row entries are given, for $i = 1, 2, 3, \ldots$, by

$$b_i = -\frac{\begin{vmatrix} a_0 & a_{2i} \\ a_1 & a_{2i+1} \end{vmatrix}}{a_1}, \qquad c_i = -\frac{\begin{vmatrix} a_1 & a_{2i+1} \\ b_1 & b_{i+1} \end{vmatrix}}{b_1}$$

$$d_i = -\frac{\begin{vmatrix} b_1 & b_{i+1} \\ c_1 & c_{i+1} \end{vmatrix}}{c_1}, \qquad e_i = -\frac{\begin{vmatrix} c_1 & c_{i+1} \\ d_1 & d_{i+1} \end{vmatrix}}{d_1}, \ldots$$

The number of right half-plane zeros is the number of sign changes in the first column of the Routh array. (We will not consider the case where a zero occurs in the first column.) In evaluating these determinants, a zero entry in the Routh array is left blank.

694 Chapter 20 Laplace Transform Applications

Show that for $Q(s) = s^3 + 3s^2 + s + 6$, the Routh array is

s^3	1	1
s^2	3	6
s	-1	
1	6	

and thus there are two right half-plane zeros.

20.36 Using the Routh algorithm of Prob. 20.35, determine the number of right half-plane zeros of the following polynomials:
(a) $s^3 + 6s^2 + 11s + 6$
(b) $s^3 + 2s^2 + 2s + 5$
(c) $s^6 + 3s^5 + 3s^4 + 4s^3 + 4s^2 + 4s + 2$
(d) $s^4 + 2s^3 + 3s^2 + 2s + 1$

20.37 Use the Routh algorithm to determine C in Prob. 20.34 so that the circuit is absolutely stable.

20.38 If $\mathbf{G}(s)$ for the circuit in the dashed box of Prob. 20.33 is

$$\mathbf{G}(s) = \frac{K}{s(s + 1)(s + 4)}$$

find K so that the overall circuit is absolutely stable.

20.39 Determine if the initial-value theorem is applicable and, if so, find $f(0^+)$ for the following values of $\mathbf{F}(s)$:

(a) $\dfrac{3(s + 2)}{(s + 1)(s + 3)}$ (b) $\dfrac{1}{s(s^2 + 1)}$ (c) $\dfrac{1}{s^2 + 2s + 5}$ (d) $\dfrac{s^2 + 4s + 7}{(s + 1)(s^2 + 2s + 5)}$

(e) $\dfrac{1}{(s + 1)(s + 2)^3}$ (f) $\dfrac{s + 2}{(s^2 + 2s + 2)(s + 1)^2}$

20.40 Use the final-value theorem to find the steady-state value of the step response of the circuit of Prob. 20.33.

20.41 Show that for the circuit of Prob. 20.33

$$\mathbf{G}(s) = \frac{\mathbf{H}(s)}{1 - \mathbf{H}(s)}$$

and

$$\mathbf{V}_3(s)/\mathbf{V}_1(s) = \frac{1}{1 + \mathbf{G}(s)}$$

For

$$\mathbf{H}(s) = \frac{K}{s^2 + as + b}$$

and $v_1(t) = u(t)$ V, find K so that the steady-state value of $v_3(t)$ is zero.

20.42 (a) If for the circuit of Prob. 20.33

$$\mathbf{H}(s) = \frac{K(1 + k_1 s)}{s^2 + as + b}$$

show that the steady-state value of $v_3(t)$ is zero if $K = b$ and $k_1 = a/b$, for both $v_1 = u(t)$ V and $v_1 = tu(t)$ V (the ramp function).

(b) Use the relation $v_2 = v_1 - v_3$ and find the steady-state response v_2 to the ramp input. Note that the final-value theorem cannot be used here because the limit of $s V_2(s)$ is infinite.

20.43 Note that

$$\mathbf{H}_1(s) = \frac{2(1 + s)}{s^2 + 2s + 2}$$

satisfies the requirement of Prob. 20.42 that the steady-state value of v_3 be zero for both the step and ramp inputs, but

$$\mathbf{H}(s) = \frac{2(1 + 2s)}{s^2 + 2s + 2}$$

does not. Use the idea of Prob. 20.42(b) to find the steady-state value of v_2 in the latter case if the input is the ramp.

20.44 Use the method of Sec. 20.7 to find the steady-state response i for the circuit of Ex. 14.6.3.

20.45 Use the method of Sec. 20.7 to find the steady-state response v for the circuit of Prob. 11.19.

20.46 Draw the straight line approximation to the Bode magnitude and phase plots for $\mathbf{H}(s)$ given by
(a) $\dfrac{10(1 + s)}{10 + s}$; (b) $\dfrac{10(1 - s)}{10 + s}$.
Obtain also the corrected magnitude response in (a).

20.47 Draw the straight line approximation to the Bode magnitude and phase plots for $\mathbf{H}(s)$ given by
(a) $\dfrac{10(s - 1)}{10 + s}$; (b) $\dfrac{100(1 + s)}{(10 + s)^2}$.

20.48 Draw the straight line approximation to the Bode magnitude and phase plots for

$$\mathbf{H}(s) = \frac{40(1 + s)}{(4 + s)(10 + s)}$$

Estimate the value of ω at which $|\mathbf{H}(j\omega)| = 1$.

20.49 Draw the straight line approximation to the Bode magnitude and phase plots for

$$\mathbf{H}(s) = \frac{100s}{s^2 + 10s + 100}$$

Correct the magnitude response at the corners.

20.50 Draw the straight line approximation to the Bode magnitude plot for

$$\mathbf{H}(s) = \frac{512(s + 1)}{s(s + 2)(s^2 + 8s + 64)}$$

Correct the magnitude response at the corners.

Determinants and Cramer's Rule

The solution of simultaneous equations, such as those often encountered in circuit theory, may be obtained relatively easily, in many cases, by the use of *determinants*. We define a determinant as a square array of numbers having a numerical value, such as

$$\Delta = \begin{vmatrix} a_{11} & a_{12} \\ a_{21} & a_{22} \end{vmatrix} \tag{A.1}$$

In this case the determinant is a 2×2 array, with two *rows* and two *columns* and a value Δ defined to be

$$\Delta = a_{11}a_{22} - a_{12}a_{21} \tag{A.2}$$

In the second-order, or 2×2, case of (A.1), the method of obtaining Δ in (A.2) may be thought of as a *diagonal rule*. That is,

$$\Delta = \begin{vmatrix} a_{11} & a_{12} \\ a_{21} & a_{22} \end{vmatrix} = a_{11}a_{22} - a_{12}a_{21} \tag{A.3}$$

or Δ is a difference of the product $a_{11}a_{22}$ of elements down the diagonal to the right and the product $a_{12}a_{21}$ of elements down the diagonal to the left.

As an example, consider

$$\Delta = \begin{vmatrix} 1 & 2 \\ -3 & 4 \end{vmatrix}$$

which is given by

$$\Delta = (1)(4) - (2)(-3) = 10$$

A third-order, or 3×3, determinant, such as

$$\Delta = \begin{vmatrix} a_{11} & a_{12} & a_{13} \\ a_{21} & a_{22} & a_{23} \\ a_{31} & a_{32} & a_{33} \end{vmatrix} \tag{A.4}$$

has three rows and three columns. It may also be evaluated by a diagonal rule, given by

$$\Delta = \begin{vmatrix} a_{11} & a_{12} & a_{13} \\ a_{21} & a_{22} & a_{23} \\ a_{31} & a_{32} & a_{33} \end{vmatrix}$$

$$= (a_{11}a_{22}a_{33} + a_{12}a_{23}a_{31} + a_{13}a_{32}a_{21})$$

$$- (a_{13}a_{22}a_{31} + a_{23}a_{32}a_{11} + a_{33}a_{21}a_{12})$$

(A.5)

Thus the value of the determinant is a difference of products of elements down the diagonals to the right and products of elements down the diagonals to the left.

An example of a third-order determinant and its evaluation is given by

$$\Delta = \begin{vmatrix} 1 & 1 & 1 \\ 2 & -1 & 1 \\ -1 & 1 & 2 \end{vmatrix}$$

$$= [(1)(-1)(2) + (1)(1)(-1) + (1)(1)(2)]$$

$$- [(1)(-1)(-1) + (1)(1)(1) + (2)(2)(1)]$$

$$= -7$$

(A.6)

A general definition of determinants may be given and used to derive a number of evaluation procedures. This is the technique usually given in elementary algebra books. However, for our purposes we shall use the diagonal rules that we have considered for second- and third-order determinants and evaluate higher-order determinants by the method of expansion by *minors*, or *cofactors*.

The *minor* A_{ij} of the element a_{ij} in the ith row and the jth column of a determinant is the determinant left after the ith row and the jth column are removed. For example, in (A.6) the minor A_{21} of the element $a_{21} = 2$ (second row, first column) is

$$A_{21} = \begin{vmatrix} 1 & 1 \\ 1 & 2 \end{vmatrix} = 2 - 1 = 1$$

The *cofactor* C_{ij} of the element a_{ij} is given by

$$C_{ij} = (-1)^{i+j} A_{ij}$$

(A.7)

In other words, the cofactor is the *signed* minor, the minor multiplied by ± 1 with the sign depending on whether the sum of the row number and column number is even or odd.

The value of a determinant is the sum of products of the elements in any row or column and their cofactors. For example, let us *expand* the determinant of (A.4) by cofactors of the first row. The result is

$$\Delta = a_{11} C_{11} + a_{12} C_{12} + a_{13} C_{13}$$

or, by (A.7),

$$\Delta = a_{11}(-1)^{1+1} A_{11} + a_{12}(-1)^{1+2} A_{12} + a_{13}(-1)^{1+3} A_{13}$$

$$= a_{11} A_{11} - a_{12} A_{12} + a_{13} A_{13}$$

Writing out the minors explicitly, we have

$$\Delta = a_{11}\begin{vmatrix} a_{22} & a_{23} \\ a_{32} & a_{33} \end{vmatrix} - a_{12}\begin{vmatrix} a_{21} & a_{23} \\ a_{31} & a_{33} \end{vmatrix} + a_{13}\begin{vmatrix} a_{21} & a_{22} \\ a_{31} & a_{32} \end{vmatrix}$$

By the diagonal rule this may be written

$$\Delta = a_{11}(a_{22}a_{33} - a_{23}a_{32}) - a_{12}(a_{21}a_{33} - a_{23}a_{31}) + a_{13}(a_{21}a_{32} - a_{22}a_{31})$$

which may be simplified to (A.5).

To illustrate expansion by minors, let us evaluate the determinant of (A.6) by applying the technique to the third column. We have

$$\Delta = 1(-1)^{1+3}\begin{vmatrix} 2 & -1 \\ -1 & 1 \end{vmatrix} + 1(-1)^{2+3}\begin{vmatrix} 1 & 1 \\ -1 & 1 \end{vmatrix} + 2(-1)^{3+3}\begin{vmatrix} 1 & 1 \\ 2 & -1 \end{vmatrix}$$

$$= (2 - 1) - (1 + 1) + 2(-1 - 2) = -7$$

A method of obtaining solutions of simultaneous equations by determinants, referred to earlier, is called *Cramer's rule*. We shall illustrate the method by applying it to two systems of equations, a second-order and a third-order system. The examples will be given in such a way that the generalization to higher-order systems should be evident. The second-order system that we shall consider is the set of equations

$$x_1 - 2x_2 = 5$$
$$6x_1 + x_2 = 4 \tag{A.8}$$

We define the determinant of the system as the determinant Δ whose first column is the coefficients of x_1, whose second column is the coefficients of x_2, etc. In the case of the system of (A.8) we have

$$\Delta = \begin{vmatrix} 1 & -2 \\ 6 & 1 \end{vmatrix} = 13 \tag{A.9}$$

Cramer's rule states that

$$x_1 = \frac{\Delta_1}{\Delta}$$

$$x_2 = \frac{\Delta_2}{\Delta} \tag{A.10}$$

$$\vdots$$

where Δ_1 is Δ with its first column replaced by the constants in the right members of the system of equations, Δ_2 is Δ with the second column so replaced, etc. In the example of (A.8), the right members of the equations are 5 and 4 so that

$$\Delta_1 = \begin{vmatrix} 5 & -2 \\ 4 & 1 \end{vmatrix} = 13$$

$$\Delta_2 = \begin{vmatrix} 1 & 5 \\ 6 & 4 \end{vmatrix} = -26$$

By Cramer's rule the solution of (A.8) is given by

$$x_1 = \frac{\Delta_1}{\Delta} = \frac{13}{13} = 1$$

$$x_2 = \frac{\Delta_2}{\Delta} = \frac{-26}{13} = -2$$

As an example of a third-order system, let us consider the equations

$$
\begin{aligned}
x_1 + x_2 + x_3 &= 6 \\
2x_1 - x_2 + x_3 &= 3 \\
-x_1 + x_2 + 2x_3 &= 7
\end{aligned}
\qquad \text{(A.11)}
$$

The solution, by (A.10), is

$$
x_1 = \frac{\begin{vmatrix} 6 & 1 & 1 \\ 3 & -1 & 1 \\ 7 & 1 & 2 \end{vmatrix}}{\begin{vmatrix} 1 & 1 & 1 \\ 2 & -1 & 1 \\ -1 & 1 & 2 \end{vmatrix}} = \frac{-7}{-7} = 1
$$

$$
x_2 = \frac{\begin{vmatrix} 1 & 6 & 1 \\ 2 & 3 & 1 \\ -1 & 7 & 2 \end{vmatrix}}{-7} = \frac{-14}{-7} = 2
$$

and

$$
x_3 = \frac{\begin{vmatrix} 1 & 1 & 6 \\ 2 & -1 & 3 \\ -1 & 1 & 7 \end{vmatrix}}{-7} = \frac{-21}{-7} = 3
$$

Appendix B

Gaussian Elimination

Simultaneous equations, such as (A.8), may be solved by a method of successively eliminating one unknown at a time, in a certain systematic manner. One such procedure, known as *Gaussian elimination,* will be given here for the special case of (A.11), repeated as

$$x_1 + x_2 + x_3 = 6$$
$$2x_1 - x_2 + x_3 = 3 \qquad \text{(B.1)}$$
$$-x_1 + x_2 + 2x_3 = 7$$

It will be clear from the procedure how to generalize it to any system.

We may systematically eliminate x_1 from all the equations of (B.1) except the first equation by subtracting twice the first equation from the second and adding the first equation to the third. We shall then have the first equation and the two new equations, neither of which contains x_1. The result is

$$x_1 + x_2 + x_3 = 6$$
$$-3x_2 - x_3 = -9 \qquad \text{(B.2)}$$
$$2x_2 + 3x_3 = 13$$

Let us now divide the second equation by -3, resulting in

$$x_1 + x_2 + x_3 = 6$$
$$x_2 + \frac{1}{3}x_3 = 3 \qquad \text{(B.3)}$$
$$2x_2 + 3x_3 = 13$$

We next eliminate x_2 from all the equations except the first two. In the simple case of (B.3), this means eliminating x_2 from the third equation. This may be done by subtracting twice the second equation from the third equation, resulting in

$$x_1 + x_2 + \quad x_3 = 6$$

$$x_2 + \frac{1}{3}x_3 = 3 \tag{B.4}$$

$$\frac{7}{3}x_3 = 7$$

We now divide the third equation by $\frac{7}{3}$, yielding

$$x_1 + x_2 + \quad x_3 = 6$$

$$x_2 + \frac{1}{3}x_3 = 3 \tag{B.5}$$

$$x_3 = 3$$

In a higher-order system we would repeat the procedure. That is, eliminate x_3 from all the equations except the first three. Eventually we reach the point, as in (B.5), where the last unknown is found (x_3 in this case). We then *back-substitute* the last unknown into the next-to-last equation to find another unknown. These two answers are then back-substituted into the next equation to get another unknown, and so forth.

To illustrate the back-substitution process, from the second equation of (B.5) we have

$$x_2 = 3 - \frac{1}{3}(x_3)$$

$$= 3 - \frac{1}{3}(3) \tag{B.6}$$

$$= 2$$

The known values of x_2 and x_3 are then substituted into the first equation of (B.5), yielding

$$x_1 = 6 - x_2 - x_3$$

$$= 6 - 2 - 3 \tag{B.7}$$

$$= 1$$

In this example, this last step completes the process, since all the unknowns have been found.

The steps in the Gaussian procedure may be put in compact form by leaving out the symbols for the unknowns, the addition operation signs, and the equal signs. That is, (B.1) may be represented by

$$\begin{bmatrix} 1 & 1 & 1 & 6 \\ 2 & -1 & 1 & 3 \\ -1 & 1 & 2 & 7 \end{bmatrix} \tag{B.8}$$

Such an entity is sometimes called a *matrix*. It has rows and columns like a determinant, but it is not necessarily a square array (the example given is 3×4) and has no number associated with it. The matrix merely represents the equations. That is, its first row means one x_1 plus one x_2 plus one x_3 equals six, etc.

We may eliminate unknowns in (B.1) by manipulating rows in (B.8). Leaving the first row intact, subtracting twice the first row from the second, and adding the first row to the third, we have the new matrix

$$\begin{bmatrix} 1 & 1 & 1 & 6 \\ 0 & -3 & -1 & -9 \\ 0 & 2 & 3 & 13 \end{bmatrix} \qquad \text{(B.9)}$$

This represents the system of (B.2).

Dividing the second row of (B.9) by -3 yields

$$\begin{bmatrix} 1 & 1 & 1 & 6 \\ 0 & 1 & \frac{1}{3} & 3 \\ 0 & 2 & 3 & 13 \end{bmatrix} \qquad \text{(B.10)}$$

This represents (B.3).

Subtracting twice the second row of (B.10) from the third row gives

$$\begin{bmatrix} 1 & 1 & 1 & 6 \\ 0 & 1 & \frac{1}{3} & 3 \\ 0 & 0 & \frac{7}{3} & 7 \end{bmatrix} \qquad \text{(B.11)}$$

which represents (B.4). Dividing its last row by $\frac{7}{3}$ gives

$$\begin{bmatrix} 1 & 1 & 1 & 6 \\ 0 & 1 & \frac{1}{3} & 3 \\ 0 & 0 & 1 & 3 \end{bmatrix} \qquad \text{(B.12)}$$

The last result is equivalent to (B.5) and may be used, as in (B.6) and (B.7), to obtain the solution.

Alternatively, we may continue the elimination process as follows. In (B.12), subtract from the second row $\frac{1}{3}$ times the last row, and subtract the last row from the first row. This results in

$$\begin{bmatrix} 1 & 1 & 0 & 3 \\ 0 & 1 & 0 & 2 \\ 0 & 0 & 1 & 3 \end{bmatrix}$$

Finally, subtract the second row from the first, obtaining

$$\begin{bmatrix} 1 & 0 & 0 & 1 \\ 0 & 1 & 0 & 2 \\ 0 & 0 & 1 & 3 \end{bmatrix} \qquad \text{(B.13)}$$

The last result represents $x_1 = 1$, $x_2 = 2$, and $x_3 = 3$, which is the solution.

This procedure of continuing the elimination process to the form (B.13) is known as the Gauss-Jordan method.

Appendix C

Complex Numbers

From our earliest training in arithmetic we have dealt with *real* numbers, such as 3, -5, $\frac{4}{7}$, π, etc., which may be used to measure distances in one direction or another from a fixed point. A number such as x that satisfies

$$x^2 = -4 \tag{C.1}$$

is not a real number and is customarily, and unfortunately, called an *imaginary* number. To deal with imaginary numbers, an *imaginary unit*, denoted by j, is defined by

$$j = \sqrt{-1} \tag{C.2}$$

Thus we have $j^2 = -1$, $j^3 = -j$, $j^4 = 1$, etc. (We might note that mathematicians use the symbol i for the imaginary unit, but in electrical engineering this might be confused with current.) An imaginary number is defined as the product of j with a real number, such as $x = j2$. In this case $x^2 = (j2)^2 = -4$, and thus x is the solution of (C.1).

A *complex* number is the sum of a real number and an imaginary number, such as

$$A = a + jb \tag{C.3}$$

where a and b are real. The complex number A has a *real part, a,* and an *imaginary part, b,* which are sometimes expressed as

$$a = \text{Re } A$$

$$b = \text{Im } A$$

It is important to note that both parts are real, in spite of their names.

The complex number $a + jb$ may be represented on a rectangular coordinate plane, or a *complex plane,* by interpreting it as a point (a, b). That is, the horizontal coordinate is a and the vertical coordinate is b, as shown in Fig. C.1, for the case $4 + j3$. Because of this analogy with points plotted on a rectangular coordinate system, (C.3) is sometimes called the *rectangular form* of the complex number A.

The complex number $A = a + jb$ may also be uniquely located in the complex

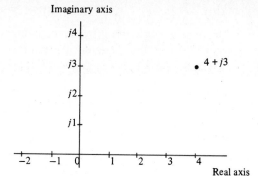

FIGURE C.1 Graphical representation of a complex number

plane by specifying its distance r along a straight line from the origin and the angle θ which this line makes with the real axis, as shown in Fig. C.2. From the right triangle thus formed, we see that

$$r = \sqrt{a^2 + b^2}$$
$$\theta = \tan^{-1} \frac{b}{a} \tag{C.4}$$

and that

$$a = r \cos \theta$$
$$b = r \sin \theta \tag{C.5}$$

We denote this representation of the complex number by

$$A = r\underline{/\theta} \tag{C.6}$$

which is called the *polar form*. The number r is called the *amplitude,* or *magnitude,* and is sometimes denoted by

$$r = |A|$$

The number θ is the *angle* or *argument* and is often denoted by

FIGURE C.2 Two forms of a complex number

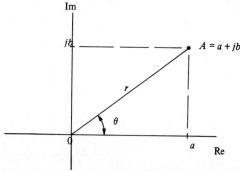

$$\theta = \text{ang } A = \text{arg } A$$

One may easily convert from rectangular to polar form, or vice versa, by means of (C.4) and (C.5). For example, the number A, shown in Fig. C.3, is given by

$$A = 4 + j3 = 5\underline{/36.9°}$$

since by (C.4)

$$r = \sqrt{4^2 + 3^2} = 5$$

$$\theta = \tan^{-1}\frac{3}{4} = 36.9°$$

The *conjugate* of the complex number $A = a + jb$ is defined to be

$$A^* = a - jb \tag{C.7}$$

That is, j is replaced by $-j$. Since we have

$$|A^*| = \sqrt{a^2 + (-b)^2} = \sqrt{a^2 + b^2} = |A|$$

and

$$\text{arg } A^* = \tan^{-1}\left(\frac{-b}{a}\right) = -\tan^{-1}\frac{b}{a} = -\text{arg } A$$

we may write, in polar form,

$$(r\underline{/\theta})^* = r\underline{/-\theta} \tag{C.8}$$

We may note from the definition that if A^* is the conjugate of A, then A is the conjugate of A^*. That is, $(A^*)^* = A$.

The operations addition, subtraction, multiplication, and division apply to complex numbers exactly as they do to real numbers. In the case of addition and subtraction, we may write, in general,

$$(a + jb) + (c + jd) = (a + c) + j(b + d) \tag{C.9}$$

and

$$(a + jb) - (c + jd) = (a - c) + j(b - d) \tag{C.10}$$

That is, to add (or subtract) two complex numbers, we simply add (or subtract) their real parts and their imaginary parts.

FIGURE C.3 Two forms of the complex number A

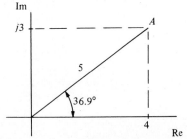

As an example, let $A = 3 + j4$ and $B = 4 - j1$. Then

$$A + B = (3 + 4) + j(4 - 1) = 7 + j3$$

This may also be done graphically, as shown in Fig. C.4(a), where the numbers A and B are represented as vectors from the origin. The result is equivalent to completing the parallelogram or to connecting the vectors A and B in head-to-tail manner, as shown in Fig. C.4(b), as the reader may check by comparing the numbers. For this reason, complex number addition is sometimes called vector addition.

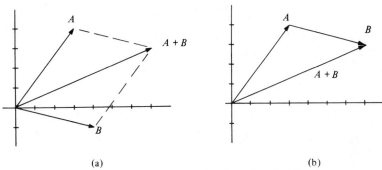

(a)   (b)

FIGURE C.4 Two methods of graphical addition

In the case of multiplication of numbers A and B given by

$$A = a + jb = r_1 \cos \theta_1 + jr_1 \sin \theta_1$$
$$B = c + jd = r_2 \cos \theta_2 + jr_2 \sin \theta_2 \tag{C.11}$$

we have

$$AB = (a + jb)(c + jd) = ac + jad + jbc + j^2 bd$$
$$= (ac - bd) + j(ad + bc) \tag{C.12}$$

Alternatively we have, from (C.11),

$$AB = (r_1 \cos \theta_1 + jr_1 \sin \theta_1)(r_2 \cos \theta_2 + jr_2 \sin \theta_2)$$
$$= r_1 r_2 [(\cos \theta_1 \cos \theta_2 - \sin \theta_1 \sin \theta_2) + j(\sin \theta_1 \cos \theta_2 + \cos \theta_1 \sin \theta_2)]$$
$$= r_1 r_2 [\cos (\theta_1 + \theta_2) + j \sin (\theta_1 + \theta_2)]$$

Therefore, in polar form we have

$$(r_1 \underline{/\theta_1})(r_2 \underline{/\theta_2}) = r_1 r_2 \underline{/\theta_1 + \theta_2} \tag{C.13}$$

and hence we may multiply two numbers by multiplying their magnitudes and adding their angles.

From this result we see that

$$AA^* = (r \underline{/\theta})(r \underline{/-\theta}) = r^2 \underline{/0} = |A|^2 \underline{/0}$$

Since $|A|^2 \underline{/0}$ is the real number $|A|^2$, we have

$$|A|^2 = AA^* \tag{C.14}$$

Division of a complex number by another, such as

$$N = \frac{A}{B} = \frac{a + jb}{c + jd}$$

results in an irrational denominator, since $j = \sqrt{-1}$. We may rationalize the denominator and display the real and imaginary parts of N by writing

$$N = \frac{AB^*}{BB^*} = \frac{a + jb}{c + jd} \cdot \frac{c - jd}{c - jd}$$

which is

$$N = \frac{(ac + bd) + j(bc - ad)}{c^2 + d^2} \tag{C.15}$$

We may also show by the method used to obtain (C.13) that

$$\frac{r_1 / \theta_1}{r_2 / \theta_2} = \frac{r_1}{r_2} / \theta_1 - \theta_2 \tag{C.16}$$

As examples, let $A = 4 + j3 = 5 / 36.9°$ and $B = 5 + j12 = 13 / 67.4°$. Then we have

$$AB = (5)(13) / 36.9° + 67.4° = 65 / 104.3°$$

and

$$\frac{A}{B} = \frac{5}{13} / 36.9° - 67.4° = 0.385 / -30.5°$$

Evidently, it is easier to add and subtract complex numbers in rectangular form and to multiply and divide them in polar form.

If two complex numbers are equal, then their real parts must be equal and their imaginary parts must be equal. That is, if

$$a + jb = c + jd$$

then

$$a - c = j(d - b)$$

which requires

$$a = c, \qquad b = d$$

Otherwise we would have a real number equal to an imaginary number, which, of course, is impossible. As an example, if

$$1 + x + j(8 - 2x) = 3 + jy$$

then

$$1 + x = 3$$
$$8 - 2x = y$$

or $x = 2$, $y = 4$.

Appendix D

Euler's Formula

To derive Euler's formula, an important result, let us begin with the quantity

$$g = \cos\theta + j\sin\theta \qquad (\text{D.1})$$

where θ is real and $j = \sqrt{-1}$. Differentiating, we have

$$\frac{dg}{d\theta} = j(\cos\theta + j\sin\theta) = jg$$

as may be seen from (D.1). The variables in the last equation may be separated, yielding

$$\frac{dg}{g} = j\,d\theta$$

Integrating, we have

$$\ln g = j\theta + K \qquad (\text{D.2})$$

where K is a constant of integration. From (D.1) we see that $g = 1$ when $\theta = 0$, which must also hold in (D.2). That is,

$$\ln 1 = 0 = 0 + K$$

or $K = 0$. Therefore, we have

$$\ln g = j\theta$$

or

$$g = e^{j\theta} \qquad (\text{D.3})$$

Comparing (D.1) and (D.3), we see that

$$e^{j\theta} = \cos\theta + j\sin\theta \qquad (\text{D.4})$$

which is known as *Euler's formula*. An alternative form found by replacing θ by $-\theta$

is

$$e^{-j\theta} = \cos(-\theta) + j\sin(-\theta)$$

which is, equivalently,

$$e^{-j\theta} = \cos\theta - j\sin\theta \qquad (D.5)$$

Evidently, (D.4) and (D.5) are conjugates.

Euler's formula provides us with a means of obtaining alternative forms of $\cos\theta$ and $\sin\theta$. For example, adding (D.4) and (D.5) and dividing the result by 2, we have

$$\cos\theta = \frac{e^{j\theta} + e^{-j\theta}}{2} \qquad (D.6)$$

Similarly, subtracting (D.5) from (D.4) and dividing the result by $2j$, we have

$$\sin\theta = \frac{e^{j\theta} - e^{-j\theta}}{2j} \qquad (D.7)$$

We may use Euler's formula also to clarify the polar representation

$$A = r\underline{/\theta} \qquad (D.8)$$

of the complex number

$$A = a + jb \qquad (D.9)$$

This was considered in Appendix C, where by (C.5) we had

$$a = r\cos\theta$$
$$b = r\sin\theta \qquad (D.10)$$

By Euler's formula we may write

$$re^{j\theta} = r(\cos\theta + j\sin\theta)$$
$$= r\cos\theta + jr\sin\theta$$

which, by (D.10), is

$$re^{j\theta} = a + jb \qquad (D.11)$$

Therefore, comparing (D.11) with (D.8) and (D.9), we have

$$A = r\underline{/\theta} = re^{j\theta} \qquad (D.12)$$

This result enables us to easily obtain the multiplication and division rules, given in (C.13) and (C.16). Clearly, if

$$A = r_1\underline{/\theta_1} = r_1 e^{j\theta_1}$$

and

$$B = r_2\underline{/\theta_2} = r_2 e^{j\theta_2}$$

then

$$AB = (r_1\underline{/\theta_1})(r_2\underline{/\theta_2})$$
$$= (r_1 e^{j\theta_1})(r_2 e^{j\theta_2})$$
$$= r_1 r_2 e^{j(\theta_1 + \theta_2)}$$
$$= r_1 r_2 \underline{/\theta_1 + \theta_2}$$

Similarly, we may obtain

$$\frac{A}{B} = \frac{r_1}{r_2}\underline{/\theta_1 - \theta_2}$$

Euler's formula is illustrated graphically in Fig. D.1. A unit vector is rotating around a circle in the direction shown, with an angular velocity of ω rad/s. Therefore in t seconds it has moved through an angle ωt as shown, and thus the vector may be specified by $1\underline{/\omega t}$ or $e^{j\omega t}$. Its real part is the projection on the horizontal axis, given by $\cos \omega t$, and its imaginary part is the projection on the vertical axis, given by $\sin \omega t$.

That is,

$$e^{j\omega t} = \cos \omega t + j \sin \omega t$$

which is Euler's formula. The projections trace out the cosine and sine waves, as shown, as the vector rotates.

FIGURE D.1 Graphical illustration of Euler's formula

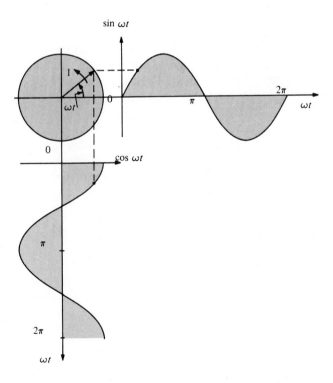

Appendix E

Computer Methods

SPICE, an acronym for Simulation Program with Integrated Circuit Emphasis, was developed in the Department of Electrical Engineering and Computer Science at the University of California at Berkeley. The program is very useful in solving simple or complex circuits and to this end is included to complement the learning and understanding of the rudiments of electric circuits. The SPICE family of circuit simulators performs dc, ac, and transient circuit analysis. The particular version chosen for illustration in this text is PSpice, from MicroSim Corporation of Laguna Hills, California, which uses the same algorithms and conforms to the input syntax of SPICE2. Although PSpice also provides a number of commands that perform operations and simulations not available in the Berkeley version of SPICE, we will restrict our discussion to the standard commands and statements that are pertinent to our study. Versions of PSpice are available for the popular personal computers such as the IBM PC family.

The general procedure for using SPICE is outlined in Section 4.8. The purpose of this appendix is to give a description of all command and data statements used in the various examples of the text. A SPICE input file or *circuit file* contains (1) title and comment statements, (2) data statements, (3) solution control statements, (4) output control statements, and (5) an end statement.

A

TITLE AND COMMENT STATEMENTS

A *title statement* is the first statement of the circuit file. It can contain any text, but it is restricted to one line. This statement is normally used to identify the circuit under investigation. An example might be

DC analysis of circuit 6.2

A *comment statement* is used for embedding descriptive statements for circuit definition or as programming aids within the circuit file. Comment lines are marked by * in the first column and may contain any text. An example is

*Data statements for Ex. 12.7.

B

DATA STATEMENTS

Data statements are used to define the elements of a circuit for the SPICE simulator. The first letter in a data statement is used to define the element type. Passive element descriptors are

R   resistor
C   capacitor

L   inductor
K   linear transformer

Independent voltage and current source descriptors are

I   current source

V   voltage source

Dependent (controlled) current and voltage source descriptors are

E   voltage-controlled
    voltage source (VCVS)
F   current-controlled
    current source (CCCS)

G   voltage-controlled
    current source (VCCS)
H   current-controlled
    voltage source (CCVS)

A subcircuit call (X) statement that can incorporate these elements into a useful subcircuit definition will be described later.

Figure E.1 shows the basic circuit connection between nodes m and n for a branch element. The reference node must be denoted as 0. The branch voltage $V(m, n)$ represents the voltage of node m with respect to node n. The voltage $V(m, 0)$ of node m with respect to the reference node can also be written as simply $V(m)$. Branch currents in SPICE are always defined as flowing from the more positive node through the element toward the more negative node. Each branch can contain any of the preceding elements, with current and voltage sources having a zero value, but resistors, capacitors, and inductors must have nonzero values.

The branches are interconnected at nodes identified by nonnegative integers that

FIGURE E.1 Branch element connected between nodes m and n

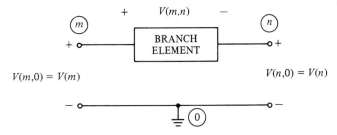

are not necessarily sequential, with the reference node being 0 (some simulators permit alphanumeric strings). Circuits cannot contain a loop of voltage sources, a *cut set* of current sources (a set whose removal will separate the circuit into two parts), or a node to which only one element is connected (dangling node). In addition, each node must have a dc path to the reference node.

Numbers associated with element values and responses in SPICE can be specified in the following formats:

Type	Examples
INTEGER	77, −56
FLOATING-POINT	7.54, −33.65
EXPONENT	100E-02, 1.775E3

Scale factors that are available for use with these numbers are

$F = 10^{-15}$	$U = 10^{-6}$	$MEG = 10^6$
$P = 10^{-12}$	$M = 10^{-3}$	$G = 10^9$
$N = 10^{-9}$	$K = 10^3$	$T = 10^{12}$

Thus, these values are all equivalent:

1.05E6	1.05MEG	1.05E3K	.00105G

Data statements for the various circuit elements take the following forms:

1. Resistor

$$R\langle NAME\rangle \ \langle(+) \ NODE\rangle \ \langle(-) \ NODE\rangle \ \langle VALUE\rangle$$

⟨NAME⟩ denotes any seven-character alphanumeric string for labeling the element.

(+) and (−) NODES define the polarity of the resistor connection. Positive current flows from (+) NODE through the resistor to (−) NODE.

⟨VALUE⟩ is the nonzero resistance value (positive or negative) in ohms.

Example E.1:   RLOAD 12 3 10K

denotes a 10-kΩ resistor labeled RLOAD connected between nodes 12 and 3 having current flowing from positive node 12 through the resistor to negative node 3.

2. Capacitor

$$C\langle NAME\rangle \ \langle(+) \ NODE\rangle \ \langle(-) \ NODE\rangle \ \langle VALUE\rangle \ [IC = \langle INITIAL \ VALUE\rangle]$$

⟨NAME⟩ denotes any seven-character alphanumeric string.

(+) and (−) NODES define the polarity of the capacitor connection. Positive current flows from (+) NODE through the capacitor to (−) NODE.

⟨VALUE⟩ is the nonzero value (positive or negative) in farads.

⟨INITIAL VALUE⟩ is optional and denotes the initial capacitor voltage at time $t = 0$ of (+) NODE with respect to (−) NODE for a transient response analysis.

Example E.2:   CEXT 2 3 10U
              CEXT 2 3 10U IC=4

denotes a 10-μF capacitor labeled EXT connected between nodes 2 and 3. If IC=4 is included, then an initial voltage of 4 V for node 2 with respect to 3 exists at t = 0. This is used only in the transient response.

3. Inductor

$$L\langle NAME \rangle \ \langle(+) \ NODE \rangle \ \langle(-) \ NODE \rangle \ \langle VALUE \rangle \ [IC = \langle INITIAL \ VALUE \rangle]$$

$\langle NAME \rangle$ denotes any seven-character alphanumeric string.

$(+)$ and $(-)$ NODES define the polarity of the inductor connection. Positive current flows from $(+)$ NODE through the inductor to $(-)$ NODE.

$\langle VALUE \rangle$ is the nonzero value (positive or negative) in henrys.

$\langle INITIAL \ VALUE \rangle$ is optional and denotes the initial inductor current at time t = 0 from $(+)$ NODE through the inductor to $(-)$ NODE for a transient response analysis.

Example E.3:   L12 100 0 10M
                 L12 100 0 10M IC=−0.5

denotes a 10-mH inductor labeled 12 connected between nodes 100 and 0 (reference node). If IC = −0.5 is included, then an initial current of −0.5 A from node 100 through the inductor to node 0 exists at $t = 0$. This is used only in the transient response.

4. Linear transformer

$$K\langle NAME \rangle \ L\langle INDUCTOR \ NAME \ A \rangle \ L\langle INDUCTOR \ NAME \ B \rangle \ \langle COUPLING \ VALUE \rangle$$

$K\langle NAME \rangle$ couples two inductors A and B using a dot convention that is determined by the node assignments of inductors A and B. The polarity is determined by the order of the nodes in the L devices and not by the order of the inductors in the K statement. The dotted terminals of inductors A and B are those connected to the positive (first labeled) nodes in their defining statements.

$\langle NAME \rangle$ denotes any seven-character alphanumeric string.

$\langle COUPLING \ VALUE \rangle$ is the coefficient of mutual coupling in the range from 0 to 1.

Example E.4:   LPRI   2 3 500M
                 LSEC   5 4 400M
                 KXFRM LPRI LSEC 0.98

denotes a linear transformer having a mutual coupling of 0.98 between LPRI and LSEC. Dotted terminals for the polarity of the coupling are the terminal of LPRI connected to node 2 and the terminal of LSEC connected to node 5.

5. Independent current and voltage sources

$$I\langle NAME \rangle \ \langle(+) \ NODE \rangle \ \langle(-) \ NODE \rangle \ [TYPE \ \langle VALUE \rangle] \ [TRANSIENT \ SPEC.]$$
$$V\langle NAME \rangle \ \langle(+) \ NODE \rangle \ \langle(-) \ NODE \rangle \ [TYPE \ \langle VALUE \rangle] \ [TRANSIENT \ SPEC.]$$

I denotes independent current source.

V denotes independent voltage source.

$\langle NAME \rangle$ denotes any seven-character alphanumeric string.

$(+)$ and $(-)$ NODES define the polarity of the source. Positive current flows from $(+)$ NODE through the source to $(-)$ NODE.

TYPE is DC (default) for a dc source and AC for a sinusoidal ac source.

$\langle VALUE \rangle$ is a dc value for DC or a magnitude and phase (in degrees) for AC. Default values are zero.

[TRANSIENT SPEC.] is used in transient analysis only and can be one of the following:

$$EXP(\langle x1 \rangle \ \langle x2 \rangle \ \langle td1 \rangle \ \langle tc1 \rangle \ \langle td2 \rangle \ \langle tc2 \rangle)$$

The EXP form causes the output current or voltage to be $\langle x1 \rangle$ for the first $\langle td1 \rangle$ seconds. Then, the output decays exponentially from $\langle x1 \rangle$ to $\langle x2 \rangle$ with a time constant of $\langle tc1 \rangle$. The decay lasts $\langle td2 \rangle$ seconds. Then, the output decays from $\langle x2 \rangle$ back to $\langle x1 \rangle$ with a time constant of $\langle tc2 \rangle$.

$$PULSE(\langle x1 \rangle \ \langle x2 \rangle \ \langle td \rangle \ \langle tr \rangle \ \langle tf \rangle \ \langle pw \rangle \ \langle per \rangle)$$

The PULSE form causes the output to start at $\langle x1 \rangle$ and remain for $\langle td \rangle$ seconds. The output then goes linearly from $\langle x1 \rangle$ to $\langle x2 \rangle$ during the next $\langle tr \rangle$ seconds. The output remains $\langle x2 \rangle$ for $\langle pw \rangle$ seconds. It then returns linearly to $\langle x1 \rangle$ during the next $\langle tf \rangle$ seconds. It remains at $\langle x1 \rangle$ for $\langle per \rangle - (\langle tr \rangle + \langle pw \rangle + \langle tf \rangle)$ seconds and the cycle repeats, excluding the initial delay of $\langle td \rangle$ seconds.

$$PWL(\langle t1 \rangle \ \langle x1 \rangle \ \langle t2 \rangle \ \langle x2 \rangle \ . \ . \ . \ . \ \langle tn \rangle \ \langle xn \rangle)$$

The PWL form describes a piecewise linear waveform. Each pair of time-output values specifies a corner of the waveform. The output at times between corners is the linear interpolation of the current at the corners.

$$SIN(\langle xoff \rangle \ \langle xampl \rangle \ \langle freq \rangle \ \langle td \rangle \ \langle df \rangle \ \langle phase \rangle)$$

The SIN form causes the output to start at $\langle xoff \rangle$ and remain for $\langle td \rangle$ seconds. Then, the output becomes an exponentially damped sine wave described by the equation

$$xoff + xampl \cdot \sin\{2\pi \cdot [freq \cdot (TIME-td) - phase/360]\} \cdot e^{-(TIME-td) \cdot tf}$$

Example E.5:   IG1 2 3 0.2A   or   IG1 2 3 DC 0.2A

denotes a dc current source labeled G1 supplying 0.2 A from node 2 to 3 through the source.

ISOURCE 0 5 AC 10 64

denotes a $10\underline{/64°}$ A ac source labeled SOURCE connected between nodes 0 and 5 with the positive terminal to node 0.

I2 0 2 EXP(10M 0M 0 0.1 1)

denotes an exponential current of $10e^{-10t}$ mA flowing from node 0 to 2 through the source in the interval $0 < t < 0.1$ s for transient analysis. Examples for PULSE, PWL, and SIN are given in Table 8.1.

6. Voltage-controlled voltage source (VCVS)

$$E\langle NAME \rangle \ \langle(+) \ NODE \rangle \ \langle(-) \ NODE \rangle \ \langle(+ \ CONTROLLING) \ NODE \rangle$$
$$+ \qquad \langle(- \ CONTROLLING) \ NODE \rangle \ \langle GAIN \rangle$$

$\langle NAME \rangle$ denotes any seven-character alphanumeric string.

$(+)$ and $(-)$ NODES define the polarity of the source. Positive current flows from $(+)$ NODE through the source to $(-)$ NODE.

$(+ \ CONTROLLING)$ and $(- \ CONTROLLING)$ are in pairs and define a set of controlling voltages that are multiplied by $\langle GAIN \rangle$. A particular node may appear more than once, and the output and controlling nodes need not be different.

Example E.6:   EBUFF 1 3 11 9 2.4

denotes a VCVS labeled BUFF connected between nodes 1 and 3, with the voltage of node 1 with respect to 3 equal to 2.4 times the voltage of node 11 with respect to 9.

7. Current-controlled current source (CCCS)

F⟨NAME⟩ ⟨(+) NODE⟩ ⟨(−) NODE⟩ ⟨(CONTROLLING V DEVICE) NAME⟩ + ⟨GAIN⟩

⟨NAME⟩ denotes any seven-character alphanumeric string.

(+) and (−) NODES define the polarity of the source. Positive current flows from (+) NODE through the source to (−) NODE. The current through ⟨(CONTROLLING V DEVICE) NAME⟩ multiplied by ⟨GAIN⟩ determines the output current.

(CONTROLLING V DEVICE) is an independent voltage source with a terminal voltage that need not be zero.

Example E.7:   F23 3 7 VOUT 1.2

denotes a CCCS labeled 23, with current flowing from node 3 through the source to node 7 having a value equal to 1.2 times that of the current flowing in independent voltage source VOUT (from + terminal to − terminal).

8. Voltage-controlled current source (VCCS)

G⟨NAME⟩ ⟨(+) NODE⟩ ⟨(−) NODE⟩ ⟨(+ CONTROLLING) NODE⟩
+   ⟨(− CONTROLLING) NODE⟩ ⟨TRANSCONDUCTANCE⟩

⟨NAME⟩ denotes any seven-character alphanumeric string.

(+) and (−) NODES define the polarity of the source. Positive current flows from (+) NODE through the source to (−) NODE.

(+ CONTROLLING) and (− CONTROLLING) are in pairs and define a set of controlling voltages that are multiplied by ⟨TRANSCONDUCTANCE⟩. A particular node may appear more than once, and the output and controlling nodes need not be different.

Example E.8:   GAMP 4 3 1 9 1.7

denotes a VCCS labeled AMP connected between nodes 4 and 3, with the current flowing from node 4 to 3 through the source equal to 1.7 times the voltage of node 1 with respect to 9.

9. Current-controlled voltage source (CCVS)

H⟨NAME⟩ ⟨(+) NODE⟩ ⟨(−) NODE⟩ ⟨(CONTROLLING V DEVICE) NAME⟩
+   ⟨TRANSRESISTANCE⟩

⟨NAME⟩ denotes any seven-character alphanumeric string.

(+) and (−) NODES define the polarity of the source. Positive current flows from (+) NODE through the source to (−) NODE. The current through ⟨(CONTROLLING V DEVICE) NAME⟩ multiplied by ⟨TRANSRESISTANCE⟩ determines the output voltage.

(CONTROLLING V DEVICE) is an independent voltage source with a terminal voltage that need not be zero.

Example E.9:   HIN 4 8 VDUMMY 7.7

denotes a CCVS labeled IN, with voltage from node 4 through the source to node 8 having a value equal to 7.7 times that of the current flowing in independent voltage source VDUMMY (from − terminal to + terminal).

10. Subcircuit call statement

X⟨NAME⟩ [NODE]* ⟨(SUBCIRCUIT) NAME⟩

⟨NAME⟩ denotes any seven-character alphanumeric string.

[NODE]* denotes a list of nodes required by the subcircuit definition.

⟨(SUBCIRCUIT) NAME⟩ is the name of the subcircuit's definition (see .SUBCKT statement below). There must be the same number of nodes in the call as in the subcircuit's definition.

This statement causes the referenced subcircuit to be inserted into the circuit, with the given nodes replacing the argument nodes in the definition. It allows defining a block of circuitry once and then the use of that block in several places.

Example E.10:   XBUFF 4 1 7 9 UNITAMP

denotes a call to a subcircuit that replaces the call statement by the content of the file UNITAMP for SPICE analysis.

C

SOLUTION CONTROL STATEMENTS

The commands to be discussed in this section are the .AC, .DC, .FOUR, .IC, .LIB, .SUBCKT, .TF, and .TRAN statements. The output control statements .PLOT and .PRINT described in the next section are used to get the results of the solution control statements.

1. AC analysis: The .AC statement is used to calculate the frequency response of a circuit over a range of frequencies. It has the form

   .AC [LIN][OCT][DEC] ⟨(POINTS) VALUE⟩ ⟨(START FREQUENCY) VALUE⟩
   +   ⟨(END FREQUENCY) VALUE⟩

   LIN, OCT, or DEC are keywords that specify the type of sweep as follows:

   LIN: Linear sweep. The frequency varies linearly from START FREQUENCY to END FREQUENCY. ⟨(POINTS) VALUE⟩ is the number of points in the sweep.
   OCT: Sweep by octaves. The frequency is swept logarithmically by octaves. ⟨(POINTS) VALUE⟩ is the number of points per octave.
   DEC: Sweep by decades. The frequency is swept logarithmically by decades. ⟨(POINTS) VALUE⟩ is the number of points per decade.

   Exactly one of LIN, OCT, or DEC must be specified.

   ⟨(END FREQUENCY) VALUE⟩ must not be less than ⟨(START FREQUENCY) VALUE⟩, and both must be greater than zero. The entire sweep may specify only one point if desired.

   Example E.11:   .AC LIN 1 100HZ 100HZ

   denotes an ac steady-state solution for a network having a frequency of 100 Hz.

   .AC LIN 101 100KHZ 200KHZ

   denotes a linear frequency response having 101 points evenly distributed in the range from 100 to 200 KHz.

2. DC analysis: The .DC statement causes a DC sweep analysis to be performed for the circuit. It has the form

.DC ⟨(SWEEP VARIABLE) NAME⟩ ⟨(START) VALUE⟩ ⟨(END) VALUE⟩
+ ⟨(INCREMENT) VALUE⟩

⟨(SWEEP VARIABLE) NAME⟩ is a name of an independent current or voltage source. It is swept linearly from ⟨(START) VALUE⟩ to ⟨(END) VALUE⟩. The increment size is ⟨(INCREMENT) VALUE⟩. ⟨(START) VALUE⟩ may be greater or less than ⟨(END) VALUE⟩; that is, the sweep may go in either direction. ⟨(INCREMENT) VALUE⟩ must be greater than zero. The entire sweep may specify only one point if desired.

Example E.12:   .DC VIN 10V 10V 1V

denotes a dc solution for a circuit with independent voltage source VIN = 10 V.

.DC IGEN 1M 10M 1M

denotes a dc sweep for independent current source IGEN being swept from 1 to 10 mA in 1-mA steps.

3. Fourier analysis: Fourier analysis performs a decomposition into Fourier components as the result of a transient analysis. A .FOUR statement requires a .TRAN statement (described below). It has the form

.FOUR ⟨(FREQUENCY) VALUE⟩ ⟨(OUTPUT VARIABLE)⟩*

⟨(OUTPUT VARIABLE)⟩ is a list of one or more variables for which the Fourier components are desired. The Fourier analysis is done by starting with the results of the transient analysis for the specified output variables. From these voltages or currents, the dc component, the fundamental frequency, and the second through ninth harmonics are calculated. The fundamental frequency is ⟨(FREQUENCY) VALUE⟩, which specifies the period for the analysis. The transient analysis must be at least $1/⟨(FREQUENCY)\ VALUE⟩$ seconds long.

Example E.13:   .FOUR 10KHZ V(5) V(6,7) I(VSENS3)

yields the Fourier components for variables V(5), V(6,7), and I(VSENS3). The fundamental frequency for the decomposition is set equal to 10 kHz.

4. Initial transient conditions: The .IC statement is used to set initial conditions for transient analysis. It has the form

.IC ⟨V(⟨NODE⟩) = ⟨VALUE⟩⟩*

Each ⟨VALUE⟩ is a voltage that is assigned to ⟨NODE⟩ for the initial node voltage at time $t = 0$ for the transient analysis.

Example E.14:   .IC V(2)=5 V(5)=−4V V(101)=10

denotes setting the initial node voltages of nodes 2, 5, and 101 to 5, −4, and 10 V, respectively, at $t = 0$.

5. Library file: The .LIB statement is used to reference a subcircuit library in another file. It has the form

.LIB ⟨(FILE) NAME⟩

⟨(FILE) NAME⟩ is the name of the subcircuit library file.

Example E.15:   .LIB OPAMP.LIB

denotes that the subcircuit library file is named OPAMP.LIB.

6. Operating (bias) point dc analysis: The .OP statement calculates all the dc node voltages and the currents in all voltage sources. It has the simple form

.OP

7. Subcircuit definition: The .SUBCKT statement is used to define a subcircuit that is called using the X statement described previously. It has the form

.SUBCKT ⟨NAME⟩ [NODE]*

The .SUBCKT statement begins the definition of a subcircuit. The definition is ended with a .ENDS statement. All statements between .SUBCKT and .ENDS are included in the definition. Whenever the subcircuit is called by an X statement, all the statements in the definition replace the calling statement.

8. Transfer function: The .TF statement produces the dc small-signal transfer function. It has the form

.TF ⟨(OUTPUT VARIABLE)⟩ ⟨(INPUT SOURCE) NAME⟩

The gain from ⟨(INPUT SOURCE) NAME⟩ to ⟨(OUTPUT VARIABLE)⟩ is calculated along with the input and output resistances. ⟨(OUTPUT VARIABLE)⟩ may be a current or a voltage: however, in the case of a current it is restricted to be the current through a voltage source.

Example E.16:   .TF V(3) IIN

produces the dc small-signal transfer function for V(3)/IIN, the input resistance seen from the terminals of independent current source IIN, and the output resistance seen at node V(3).

9. Transient analysis: The .TRAN statement causes a transient analysis to be performed for the circuit. It has the form

.TRAN ⟨(PRINT STEP) VALUE⟩ ⟨(FINAL TIME) VALUE⟩ [UIC]

The transient analysis calculates the behavior of the circuit over time, starting at $t = 0$ and going to ⟨(FINAL TIME) VALUE⟩. ⟨(PRINT STEP) VALUE⟩ is the time interval used for plotting or printing the results of the analysis. The keyword UIC (use initial conditions) causes the initial conditions set for capacitors and inductors with the IC specification to be used.

Example E.17:   .TRAN 1NS 100NS UIC

produces a transient analysis in the interval from 0 to 100 ns, with plotted or printed output on 1-ns intervals.

D

OUTPUT CONTROL STATEMENTS

The output control statements for plotting and printing are the .PLOT, .PRINT, and .WIDTH statements.

1. Plot: The .PLOT statement allows results from dc, ac, and transient analysis to be output in the form of *line printer* plots. It has the form

.PLOT [DC][AC][TRAN][OUTPUT VARIABLE]*
+   ([⟨(LOWER LIMIT) VALUE⟩, ⟨(UPPER LIMIT) VALUE⟩])

DC, AC, and TRAN are the analysis types that can be output. Exactly one analysis type must be specified. [OUTPUT VARIABLE]* is a list of the output variables desired for plotting. A maximum of eight output variables are allowed in one .PLOT statement.

The range and increment of the x-axis are fixed by the analysis being plotted. The range of the y-axis can be set by adding (⟨(LOWER LIMIT) VALUE⟩, ⟨(UPPER LIMIT) VALUE⟩) to output variables with the same y-axis range. Each occurrence defines one y-axis with the specified range. All output variables that come between it and the next range to the left are put on its corresponding y-axis. If no y-axis limits are specified, the program automatically determines the plot limits.

Example E.18:   .PLOT DC V(2) V(3, 5) I(R2)

plots the dc response for V(2), V(3, 5), and I(R2), the current in R2.

.PLOT AC VM(3) VP(3) IR(C1) II(C1)

plots the magnitude and phase of V(3) and the real and imaginary parts of the current through C1.

.PLOT TRAN V(5) V(2, 3) (0, 5V) I(R1) I(VCC) (−5MA, 5MA)

plots the transient response of V(5) and V(2, 3) between the limits of 0 and 5 V and I(R1) and I(VCC) between the limits of −5 and 5 mA.

2. Print: The .PRINT statement allows results from dc, ac, and transient analysis to be output in the form of tables. It has the form

.PRINT [DC][AC][TRAN][(OUTPUT VARIABLE)]*

DC, AC, and TRAN are the analysis types that can be output. Exactly one analysis type must be specified. [(OUTPUT VARIABLE)]* is a list of output variables desired. There is no limit to the number of output variables. The output format is determined by the specification of the .WIDTH command.

Example E.19:   .PRINT DC V(1) I(R12)

prints the dc values for V(1) and I(R12).

.PRINT AC VM(1, 5) VP(1, 5) IR(L2) II(L2)

prints the magnitude and phase of V(1, 5) and the real and imaginary parts I(L2).

.PRINT TRAN V(7) I(L4) I(VCC) V(3, 1)

prints the transient response for V(7), I(L4), I(VCC), and V(3, 1).

3. Width: The .WIDTH statement sets the width of the output. It has the form

.WIDTH OUT = ⟨VALUE⟩

⟨VALUE⟩ is the number of columns and must be either 80 or 132. The default value is 80 columns.

Example E.20:   .WIDTH OUT = 132

E
END STATEMENTS

End statements for subcircuit files and circuit files are .ENDS and .END, respectively.

1. End of subcircuit definition: The .ENDS marks the end of a subcircuit definition (started by a .SUBCKT statement). It has the form

    .ENDS [(SUBCIRCUIT) NAME]

    It is good practice to repeat the subcircuit name, although this is not required.

2. End of circuit: The .END statement marks the end of the circuit. It has the form

    .END

Appendix F

Answers to Selected Odd-Numbered Problems

CHAPTER 1

1.1 (a) 32 mC; (b) 4 mA
1.3 5π W
1.5 (a) $24e^{-2t}$ W, $12(1 - e^{-2})$ J;
(b) $-12e^{-2t}$ W, $6(e^{-2} - 1)$ J;
(c) $-8e^{-2t}$ W, $4(e^{-2} - 1)$ J
1.7 (a) 86,400 J; (b) 14,400 C
1.9 (a) 14,400 J; (b) 1200 C
1.11 3 C, 7.46 W
1.13 $\frac{5}{8}$ W
1.15 22 μJ

CHAPTER 2

2.1 30 mA
2.3 9.6 kJ
2.5 1 A, 2 A, 9 V
2.7 0.3 A, 13 V; Circuit: 6-V source and 20-Ω resistor
2.9 12.5 V, 37.5 V
2.11 (a) 3; (b) 2
2.13 12 Ω, 4 A
2.15 4, 6, 8, 12, 18, 24, 36 Ω
2.17 5 A, -3 A
2.19 3 A
2.21 7 A, 35 V, 8 V
2.23 9 A, 52 V
2.25 10 A, 2.5 A, 10 V
2.27 37.5 μA
2.29 4.5 kΩ

CHAPTER 3

3.1 -2 V, 2 W
3.3 3 A
3.5 (a) 3 A, 6 V; (b) -1 A, -6 V
3.7 -3 V
3.13 6 cos $2t$ V
3.15 -12 V
3.17 -3 A
3.19 2 sin $3000t$ mA
3.21 (a) $v_0 = [\mu/(R_1 + R_2)](R_2 v_1 + R_1 v_2)$, where $\mu = 1 + R_f/R$; (b) 17 V

CHAPTER 4

4.1 10 V, 6 V
4.3 8 A
4.5 10 V, 1 A
4.7 3 V
4.9 -2 V, -4 V
4.11 2 V
4.13 4 cos $3t$ V
4.15 4 mA
4.17 4 W
4.19 8 V
4.21 8 V (loop analysis)
4.25 -2 cos $2t$ A
4.27 (a) 3 kΩ; (b) $\frac{2}{3}$ kΩ
4.33 $v = 3$ V, $R_{in} = 1.2$ kΩ
4.35 $v_1 = 5, 6, 7, 8, 9, 10$ V

CHAPTER 5

5.1 7 A, 35 V, 8 V
5.3 $4 \cos 4t$ mA
5.5 3 A
5.7 $4 + (-4/3) + 16/3 = 8$ V
5.9 $-4 - 2 = -6$ A
5.11 $v_{oc} = 20$ V, $R_{th} = 4\ \Omega$, $i = 2.5$ A
5.13 $i_{sc} = 3$ A, $R_{th} = 4\ \Omega$, $i = 1$ A
5.15 $v_{oc} = 12$ V, $R_{th} = 8\ \Omega$, $v = 4$ V
5.17 $i_{sc} = -6 \cos 2t$ mA, $R_{th} = 3$ kΩ,
      $i = -2 \cos 2t$ mA
5.19 4.5 W
5.21 2.25 W
5.23 3 Ω, 3 W
5.25 $v_{oc} = 14$ V, $R_{th} = 14\ \Omega$, $R = 14\ \Omega$,
      $P_{max} = 3.5$ W

CHAPTER 6

6.1 2 V
6.3 2 A
6.5 2.5 A
6.9 -11, -20 V

CHAPTER 7

7.1 4 ms, 100 V
7.3 (a) 0; (b) 20 μA; (c) $20e^{-2t}$ μA;
      (d) $3 \cos 100\,t$ mA
7.5 (a) $10t + 4$ V; (b) $10t^2 + 4$ V;
      (c) $-5e^{-t} + 9$ V; (d) $3 \sin 2t + 4$ V
7.7 $12e^{-2t}$ W
7.9 2 J at 1 s
7.11 0, -1 A
7.13 (a) 10, 0.1 μF; (b) one answer: 5 parallel
      sets of 2 in series; (c) one answer: 2
      parallel sets of 5 in series
7.15 20, 0, -60 V
7.17 $4e^{-10t}$ V
7.19 (a) 18 J; (b) 9 J
7.21 100 V, -100 V, -20 A/s
7.23 -2 A, 7.5 A/s
7.25 (a) 50, 0.5 mH; (b) one answer: 2 in series
      with 4 parallel sets of 2 each
7.27 1 nF, $10^{12}\ \Omega$
7.29 2 V or 10 V

CHAPTER 8

8.1 (a) 100 V; (b) $100e^{-100t}$ V; (c) 5 mJ
8.3 $6e^{-2t}$ V
8.5 $-6e^{-9t}$ A
8.7 (a) $V_0 e^{-t/RC}$; (b) $[RV_0/(R + R_1)]e^{-t/RC}$
8.9 $32e^{-16t}$ V
8.11 $2 + 5e^{-2t}$ A, $24 - 30e^{-2t}$ V
8.13 $14e^{-3t} - 7e^{-2t}$ A, $42e^{-2t} - 48e^{-3t}$ V
8.15 $e^{-9t} + 17$ V
8.17 $5 - 3e^{-8t} + 6e^{-t}$ A
8.19 $10e^{-3t} - 8e^{-2t}$ V
8.21 $\dfrac{1}{24}(1 - e^{-5t})u(t)$ A
8.23 $(3e^{-3t} + 6)u(t)$
      $+ (3e^{-3(t-1)} - 6e^{-3t} - 6)u(t - 1)$ V
8.25 $8(1 - e^{-2t})$ V

CHAPTER 9

9.3 $6e^{-t} - e^{-6t}$ A
9.5 $18e^{-t} - 3e^{-6t}$ V
9.7 $(2t + 4)e^{-2t}$ A
9.9 $e^{-5t}(\sin 5t + 5 \cos 5t)$ A
9.11 $8e^{-t} - 2e^{-4t} + 2$ A
9.13 $8 - 2e^{-4t}$ V, $10 - 4e^{-2t}$ V
9.15 (a) $0.5e^{-2t} - 1.5e^{-6t} + 1$ A;
      (b) $(-1 + 4t)e^{-4t} + 1$ A;
      (c) $e^{-4t}(-\cos 4t + \sin 4t) + 1$ A
9.17 (a) $8 - e^{-2t}(8 \cos t + 6 \sin t)$ V;
      (b) $15e^{-t} - e^{-2t}(15 \cos t + 5 \sin t)$ V
9.19 (a) $e^{-2t}(-12t + 2) - 2$ V;
      (b) $e^{-2t}(4 - 6t) - 4 \cos 2t - \sin 2t$ V
9.21 (a) $-6e^{-2t} + 2e^{-3t} + 4$ V;
      (b) $-33e^{-2t} + 30e^{-3t} + 3 \cos 2t$
      $+ 15 \sin 2t$ V;
      (c) $6(e^{-t} - e^{2t})$ V
9.23 $10 - 10(1 + 1000t)e^{-1000t}$ V

CHAPTER 10

10.1 (a) Leads by 60°; (b) lags by 6.9°
10.3 2 kΩ, 0.5 H
10.5 $(V_m/\sqrt{R^2 + \omega^2 L^2})$
      $\times \sin[\omega t - \tan^{-1}(\omega L/R)]$
10.7 (a) $v_1 = 6e^{j(8t - 53.1°)}$ V,
      $v = 6 \cos(8t - 53.1°)$ V;
      (b) $v = 6 \sin(8t - 53.1°)$ V

10.9 (a) $5\sqrt{2}\cos(10t + 135°)$;
  (b) $13\cos(10t + 112.6°)$;
  (c) $5\cos(10t - 36.9°)$;
  (d) $-10\cos 10t$; (e) $5\sin 10t$

10.11 (a) $10\underline{/111.9°}\Omega$; (b) $(1/\sqrt{2})\underline{/15°}$ kΩ;
  (c) $a\underline{/\alpha}$ Ω

10.13 5 Ω, -5 Ω, 0.1 Ω, 0.1 Ω,
  $\sqrt{2}\cos(2t + 45°)$ A

10.15 (a) $2\cos(4t - 36.9°)$ A;
  (b) $2.5\cos 2t$ A

10.17 $\dfrac{1}{32}$ F, $24\cos^2 8t$ W

10.19 $2\cos(t - 53.1°)$ A,
  $0.5\cos(t - 53.1°)$ A

10.21 $(5/\sqrt{2})\cos(4t - 45°)$ V

10.23 $\sqrt{5}\cos(10,000t + 116.6°)$ V

10.25 $e^{-6t} - 3e^{-2t} + 2\sqrt{5}\cos(6t - 26.6°)$ A

10.27 $8\cos(2t - 36.9°)$ V

CHAPTER 11

11.1 $16\sin 2t$ V

11.3 $9.6\cos(4t - 53.1°)$ V

11.5 $3\sqrt{2}\cos(4t - 135°)$ V

11.7 $2\cos(2t + 53.1°)$ V

11.9 $5\cos(2t + 53.1°)$ A

11.11 $10\cos(2t + 36.9°)$ V

11.13 $5\sqrt{2}\cos(6t - 36.9°)$ V

11.17 $-2\sin 2t$ V

11.19 $6\sqrt{2}\cos(2t + 135°)$ V

11.21 $1 - 2\cos(t - 36.9°)$
  $+ 3\cos(2t + 112.6°) + \sin 3t$ A

11.23 $\mathbf{V}_{oc} = \frac{18}{17}(3 - j5)$ V,
  $\mathbf{Z}_{th} = (19 - j9)/17$ Ω,
  $i = 2\sqrt{2}\cos(2t - 45°)$ A

11.25 $\mathbf{V}_{oc} = \dfrac{10}{3}$ V, $\mathbf{Z}_{th} = -j\frac{4}{3}$ Ω,
  $v = 2\cos(2t + 53.1°)$ V

11.27 $\frac{1}{2}\cos(1000t - 135°)$ V

11.29 $0.5\cos(3t + 126.9°)$ A

11.31 $\mathbf{V}_c = x + jy$, where $x^2 + (y + \frac{1}{2})^2$
  $= (\frac{1}{2})^2$, $L = 1$ H, $v_c = \sin 2t$ V

11.33 $v = 2.82\cos(4t - 135°)$ V

11.35 $v = 5.65\cos(2t - 135°)$ V

CHAPTER 12

12.1 10 mW

12.3 15 W

12.5 3.4 W

12.7 (a) Load = 1 Ω, $P = 4.25$ W;
  (b) load = $(4 + j3)/5$ Ω, $P = 4.78$ W

12.9 40 W

12.11 3 W

12.15 (a) $\sqrt{6}$ A; (b) 3 A; (c) $I_m/2$ A; (d) $I_m/\sqrt{2}$ A

12.17 3 A, 0.707 leading, $\frac{2}{3}$-H inductor

12.19 $\frac{1}{160}$-F capacitor

12.21 $12 + j16$ VA, 0.6 lagging

12.23 0.5 A, 0.8 lagging

12.25 4 W

12.27 3 W

CHAPTER 13

13.1 $10\underline{/-90°}$ A rms

13.3 $\sqrt{3}\underline{/30°}$ Ω, 1.8 kW

13.5 12.65 A rms

13.7 $25\sqrt{2}$ A rms, 11.25 kW

13.9 25.9 μF

13.11 $200\sqrt{3}$ V rms, 120 A rms, 57.6 kW

13.13 20 A rms

13.15 $12 + j12$, $-22 - j16$, $10 + j4$ A rms

13.17 $10\sqrt{3}$ A rms, 7.2 kW

13.23 500 W

13.25 $500\sqrt{3}$, $1000\sqrt{3}$, $1500\sqrt{3}$ W

13.29 $13.83\underline{/26.6°}$, $19.03\underline{/-140.39°}$,
  $6.37\underline{/68.9°}$ A

13.31 $12.74\underline{/7.43°}$ A

CHAPTER 14

14.1 (a) $5\underline{/-30°}$; (b) $1\underline{/-45°}$;
  (c) $10\underline{/-53.1°}$; (d) $5\underline{/0°}$

14.3 (a) $-1 \pm j2$; (b) -3, -4, $\pm j1$; (c) 0

14.5 $-16e^{-4t}\cos 2t$ V

14.7 $-j\frac{8}{5}$ Ω, $8\,e^{-2t}\sin t$ V

14.9 $1/[2(s^3 + 2s^2 + 2s + 1)]$,
  $\sqrt{2}\cos(t - 135°)$ V

14.11 $8/(2s^2 + 3s + 4)$,
  $4\sqrt{2}\,e^{-t}\cos(t + 45°)$ V

14.13 $\mathbf{V}_{oc} = 8(17 - j6)/5$ V,
  $\mathbf{Z}_{th} = 12(2 - j1)/5$ Ω
  $i = \sqrt{5}\,e^{-2t}\cos(4t - 26.6°)$ A

14.15 $1/[2(s^3 + 2s^2 + 2s + 1)]$, $-e^{-2t}$ V

14.17 $6e^{-t} - 6e^{-3t} - 12te^{-t}$ V

14.19 $6/(s^2 + 3s + 3)$, $-12e^{-2t}\sin t$ V

14.21 $8 + e^{-3t}\sin 6t - 8e^{-3t}$ V

14.23 $2/(s + 1)^2$, $-6e^{-2t}\sin t$ V

14.25 $y_{11} = y_{22} = \dfrac{s^2 + 8}{4s}$,

$\qquad y_{12} = y_{21} = \dfrac{4 - s^2}{4s}$

14.27 $\mathbf{h}_{11} = \dfrac{25s}{s^2 + 250}$,

$\qquad \mathbf{h}_{12} = -\mathbf{h}_{21} = \dfrac{s^2}{s^2 + 250}$,

$\qquad \mathbf{h}_{22} = \dfrac{10s}{s^2 + 250}$, $\mathbf{g}_{11} = 10/s$,

$\qquad \mathbf{g}_{12} = -\mathbf{g}_{21} = -1$, $\mathbf{g}_{22} = 25/s$

14.29 $A = 1$, $B = 25/s$, $C = 10/s$,

$\qquad D = (s^2 + 250)/s^2$

14.31 $\mathbf{y}_{11} = s + 1$, $\mathbf{y}_{12} = -s$,

$\qquad \mathbf{y}_{21} = \frac{1}{4} - s$, $\mathbf{y}_{22} = s + \frac{1}{2}$,

$\qquad \mathbf{V}_2/\mathbf{V}_1 = \dfrac{4s - 1}{4s + 6}$

14.33 $-10,000/9$

14.35 -67.295

14.39 $\dfrac{s^2 - 6s + 12}{s^2 + 6s + 12}$

CHAPTER 15

15.1 $2s/(s^2 + 2s + 100)$

15.3 $2s/(s^2 + 2s + 100)$

15.9 $\mathbf{H} = 4s/(s^2 + 2s + 7)$, 2, 2 rad/s,
$\qquad \sqrt{7}$ rad/s

15.11 $\mathbf{H} = (2/3)/(s^3 + 2s^2 + 2s + 1)$,
$\qquad |\mathbf{H}(j\omega)| = (2/3)/\sqrt{1 + \omega^6}$

15.13 $R_1 = 1/2$, $R_2 = 1/\sqrt{2}$, $R_3 = 1\ \Omega$

15.17 $R_1 = 79.6\ \text{k}\Omega$, $R_2 = 159.2\ \text{k}\Omega$, all
$\qquad$ other R's $= 15.92\ \text{k}\Omega$

15.19 $-80s/(s^2 + 5s + 400)$,
$\qquad \omega_0 = 20$ rad/s, $B = 5$ rad/s, R's are
$\qquad 1.25$, 40 kΩ

15.23 1.06, 1.325, 5.3 MΩ

CHAPTER 16

16.1 $-50 \sin 100t$ V

16.3 (a) $20 \cos 2t - 36 \sin 4t$ V,
$\qquad -30 \cos 2t + 60 \sin 4t$ V;
$\qquad$ (b) $-30 \cos 2t$ V

16.5 $2e^{-2t}$ V

16.7 $(5\sqrt{2}/6) \cos (2t - 135°)$ V

16.9 (a) 116 J; (b) 52 J

16.11 $\dfrac{5}{\sqrt{2}} \cos (2t + 135°)$ A

16.13 $8 \cos (t - 53.1°)$ A

16.15 $15 \cos (4t - 36.9°)$ A

16.17 $s/(3s + 2)$

16.19 1200 W

16.21 1200 W

16.23 50

16.25 $6 \cos 100t$ V

16.27 $40\ \underline{/0.1°}$ V

CHAPTER 17

17.1 (a) $a_0 = 3$, $a_n = 0$,
$\qquad b_n = \dfrac{(-1)^n - 1}{n\pi}$, $\omega_0 = 1$

$\qquad$ (b) $a_n = \dfrac{2(-1)^n}{\pi(n^2 + 1)} \sinh \pi$,
$\qquad n = 0, 1, 2, \ldots$; $b_n = -2na_n$,
$\qquad n = 1, 2, 3, \ldots$; $\omega_0 = 1$;

$\qquad$ (c) $\dfrac{1}{3} + \dfrac{4}{\pi^2} \displaystyle\sum_{n=1}^{\infty} \dfrac{(-1)^n}{n^2} \cos n\pi t$

17.5 $\displaystyle\sum_{n=1}^{\infty} b_{2n-1} \sin \dfrac{(2n - 1)\pi t}{8}$; $b_{2n-1} =$

$\qquad \dfrac{4}{(2n - 1)\pi}\left[1 + \dfrac{\dfrac{8 \sin (2n - 1)\pi}{8} \cos \dfrac{(2n - 1)\pi}{4}}{(2n - 1)\pi}\right]$

17.9 $\dfrac{4}{\pi} + 2 \sin 2(t - \pi/6)$

$\qquad - \dfrac{8}{\pi} \displaystyle\sum_{n=1}^{\infty} \dfrac{1}{4n^2 - 1} \cos \left(4nt - \dfrac{2\pi}{3}\right)$

17.11 $\dfrac{1}{2} + \dfrac{2}{\pi} \displaystyle\sum_{n=1}^{\infty} \dfrac{(-1)^{n+1}}{(2n - 1)} \cos \dfrac{(2n - 1)\pi t}{2}$

17.13 $\dfrac{16}{\pi}\left[\dfrac{1}{2} + \displaystyle\sum_{n=1}^{\infty} \dfrac{\sin(4nt - \tan^{-1}\frac{1}{n})}{(1 - 4n^2)\sqrt{n^2 + 1}}\right]$ V

17.15 $2 - \displaystyle\sum_{n=1}^{\infty} \dfrac{8 \sin\left(n\pi t - \tan^{-1}\dfrac{2n\pi}{2 - n^2\pi^2}\right)}{n\pi\sqrt{4 + n^4\pi^4}}$ V

17.19 $\dfrac{1}{3} + \dfrac{4}{\pi^2} \displaystyle\sum_{n=1}^{\infty} \dfrac{(-1)^n}{n^2} \cos n\pi t$

17.21 $\displaystyle\sum_{n=1}^{\infty} \dfrac{128}{\pi^2(2n - 1)^2}\left[\dfrac{2 + (2n - 1)^2 \pi^2}{4 + (2n - 1)^2 \pi^2}\right]$ W

17.23 $\dfrac{16}{\pi} \displaystyle\sum_{n=1}^{\infty} \dfrac{1}{2n - 1} \sqrt{\dfrac{1 + (2n - 1)^2 \pi^2}{4 + (2n - 1)^2 \pi^2}} \sin \theta$ A;

$$\theta = (2n-1)\pi t + \tan^{-1} \frac{(2n-1)\pi}{2 + (2n-1)^2 \pi^2}$$

$$\frac{16}{\pi} \sum_{n=1}^{\infty} \frac{\sin\left[(2n-1)\pi t - \tan^{-1} \frac{(2n-1)\pi}{2}\right]}{(2n-1)\sqrt{4 + (2n-1)^2 \pi^2}} \text{ V}$$

left:

$$\frac{128}{\pi^2} \sum_{n=1}^{\infty} \frac{1 + (2n-1)^2 \pi^2}{(2n-1)^2 [4 + (2n-1)^2 \pi^2]} \text{ W}$$

right:

$$\frac{128}{\pi^2} \sum_{n=1}^{\infty} \frac{1}{(2n-1)^2 [4 + (2n-1)^2 \pi^2]} \text{ W}$$

17.25 $\displaystyle\sum_{n=-\infty}^{\infty} \frac{8}{\pi(1-4n^2)} e^{j4nt}$

17.27 (a) $|c_0| = \frac{3}{2}$, $\phi_0 = 0$;
$|c_{-n}| = |c_n| = 0$, n even; $= 1/|\pi n|$,
n odd; $\phi_n = -90°$, $n = 1, 3, 5, \ldots$

17.31 For $A = 1$,
(a) $f(t) = 1.273 \sin(2\pi t) + 0.423 \sin(6\pi t)$
$+ 0.253 \sin(10\pi t)$
$+ 0.179 \sin(14\pi t)$
$+ 0.138 \sin(18\pi t)$

(c) $f(t) = 0.642 + 0.418 \sin(2\pi t - 92°)$
$+ 0.084 \sin(4\pi t - 94°)$
$+ 0.037 \sin(6\pi t - 95°)$
$+ 0.021 \sin(8\pi t - 97°)$
$+ 0.013 \sin(10\pi t - 99°) + \cdots$

17.33 $f(t) = 6.379 + 1.986 \sin(2\pi t + 10.8°)$
$+ \sin(4\pi t + 8.10°)$
$+ 0.670 \sin(6\pi t + 8.40°)$
$+ 0.504 \sin(8\pi t + 9.44°)$
$+ 0.404 \sin(10\pi t + 10.8°) + \cdots$

CHAPTER 18

18.1 (a) $e^{-j\omega/2} \text{Sa}(\omega/2)$

(b) $\dfrac{a + j\omega}{(a + j\omega)^2 + b^2}$

(c) $\dfrac{1 - e^{-(a+j\omega)}}{a + j\omega}$

18.3 $\dfrac{-2j\omega}{a^2 + \omega^2}$

18.5 (a) $\dfrac{1}{\sqrt{a^2 + \omega^2}}$; $\tan^{-1} \dfrac{\omega}{a}$

(b) $\dfrac{b}{\sqrt{(a^2 + b^2 - \omega^2)^2 + 4a^2\omega^2}}$
$-\tan^{-1} \dfrac{2a\omega}{a^2 + b^2 - \omega^2}$;

(c) $\dfrac{b}{\sqrt{(a^2 + b^2 - \omega^2)^2 + 4a^2\omega^2}}$;
$180° + \tan^{-1} \dfrac{2a\omega}{a^2 + b^2 - \omega^2}$;

(d) $\dfrac{2|\omega|}{\omega^2 + a^2}$; $\begin{array}{l} -90°, \ \omega > 0 \\ 90°, \ \omega < 0 \end{array}$

18.7 $\dfrac{-1 + j\omega}{\sqrt{2}(5 - \omega^2 + j2\omega)}$

18.9 (a) $\dfrac{-j3}{13 - \omega^2 + j4\omega}$

(b) $\dfrac{2 + j\omega}{13 - \omega^2 + j4\omega}$

18.11 $\dfrac{8e^{j\omega/2} \cos\dfrac{5\omega}{2} \sin \omega}{\omega}$

18.13 (a) $\dfrac{a}{\pi(t^2 + a^2)}$

(b) $\dfrac{-2at}{\pi(t^2 + a^2)^2}$

(c) $\dfrac{a}{\pi[(t-2)^2 + a^2]}$

18.15 (a) $-e^t u(-t)$ (b) $-te^t u(-t)$

18.17 (a) $\dfrac{1}{2 + j\omega}$

(b) $(e^{-t} - e^{-2t})u(t)$ V

18.19 $\dfrac{2 + j\omega}{1 + j\omega}$; $y' + y = x' + 2x$

18.21 $\dfrac{1}{4} J$

18.23 $\dfrac{1}{3} J$

CHAPTER 19

19.1 (a) $\dfrac{2e^{-3s} - 1}{s}$ (b) $\dfrac{e^{-as} - e^{-bs}}{s}$

19.3 $\dfrac{s}{s^2 + 1}$

19.5 (a) $(2e^{-4t} + 3t)u(t)$

(b) $(1 - \cos 2t)u(t)$

(c) $2(\cos 3t + \sin 3t)u(t)$

(d) $(e^{-t} + t - 1)u(t)$

19.7 (a) $e^{-t}(6 \cos 3t - 2 \sin 3t)u(t)$

(b) $e^{-2t}(2 \cos 3t + 2 \sin 3t)u(t)$

(c) $te^{-2t}u(t)$

(d) $e^{-t}(2 \cos t - 8 \sin t)u(t)$

19.9 (a) $\dfrac{1 - e^{-s}}{s}$

(b) $\dfrac{1 - e^{-2(s+2)}}{s + 2}$

(c) $\dfrac{e^{-2s}(2s + 1)}{s^2}$

(d) $\dfrac{e^{-2(s + 1)}}{s + 1}$

19.11 $\dfrac{e^{-2s}(2s + 1)}{s^2}$; $\dfrac{(1 - e^{-s})^2}{s^2}$

19.13 (a) $\dfrac{1}{2} \sin 2t \,[u(t) + u(t - \pi)]$

(b) $(t - 4)u(t - 4) - (t - 7)u(t - 7)$

(c) $-(5 \cos 3t + 2 \sin 3t)u(t - \pi)$

(d) $e^{-t}(2 \cos 2t + \sin 2t)$
$\times [u(t) - e^{\pi/2}u(t - \pi/2)]$

19.15 $t \cos t\, u(t)$

19.17 $(\cos t + \sin t)u(t)$

19.19 (a) $\dfrac{(ae^{-at} - be^{-bt})u(t)}{a - b}$

(b) $(2e^{-t} - e^{-2t})u(t)$

(c) $(-4 + 5e^{-t} + e^{-3t})u(t)$

(d) $(2 - t + 3 \cos t - 2 \sin t)u(t)$

19.21 (a) $e^{-t}(1 + \sin 2t)u(t)$

(b) $e^{-t}(1 + t - \cos t - \sin t)u(t)$

(c) $[-2 \sin t + 2e^{-2t}(2 \cos 3t - \sin 3t)]u(t)$

(d) $[4e^{-2t} + e^{-t}(-4 \cos 2t + \frac{13}{4} \sin 2t$
$\quad - 5t \sin 2t - \frac{5}{2}t \cos 2t)]u(t)$

19.23 (a) $\dfrac{1}{s(1 + e^{-s})}$ (b) $\dfrac{1 + e^{-\pi s}}{(1 - e^{-\pi s})(s^2 + 1)}$

19.25 $\dfrac{s}{s^2 + 1}$

19.27 $\dfrac{8}{\pi} \displaystyle\sum_{n=1}^{\infty} \dfrac{1}{2n - 1} \sin(2n - 1)\pi t$

19.31 $t \sin 2t$ A

19.33 (a) $2e^{-t} - e^{-3t}$ (b) $3 \cos 2t + \cos t$

19.35 $e^{-2t}(4 \cos t + 3 \sin t)$

CHAPTER 20

20.1 (a) $8 - e^{-2t}(6 \sin t + 8 \cos t)$ V

(b) $15e^{-t} - e^{-2t}(5 \sin t + 15 \cos t)$ V

20.3 $2 + 6e^{-8t} - 8e^{-5t}$ V

20.7 (a) $6e^{-3t/2} - 4e^{-2t}$ A

(b) $(2 + 6t)e^{-2t}$ A

(c) $e^{-t}(2 \cos t + 8 \sin t)$ A

20.9 $(4 + 4t + t^2)e^{-t}$ V

20.11 $(4 + 24t)e^{-8t} + 2 \sin 8t$ V

20.13 (a) $2 \sin 2t$ V

(b) $2 \cos 2t$ V

(c) $2 \cos 2t + 2 \sin 2t$ V

20.15 $e^{-t} - e^{-3t}$ A

20.17 (a) $(4 - 5e^{-t} + e^{-5t})u(t)$;
$\quad 5(e^{-t} - e^{-5t})u(t)$

(b) $[1 - e^{-2t}(\cos 3t + \frac{2}{3} \sin 3t)]u(t)$;
$\quad \frac{13}{3}e^{-2t} \sin 3t\, u(t)$

20.19 $e^{-2t}(1 - \cos t)u(t)$

20.21 $\dfrac{2}{(s + 3)^2 + 1}$; $2e^{-3t} \sin t\, u(t)$

20.23 $p(t) + p(t - 2) + p(t - 4) + \cdots$,
where
$p(t) = (1 - e^{-t})u(t) - [1 - e^{-(t-1)}]u(t - 1)$

20.25 $\dfrac{1}{s^2 + \sqrt{2}s + 1}$; $\dfrac{1}{\sqrt{1 + \omega^4}}$

20.27 $\mathbf{H} = \dfrac{\dfrac{\mu}{2R}}{s^2 + \left(2 - \mu + \dfrac{1}{R}\right)s + \dfrac{1}{R}}$

(a) $2e^{-t} \sin t\, u(t)$ A
$\quad [1 - e^{-t}(\cos t + \sin t)]u(t)$ A

(b) $(e^{-t} - e^{-2t})u(t)$ A
$\quad \left(\dfrac{1}{2} - e^{-t} + \dfrac{1}{2}e^{-2t}\right)u(t)$ A

(c) $4te^{-2t}u(t)$ A
$\quad (1 - e^{-2t} - 2te^{-2t})u(t)$ A

20.29 $\mathbf{V}_{oc} = \dfrac{2s^2 + 10}{s + 1}$, $\mathbf{Z}_{th} = \dfrac{s^2 + s + 1}{s + 1}$

$i = (2 + 6t)e^{-2t}$ A

20.37 $C > \dfrac{1}{12}$ F

20.39 (a) 3, (b) 0, (c) 0, (d) 1, (e) 0, (f) 0

20.41 b

20.43 $t + 1$

20.45 $6\sqrt{2} \cos(2t + 135°)$ V

Index

Impedance(s) (*cont.*)
  phase, 396
  reflected, 520–21
  sinusoidal excitations and, 426–27,
    429
  Thévenin, 656
  transfer, 442
Impedance scaling, 486–88
Impulse function, 603, 604, 618–25
Impulse response, 623
  Laplace transforms and, 662–64,
    665–67
Independent equations:
  current, 154–57
  voltage, 151–54
Independent sources:
  current, 13
  voltage, 12–14
Inductance, 174, 645
  mutual, 503–4, 506–12
  self-, 503, 506
Inductive reactance, 314
Inductors, 2
  current changes in, 174–77
  duality of, 184
  energy storage in, 177–78
  linearity of, 184–85
  parallel connections of, 180–81
  practical (real), 183
  series connections of, 179–80, 181
Initial-value theorem, 672–74
Instantaneous power, 10, 31–32, 34,
    39
  peak (maximum), 358
Insulators (dielectrics), 164
Integral equations, 206
Integrated circuits, 49–50
Integrating factor method for equa-
    tions, 213–15
Integrator circuits, 227
Integrodifferential equations, Laplace
    transforms and, 637–40
Internal resistance, 131
International System of Units (SI),
    3–5
Inverse Laplace transforms, 605,
    625–33
Inverter circuits, 64
Inverting input terminals, 59
Isolation, dc, 516

J

Joule, James Prescott, 4, 357
Joules (J), 4, 357
Joule's law, 357

K

Kilograms (kg), 3, 4
Kilohms (kΩ), 19

Kirchhoff, Gustav Robert, 23, 72
Kirchhoff's current law (KCL), 23,
    72, 146, 152, 158–59, 172, 180,
    196, 197, 206, 217, 218, 225,
    272, 277–78, 307, 392, 650
  dependent sources and, 58, 60–61,
    62, 63–64
  impedance combinations and,
    317–18
  linear circuits and, 113–15
  in mesh analysis, 88
  in nodal analysis, 73, 74–75, 76,
    80–82, 84, 329, 331
  parallel resistance and, 35–38,
    41–42
  phasors and, 317–18
  series resistance and, 30
  sinusoidal excitations and, 426
  statements of, 24–26, 28–29
Kirchhoff's law of radiation, 72
Kirchhoff's voltage law (KVL), 23,
    46, 72, 146, 150, 151–52, 153,
    155, 171, 179, 196, 206, 217,
    218, 231, 272, 273, 277–78,
    307, 650
  dependent sources and, 58, 60–61,
    62
  impedance combinations and,
    317–18
  linear circuits and, 113–15, 122
  in mesh analysis, 86, 87, 88,
    90–92, 336
  in nodal analysis, 73, 74, 329
  parallel resistance and, 35, 40–41,
    43
  phasors and, 317–18
  series resistance and, 30, 31, 32
  sinusoidal excitations and, 426
  statements of, 26–28
  transformers and, 508–10, 515–19,
    520

L

Ladder networks, 114
Laplace, Pierre Simon, Marquis de,
    602, 603
Laplace transform(s), 604–6, 645
  Bode plots and, 678–84
  convolution theorem and, 614–17
  definition of the, 603–4
  differentiation theorems and,
    634–36
  final-value theorem and, 672,
    674–75
  Fourier transforms compared to,
    595, 603, 604, 605, 607, 637,
    658
  impulse function and, 603, 604,
    618–25
  impulse response and, 662–64,
    665–67

initial-value theorem and, 672–74
integrodifferential equations and,
    637–40
inverse, 605, 625–33
linearity of, 607–10
network functions and, 657–71,
    678–88
quadratic factors and, 685–88
solving electric circuits with,
    646–56
stability and, 668–71
steady-state sinusoidal response
    and, 676–78
step response and, 662, 664,
    665–66
translation theorems and, 610–13
Leakage current, 182–83
Leakage flux, 504
Leakage resistance, 182–83
Lenz, Heinrich F. E., 508
Lenz's law, 508
Leyden jar, 1
Linear capacitors, 165
Linear circuit(s):
  Cramer's rule and, 118
  definition of a, 113
  Kirchhoff's current law (KCL)
    and, 113–15
  Kirchhoff's voltage law (KVL)
    and, 113–15, 122
  Ohm's law and, 112, 115, 118,
    123
  superposition of, 117–22, 123,
    337–39
Linear element, 113
Linear equations, 113–14, 118
Linearity:
  of capacitors, 184–85
  of inductors, 184–85
  of Laplace transforms, 607–10
Linear resistors, 19–20, 112
Linear transformers, 503
  circuits with, 515–19
Line spectra (discrete amplitude spec-
    tra), 569
Line voltage, 395
Links, of graphs, 150
Loop, definition of, 86–87
Loop analysis. *See* Mesh analysis
Loosely coupled transformers,
    515
Losses: 490–91
  core, 183
  ohmic, 183
Low-pass filters, 475
Lumped-parameter circuits, 23

M

Magnetic fields, 174
Magnetic flux, 174–78, 419, 503,
    504

TABLE OF SELECTED LAPLACE TRANSFORM PAIRS

$f(t) = \mathcal{L}^{-1}[F(s)]$	$F(s) = \mathcal{L}[f(t)]$
$\delta(t)$	1
$u(t)$	$\dfrac{1}{s}$
e^{-at}	$\dfrac{1}{s+a}$
t	$\dfrac{1}{s^2}$
$\sin kt$	$\dfrac{k}{s^2+k^2}$
$\cos kt$	$\dfrac{s}{s^2+k^2}$
te^{-at}	$\dfrac{1}{(s+a)^2}$
$e^{-at}\sin kt$	$\dfrac{k}{(s+a)^2+k^2}$
$e^{-at}\cos kt$	$\dfrac{s+a}{(s+a)^2+k^2}$
$t^n;\ n = 1, 2, 3, \ldots$	$\dfrac{n!}{s^{n+1}}$
$\dfrac{t^{n-1}e^{-at}}{(n-1)!};\ n = 1, 2, 3, \ldots$	$\dfrac{1}{(s+a)^n}$
$2\lvert K \rvert e^{-\alpha t}\cos(\beta t + \arg K)$	$\dfrac{K}{s+\alpha-j\beta} + \dfrac{K^*}{s+\alpha+j\beta}$
$\dfrac{1}{b-a}(e^{-at} - e^{-bt});\ a \neq b$	$\dfrac{1}{(s+a)(s+b)}$
$\sin(kt + \theta)$	$\dfrac{s\sin\theta + k\cos\theta}{s^2+k^2}$
$\cos(kt + \theta)$	$\dfrac{s\cos\theta - k\sin\theta}{s^2+k^2}$